T0337990

Applied Engineering Analysis

Applied Engineering Analysis

Tai-Ran Hsu
San Jose State University, San Jose, USA

Registered Office(s)
John Wiley & Sons, Inc., 111 River Street, Hoboken, NJ 07030, USA
John Wiley & Sons Ltd, The Atrium, Southern Gate, Chichester, West Sussex, PO19 8SQ, UK

Editorial Office
The Atrium, Southern Gate, Chichester, West Sussex, PO19 8SQ, UK

For details of our global editorial offices, customer services, and more information about Wiley products visit us at www.wiley.com.

Wiley also publishes its books in a variety of electronic formats and by print-on-demand. Some content that appears in standard print versions of this book may not be available in other formats.

Library of Congress Cataloging-in-Publication Data

Names: Hsu, Tai-Ran, author.
Title: Applied engineering analysis / by Tai-Ran Hsu.
Description: Hoboken, NJ : John Wiley & Sons, 2017. | Includes
    bibliographical references and index. |
Identifiers: LCCN 2017017185 (print) | LCCN 2017037252 (ebook) | ISBN
    9781119071181 (pdf) | ISBN 9781119071198 (epub) | ISBN 9781119071204
    (cloth)
Subjects: LCSH: Engineering mathematics—Textbooks.
Classification: LCC TA332.5 (ebook) | LCC TA332.5 .H78 2017 (print) | DDC
    620.001/51—dc23
LC record available at https://lccn.loc.gov/2017017185

Cover design by Wiley
Cover image: (Sir Isaac Newton) © duncan1890/Gettyimages; (Hand) © Colin Anderson/Gettyimages

Set in 10/12pt WarnockPro by SPi Global, Chennai, India

My wife Grace Su-Yong who has supported me for decades with endless encouragement and love, my three loving adult children, Jean, Euginette, and Leigh, for their consistent support, and above all, my motivated students who inspired me to write this book.

莊子·天下篇·第七節：

一尺之棰，
　　日取其半，
　　　　萬世不竭

An excerpt from Chapter 7 of "Under Heaven" by Zhuang-Zi, 350 BC:

"A foot-long stick,

cut in half every day,

the last cutting to eternity."

An ancient Chinese wisdom on the concept of infinitesimal calculus

Contents

Preface

This book is designed to be a textbook for a one-semester course in engineering analysis for both junior and senior undergraduate classes or entry-level graduate programs. It is also designed for practicing engineers who are in need of analytical tools to solve technical problems in their line of duties. Unlike many textbooks adopted for class teaching of engineering analysis, this book introduces fewer additional mathematical topics beyond the courses on calculus and differential equations, but has heavy engineering content. Another unique feature of this book is its strong focus on using mathematics as a tool to solve engineering problems. Theories are presented in the book to show students their connection with practical issues in problem-solving. Overall, this book should be treated as an engineering, not advanced engineering mathematics textbook.

Mathematics and physics are two principal pillars of engineering education of all disciplines. Indeed, courses in mathematics and physics dominate the curricula of lower division engineering education in both the Freshman and Sophomore years in most engineering programs worldwide. Many engineering schools offer a course on engineering analysis that follows classes on precalculus, calculus, and differential equations. Engineering analysis is also offered at the entry level of graduate studies in many universities in the world.

The widespread acceptance of engineering analysis as a core curriculum by many educators is attributed to their conviction that students need to synergistically integrate all of the mathematical subjects that they learned earlier and apply them in solving engineering problems. However, the pedagogy of engineering analysis and its outcome has rarely been discussed in open forums. Many universities offer a course on engineering analysis as a terminal mathematics course with additional advanced mathematics subjects. Consequently, all textbook vendors with whom I have had contact in the last 30 years have consistently published books on advanced engineering mathematics, as textbooks for my course on engineering analysis. Upon close inspection, almost all have little direct relevance to the engineering profession. Additionally, all of these books are close to, or exceed, 1000 printed pages, with overwhelming coverage of detailed and elegant mathematical treatments to mostly mathematical problems. I have also observed that in all of the advanced engineering mathematics books such as those cited in the bibliography of this book, less than 10 percent of the pages have applications to engineering problems. Consequently, a textbook that is designed to teach students to solve engineering problems using mathematics as a tool is truly needed in classes on engineering analysis or those with similar objectives.

Many science and engineering educators are of the opinion that most engineering problems in the real world are of a physical nature. The disconnect in teaching mathematics and physics, as it occurs in lower division engineering education, has resulted in the inability of students to use mathematics as a tool to solve such engineering problems. Many students in my engineering analysis classes are skillful in manipulating mathematics in their assigned problems, including performing integrations and solving differential equations either using classical solution techniques learned previously, or using modern tools such as electronic calculators and computers. However, they are not capable of deriving appropriate equations for solving particular genuine engineering problems. Even more, students cannot apply integrations to determine simple design engineering properties such as areas, volumes, and centroids of solids of given geometry. The situation has worsened in recent years with rapid advances in information technology, which offer students ready access to turnkey software packages such as finite-element and finite-difference codes. This results in obtaining the solutions of engineering problems, in which insight, knowledge, and experience are sacrificed for numbers with seven-decimal point accuracy and fancy graphics. Unfortunately, most of these student users do not know what these numbers and graphs mean as solutions to the problems. These readily available commercial computer codes have actually further prevented engineering students from understanding fundamental engineering principles, worsening an already serious dissociation occurring early in mathematics and physics education.

An encouraging sign in recent years, however, has been the emergence of a consensus among visionary educators that students should relate their mathematics to the engineering subjects they will encounter in their upper division classes, and develop them as tools to solve real-world problems. It was with this conviction that I was motivated to write this book.

The present book intends to develop the analytical capability of students in engineering education. I am convinced that upper division students are not short of exposure to mathematics; what is lacking is the opportunity that they get to use what they learned in solving engineering problems. Consequently, no advanced mathematical subjects need to be added to this book. Rather, I have placed strong emphasis on how students will learn to apply the mathematics that they learned in previous years to solve engineering problems. Another aspect of this book is to include sufficient materials to fit the 3 hours per week in a 15-week timeframe that most engineering schools provide for this course. The topics to be covered were carefully chosen to ensure proper balance between breadth and depth, with lower division mathematics and physics courses as prerequisites.

There are 12 chapters in this book. Chapter 1 offers an overview of engineering analysis, in which students will learn the need for a linkage between physics and mathematics in solving engineering problems. Chapter 2 provides students with basic concepts of mathematical modeling of physical problems. Mathematical modeling often requires setting up functions and variables that represent physical quantities in practical situations. It may involve all forms of mathematical expressions ranging from algebraic equations to integrations and differential equations. Students are expected to apply their skills to determine physical quantities such as areas, volumes, centroids of plane subjects, moments of inertia, and so on as required in many engineering analyses. Special functions and curve-fitting techniques that can model specific engineering effects and phenomena are also presented. Chapter 3 refreshes the topics on vectors and vector calculus, which are viable tools in dealing with complicated engineering problems of different disciplines. Application of vector calculus, in particular, to rigid body dynamics is illustrated. Chapter 4 relates to the application of linear algebra and matrices in the formulation of modern-day analytical tools, and the solution techniques for very large numbers of simultaneous equations such as in the finite-element analysis. Chapter 5 deals with Fourier series, which are used to represent many periodic phenomena in engineering practices. Chapter 6 relates to Laplace transformation for functions that represent physical phenomena covering half of the infinite space or time domain, such as in the case of indeterminate beams subjected to distributed loads in various sections of the spans, or with discrete concentrated forces. Chapters 7 and 8 deal with the derivation, not just solution techniques, of first- and second-order ordinary differential equations with applications in fluid dynamics and heat transfer by conduction and convection in solids interfaced with fluids with applications in heating, cooling, and refrigeration of small solids. Chapter 8 presents the principles and mathematical modeling of free and forced vibrations, as well as resonant and near-resonant vibrations of solids with elastic restraints. Chapter 9 deals with the solutions of partial differential equations, in which equations for heat conduction and mechanical vibrations in solid structures are introduced. This chapter also offers solution methods such as the separation of variables technique and integral transform methods involving Laplace and Fourier transforms with numerical illustrations. Chapter 10 offers numerical solution methods for solving nonlinear and transcendental equations and differential equations and integrals, with examples that will facilitate learning of these techniques. Special descriptions of the overviews of popular Mathematica and MatLAB software are included in this chapter with a special article on the use of MatLAB in one of the appendices of the book. Chapter 11 introduces the principle and mathematical formulation of the finite-element method, which intends to make readers intelligent users of this versatile and powerful numerical technique for obtaining the solutions to many engineering and scientific problems with complicated geometry, loading, and boundary conditions. The book ends with a special chapter on statistics for engineering analysis as Chapter 12, in which the readers will learn the common terminologies in the science of statistics with physical meanings. This chapter will usher the readers to the common practice of statistical process control (SPC) currently adopted by industries involved in mass production. This chapter also includes probabilistic design methods for structures and mechanical systems that would not be otherwise handled using traditional deterministic techniques.

I have taught engineering analysis to senior undergraduate and entry-level graduate students in two major universities in the U.S. and Canada for over 30 years. I have found that most students are not accustomed to the application of mathematics to solving descriptive engineering problems that often require the derivation of equations and mathematical formulae for solutions. I attribute this to a major psychological barrier that many engineering students need to overcome before they can be effective analysts. Consequently, I have included many examples and problems that are descriptive in nature in this book. These problems are drawn from several engineering disciplines and many of them require numerical solutions that relate to theoretical concepts. Most of these problems can be solved using pocket electronic calculators.

It is by no means a trivial job to develop a book of this breadth and depth single-handedly. I wish to thank a number of students who helped me in shaping its content. In particular, my appreciation goes to a former student, Vaibhav Tank, for his contribution of the application of MatLAB software in problem-solving. Such dedicated students made an effort to develop this book into a memorable pleasure for me, and gave me a feeling of accomplishment in engineering education.

Tai-Ran Hsu
San Jose, California

Suggestions to instructors

This book is written to be a textbook for upper division undergraduate and entry-level graduate classes. It is also intended to be a reference book for practicing engineers to refresh their experience and skills in using mathematics for their engineering analyses, or to upgrade their understanding of contemporary analytical tools using digital technology in numerical methods as well as the use of commercial software packages such as MatLAB and the finite-element method for advanced engineering analyses.

The content of this book is designed for 3 hours per week in a 15-week long semester at both undergraduate and graduate levels. With significant omission of materials, the book can also be used for classes with well-prepared students for 10-week long quarters.

This textbook would be more effective for students with the following academic experience and backgrounds:

1) Undergraduate students in good upper division academic standing with sound knowledge and experience in college mathematics that includes calculus, differential equations, fundamental physics, and concurrent learning of engineering subjects in solid and fluid mechanics and heat transfer.

2) Students with working experience in computer software packages such as Microsoft Office, in particular the MS Excel, and other software packages such as Mathematica and MatLAB over those who do not have such experience.

Teaching a course on engineering analysis with sufficient breadth and depth such as that presented in this textbook in the aforementioned time frames could be a challenge for instructors. This situation may be further compounded by the likelihood of having students who completed their prerequisite mathematics in lower division coursework which often focuses on drilling rather than applications. The following Schedules A and B offer suggested topics from the book that the instructor may cover in either one semester or one quarter. The instructor would use his or her discretion to assign the unlisted sections in the following tables either as omissions or as assigned reading materials to the students.

Schedule A: For 15-week, 3 hours/week-long semesters

Week no.	Undergraduate classes	Graduate classes
1	Sections 1.3, 1.4, 2.2.3, 2.2.4, 2.3, 2.5	Sections 1.3, 1.4, 2.2.3, 2.2.4, 2.3, 2.4
2	Sections 3.4, 3.5, 3.6,	Sections 3.5, 3.6, 3.7, 4.3, 4.4
3	Sections 4.2.2, 4.3–4.7	Sections 4.5–4.8
4	Sections 5.2–5.5	Sections 5.2–5.5, 6.2, 6.4, 6.5
5	Sections 6.2–6.4	Sections 6.6, 6.7, 7.2
6	Sections 6.5, 6.6, 7.2, 7.3	Sections 7.3–7.5
7	Sections 7.4, 7.5	Sections 8.2, 8.3. Assigned reading 8.4–8.6
8	Sections 8.2–8.4	Sections 8.7–8.9
9	Sections 8.5–8.6	Sections assigned reading 9.2., 9.3, 9.4
10	Sections 8.7, 8.8, 8.9	Sections 9.5.2, 9.6
11	Sections 9.1, 9.3.1, 9.4	Sections assigned 9.7, 9.8
12	Sections 9.5.1, 9.6.1, 9.7	Sections assigned reading 10.2., 10.3–10.6
13	Sections 10.2–10.4	Sections 11.3–11.6
14	Sections 10.5–10.6, 11.2–11.4	Sections assigned reading 11.7, 11.8. 12.4–12.7
15	Sections 11.6–11.8, 12.1–12.6, 12.8–12.10	Sections 12.8–12.10

Schedule B: For 10-week, 3 hours/week-long quarters

Week no.	Undergraduate classes	Graduate classes
1	Sections 1.3, 1.4, 2.3, 2.5, 3.3	Sections 1.3, 1.4, 2.3–2.5, 3.4
2	Sections 3.4–3.6	Sections 3.5–3.7
3	Sections 4.3–4.5, 4.6	Sections 4.5–4.8
4	Sections 4.8, 5.3–5.5	Sections 5.2–5.5, 6.2–6.4, 6.6, 6.7
5	Sections 6.2–6.7	Sections 7.3, 7.4, 7.5
6	Sections 7.3–7.5	Sections 8.4–8.9
7	Sections 8.2, 8.4, 8.5	Sections 9.3, 9.4–9.6
8	Sections 8.6–8.9	Sections 10.3–10.4, 10.5, 10.6
9	Sections 9.1, 9.4–9.6, 9.8, 10.3, 10.4, 10.5.2	Sections 11.2–11.8
10	Sections 10.6.2, 11.3, 11.8, 12.4, 12.5, 12.6, 12.8–12.10	Sections 12.4–12.10

Instructors may, of course, use their discretion in selecting the topics other than those suggested in the above tables to suit their own preferences and schedules.

The author would like to recommend using a black or white board in addition to slide projections for examples that are offered by the textbook. From his own experience, the author found it challenging to have students who develop a mindset to learn to use their skills in math drilling to the applications in solving a wide range of problems in a single 15-week semester. In pursuit of this goal, much effort needs to be made in extra tutoring and advising students outside the classroom to help them acquire this new experience.

Further, a few additional suggestions for the successful teaching of this class include: offering quizzes and examinations in open book format, for which students may bring any reference materials they wish instead of memorizing all formulae and equations required for solving the problems; students are also encouraged to form "study groups" on their own initiative, and above all, bringing "models" for classroom demonstrations, such as wine bottles for determining their volume content, tightening a bolt to a fixture to demonstrate the principle of "cross product of vectors," etc. These simple classroom demonstrations would not only help students in understanding the subject matter, but also reinforce students' appreciation of the value of applied engineering analysis by using mathematics as a tool for solving real-world problems.

A final remark on stimulating student's interest in learning engineering analysis is to encourage and reward them for using the available online solution methods in mathematical operations in their homework assignments and quizzes and exams at all times.

About the companion website

Don't forget to visit the companion website for this book:

www.wiley.com/go/hsu/applied

There you will find valuable material designed to enhance your learning, including:

1) Solutions manual
2) PowerPoint slides

Scan this QR code to visit the companion website

1

Overview of Engineering Analysis

Chapter Learning Objectives

- Learn the concept and principles of engineering analysis, and the vital roles that engineering analysis plays in professional engineering practices.
- Learn the need for the application of engineering analysis in three principal functions of professional engineering practice: creation, problem solving, and decision making.
- Learn that engineers are expected to solve problems that relate to protection of properties and public safety and also to make decisions.
- Appreciate the roles that mathematics plays in engineering analysis, and acquire the ability to use mathematical modeling in problem solving and decision making in dealing with real physical situations.

1.1   Introduction

Engineering analysis involves the application of scientific principles and approaches that often use mathematical modeling as a tool to reveal the physical state of an engineering system, a machine or device, or structure under study. Applications of engineering analysis also include electrical circuit design, derivation of algorithms for computer programming, and so on. It is an integral part of the professional practice of all engineering disciplines.

Engineering analysis provides engineers with viable tools in their professional practice, which in general involves creating new products or engineering systems; solving technical problems that either are required to sustain current operation and maintenance of engineering processes or relate to developing new products or engineering systems; and making logical decisions in dealing with complex professional activities, such those involved in design processes and concerning proper actions required to deal with many critical problems that often involve serious consequences to public welfare and safety.

Engineers who apply engineering analysis to these professional activities must realize that mathematical modeling is widely used as a "working tool" for their analyses, and the problems that they are required to deal with are fundamentally of a physical nature. Consequently, engineering analysis does not involve simply finding suitable mathematical formulas or equations with mathematical solutions for the problems. Engineers must have thorough understanding of the physical nature of the problem and be able to identify the appropriate mathematical modeling as a tool in finding the solutions. It also requires engineers to recognize that mathematics serves as the "servant" to its "master"—the corresponding physics.

Applied Engineering Analysis, First Edition. Tai-Ran Hsu.
© 2018 John Wiley & Sons Ltd. Published 2018 by John Wiley & Sons Ltd.
Companion Website: www.wiley.com/go/hsu/applied

This book will primarily focus on application of engineering analysis to problems primarily relating to mechanical engineering. However, the principles of mathematical modeling derived for these applications may also be applicable to other engineering disciplines.

1.2 Engineering Analysis and Engineering Practices

Practicing engineers are often expected to undertake the following functions in their professional practices.

1.2.1 Creation

There is a common understanding that engineers originate many devices and engineering systems that benefit humankind and as such are principal contributors to those aspects of our contemporary civilization that improve the quality of lives and living standards for all. This belief in the creativity of engineers is exemplified and supported by this quotation from Albert Einstein, underlining that "creativity" is indeed a part of engineers' professional practice

> Scientists investigate that which already is. Engineers create that which has never been. (http://thinkexist.com/quotation/scientists-investigate-that-which-already-is/761229 .html)

Table 1.1 lists the 20 greatest achievements by engineers in the last century as selected by the American Academy of Engineering. One will readily observe that all these achievements have substantially enhanced people's quality of life. Engineering analysis offers the necessary means for engineers to develop and produce all the products and systems listed in Table 1.1.

1.2.2 Problem Solving

Engineers are constantly required to solve technical problems that can vary in many different ways. Some of the problem areas that engineers often encounter include ambiguity in the design of new products or engineering systems, improper ways of manufacturing and production of products, inferior quality of products, run-away cost control in design and/or production, resolution of customers' complains and grievances, and addressing public grievances and mistrust. Many of these problems require immediate solutions, whereas others may require and be afforded more time to resolve. Some problems that engineers contribute toward solving

Table 1.1 Greatest engineering achievements of the 20th century.

• Electrification	• Highways
• Automobile	• Spacecraft
• Airplane	• Internet
• Water supply and distribution	• Imaging
• Electronics	• Household appliances
• Radio and television	• Health technologies
• Agricultural mechanization	• Petroleum and petrochemical technologies
• Computers	• Laser and fiber optics
• Telephone	• Nuclear technologies
• Air conditioning and refrigeration	• High-performance materials

American Academy of Engineering (http://www.greatachievements.org/)

may have major impacts on the wellbeing and safety of the general public: a real-life case is the devastating rupture of a small section of natural gas pipelines in San Bruno, California in the autumn of 2010. This accident destroyed 20 family houses in the neighborhood and caused the loss of 10 human lives. The owner and operator of the gas pipelines, Pacific Gas and Electricity, has since been involved in major litigation by the victims. Large numbers of engineers have spent countless hours of time in the search to identify the causes of the rupture of the pipeline, and from these are expected to devise solutions to avoid recurrences.

Problems such as this relating to rupture of pressurized pipelines can be investigated by engineering analysis using scientifically sound mathematical modeling based on fracture mechanics analysis, such as proposed by (Hsu and Bertel, 1976; Hsu *et al.*, 1988).

1.2.3 Decision Making

Engineering is a profession, not just an occupation. The welfare of others depends greatly on the expertise and judgment exercised by engineers as reflected in the quality of what they design and manufacture. Decisions made by engineers represent a public responsibility (Pearsall and Hadley Cocks, 1996). Engineers are often required to investigate serious issues of various natures and degrees of complexity; many of the decisions that engineers make in resolving these issues are based on the use of proper engineering analysis. Incorrect or inadequate analyses by engineers may result in loss of productivity, compromise of the value of products or systems, and—most seriously—have grave consequences in terms of loss of both property and human lives. Engineering analyses used in decision-making processes may sometimes involve simple scientific principles and simple mathematical manipulations, but the analysis of many problems requires sophisticated advanced analytical tools and modern computational power.

The following represent some common classes of problems that require decision-making in mechanical engineering practice.

1) **Decisions relating to design of new products or systems:** Decisions are required in configuring new products or engineering systems and selecting design methodology. Decision on the latter activity may involve selecting the empirical formulas established by a company or those available in published design handbooks (Avallone *et al.*, 2006; Kreith, 1998; Bishop, 2002; Gad-el-Hak, 2002; Whitaker, 1996; Gibilisco, 1997). On other occasions engineers may decide to derive their own methodology for the design analysis. Important decisions are also required in selecting materials for the product or engineering systems at this stage of the design process.
2) **Decisions on manufacturing and production of new products and engineering systems:** Once the design of new products or engineering systems is completed, engineers need to make the further major decisions relating to the manufacturing and production of the products or engineering systems.

Decisions on optimal ways of fabricating the parts and components of the target products or engineering systems may include the choice of suitable and cost-effective machine tools, fabrication processes, and assembly and packaging of components into the finished product or engineering system. Often, many of these parts can be purchased from suppliers, in which case the engineers need to decide what parts to use and select the suppliers.

Engineers are frequently also required to make decisions of a nontechnical nature. Such decisions may include quality control and assurance of the products or engineering systems, as well as their maintenance requirements. At times engineers must decide how to deal with unexpected demand for change orders coming from the customers, as well as due to unexpected causes relating to the production.

On relatively rare but critical occasions, engineers are expected to be involved in making decisions in "real-time" technical risk assessment to their superiors or clients. These may often involve potentially grave consequences in terms of serious financial and public safety implications. Two hypothetical cases are presented here to illustrate the critical decision making processes that may be encountered by engineers.

Case 1. Decision on what action to take when a small crack appears on a pipeline
We use the Trans-Alaska Pipeline as an example. Part of this mighty engineering undertaking is shown in Figure 1.1.

Figure 1.1 The Trans-Alaska Pipeline. (Courtesy of Alyeska Pipeline Service Company, Anchorage, Alaska, USA)

This vital long pipeline transports crude oil from Prudhoe Bay at the north shore of the state of Alaska to the port of Valdez in the south shore of the state (https://en.wikipedia.org/wiki/Trans-Alaska_Pipeline_System). The 800 mile-long (287 km-long) pipeline was built between 1974 and 1977. A large portion of the pipeline was built over hundreds of miles of uninhabited fragile permafrost land. The pipeline is supported by specially designed supports that have foundations rooted deep in the ground below the permafrost layer, and that elevate the entire pipeline to allow free passage for wildlife over this vast uninhabited land. It costs millions of dollars to build but it also earns millions of dollars revenue daily for the owner with a daily crude oil transportation capacity of 2136 million barrels. The oil inside the pipeline is subjected to high pressure maintained by 11 pumping stations along the way.

One scenario is "what would the field engineer do if he or she detects a fine crack on the surface of the pipeline, either by visual inspection or by online instrumentations?" A sensible action would be for the engineer to report this discovery to his or her superior without delay. Upon the receipt of the report of the finding of the field engineer, the management (or the owner) of the pipeline has the two options: (1) Do nothing and continue the routine operations as usual, but run a risk of possible major pipeline rupture after the "small" crack grows beyond a "critical length" under the internal pressure of the pipeline. (2) Order an immediate shut-down of the pipeline operations and invoke a thorough investigation to find the cause or causes of the small crack and assess the possibility of its stability against further growth under the normal operating conditions of the pipeline.

Either of these actions may have serious consequences: Option (1) could lead to a devastating rupture of the pipeline by miles should the crack grow to a critical length over the time of continuous operations. This would not only require costly replacement of the ruptured pipeline section but would also jeopardize the public safety with possible losses of human life and wild animals' lives. Option (2) in contrast appears prudent as far as safe operations are concerned. However, such action requires the shut-down of the pipeline operations and that would result in substantial loss of income revenue to the company.

Case 2. Decision on proper actions to be taken when cracks appear on the surface of an aircraft
This scenario is similar to that in Case 1 but involves even more grave consequences and the need for a "real-time" decision.

It is common practice for commercial airline pilots or technicians to conduct visual inspections of their aircraft prior to scheduled flights. Figure 1.2 shows a commercial airplane docked at a gate in a civilian airport, ready for flight.

Figure 1.2 A commercial airplane ready for a scheduled flight. (Courtesy of San Francisco International Airport.)

The purpose of walk-through visual inspections by field inspectors is to ensure that the airplane is sound in structure and is ready for a safe flight. However, the inspection may reveal the appearance of a small crack in some part of the aircraft—the wings, stabilizer and rudder, engines, or fuselage. Such finding pose the crucial question "what should the flight manager do after the field inspector has reported such finding to him or her?"

As in Case 1, the manager may choose one of two options: (1) Ignore the report by the field inspector and allow the aircraft to take off as scheduled, but run the risk of a mid-air failure if the "small crack" found by the field inspector grows to a "critical length" and propagates through the aircraft structure, resulting in the airplane coming down. (2) Abort the scheduled flight and tow the evacuated aircraft to a hangar for detailed investigation of the cause or causes of the crack and assess its stability under conceivable in-flight conditions and situations. The latter decision is prudent but would lead to a substantial loss of revenue to the airline in addition to unavoidable complaints by the ticketed customers about the disruption in their travel plans.

In either of the above hypothetical cases, the persons who make the ultimate decisions on the optional actions rely heavily on the "risk assessments" by their engineers and technical staff, from which they would recommend the optimal decisions on the follow-up actions. Real-time risk assessments submitted by engineers to their superiors obviously cannot—and should not—be reached arbitrarily. What engineers need in such decision-making process are the "appropriate tools" that are available to them in such process. Whatever tool they use should be derived from logical "scientific principles" and employ sound engineering analysis.

1.3 "Toolbox" for Engineering Analysis

The "**tools**" that are available in engineers' tool box for their **creation** of new products or engineering systems, **problems solving** and making critical **decisions** are the *mathematics* derived from the *laws of physics*. For instance, the tools that engineers could use in making real-time risk assessments and from there making reasoned judgments on whether the surface cracks appearing in the pipeline or airplane in the two scenarios in presented are "stable" (i.e., safe) or "unstable" (i.e., leading to structural failure) may be found among the theories and principles of *fracture mechanics*, such as described in the literature (Landes et al., 1979; Kim and Hsu, 1982; Hsu, 1986; Luxmoore et al., 1987).

While using mathematics as the principal tool for engineering analysis, engineers have to recognize this tool as a "means" for obtaining the solutions of the many problems that they have to deal with. These problems are actually the "ends" in the process. It is fundamentally important to realize that the "means" should be made to serve the "ends"—not the other way round. In the real world, engineers have problems on hand first and then seek appropriate mathematical tools to solve the problems. Often the search for the appropriate mathematical tools in engineering analysis is nonmathematical, as is illustrated in the following typical statements of the problems

"A coat hanger that hangs garments of a specified range of weights and sizes."

"A diving board that can spring a person weighing up to 100 kilograms to a height of 20 centimeters."

"An underwater electric light fixture that is safe to swimmers against electrocution and has a long working life of 2000 hours."

"A washing machine that will function for 20 000 cycles."

"A wine bottle to contain a volume of 750 milliliters of wine."

"The rate of growth of a surface crack is 0.1 mm per hour on a pressurized pipe with given geometry and loading."

There are innumerable physical statements of problems involving complex engineering systems, such as in the design and production of special-purpose bicycles, computer disk drives for mass data storage, engines for automobiles and airplanes, and so on. Some of these physical statements are expressed in the specifications describing the product, but many others will be long documents that may fill several freight craters, as in the case of airplanes.

We thus see that *engineering problems are of "physical" nature not mathematical*. Accordingly, solutions to engineering problems also need to be physical rather than mathematical. Unfortunately, there is rarely a direct link between the physical engineering problem and the answer that is also of physical nature. In fact, solutions to most engineering problems can be obtained only by engineering analysis that involves mathematical manipulation, as illustrated in Figure 1.3.

Figure 1.3 illustrates how most engineering problems are solved by **engineering analysis**, which involves first translation of the physical conditions of the engineering problem into mathematical form, as indicated in the dashed box "Mathematical interpretation". Only then can a mathematical analysis be performed, leading to a logical solution. However, the results of all mathematical or analytical analyses are in the form of numbers, charts, or graphs, which cannot be directly related to the sought solutions to the physical problem. Consequently, one needs to translate whatever mathematical solutions are obtained from the mathematical analysis into the physical sense, as indicated by the dotted box "Physical interpretation of results." These general

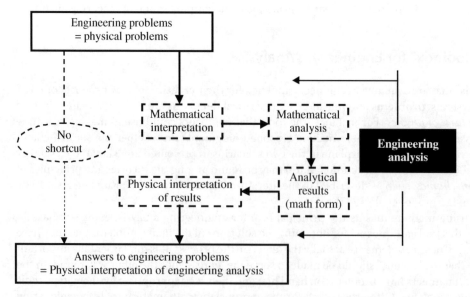

Figure 1.3 Role of engineering analysis in the solution of engineering problems.

steps in the engineering analysis, although apparently complex, are necessary in solving most engineering problems that are of physical nature.

The flow chart in Figure 1.4 illustrates the major steps involved in a typical engineering design process. Depending on the complexity of the case, some of these steps in this flow chart may be skipped, but the design of most engineering systems involves all of these steps in the design process.

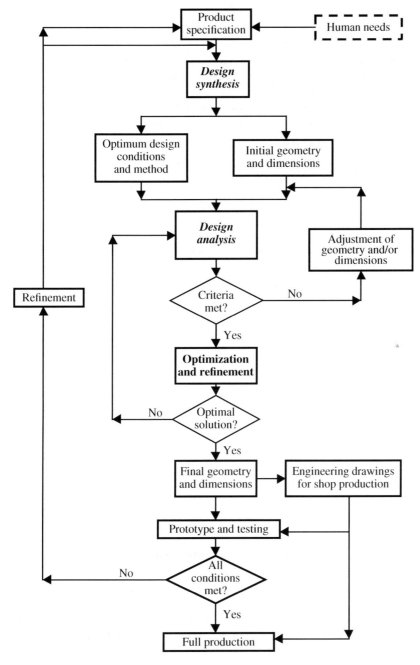

Figure 1.4 General engineering design procedures.

The three principal components in a design process include "design synthesis," "design analysis," and "design optimization" (Hsu and Sinha, 1992).

"Design synthesis" involves "brainstorming" to search for all possible solutions that might satisfy the design objectives. Determination of design parameters may include considering plausible configurations, materials, design, fabrication and assembly methods, design verifications (testing, etc.), and so on, along with the pros and cons of each of the listed options. The engineer(s) will then conduct an elimination process on each of the listed options based on the relative merits of each listed option and thereby reach an optimal, but not necessarily *the* best, decision on the required design parameters.

Design synthesis is usually followed by "design analysis," in which all the design parameters established from the synthesis procedure are considered using whatever methodology has been chosen for reaching quantitative decisions. The methodology, or the "tools," involved in the design analysis often involves mathematical formulations; these can be as simple as empirical or algebraic equations to complex sophisticated numerical analyses such as the finite element method (FEM) described in Chapter 11, or the finite difference method (FDM) using digital computer methods as presented in Chapter 10. The remaining procedures included in Figure 1.4 relate to the logical sequences from which optimal design of a product or engineering system is reached by meeting all required conditions and set criteria.

One will readily observe that engineering design is an "open-ended" problem that requires design engineers to make a number of assumptions and hypotheses about many items of information and conditions that are normally available in the case of class assignments in the school but are missing, though required, in reality. Solution to these problems involves "iterative" procedure as illustrated in Figure 1.4. The solution that design engineers obtain are "optimal" but not necessarily the "best" solution that many might expect.

Another aspect of engineering design analysis is that "design is a process of creating something from nothing." In this, the design engineer needs to assign values of missing information relating to the geometry and dimensions of essential components on the basis of his or her professional intuition and experience. The validity and suitability of these assumed values are to be confirmed at the end of the design process, as shown in the flowchart in Figure 1.4.

1.4 The Four Stages in Engineering Analysis

Engineers create new products or engineering systems through design processes that involve engineering analysis as described. "Design analysis" as presented in Figure 1.4 is an important part of an engineering design process. While the complexity of the problems and thus the "tools" used in engineering analysis may vary from case to case, these tools play major roles in all types of design processes.

In general, engineering analysis involves the following four stages.

Stage 1. Identification of the physical problem
The engineer must first have a clear understanding of the problem for analysis, whether it is an original design problem or an existing problem that requires a solution or decision. For many design problems, this stage of work relates to understanding of the specification of the product or engineering system.

The items listed below are some pertinent items of information that are normally required in the design of a product or an engineering system. This information is normally supplied by the customer, or come from other sources. Essential information required in this stage of analysis may include, but is not limited to the following:

1) The design objective such as the intended application(s).
2) Options and/or restraints in configurations in geometry and size of the product or system to be designed.
3) Selection of materials for all components.
4) Expected loading conditions, including normal and overloading situations.
5) Other conditions and constraints, on space, environment, cost, government regulations, etc.

Stage 2. Idealization of actual physical situations for mathematical analysis

After being made fully aware of the specific conditions and requirements for the intended product or engineering system, the engineer will search for the analytical tools that is required to do the work. There are times when the engineer will realize that many of the stipulated requirements and constraints cannot be met with the available analytical "tools" that he or she is aware of or has the necessary access to. The engineer would thus be compelled to create idealizations of many of the physical aspects of the problem through assumptions about the conditions that were stipulated in Stage 1 of the analysis so that he or she can handle the analyses using the available analytical tools. Idealization may be required in the following specific areas:

1) *The geometry.* Most engineering analytical tools, such as mathematical formulations and solutions offered in much published literature and many handbooks (such as referenced in Sections 1.2 and 1.3) are available for typical solid geometries of beams, rectangular or circular plates, cylinders, etc., whereas the geometry of machine components in reality may not fit any of these typical geometries. Idealization of the geometry of the products in real-world design analysis is thus necessary to allow use of the available analytical tools.
2) *Loading conditions.* Again, most existing engineering solutions available in textbooks or handbooks are applicable only for simple loading conditions. For example, classical solutions of beam analysis are available for beams subject to concentrated forces or uniformly distributed forces. Cases with beams subject to other types of loading require special derivation of equations and their solution. Engineers are expected to idealize the expected loadings to the components in the current design analysis so as to enable use of the available analytical tools.
3) *Boundary or support conditions.* Many end fixtures of machine components are neither completely fixed nor freed to rotate. For example, for a beam "bolted" to end supports, this joint, in reality is neither a "hinged" support that allows the beam complete freedom to rotate nor is it rigidly fixed, the cases offered by almost all textbooks and handbooks. In such cases, engineers need to "idealize" the boundary or end conditions in their analysis by using simply supported conditions with "frictionless hinge and rollers," or rigidly held ends with no rotations, as in the cases of beams subjected to bending. Approximate bolted end supports are usually handled as idealized situations between the two extreme cases of simply supported and rigidly held end conditions.

One must realize that idealization of the physical requirements of the problems via a number of assumptions, although necessary for the analysis, will compromise the accuracy and credibility of the analytical solutions and prevent them from being entirely realistic. This shortcoming is usually compensated by "design margins" set by the design engineers, as will be illustrated by examples in Section 1.5.

Stage 3. *Mathematical modeling and analysis*

At this stage of analysis, the engineer is ready to apply whatever available analytical tools he or she has acquired either from what he or she previously learned from his or her schools or from available engineering handbooks such as that of Avallone *et al.* (2006) to solve the problem. Major tasks at this stage of analysis may involve the following:

1) Developing a suitable mathematical model based on the idealized physical characteristics of the problem.
2) Deriving applicable mathematical expressions; these may be in simple algebraic form or may be published or self-derived differential equations, or empirical formulas developed by the engineers themselves, or empirical equations developed by their employer in handling similar problems in the past.
3) Establishing a mathematical representation of loading and boundary conditions.
4) Solving the equations for numerical solutions with idealized loading and boundary conditions.

Stage 4. *Interpretation of results*

This is a critical stage of any engineering analysis. Whatever analytical method the engineer uses in the analysis, the results are most likely to be in the forms of raw numbers or in graphs or charts. A major effort at this stage of the analysis is to invest these forms of solution with a physical sense required for the solution to the physical problem.

Some of these four stages in engineering analysis may well take longer time and more effort to accomplish than others. Some analytical cases may involve only the use of formulas available in textbooks or handbooks; others may require tedious derivation of equations and computations using powerful digital computers running appropriate numerical analysis software. There are still cases that are too complicated to be handled by closed-form solutions from the required analysis. In these cases, qualitative solutions based on designers' own experience and interpretation may, necessarily, prevail.

Figure 1.5 shows how the four stage of engineering analysis fit into general design process illustrated in Figure 1.4.

1.5 Examples of the Application of Engineering Analysis in Design

Two examples will illustrate the application of engineering analysis in the four stages described in Section 1.4. The first example relates to the simple design analysis of a low-cost but common consumer product—a coat hanger; the other relates to the design of a bridge structure with aspects that have implications for public safety.

Example 1.1 Design a simple coat hanger involving the four stages in engineering analysis

Stage 1. *The problem*

The objective of this analysis is to assess whether a coat hanger designed for a customer is strong enough to hang an overcoat weighing up to 6 pounds (lb_f). The geometry and dimensions of this hanger are illustrated in Figure 1.6.

The customer requires the hanger be made of plastic for low cost and light weight with an allowable tensile strength of 500 psi from a materials handbook. Other requirements include the length of the hanger to be limited to 17 inches as shown in Figure 1.6, and the cost of producing each hanger from a batch of 30 000 units be kept to less than 50 cents.

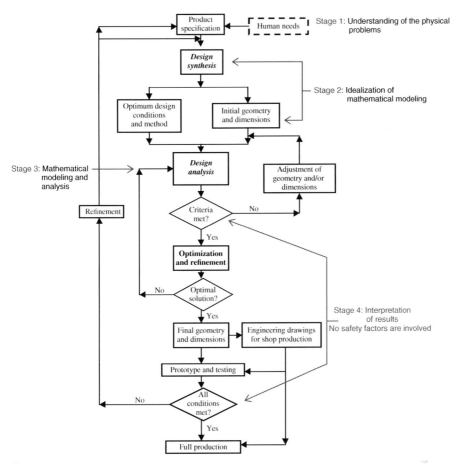

Figure 1.5 Four-stage engineering analysis in general engineering design process.

Figure 1.6 Coat hanger with specified geometry and dimensions.

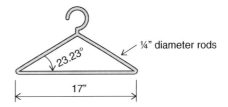

Stage 2. Idealization of actual physical situations for subsequent mathematical modeling and analysis

The engineer recognizes that the principal loading on the coat hanger in Figure 1.6 would cause the two inclined support plastic rod members to bend, leading to a decision to use the mathematical formulas for "beam bending" in the analysis. Solutions for the strength of the beam bending under different loading and end conditions are available in many textbooks or handbooks (such as Avallone *et al.*, 2006 or Kreith 1998). However, the geometry of the hanger as shown in Figure 1.6, with rounded corners and a hook, is too complicated to match any structural geometries that are available in any textbook or reference book. Consequently, the engineer has to make the necessary idealizations of the geometry and loading conditions on the load-bearing members of the hanger in the analysis so as to be able to make use of the available mathematical formulas and their solutions available in reference works.

1) *On the geometry:*

From 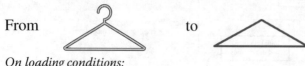 to

2) *On loading conditions:*

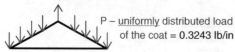 P – <u>uniformly</u> distributed load of the coat = 0.3243 lb/in

3) *On boundary (end) conditions:*

Stage 3. Mathematical modeling and analysis

The problem required to be answered in the present case is "will the coat hanger depicted in Figure 1.6 withstand the specified maximum weight of a coat up to 6 lb$_f$?" This is a physical problem that requires a physical statement in the answer.

The answer to this question may be obtained by following the principles of structure design by keeping the maximum stress induced in the coat hanger by the expected maximum load (the weight of a coat up to 6 lb$_f$) below the allowable limit. In the present case, keeping the maximum tensile strength of the hanger material below the specified value of 500 psi.

With the idealization made in Stage 2, the maximum stress in the coat hanger can be computed from the formulas of "simple beam theory" available from handbooks (Avallone *et al.*, 2006; Kreith, 1998) as follows.

First using the actual loading on the load-bearing member as

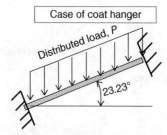

and subsequently converting to an case available in the handbook consulted, as illustrated in Figure 1.7.

Uniformly distributed load $w = P \cos(23.23°)$ **Figure 1.7** Loading on the idealized coat hanger member.

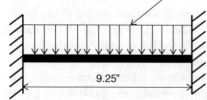

The maximum normal bending stress induced by the distributed load in Figure 1.7 can be obtained with the following formula:

$$\sigma_{n,max} = \frac{M_{max}c}{I} \tag{1.1}$$

In Equation 1.1, M_{max} is the maximum bending moment in the beam, c is the radius of the beam of circular cross-section, and I is the section moment of inertia of the beam.

We may compute the following:

The section moment of inertia of the frame rod:

$$I = \frac{\pi d^4}{64} = 1.9165 \times 10^{-4} \text{ in}^4$$

The maximum bending moment:

$$M_{max} = \frac{p\ell^2}{8} = \frac{0.298 \times (9.25)^2}{8} = 3.1872 \text{ lb-in}$$

The radius of the frame rod: $c = 0.125$ inch.
The maximum bending stress is computed to be

$$\sigma_{n,max} = \frac{3.1872 \times 0.125}{1.9165 \times 10^{-4}} = 2078.8 \text{ psi}$$

It occurs at both ends of the inclined rod members in Figure 1.6, and it is a tensile stress at the top face of the rod members.

Stage 4: Interpretation of results

As described in Section 1.4, the results obtained from Stage 3 of the analysis are usually in the form of raw numbers or graphics. In the current analysis, we have computed the maximum normal bending stress $\sigma_{n,max}$ to be 2078.8 psi that occurs at both ends of the rod members of the hanger. This "number" needs to be interpreted to the physical statement about "will the coat hanger withstand the specified maximum weight of a coat up to 6 lb$_f$?", which is the ultimate answer required from the current design analysis.

We know from the specification of this example that the allowable tensile strength of the material is $\sigma_{all} = 500$ psi as specified in Stage 1 of the analysis. Physically, it means that the coat hanger material cannot sustain any tensile stress larger than the specified 500 psi in any part of this product. The maximum normal bending stress (which turns out to be in tension, at the top surface of the inclined rod members) at 2078.8 psi is much greater than the allowed limit of 500 psi, which means that the hanger will not have the strength to withstand the weight of a coat at 6 lb$_f$.

The conclusion reached at the end of this analysis means that the specified design of the coat hanger with its geometry and dimensions indicated in Figure 1.6 will not be strong enough to carry the intended weight of the coat. The customer must be informed of this finding and advised either to adjust the dimensions of the hanger, for example, by increasing the diameter of the members of the coat hanger or to implement some other change such as replacing the plastic material for the hanger with a stronger material such aluminum or steel. These adjustments will increase the cost of production. Alternatively, the customer will need to reduce the maximum weight of the garment to be supported. In that case, the engineer will use a similar analytical approach to determine the limit on the weight of the garment for the new design of the coat hanger.

Example 1.2   Design analysis on a bridge cross a narrow creek
This example demonstrates the use of the four stages in engineering analysis on the design of a bridge structure such as that illustrated in Figure 1.8. The bridge is designed and constructed to handle limited local traffic over a narrow creek. The span of the bridge is 20 feet, and the maximum load designed for the bridge is 10 tons (or 20 000 lb_f).

The bridge is supported by two I-beams made of steel, as illustrated in Figure 1.9. The nomenclature of the I-beam cross-sections is shown at the right of the figure.

The engineering analysis on the strength of the bridge begins with a summary of the description of the problem as indicated in the following Stage 1 of the analysis.

Stage 1. Definition of the problem
The analysis is intended to satisfy the following physical requirements:

1) A bridge over a narrow creek.
2) Distance of crossing, i.e., the length of the bridge is 20 feet.
3) Width of the bridge is 8 feet.
4) Maximum designed load is 10 tons (or 20 000 lb_f)
5) Steel is chosen as the load-carrying structural material.
6) Maximum tensile strength for steel = 75 000 psi (from a handbook).
7) Maximum shearing strength for steel = 25 000 psi (from the same handbook).
8) Maximum allowable deflection of the structure is not critical in this case.

The relatively simple design of the bridge structure as illustrated in Figure 1.9 requires the two steel I-beams to bear the load of the crossing traffic, which allows the engineer to use simple formulas for beam bending available in a popular handbook (e.g., Young *et al.*, 2012).

Figure 1.8 A bridge across a narrow creek. (Courtesy of HC Bridge Company, Wilmette, Illinois, USA)

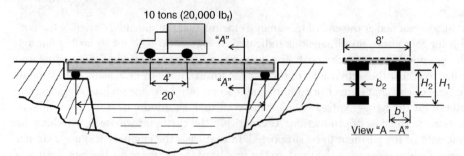

Figure 1.9 Specifications of a bridge over a narrow creek.

Figure 1.10 An idealized structural support of a bridge.

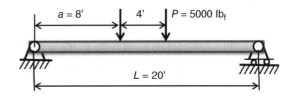

However, the desirable convenience of using existing formulas requires several idealizations of the real-world situation with regard to the geometry, loading, and end support conditions as described for Stage 2 in Section 1.4. It also requires caution in the "interpretation of results" in Stage 4 of the analysis using proper values of "safety factors," because of the high potential risk of loss of property and human life resulting from a structural failure of the bridge. The "safety factor" is often used as a "design allowance factor" in engineering analysis involving structures. It is intended to make up the likely discrepancies between the solutions for the "real" structure and those obtained from "idealized conditions."

Stage 2. Mathematical idealization

As mentioned in Section 1.3, engineers are often required to assign values to information that is required for the engineering analysis but is missing. In this situation, we need to make the following "logical" assumptions on the missing conditions.

1) The maximum load applied to the bridge is carried equally by two identical I-beams as shown on the right of Figure 1.9. The weights of the beams and other materials are neglected. (This is obviously a serious violation of reality. We need to make this idealization, nonetheless, in order to simplify the subsequent mathematical analysis.)
2) We have tentatively selected 12-inch I-beams for the load-bearing components. The exact dimensions for these beams are presented in manufacturer's handbook, which indicates: $H_1 = 12$ inches, $H_2 = 10.92$ inches, $b_1 = 5$ inches, and $b_2 = 0.35$ inches (these dimensions are not the choice of the design engineer).
3) Each I-beam carries half of the total load, i.e., 10 000 lbf.
4) Assume that each of the four wheels of the vehicle crossing the bridge carries an equal amount of load. (This may be another questionable assumption, but it is a necessary condition for the analysis.)
5) The worst condition to be considered for the integrity of this bridge structure is when the vehicle is at the mid-span of the bridge.
6) In order to use the formula for beam bending with "simply supported ends" available in the handbooks, we need to fix the two I-beams with one end of each of the beams hinged to a fixed anchor and the other resting on "rollers."

Stage 3. Mathematical analysis

Based on the assumptions and idealizations made in Stage 2 of the analysis, we can now use the simple beam theory for the case illustrated in Figure 1.10 for the current analysis.

In Figure 1.10, the concentrated forces, $P = 5000$ lbf represent the load transmitted to the bridge by the wheels of the vehicle. The objective of the current analysis is to assess whether the proposed structure of the bridge would be strong enough to sustain the expected load: expressions for the induced bending stresses and deflection due to the loading conditions illustrated in Figure 1.10 are available in handbooks (Young *et al.*, 2012; Avallone *et al.*, 2006; Kreith, 1998) or any textbook on strength of materials, from which we have the following formula for the induced bending stresses and deflection. The maximum induced normal tensile bending stress at the bottom surface of the I-beam's edge is

$$\sigma_{n,max} = \frac{M_{max}c}{I} \tag{1.2}$$

and the maximum shear stress at the center of the I-beam cross section is

$$\sigma_{s,max} = \frac{VQ(0)}{Ib_2} \tag{1.3}$$

where

M_{max} = maximum bending moment induced by the load.

c = half the depth of the I-beam cross section (6 inches in the present case).

I = section moment of inertia of the beam cross-section.

V = maximum vertical shearing force induced by the load.

b_2 = width of the rib of the I-beam

$Q(0)$ = static moment at the center of beam cross-section.

The maximum deflection of the beam is

$$y_{max} = \frac{Pa}{6EI} \left(\frac{3L^2}{4} - a^2 \right) \tag{1.4}$$

at the mid-span of the bridge, where E = Young's modulus of the beam material ($=30 \times 10^6$ psi for steel, from handbooks).

One may determine the following parameters required for determining the stresses and deflection of the beam structure illustrated in Figure 1.10: $M = Pa = 480\,000$ in-lb$_f$, $c = 6$ in, $I = 215.4$ in^4; $V = 5000$ lb$_f$, $b_2 = 0.35$ in, $Q(0) = 20.688$ in^3. The maximum bending and shearing stresses and the deflection are obtained by substituting these parameters into Equations 1.2, 1.3, and 1.4, respectively. The corresponding maximum values of the induced normal and sharing stresses and deflection in each of the two beams subjected to the specified loads are

Maximum normal stress: $\sigma_{n,max} = 13\,370$ psi (tensile at the bottom of the I-beams).

Maximum shear stress: $\sigma_{s,max} = 1372$ psi.

Maximum deflection: $y_{max} = 0.42$ in.

Stage 4. Interpretation of results with safety factors (SF)

The magnitude of the maximum normal and shearing stresses and deflection of the beam computed in Stage 3 of the analysis need to be interpreted into physical terms to be of practical value. For the present case, a critical question that engineers need to answer is "would the bridge structure with the selected I-beams be strong enough to sustain the maximum intended load of 10 tons?" We will thus need to translate the computed numerical results of the analysis with the maximum normal bending stresses $\sigma_{n,max} = 13\,370$ psi, the maximum shearing stress $\sigma_{s,max} = 1372$ psi, and the maximum deflection $y_{max} = 0.42$ in into physical statements to form the answers to the above questions.

A conventional way of translating these computed stresses into physical meanings is to compare these values with the "maximum strength" of the materials of the structure specified by material handbooks, as in Example 1.1. The maximum strength of the material is a material property that can be taken as either the measured yield stress (σ_y) or ultimate tensile strength (σ_u). Values of both σ_y and σ_u for common materials are available in handbooks (Avallone *et al.*, 2006). However, because the maximum induced stresses and deflection of the structure are obtained by an analysis that is based on the assumption of an idealized situation in Stage 2 of this analysis, which may not be sufficiently realistic to reflect the real situation, a factor, called the "safety factor" (SF) with a value greater than 1.0 is used in setting the allowable stress defined by the expression (1.5):

$$\sigma_a = \frac{\text{Measured } \sigma_y \text{ or } \sigma_u \text{ of the structure material}}{\text{SF}} \tag{1.5}$$

The safety factor SF may be defined in alternative way by the expression (1.6):

$$\text{Safety factor (SF)} = \frac{\text{Ultimate tensile strength of the material } \sigma_u}{\text{Maximum tensile stress in the structure } (\sigma_{max})} \tag{1.6}$$

We observe from the definition of SF in Equation 1.6 that a "discounting" of the ultimate tensile strength of the material (σ_u) is applied in order to make up for the less-than-realistic maximum induced stresses and deflection obtained in the analysis.

With the maximum tensile stress, $\sigma_{max} = 13\,370$ psi obtained from Stage 3 of the analysis, a safety factor, SF $= 75\,000/13\,370 = 5.61$ is obtained from Equation (1.6). The induced maximum deflection of 0.42 inch in the mid-span of the bridge by the expected maximum loads is not considered to be excessive for this type of application. The SF of 5.61 in this analysis is considered to be within a "safe" margin for the bridge structure, as will be elaborated in the subsequent section. Hence, the selected I-beam structure is considered to be strong enough to withstand the designed maximum load of 10 tons.

1.6 The "Safety Factor" in Engineering Analysis of Structures

"Safety factor" or "factor of safety" is common terminology in machine design. The definition of safety factor may vary according to occasion (Shigley and Mischke, 1989; Norton, 2013; Hagen, 2014). In engineering analysis that involves the structure of machines or engineering systems, the definitions of SF as expressed in Equations 1.5 and 1.6 are frequently used in interpreting the analytical results from analyses into the physical states of an engineering problem.

The *safety factor* (SF), as defined in Equations (1.5) and (1.6) has numerical values great than 1.0, that is, SF > 1.0. A value SF $= 2$ means that only half of the ultimate tensile strength of the material is used in the design.

There is no established rule for setting values of SF in engineering analysis. The definition of SF in Equations 1.5 and 1.6, however, implies that SF is used to "compensate" the discrepancies between analytical results obtained from the idealized state of the structures and those from their actual state. One would thus expect the value of SF be established on factors depending on the nature of the problems in the analysis. Primary factors that affect setting of the values of SF may include the following: (1) The degree and number of idealizations on the real conditions made in Stage 2 in the analysis. The greater the idealization of the problem, the higher the SF values used in Stage 4 of the analysis. (2) The credibility and sophistication of mathematical analysis used in Stage 3 of the analysis. The more sophisticated the analysis in Stage 3, the lower the value of SF. (3) The reliability of material properties used in the analysis. The greater the amount of testing on the materials and components, the lower the value of SF. (4) The liability of the product or engineering system to public safety. (5) The environmental conditions for which the products or engineering systems are designed. Table 1.2 shows recommended typical safety factors for engineering analysis with various applications (http://www.engineeringtoolbox.com/factors-safety-fos-d_1624.html).

The SF values presented in Table 1.2 may be used in interpretation of analytical results in Stage 4 of engineering analysis involving structural design of machines or engineering systems.

Table 1.2 Typical safety factors for engineering analyses.

Equipment	Recommended safety factor (SF)
Aircraft components	1.5–2.5
Boilers	3.5–6.0
Bolts	8.5
Cast iron wheels	20
Engine components	6.0–8.0
Heavy duty shafting	10.0–12.0
Lifting equipment	8.0–9.0
Pressure vessels	3.5–6.0
Turbine components—static	6.0–8.0
Turbine components—rotating	2.0–3.0
Springs, large heavy duty	4.5
Structural steelwork in buildings	4.0–6.0
Structural steelwork in bridges	5.0–7.0
Wire ropes	8.0–9.0

The Engineering ToolBox (http://www.engineeringtoolbox.com/factors-safety-fos-d_1624.html)

For instance, the value SF = 5.61 obtained in Example 1.2 may be translated into a "safe" design because this value is within the range of SF recommended in Table 1.2.

Table 1.2 also reveals an interesting fact that higher SF values are used for the design analysis of pressure vessels with a range of 3.5 to 6. The adoption of a high value of SF in such case is mainly due to serious concern about public liability in the event of structural failure. Consequently, the Unfired Pressure Code (http//en.wikipedia.org/wiki/ASME_Boiler_and_pressure_vessel_code) developed and published by the American Society of Mechanical Engineers (ASME) uses a value of SF = 4. A safety factor of 4 in this case means that pressure vessels designed by the ASME code use only 25% of the maximum measured strength of the vessel materials. The "unused" 75% of the vessel material allows engineers to use simple design formulas such as $\sigma_{\theta\theta} = pr/t \leq \sigma_a = 0.25\sigma_u$, in which $\sigma_{\theta\theta}$ is the hoop stress of a cylindrical pressure vessel' p, r, and t are the internal pressure, inner radius, and required thickness of the vessel respectively; and σ_u is the ultimate tensile strength of the vessel material. The physical implication of using such a high value of SF means an "underuse" of the material strength in the design with corresponding use of more than the necessary amount of materials. However, potential grave public liability takes priority over material usage, and the overweight of the pressure vessel associated with high SF in the design analysis is not a primary design consideration.

In contrast to the case of design analysis of pressure vessels, in which high SF values are used because of the potentially serious public liability in the event of the vessel bursting, very low SF values are used in aircraft component design analysis, with SF = 1.5–2.5 as indicated in Table 1.2. One might wonder why such a low SF value is assigned for aircraft design analysis given that the possible crash of a commercial airplane due to structural failure would generate even more serious public liability than failure of a pressure vessel. The reason for using low SF values in aircraft components design is to use the "good" portion of the material strength to minimize the weight of the aircraft's structure so as to allow higher payloads carried by the aircraft to maximize revenue earned by the owners. The risk of using low SF values in aircraft design analysis is justified by using highly sophisticated analytical tools such as the finite-element method in the design analysis, as will be presented in Chapter 11, coupled with extensive materials and components testing and with stringent quality inspections and control. All of these activities permit the least requirements of "idealization" of conditions in Stage 2 of the analysis, which means that the analytical results are closer to the realistic situations.

Comparison of these two analytical cases typifies the important roles that the safety factor plays in major engineering systems. It also underlines the significance of the safety factor in the design analysis of machine and structure analyses in engineering practice.

1.7 Problems

1.1 Why do engineers "create" but not scientists?

1.2 Give one example for each of the three major functions of professional engineers in their routine work.

1.3 Why is engineering a profession and not merely an occupation?

1.4 Describe the role of mathematical modeling in engineering analysis and also the major tasks involved in the four stages of engineering analysis.

1.5 Use no more than 25 words in answering each of the following questions:
   a) Why is Stage 2 a part of an engineering analysis?
   b) Why is Stage 4 necessary in an engineering analysis?
   c) What is the reason for including safety factors in engineering analysis involving structure design?
   d) How are safety factors determined, and on what bases?
   e) Why the use of higher safety factors in the analysis means a smaller portion of the material's strength is used?

1.6 What would you do in the four stages of engineering analysis in the design of
   a) A diving board for a swimming pool 2 m from the water level with a maximum deflection of 5 cm by a diver weighing 80 kg$_f$.
   b) A bar handle bolted to the chassis of a bus; the bar should be strong enough to support a person weighing 200 lb in a sudden stop from 20 mile/hour cruising.

1.7 Comment on the validity of the idealizations made in Stage 2 in Examples 1.1 and 1.2 in Section 1.5.

1.8 Conduct an engineering analysis on the bridge structure given in Example 1.2, but include the weight of the steel structure plus the weight of the concrete road surface, 20 ft long × 12 ft wide and 6 in thick.

1.9 Conduct an analysis to find the variation of bending stresses with the truck at four different locations on the bridge.

1.10 Study the principles of fracture mechanics on your own initiative and indicate what judgment you would offer to your superiors as to what decisions they should make in dealing with the situations described in Case 1 and Case 2 in section 1.2.3.

2

Mathematical Modeling

Chapter Learning Objectives

- Learn what mathematical modeling is and its application in engineering analysis.
- Learn the physical representation of mathematical entities such as functions, variables, derivatives, and integrals in engineering analysis.
- Understand continuous functions and functions with discrete values.
- Understand curve fitting by polynomial functions.
- Appreciate the application of derivatives in engineering analysis.
- Appreciate the application of integrals in engineering analysis.
- Learn special functions in engineering analysis.
- Understand differential equations in engineering analysis and how they are derived.

2.1　Introduction

As was indicated in Section 1.1 in Chapter 1, "mathematical modeling" is used as a principal tool in engineering analysis. It is reasonable to ask "What is mathematical modeling?" Different answers to the question will be given by different people. Our definition of mathematical modeling is that it is "an act of translating back and forth between mathematics and physical situations."

An analogy to the above definition might be a beautiful melody being developed in the mind of a composer with his or her subsequent expressing this melody in symbolic notes on the stave, as illustrated in Figure 2.1

In engineering analysis, however, the role that mathematical modeling plays is significantly more complicated than that implied in Figure 2.1. Figure 2.2 illustrates that mathematical modeling involves most of the tasks stipulated in Stages 2 and 3 of engineering analyses as described in Section 1.4 in Chapter 1.

There are essentially four different forms of math tools available to engineers in their engineering analyses.

Form 2.1　*Empirical formulas*

Empirical formulas are developed to describe physical situations that are too complicated to be expressed in closed-form mathematical expressions. These formulas contain factors that can only be determined by experiment. Many of these formulas are derived from engineers' own experience, or their employers.

Applied Engineering Analysis, First Edition. Tai-Ran Hsu.
© 2018 John Wiley & Sons Ltd. Published 2018 by John Wiley & Sons Ltd.
Companion Website: www.wiley.com/go/hsu/applied

Figure 2.1 Translation of a melody into music notes—an analogy to mathematical modeling in engineering.

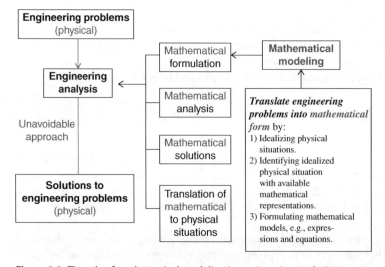

Figure 2.2 The role of mathematical modeling in engineering analysis.

There are types of empirical formulas that are frequently used in engineering analysis. The following are two examples that engineers may use in their analyses.

1. Estimation of the pressure drop in fluid flow in pipes

Pressure drop (ΔP) is a primary parameter for engineers in their design of fluid flow in pipes and tubes and occurs in many thermal/fluid engineering systems analyses. Fluid flow is possible only if the supply pumping energy is sufficient enough to overcome the pressure drop along its flow path. Pressure drop in pipe flow is primarily induced by friction between the pipe (or tube) walls and the contacting fluids.

Several empirical formulas are available for estimating the pressure drop in straight pipe flow, such as that shown in Equation 2.1 (Moody, 1944):

$$\Delta P = f \frac{\rho v^2}{2} \frac{L}{D} \tag{2.1}$$

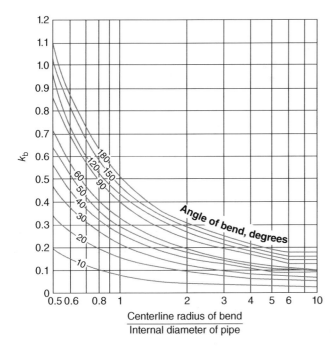

Figure 2.3 Head loss of fluids passing pipe bends.

where ρ = mass density of the fluid, v = average velocity of fluid flow, L = length of the pipe, D = diameter of the pipe, and f = friction factor from Moody chart (http://en.wikipedia.org/wiki/Moody_chart).

The friction factor f in Equation 2.1 with the values available from the Moody chart is derived from experiment.

Additional pressure drop occurs when the fluid flows through bends in a pipeline (see "Bends, flow and pressure drop in" at http://www.thermopedia.com/content/577/). This additional pressure drop in the bends is induced by a radial pressure gradient created by the centrifugal force acting on the fluid in the flow. The following empirical formula is used for estimating the pressure drop in pipe bends:

$$\Delta P = f \frac{\rho v^2}{2} \frac{\pi R_{\mathrm{b}}}{D} \frac{\theta}{180°} + \frac{1}{2} k_{\mathrm{b}} \rho v^2 \tag{2.2}$$

where f is the same Moody friction factor as shown in Equation 2.1, R_{b} = radius of the pipe bend, θ = the angle of the bend, and k_{b} = the bend loss coefficient.

Numerical values of the bend loss coefficient k_{b} in Equation 2.2 are available in handbooks (e.g., Friend and Idelchik, 1989), or from plots such as that in Figure 2.3.

2. Empirical formulas in convective heat transfer analysis

Heat transfer by convection, in particular, forced convective heat transfer is used in the design analysis of many items of heat transfer equipment such as tubular and compact heat exchangers and steam generators. Physical situations in such analyses involve convective heat transfer between the hot and cold fluids separated by tubular containing surfaces such as illustrated in Figure 2.4. Effective heat transfer between the two fluids is achievable with motion of the fluids.

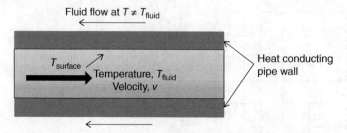

Figure 2.4 Convective heat transfer between two fluids at different temperatures.

The pressure drop (ΔP) appearing in Equations 2.1 and 2.2 thus plays a significant role in this type of analysis.

The amount of heat transfer from the hot fluid to the cold in Figure 2.4 is characterized by a heat transfer coefficient h, which appears in the newtonian cooling law that governs convective heat transfer (Kreith and Bohn, 1997):

$$q_c = h A (T_{surface} - T_{fluid}) \tag{2.3}$$

where q_c = heat transferred from the temperature at the wall surface ($T_{surface}$) of the tube to the temperature of the fluid (T_{fluid}) flowing inside the tube. A = the contacting surface area of the fluid and the tube wall.

Numerical values of the heat transfer coefficient h in Equation 2.3 are too complicated to be determined by closed-form formulas. The following expression in Equation 2.4 is used for the evaluation of the heat transfer coefficient h in Equation 2.3:

$$\mathrm{Nu} = \alpha (\mathrm{Re})^{\beta} (\mathrm{Pr})^{\gamma} f \left(\frac{x}{d_e} \right) \tag{2.4}$$

where Nu = average Nusselt number = (hD/k), Re = Reynolds number = $(\rho Dv/\mu)$, and Pr = Prantl number = $(\mu c_p/k)$, and the coefficient α and indices β and γ are constants determined by experiment. Other symbols in Equation 2.4, ρ, k, c_p, and μ are the respective mass density, thermal conductivity, specific heat under constant pressure and dynamic viscosity of the fluid inside the pipe, and D is the average pipe diameter.

The function $f(x/d_e)$ in Equation 2.4 relates to the aspect ratio of the pipe cross-section, with x indicating the location of the cross-section, and the hydraulic diameter located at x is expressed by

$$d_e = \frac{4A}{p}$$

in which A is the cross-sectional area of the fluid flow and p is the wet perimeter.

This hydraulic diameter d_e is used for either partially filled or fully filled flow in the cross-section of a pipe of a cross-sectional given geometry.

Thus, the value of heat transfer coefficient, h can be obtained from the Nusselt number in terms of the Reynolds and Prantl numbers computed with experimentally determined coefficients and parameters in Equation 2.4.

Form 2.2 *Algebraic equations*

Many textbooks, handbooks (Avallone et al., 2006; Kreith, 1998; Bishop, 2002; Gad-el-Hak, 2002; Whitaker, 1996; Gibilisco, 1997), and technical papers offer simple algebraic formulas and expressions for engineering analysis. Typical examples are reported as "stresses and deflection of beam structures subjected to bending loads, dynamic forces analysis" from engineering mechanics (Meriam and Kraige, 2007), and "temperature and heat flow in solids or fluids" from heat transfer textbooks (Kreith and Bohn, 1997; White, 1994; Janna, 1993). Equations 1.1 to 1.4 given in Chapter 1 are typical algebraic equations used in mathematical modeling for simple structure design analysis.

Form 2.3 *Differential and integral equations*
These equations are derived from the laws of physics for applications in engineering analysis. The following are typical examples of differential and integral equations for such applications

1. *Differential equation in diffusion analysis*
The following differential equation (2.5) is used to determine the concentration of a foreign substance such boron ions, represented by $C(x,t)$, diffusing into another base substance such as silicon substrate—a case of analysis of doping of semiconductor (Hsu, 2008):

$$\frac{\partial C(x,t)}{\partial t} = D\frac{\partial^2(x,t)}{\partial x^2} \tag{2.5}$$

in which $C(x,t)$ is the concentration of the foreign substance at depth x in the base substance at time t; D is the diffusivity of the specific foreign substance in the base substance, a given material characteristic available from handbook.

2. *Integral equation in mathematical modeling of engineering analysis*
Often, engineers need to solve for the function $f(x)$ that represents the solution of the problem embedded in an equation that takes the form of an integral as presented in Equation 2.6. (Spiegel, 1963):

$$\int_0^\infty f(x)\cos ax\,dx = \begin{cases} 1-\alpha & 0 \le \alpha \le 1 \\ 0 & \alpha > 1 \end{cases} \tag{2.6}$$

The solution of this integral equation is

$$f(x) = \frac{2(1-\cos x)}{\pi x^2}$$

Form 2.4 *Numerical solution methods*
There are ample occasions when mathematical modeling for engineering analysis become too complicated, with the requirement for solving nonlinear equations such as transcendental equations in solutions of partial differential equations, as will be presented in Chapter 9, as well as nonlinear differential and integral equations. Closed-form solutions will not be available for these cases. Numerical analysis techniques that require the use of digital computer with algorithms developed specifically for these types of problems are then the only viable technique for solution. There are a number of sources that engineers may use for their numerical analysis, for example[Burden and Faires (2011) and Sauer (2011). Some numerical methods for engineering analysis will be presented in Chapters 10 and 11.

As demand arises for sophisticated engineering analysis on increasing numbers of cases involving complex physical conditions in the geometry, loading, and boundary conditions—such as thermofracture, thermohydraulic, and thermomechanical analyses that require advanced engineering analysis—the forms 1 to 3 of mathematical modeling just outlined are no longer adequate. The finite-element method (FEM) and finite-difference method (FDM) are the appropriate tools for such analyses. Both of these methods are based on discretizing bodies of complex geometry into a finite number of elements of specific simple geometry interconnected at nodes of the elements. Analyses are performed on these elements rather than on the whole body. Publications describing the principles of these advanced engineering analysis techniques are available in the literature (Turner et al., 1959; Zienkiewicz, 1971; Bathe and Wilson, 1976; Hsu, 1986; LeVeque, 2007). There are several commercial software packages available to engineers for advanced engineering analysis of virtually every discipline. Chapter 11 will cover the basic principles and formulations of stress analysis using the finite-element method.

2.2 Mathematical Modeling Terminology

We will begin mathematical modeling by reviewing the physical meanings of some of the terminology that is frequently used in engineering analyses. It is important for engineers to relate many of the terms that they encountered in previous mathematical courses with the physical meanings these terms represent in solving engineering problems. In this section, we will learn what roles such common terms as "function" and "variable" play in engineering analysis, as well as the physical meanings of "differentiation" and "derivative" in the analysis. We will also learn what "integration" is in a physical sense, and how this mathematical operation can help in solving many engineering problems.

2.2.1 The Numbers

2.2.1.1 Real Numbers
Real numbers can either be integers or rational numbers—a, b, a/b, and so on, with the numbers a and b being constants. Rational numbers can be integers or fractional numbers.

2.2.1.2 Imaginary Numbers
These are the numbers that are multiplies of square root (-1) and that, along with real numbers, compose the complex numbers. Imaginary numbers do not have physical meaning but they appear in some mathematical expressions and solutions.

2.2.1.3 Absolute Values
The absolute value of a number recognizes the value ("magnitude") but not the sign attached to the value. Mathematically, the absolute value can be expressed as

$$|x| = x \qquad \text{meaning } x = 0 \text{ or a positive value}$$
$$= -x \qquad \text{meaning } x \text{ is negative, i.e. } x < 0$$

2.2.1.4 Constants
For a constant the value of the number does not change and is always fixed.

2.2.1.5 Parameters
When the value of a number is treated as a "constant" under specific conditions or circumstances we refer to it as a parameter.

2.2.2 Variables

The value of a variable varies with physical conditions, mainly in terms of spatial position or with time. There can be more than one variable involved in engineering analysis.

Two types of variables are commonly involved in engineering analysis:

1) *Spatial variables.* These designate positions in a given space, e.g., (x,y,z) in a space defined by a rectangular coordinate system, or (r,θ,z) in a cylindrical polar coordinate system.
2) *Temporal variable.* The variable representing variation of a physical quantity or system with time t.

Both these types of variables are called "independent variables" because variation of any of these individual variables x, y, and z, or r, θ, z, and t do not affect the values of the other variable presented in the same cases.

2.2.3   Functions

Functions are normally used to represent the physical properties in engineering analyses. Specific functions may be represented with the same symbols used to denote various physical quantities in the analysis. The functions or physical quantities may be represented in mathematical modeling with Latin or Greek letters, according to convention; typical notation includes the following:

Mass (m), weight (W), length (L), area (A), and volume (V) of a solid.
Forces applied to a solid (F).
Temperature (T) of substances, e.g., temperature of a solid or fluid.
Velocity of a rigid solid body or a fluid in motion (V).
Distance that a rigid body travels in a straight or curved path (S).
Stress (σ) and strain (ε) in deformed solids.

The typical mathematical expression of functions such as those designating the physical quantities listed above, usually involve associated variables. For example, the function $F(x)$ may represent the physical quantity of force F acting on a beam at a distance x from a reference point. The value of F depends on the value of the variable x in this case.

Change of the value (or magnitude) of function F associated with change in the values of the variable (or variables) of the function can take three different forms.

2.2.3.1   Form 1. Functions with Discrete Values
In general, the value of a function is determined by the values of all variables associated with the function. The values of all the variables may vary independently of all other variables associated with the function. The values of functions with discrete values are determined by the values of discrete variables. For example, in a hypothetical case that involves a family of four (father, mother, and two children) standing on a beam as illustrated in Figure 2.5a, the load to the beam involves four equivalent concentrated forces corresponding to the weights of the members of this family defined by where the individual members stand. The function $W(x)$ represents the loading forces to the beam with the variable x being the location of the beam, at which the load W is accounted for. Thus, the function $W(x)$ has four discrete values W_1 at $x=x_1$, W_2 at $x=x_2$, W_3 at $x=x_3$, and W_4 at $x=x_4$ as shown in Figure 2.5b.

2.2.3.2   Form 2. Continuous Functions
Because the value of a function is determined by the value of the associated variable or variables, the value of the function can vary *continuously* with the variation of the associated variable(s), as illustrated in Figure 2.6 for the ambient temperature variation at a place over a period of a day; in the function T will represent the ambient temperature at the time of day, which is denoted by t. The value of the function $T(t)$ varies continuously with continuous variation of the value of the variable t.

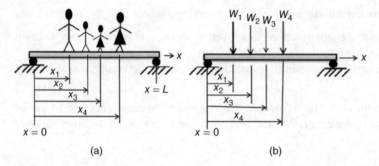

Figure 2.5 Discrete forces applied to a beam structure. (a) People standing on a beam. (b) Equivalent forces applied to the beam.

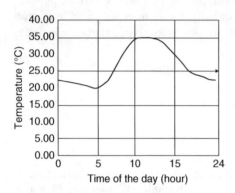

Figure 2.6 Typical diurnal variation of the ambient temperature of a location.

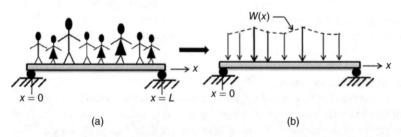

Figure 2.7 Discrete and approximated continuous functions. (a) People standing close to each other. (b) Approximate continuous variation of forces.

Often a function of discrete values may be approximated by a continuous function if the values of the discrete function are small increments of the associated variable, such as in the case illustrated in Figure 2.7, where the weight W applied to the beam by people standing close together on the beam is as shown in Figure 2.7a. The function representing the equivalent forces W applied to the beam may be approximated by an approximated continuous function $W(x)$ as shown in Figure 2.7b.

2.2.3.3 Form 3. Piecewise Continuous Functions

The magnitudes of the physical quantities associated with many physical phenomena in engineering practice vary in ways that can be represented by either continuous functions, as illustrated in Figure 2.6, or by functions of discrete values as illustrated in Figure 2.5. However, there are other cases; one may observe, for example, that the instantaneous water level (y) in

Figure 2.8 A typical water cooler tank.

Serving number:

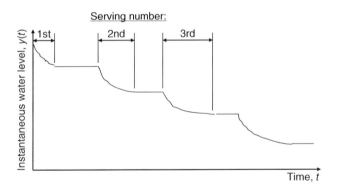

Figure 2.9 A qualitative representation of water level in the drinking water tank at different times.

the office water cooler tank shown in Figure 2.8 may be represented by both discrete values and continuous variation with the variable time (t). Graphical representation of this physical phenomenon is shown in Figure 2.9, which shows how the water level drops intermittently over time after each serving of water. It illustrates a continuous drop of water level while the release tap is open, followed by a period of constant water level before the next serving.

In many cases, functions may involve more than one independent variable, e.g., a function $F(x,y)$, in which the value of function F varies with independent variations of the two independent variables, x and y.

In summary, we observe the following properties of a function:

1) Functions are dependent variables themselves because the value of a function changes depending on the values of its associated variables.
2) The independent variables are spatial variables or temporal variables, or both.

2.2.4 Curve Fitting Technique in Engineering Analysis

Curve fitting in engineering analysis is undertaken to derive a continuous function that will best fit either a set of data points from experiments or functions with discrete values such as that illustrated in Figure 2.7b. This technique is frequently used to describe the geometric profile of solids in engineering analyses. The continuous functions that are derived using a curve fitting technique can also be used for "interpolation" and "extrapolation" of the value of the function within and outside the range of the variables used in defining this function.

A number of techniques are available for performing curve fitting. Popular techniques used in engineering analysis include (1) least-squares technique, (2) spline curve fitting technique, and (3) polynomial curve fitting technique.

The least-squares curve fitting technique (https://en.wikipedia.org/wiki/Least_square) results in more accurate fitting of large numbers of widely scattered data, whereas the function that is derived using the spline fitting technique (https://en.wikipedia.org/wiki/Spline_(mathematics)) passes through all the data points with smooth transitions. Use of the polynomial function for curve fitting is the simplest of all curve fitting techniques, and will be described in the next subsection.

2.2.4.1 Curve Fitting Using Polynomial Functions

This technique involved the derivation of a polynomial function of order n that will pass through all the given data or sample points. Figure 2.10 shows a data set of five temperatures measured in a production process.

We will use this example to derive a polynomial function that will pass through all measured data points by following a procedure that begins with the assumption of a polynomial function of order n in the general form

$$Y(x) = A_0 + A_1 x + A_2 x^2 + A_3 x^3 + \cdots + A_n x^n \tag{2.7}$$

in which the order of the polynomial function $n = N - 1$, with N being the total number of available data (sample) points. The unknown coefficients $A_0, A_1, A_2, A_3, \ldots, A_n$ are determined by a set of simultaneous equations established from the given coordinates of the sample points.

The coordinate system, x–y used in Equation 2.7 is shown in Figure 2.11.

Because the assumed polynomial function in Equation 2.7 is expected to pass through all the sample points with specified coordinates (x_i, Y_i) as indicated in the right portion of Figure 2.11, the following five simultaneous equations will be established for the five given sample points in Figure 2.10:

$$Y_1 = A_0 + A_1 x_1 + A_2 x_1^2 + A_3 x_1^3 + A_4 x_1^4$$
$$Y_2 = A_0 + A_1 x_2 + A_2 x_2^2 + A_3 x_2^3 + A_4 x_2^4$$
$$Y_3 = A_0 + A_1 x_3 + A_2 x_3^2 + A_3 x_3^3 + A_4 x_3^4$$
$$Y_4 = A_0 + A_1 x_4 + A_2 x_4^2 + A_3 x_4^3 + A_4 x_4^4$$
$$Y_5 = A_0 + A_1 x_5 + A_2 x_5^2 + A_3 x_5^3 + A_5 x_5^4$$

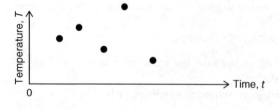

Figure 2.10 Measured temperatures from a fabrication process.

Figure 2.11 Coordinate system for polynomial function curve fitting.

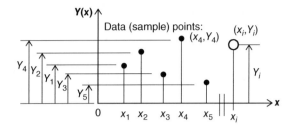

The five unknown coefficients, $A_0, A_1, \ldots, A_4$ can be determined by solving these five simultaneous equations.

Example 2.1

Derive a polynomial function that will pass the following three (3) data (sample) points: (1, 1.943), (2.75, 7.886), and (5, 1.738).

Solution:

We first note that the total number of sample point is 3, i.e., $N = 3$, giving 2 as the highest order of the assumed polynomial function: $n = N - 1 = 3 - 1 = 2$.

We assume a polynomial function of order of 2 in the form

$$Y(x) = A_0 + A_1 x + A_2 x^2 \tag{a}$$

The assumed polynomial function in Equation (a) will fit (i.e., pass through) the three given data (sample) points with given coordinates:

$$Y_1 = 1.943 \text{ at } x_1 = 1, \qquad Y_2 = 7.886 \text{ at } x_2 = 2.75, \qquad Y_3 = 1.738 \text{ at } x_3 = 5$$

Consequently, by substituting the coordinates of the sample points into Equation (a), we will obtain three simultaneous equations:

$$\text{For } i = 1: \quad A_0 + A_1(1) + A_2(1)^2 = 1.943 \tag{b1}$$

$$\text{For } i = 2: \quad A_0 + A_1(2.75) + A_2(2.75)^2 = 7.886 \tag{b2}$$

$$\text{For } i = 3: \quad A_0 + A_1(5) + A_2(5)^2 = 1.738 \tag{b3}$$

Solving the simultaneous equations in Equations (b1), (b2), and (b3), we obtain the values of the three unknown coefficients as

$$A_0 = -5.6663; \qquad A_1 = 9.1414; \qquad A_2 = -1.5321$$

Hence the polynomial function that will pass through the three given data points takes the form:

$$Y(x) = -1.5321x^2 + 9.1414x - 5.6663 \tag{c}$$

The function in Equation (c) is shown graphically in Figure 2.12.

Having obtained this continuous function that includes the three data points, we may use it for interpolating the value of the function between the limits of the sample points, and also for extrapolating for the values outside the limits of the sample range, as illustrated in Figure 2.12.

2.2.5 Derivative

The values of most engineering quantities vary with *position* in the space, defined by the spatial variables, or/and with *time*, the temporal variable. For instance, the pressure in a moving fluid varies with position, and it also often varies with time if conditions change.

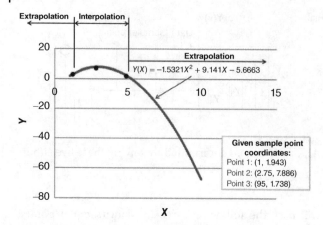

Figure 2.12 The fit of the derived function to the three given data (sample) points.

In most cases, variations of the value of functions occur *continuously* within the ranges of the associated variables, as illustrated in Figures 2.6 and 2.9.

2.2.5.1 The Physical Meaning of Derivatives

We define a derivative to be the *rate* of change of the value of a continuous function with respect to one of its associated variables.

Take for example the situation of a pressure chamber used in fabricating a product as illustrated in Figure 2.13. The process requires the product to be subjected to hydrostatic pressure. The chamber can be pressurized by regulating the pumping power and the pressure relief valve of the chamber. The rate of change of the pressure in the chamber varies as illustrated in Figure 2.14 in which the pressure at the closed circle points are the nine measured pressure readings at specific instances from the meter attached to the chamber. The rate of change of the pressure, P of the fluid at time t is expressed mathematically as $(\Delta P/\Delta t)$.

We may observe from the measured pressure in the chamber as shown in Figure 2.14 that the rate of pressure variations represented by $\Delta P/\Delta t$ varies at different time periods Δt at which the measurements of the pressures in the process chamber were taken during the fabrication process. For example, the rate of pressure variation in the period designated Δt_2 is $\Delta P_2/\Delta t_2$ as shown in the figure. The rate of change of the function P in other periods Δt_i can be obtained using similar expressions, generalized as

$$\text{Rate of change of the value of the function } P = \frac{\Delta P_i}{\Delta t_i} \tag{2.8}$$

with $i = 1, 2, 3, \ldots, 6$ in Figure 2.14.

At times the value of a function that represents a physical situation may vary continuously with the continuous variation of the associate variable, such as in the case of the pressure in

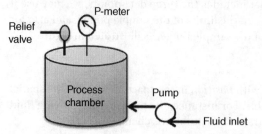

Figure 2.13 A pressurized process chamber.

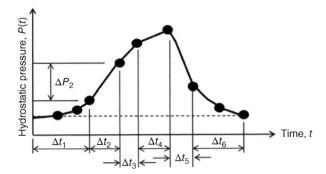

Figure 2.14 Variation of pressure in the process chamber.

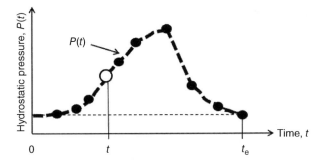

Figure 2.15 Pressure variations in the process chamber in Figure 2.13.

a pressurized process chamber in Figure 2.13. One may readily conceive that the continuous variation of the pressure P in the chamber with time t cannot be represented by the connection of straight lines of all the measured pressures. The real situation would be more like that represented by the dashed curve in Figure 2.15. The rate of change of the function P can no longer be evaluated by Equation 2.8 because the value of P is changing at all times, not in distinct increments over a finite increment of the variable Δt as was represented in Figure 2.14. A different mathematical expression of $P(t)$ derived by a curve fitting technique such as presented in Section 2.2.4 is thus required.

2.2.5.2 Mathematical Expression of Derivatives

Derivatives are used to evaluate the continuous change of the value of a function with infinitesimally small increments in the associate variables. The increments of the variable are so small that we may use the "approximation" of letting $\Delta t \to 0$ in our mathematical formulations.

In Figure 2.16 we illustrate the function y with its values "continuously" varying with an independent variable x; expressed mathematically, $y = f(x)$.

Let us pick two arbitrary points A and B on the curve that graphically represents the function $y = f(x)$. The function's values at these two points are $y_0 = f(x_0)$ at point A and $y_1 = f(x_1)$ at point B, where x_0 and x_1 are the corresponding values of variables associated with function values y_0 and y_1, respectively.

Let $\Delta x =$ the increment of the variable x, then the corresponding change of the function values associated with this increment Δx is $\Delta y = y_1 - y_0 = f(x_1) - f(x_0)$. But we also have the relationship $x_1 = x_0 + \Delta x$ where Δx is the increment of the variable x as shown in Figure 2.16, which leads

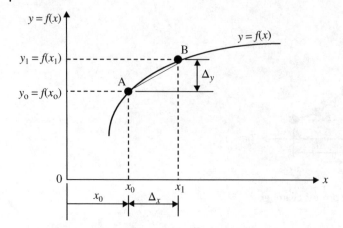

Figure 2.16 Continuous variation of a continuous function.

to the expression for the corresponding change of the value of the function:

$$\Delta y = f(x_0 + \Delta x) - f(x_0)$$

The rate of change of $f(x)$ with respect to x is expressed as

$$\frac{\text{Change of value of function } f(x)}{\text{Change of value of } x\text{-coordinate}} = \frac{\Delta y}{\Delta x}$$

$$= \text{average rate of change within increment } \Delta x$$

This expression can be interpreted as the rate of change of the value of the function between point A and B in Figure 2.16; alternatively, it is the *slope of the curve* represented by $y = f(x)$ between point A and B as illustrated in Figure 2.16.

However, it must be realized Figure 2.16 that the average rate change between points A and B is not equal to the exact rate change at intermediate points between the two bounds A and B if Δx is *large*. The discrepancy becomes smaller with smaller Δx.

A more precise expression for the rate change of function $f(x)$ between x_0 and x_1 is obtained by reducing the size of the increment Δx. In other words, the rate of change of $y = f(x)$ can be represented accurate only if Δx is very small. The increment Δx is made so small that $\Delta x \to 0$. This, in reality, means the variation of the function with respective to the variation of x is *continuous*. Mathematically, it can be expressed as

$$\lim_{\delta x \to 0} \frac{\Delta y}{\Delta x} = \lim_{\delta x \to 0} \frac{f(x_0 + \Delta x) - f(x_0)}{\Delta x} = \frac{dy(x)}{dx}\bigg|_{x=x_0} = \text{the derivative}$$

In general, the derivative of function $y = f(x)$ with respect to the variable x can be expressed in the following ways:

$$\frac{dy}{dx} = \frac{dy(x)}{dx} = \frac{df(x)}{dx}$$
$$= f'(x) = y'(x)$$
$$= \lim_{\delta x \to 0} \frac{f(x + \Delta x) - f(x)}{\Delta x} \tag{2.9}$$

Engineers should realize that not all functions, as represented by $y(x)$ in Equation (2.9), are differentiable. A function is differentiable only if the derivative exists.

Examples of derivatives of functions of various kinds are available in the literature (Ayres, 1964; Spiegel, 1963).

2.2.5.3 Orders of Derivatives

A function, $y(x)$, may be differentiated with its variable x more than once to give various "orders" of derivatives.

The derivative $dy(x)/dx$ in Equation (2.9) is referred to as the first-order derivative of the function $y(x)$, and may be itself a function or a constant.

The derivative of the derivative of the first order $(dy(x)/dx)$ is called the second-order derivative. Mathematically, it is expressed as

$$\frac{d^2y(x)}{dx^2} = \frac{d}{dx}\left(\frac{dy(x)}{dx}\right) = \text{the second-order derivative of function } y(x)$$

$$= \text{the rate of change of the first-order derivative}$$

There are also higher-order derivatives for some functions, such as the third- and fourth-order derivatives are shown below:

$$\frac{d^3y(x)}{dx^3} = \frac{d}{dx}\left(\frac{d^2y(x)}{dx^2}\right) = \text{the third-order derivative of function } y(x)$$

$$\frac{d^4y(x)}{dx^4} = \frac{d}{dx}\left(\frac{d^3y(x)}{dx^3}\right) = \text{the fourth-order derivative of function } y(x)$$

Engineering analyses, especially mechanical and civil engineering analyses, rarely involve derivatives of higher order than 4.

2.2.5.4 Higher-order Derivatives in Engineering Analyses

Derivatives of a function of different orders have different physical meanings in engineering analysis. Figure 2.17 illustrates a beam in a deflected state due to applied bending loads, with the deflected shape of the beam depicted by the dashed line.

The following physical quantities required in beam stress analysis are given by derivatives of various orders of the beam deflection at location x, expressed as $y(x)$:

$$\frac{dy(x)}{dx} = \theta \text{ is the slope of the deflection curve of the beam at location } x \qquad (2.10a)$$

$$EI\frac{d^2y(x)}{dx^2}\bigg|_x = M(x) \text{ is the bending moment in the beam at } x \qquad (2.10b)$$

where E and I are the Young's modulus of the beam material and the moment of inertia of the beam cross-section respectively.

$$C\frac{d^3y(x)}{dx^3}\bigg|_x = V(x) \text{ is the shear force in the beam at } x \text{ } (C \text{ is a constant)} \qquad (2.10c)$$

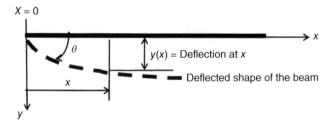

Figure 2.17 A deflected beam subject to bending load.

The deflection $y(x)$ of a beam of constant cross-section subjected to a distributed load $Q(x)$ can be obtained by solving the differential equation (Equation 2.11):

$$\frac{d^4 y(x)}{dx^4} = \frac{Q(x)}{EI}$$

(2.11)

We will illustrate solving for the deflection and the induced stresses of beams subjected to complex loading and end conditions using Equations 2.10 and Equation 2.11 in a later part of this chapter.

2.2.5.5 The Partial Derivatives

There are engineering problems that involve more than one independent variable in the analysis. The value of the function varies independently with each of the associated independent variables. Figure 2.18 illustrates such a case: that the cantilever beam shown in Figure 2.18a is subjected to a moving load at a variable velocity $v(t)$, where t is the time. Clearly, the load on the beam varies not only with the position x but also with the time t, as expressed in Figure 2.18b.

The derivatives of the loading function P given in Figure 2.18b are expressed in terms of each of the two independent variables x and t, but not with both variables included in one derivative. The rate of change of the loading function P needs to be evaluated separately with respect to each of these two variables. We have then two derivatives representing the rate change of the function $P(x,t)$:

$$\frac{\partial P(x, t)}{\partial x} \quad \text{and} \quad \frac{\partial P(x, t)}{\partial t} \quad \text{for the first-order partial derivatives}$$

and

$$\frac{\partial^2 P(x, t)}{\partial x^2} \quad \text{and} \quad \frac{\partial^2 P(x, t)}{\partial t^2} \quad \text{for the second-order partial derivatives}$$

Note that the differential operators $\partial/\partial x$ and $\partial/\partial t$ are used in the above derivatives to indicate these are derivatives of a function with respect to the "partial" not the full rate of change of the function values with both variables associated with the function.

2.2.6 Integration

2.2.6.1 The Concept of Integration

Contrary to differentiation, to obtain the derivatives of functions, which account for the rate of change of the value of a function within selected infinitesimally small increments of the associated variable, integration sums the areas of infinitesimally small elements of area under the graph of the curve $y = f(x)$ with infinitesimally small increments of the associated variable x. Thus, the integral accounts for the total area bounded by the curve defined by the function at

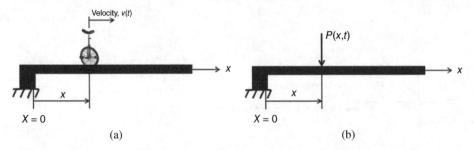

Figure 2.18 A cantilever beam subject to a moving load. (a) Moving load on the beam. (b) The loading function of the beam.

Figure 2.19 Illustration of the concept of integration.

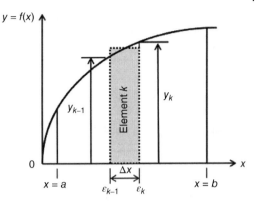

$x = a$ and $x = b$ with infinitesimally small increments of variable x, or $\Delta x \to 0$, as illustrated in Figure 2.19, where a and b are the limits of the integration. There are two types of integrals: definite integrals and indefinite integrals. Definite integrals involve a definite range of the associated variable in the determination of the area bounded by the curve $y(x)$ between these limits; indefinite integrals have no specified range of variable in the evaluations.

2.2.6.2 Mathematical Expression of Integrals

Figure 2.20a represents a measured continuous increase of pressure in the process chamber illustrated in Figure 2.13. We denote the pressure increase with time t as $P(t)$. We readily evaluate the rate of the pressure increase at time t_k to be $dP(t)/dt|_{t=t_k}$ according to Equation 2.9. The formula for the area bounded by the curve of function $P(t)$ can be derived by referring to Figure 2.20b.

Referring to the general case of small element k bounded by function $y = f(x)$ in Figure 2.19, the area of the element is $A_k \approx y_{i-1}\Delta x \approx y_k \Delta x$ with $y_{i-1} \approx y_k$ because the two are located very closely together as $\Delta x \to 0$. The same may be applied to the formulation of element area A_k in Figure 2.20b as $A_k \approx P_k \Delta t_k$. The total area bounded by the curve of the function $P(t)$ between values of the variable of t_a and t_b is the sum of all the element areas:

$$A = \sum_{k=1}^{n} A_k = \sum_{k=1}^{n} P_k \Delta t_k \tag{2.12}$$

We can see from Figure 2.20b that the rectangular-shaped elements shown in Figures 2.19 and 2.20 will more accurately represent the curve represented by the functions $y = f(x)$ in Figure 2.19 and $P(t)$ in Figure 2.20b as the corresponding increments Δx and Δt_k become small; and the

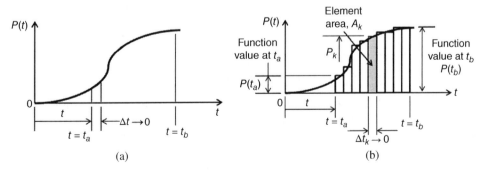

Figure 2.20 Area bounded by a continuous function $P(t)$. (a) A continuous function $P(t)$. (b) The elements of area bounded by $P(t)$.

true representation of the real continuous functions will be obtained as the increments become infinitesimally small, i.e., as $\Delta x \to 0$ in Figure 2.19 and $\Delta t_k \to 0$ in Figure 2.20b. Consequently, we have the following expression for the definite integral for continuous functions $f(x)$ (as in Figure 2.19):

$$A = \lim_{\delta n \to 0} \sum_{k=1}^{n} y_k \, \Delta x_k = \int_a^b f(x) \, dx \qquad (2.13)$$

Similarly, the integral for the situation depicted in Figure 2.20 can be expressed as

$$A = \int_{t_a}^{t_b} P(t) \, dt$$

The results of integration of many different forms of functions can be found in references such as Zwillinger (2003).

If we let $F(x)$ be a function whose derivative, $F'(x) = f(x)$, then the function $F(x)$ is an "*Anti-derivative*" or "*indefinite integral*" of *f(x)*. Mathematically, this relationship is expressed as

$$F(x) = \int F'(x) \, dx = \int f(x) \, dx$$

The function $f(x)$ in the integral is called the *integrand*.

Integration of functions has many applications in engineering analysis. It is used to evaluate areas and volumes enclosed by curves represented by functions. Integration is also used to find geometric quantities such as centroids of plane areas and centers of gravity of solids, as well as moment of inertias for cross-sections of beams and mass moment of inertia of solids involved in curvilinear motion. These geometric properties are frequently involved in many computer-aided design analyses.

2.3 Applications of Integrals

2.3.1 Plane Area by Integration

In Section 2.2.6, we derived the formula for integrals in Equation 2.13 for the area bounded by the curve represented by the function $f(x)$. We will use the same formula for determining the plane areas that can be described by continuous functions.

Example 2.2
Find the area of the right triangle in Figure 2.21 using an integration method.

Solution:
In this rather special case, we may take advantage of the right angle between the two shorter sides of the triangle by setting the x–y coordinates. This would leave the inclined side of the triangle (the hypotenuse) to be the "curve" under which the area is to be evaluated as shown in Figure 2.21. Thus the "curve" is actually a straight line and the mathematical expression to describe this straight line and thus the "curve" is $y(x) = -2x + 4$. Consequently, the area that is enclosed by this straight line is

$$A = \int_0^2 y(x) \, dx = \int_0^2 (-2x + 4) \, dx = (-x^2 + 4x)|_0^2 = 4$$

by using Equation 2.13.

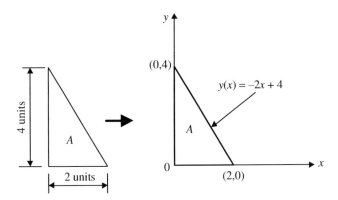

Figure 2.21 Plane area of a right triangle.

Example 2.3

Determine the plane area of a quarter-circular plate with dimensions as shown in Figure 2.22.

Solution:

Since the curved edge of the plate shown in Figure 2.22b fits in a circle, and the equation of a circle is $x^2 + y^2 = R^2 = (6)^2 = 36$, from which we may establish the following functions:

$$x(y) = \sqrt{36 - y^2} \qquad (2.14a)$$

$$y(x) = \sqrt{36 - x^2} \qquad (2.14b)$$

The area A under the arc between $x = 0$ and $x = 6$ in Figure 2.22b can thus be determined by using formula in Equation 2.13 as

$$A = \int_0^6 y(x)\,dx = \int_0^6 \sqrt{36 - x^2}\,dx$$

$$= \frac{1}{2}\left(x\sqrt{36 - x^2} + 36\sin^{-1}\frac{x}{6}\right)\Big|_0^6$$

$$= 9\pi \quad \text{or} \quad 28.26 \text{ cm}^2$$

Example 2.4

Find the area of the plate in Figure 2.23 with a curved edge that fits a cosine function.

Solution:

The curved edge of the plate in Figure 2.23b fits a function $y(x) = 25 \cos[(\pi/80)x]$. Thus the area of the plate can be obtained using the integral shown in Equation (2.13):

$$A = \int_0^{40}\left(25\cos\frac{\pi}{80}x\right)dx = 636.94 \text{ cm}^2$$

Figure 2.22 Plane area of a quarter circular plate. (a) Plate geometry. (b) The plate in an x–y coordinate system.

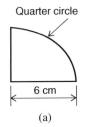

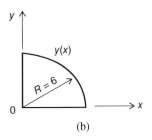

(a)   (b)

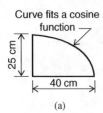

Curve fits a cosine function

25 cm

40 cm

(a)

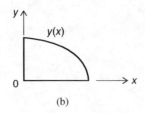

y

$y(x)$

0 → x

(b)

Figure 2.23 Plate with a curved edge. (a) Plate geometry. (b) The plate in an x–y coordinate system.

Example 2.5

Find the area of a plate with a curved edge that fits an ellipse as shown in Figure 2.24a.

Solution:

The function $y(x)$ that represents an ellipse may be derived from the equation of an ellipse:

$$\frac{x^2}{a^2} + \frac{y^2}{b^2} = 1$$

With $a = 3$ m and $b = 6$ m, we will have the function $y(x) = 2\sqrt{9 - x^2}$. Thus, by substituting this function into Equation (2.13), we obtain the area of the plate as

$$A = \int_0^3 (2\sqrt{9 - x^2})\,dx = \frac{9\pi}{2} = 14.13 \ \text{m}^2$$

Example 2.6

Determine the area of a plate with the geometry and overall dimensions shown in Figure 2.25a.

Solution:

The curved edge of the plate as shown in Figure 2.25 was the work of a designer. It does not fit to any common function as were the cases in Examples 2.2 to 2.5. Consequently, one needs to use a curve fitting technique to develop a continuous function to describe this curved edge.

We may place the plate in an x–y coordinate system as shown in Figure 2.25b, in which five points, A, B, C, D, and E, are used to define the geometry of the plate with the coordinates of each point being A(1,0), B(1, 1.943), C(2.75, 7.886), D(5,1.738), E(5,0).

We may use the polynomial curve fitting technique described in Section 2.2.4 with these measured data points to produce a polynomial function with the three data points B, C, and D similarly to what was done in Example 2.1. This gives the function that fits the curved portion of the plate as

$$y(x) = -1.5321x^2 + 9.1414x - 5.6663$$

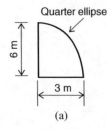

Quarter ellipse

6 m

3 m

(a)

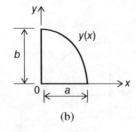

y

$y(x)$

b

0 a → x

(b)

Figure 2.24 Plate with a curved edge that fits an ellipse. (a) Plate geometry. (b) The plate in an x–y coordinate system.

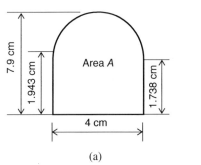
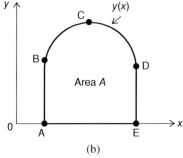

(a)   (b)

Figure 2.25 Plate with curved edge by a designer. (a) Geometry of the plate. (b) The plate in an *x–y* coordinate system.

The area of the plate can thus be determined from the integral

$$A_1 = \int_1^5 y(x)\,dx$$

$$= \int_1^5 (-1.5321x^2 + 9.1414x - 5.6663)\,dx$$

$$= 23.7 \text{ cm}^2$$

The area of the other part, the trapezoid defined by AEDB is

$$A_2 = \frac{(1.738 + 1.943)4}{2} = 7.362 \text{ cm}^2$$

Hence the total plane area of the plate is $A = A_1 + A_2 = 31.062\,\text{cm}^2$.

2.3.1.1 Plane Area Bounded by Two Curves

Integration can also be used to determine areas enclosed by two functions $f(x)$ and $g(x)$ as illustrated in Figure 2.26.

The area defined by the two functions in Figure 2.26 between the limits $x = a$ and $x = b$ is

$$A = \int_a^b f(x)\,dx - \int_a^b g(x)\,dx$$

$$= \int_a^b [f(x) - g(x)]\,dx \tag{2.15}$$

Figure 2.26 Plane area between two curves.

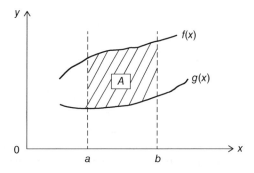

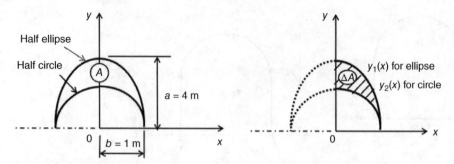

Figure 2.27 Area between a half ellipse and a half circle. (a) Plane area defined by two curves. (b) Symmetry of the geometry about the *y*-axis.

Example 2.7

Determine the plane area between a half ellipse and a half circle as shown in Figure 2.27a.

Solution:

Making use of Figure 2.26 and Equation 2.15, the area between the two curves in Figure 2.27a may be determined defining these two curves in the x–y coordinate systems and using the symmetry of the plane about the y-axis as shown in Figure 2.27b.

The function $y_1(x)$ in Figure 2.27b may be expressed using the equation of ellipse:

$$\frac{x^2}{b^2} + \frac{y^2}{a^2} = 1$$

from which we obtain

$$y(x) = \frac{a}{b}\sqrt{b^2 - x^2}$$

With $a = 4$ m and $b = 1$ m from Figure 2.27a, we have the function $y_1(x) = 4\sqrt{1 - x^2}$.

The other function, $y_2(x)$ may be derived from the equation of circle: $x^2 + y^2 = r^2$, from which we obtain $y(x) = \sqrt{r^2 - x^2}$. With the radius of the circle being $r = 1$ m, we have the function $y_2(x) = \sqrt{1 - x^2}$.

By substituting both $y_1(x)$ and $y_2(x)$ into Equation 2.15, we get the area between these two functions:

$$\Delta A = \int_0^1 [y_1(x) - y_2(x)]\,dx$$

$$= \int_0^1 4\sqrt{1 - x^2}\,dx - \int_0^1 \sqrt{1 - x^2}\,dx$$

$$= 3\int_0^1 \sqrt{1 - x^2}\,dx = 4.71 \ \text{m}^2$$

which leads to a total area between the two functions of $2 \times \Delta A = 9.41$ m^2.

2.3.2 Volumes of Solids of Revolution

This type of solid has a geometry that is symmetrical about one axis ("the axis of symmetry" in Figure 2.28). It is common in mechanical engineering applications. Typical solids of revolution include cylinders, cones, conical frustums, nozzles, etc. These solids are made up by revolving a plane area about a line, which is often called the *axis of rotation*.

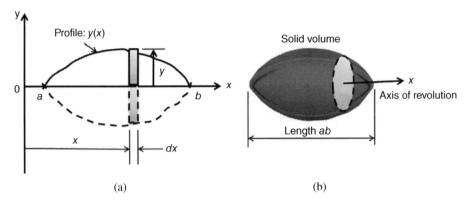

Figure 2.28 Solid of revolution. (a) Exterior profile of the solid. (b) The same solid with axis of revolution.

Often, engineers need to find the volume of a solid of revolution in order to determine the mass or weight of the solid in many design analyses. The volumes of such solids can be obtained using the special definite integrals described here.

Figure 2.28a indicates a function $y(x)$ that represents the profile of the solid of revolution shown in Figure 2.28b. The gray element of volume shown in Figure 2.28a corresponds in position to the cross-section of the solid normal to the axis of revolution that is shown in Figure 2.28b.

The volume of the solid in Figure 2.28b may be obtained by summing up the volumes of small elements in Figure 2.28a. These elements may be viewed as thin "disks" with radius $y(x)$ and thickness Δx. This is expressed mathematically as

$$\Delta V_i = (A_i)\,dx = (\pi r^2)\Delta x$$
$$= \pi[y(x)]^2\Delta x$$

where A_i and r are the cross-sectional area and the radius of the "thin" disk element, respectively.

The volume of the entire solid is the sum of the volumes of all these "thin" disk elements along the axis of revolution, or

$$V = \sum_{i=1}^{n} \Delta V_i = \sum_{i=1}^{n} [\pi(y(x))^2]\,\Delta x$$

in which i designates the disk element number and n is the total number of the disk elements in the subdivided solid.

Because the function $y(x)$ that represents the profile of the solid is a continuous function as shown in Figure 2.28a, we may express the volume of entire solid with the condition that $\Delta x \to dx$ with $dx \to 0$. The summation sign in the above expression for the volume V is thus replaced by the integral with the form

$$V = \int_a^b [\pi y(x)^2]\,dx$$
$$= \pi \int_a^b [y(x)^2]\,dx \tag{2.16}$$

The volume of a solid of revolution about the y-axis, such as shown in Figure 2.29, can be determined in a similar manner.

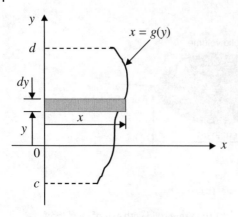

Figure 2.29 Solid volume of revolution about the y-axis.

The area of the shaded element can be expressed as $dA = \pi x^2$ and the volume of revolution of the shaded element is $dv = \pi x^2 \, dy$. The total volume of the solid of revolution about the y-axis can be expressed as

$$V = \pi \int_c^d x^2 \, dy = \pi \int_c^d [g(y)]^2 \, dy \tag{2.17}$$

Example 2.8
Determine the volume of the right cone shown in Figure 2.30 by using an integration method.

Solution:
The right solid cone in Figure 2.30 can be shaped by the rotation of a straight line described by a function $y(x) = 0.5x$ about the x-axis as illustrated in Figure 2.31.

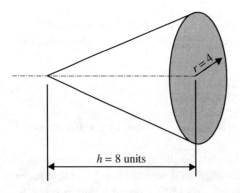

Figure 2.30 A right solid cone.

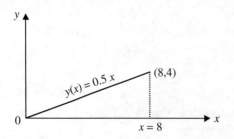

Figure 2.31 Function describing a cone.

The volume of the solid of revolution of the function $y(x)$ about the x-axis is determined using Equation (2.16):

$$V = \pi \int_0^8 [y(x)]^2 \, dx = \pi \int_0^8 (0.5x)^2 \, dx = 42.67\pi$$

One may find from engineering handbooks that the volume of a right cone is equal to

$$V = \frac{1}{3}\pi r^2 h = \frac{1}{3}\pi(4)^2 \times 8 = 42.67\pi$$

which is identical to the result determined using the integration method.

Example 2.9

Find the volume generated by revolving the first-quadrant area bounded by the parabola $y^2 = 8x$ and its latus rectum $(x = 2)$ about the x-axis as illustrated in Figure 2.32.

Solution:

Using Equation 2.16 we will find the volume of revolution V to be

$$V = \pi \int_a^b [f(x)]^2 \, dx = \pi \int_0^2 (8x) \, dx = 4\pi x^2 |_0^2 = 16\pi$$

Example 2.10

Use the integration method to determine the volume of a wine bottle as shown in Figure 2.33a. The dimensions of the bottle are shown in Figure 2.33b.

Solution:

Figure 2.33b shows that the wine bottle is made up of three sections: two straight sections and one with curved cross-section. We may readily determine the volumes of the two straight sections with the expression $V = [(\pi d^2)/4]L$, in which d and L are the respective diameters and lengths of the straight sections of the wine bottle. The volume of the curved part of the bottle requires the use of the integral expression in Equation 2.17.

Let us use the coordinate system (r,z) established as in Figure 2.29, with the radial coordinate r coinciding with the x-coordinate in Figure 2.29 and the axial coordinate z in place of the y-axis in the same figure. Letting $V_1 =$ the volume of the top straight-sided section and $V_2 =$ the volume of the bottom straight sided section:

$$V_1 = \frac{\pi}{4}d_1^2 L_1 = 0.785(2)^2 \times 10 = 31.4 \text{ cm}^3$$

$$V_2 = \frac{\pi}{4}d_2^2 L_2 = 0.785(8)^2 \times 14 = 703.36 \text{ cm}^3$$

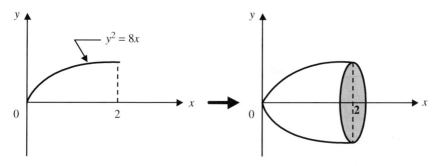

Figure 2.32 A solid volume of revolution of a parabolic curve.

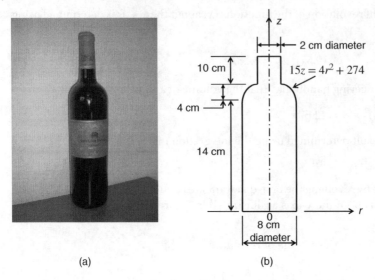

(a)   (b)

Figure 2.33 A wine bottle. (a) The physical bottle. (b) The profile and dimensions of the bottle.

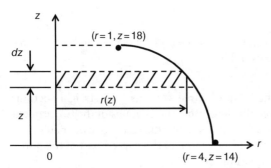

Figure 2.34 Volume of revolution of the curved section of the wine bottle.

We will use the polynomial curve fitting technique described in Section 2.2.4 to fit the profile of the curved portion of the wine bottle with three measure diameters along the z-axis, the axis of revolution, as shown in Figure 2.34.

Using the formulas derived in Section 2.2.4, we can determine the volume of this intermediate section from the cross-hatched volume element in Figure 2.34 to be

$$dV_3 = \pi[r(z)]^2 \, dz$$

where the fitted curved profile of this section of the wine bottle can be represented by a polynomial function $15z = -4r^2 + 274$. Consequently, the total volume in this portion of the wine bottle under the above function is computed as

$$V_3 = \int_{14}^{18} \pi[r(z)]^2 \, dz = \pi \int_{14}^{18} r^2 \, dz = \pi \int_{14}^{18} \frac{274 - 15z}{4} \, dz = 106.76 \ \text{cm}^3$$

The total volume of the wine bottle is then equal to the sum of the volumes of the three sections: $V_1 + V_2 + V_3 = 841.52 \, \text{cm}^3$.

The total volume as computed using the integration method was larger than the $750 \, \text{cm}^3$ shown in the label of the wine bottle. The discrepancy was mainly due to the discounting of the volume of the "punt" of the wine bottle in the above computation. Figure 2.35 shows the punt of a typical wine bottle.

Figure 2.35 The "punt" at the bottom of a wine bottle.

2.3.3 Centroids of Plane Areas

The centroid of a solid of plane geometry is the point that coincides with the center of gravity of the solid. In engineering analysis, we often consider that the mass or weight of machine components of plane geometry (e.g., a *gear* or *plate*) is "concentrated" at the centroid in a typical dynamic analysis of a rigid body of planar geometry in motion (Meriam and Kraige. 2007). Other examples of involving determining the location of centroids of solids of plane geometry include the moving couplers in mechanisms such as illustrated in Figure 2.36. The coupler that is attached to a 4-bar linkage is designed to have its tip A follow a prescribed trajectory with the rotating crank. The induced dynamic force in this oscillating coupler can be a major load to the linkage, and the location of the center of gravity of this coupler of solid plane geometry must be determined in the subsequent stress analysis for the entire mechanism. Figure 2.37 shows an assembly of a mechanism involving a cam and follower that is common in many mechanical systems. The cam, which is usually made in plane geometry, rotates about an axle. The rotary motion of the cam pushes the follower to move up and down according to the profile of the cam. In both of these systems—the coupler of a 4-bar linkage and the cam–follower assembly—rapid motion or rotation of the coupler and the cam is common in practice. The dynamic forces induced by these motions can be significant, and the location of the points of application of force is critical in the design analyses. Identification of the centroids of these planar solids is thus an important part of the design analysis.

A solid of arbitrary plane geometry is shown in Figure 2.38, for which the centroid of the solid plane is located at the coordinates $(\bar{x}, \bar{y})$.

The area of a small element of the plane, such as the one shaded in dark gray in Figure 2.38, is expressed as $dA = dx\, dy$. We will define the area moment of the element area dA about the x-axis to be $dM_x = y(dA) = y\, dx\, dy$, and the area moment about the y-axis to be $dM_y = x\, dA = x\, dx\, dy$.

Figure 2.36 Mechanism with a coupler.

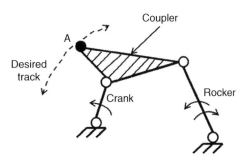

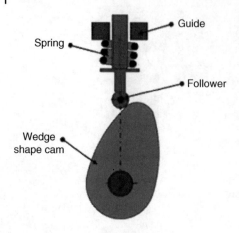

Figure 2.37 Cam with follower.

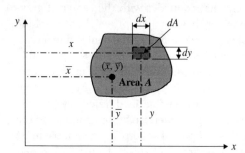

Figure 2.38 Centroid of a plane.

The total area moments can thus be obtained by summing the area moments over all area elements in the entire plane solid:

Area moment about the x-axis:

$$M_x = \int_A dM_x = \int_A y\,dA = \int_x \int_y y(dy\,dx) \tag{2.18a}$$

Area moment about the y-axis:

$$M_y = \int_A dM_y = \int_A x\,dA = \int_y \int_x x(dx\,dy) \tag{2.18b}$$

The coordinate of the centroid of the entire plane can thus be determined by the relations

$$\bar{x} = \frac{M_y}{A} \tag{2.19a}$$

$$\bar{y} = \frac{M_x}{A} \tag{2.19b}$$

where the area of the plane is

$$A = \int_A dA = \int_y \int_x dx\,dy \tag{2.20}$$

Figure 2.39 Solid of plane geometry with straight edges.

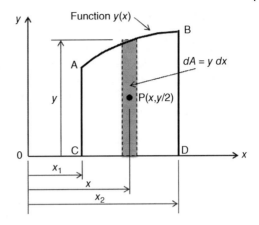

2.3.3.1 Centroid of a Solid of Plane Geometry with Straight Edges

In the case of solids of plane geometry with at least one straight edge, such as the edge C–D of the plane ABCD shown in Figure 2.39, we may set the straight edge to coincide with one of the coordinates as illustrated in the same figure. The area moments for such case can be expressed as

$$M_x \int_{x_1}^{x_2} \left(\frac{1}{2}y\right) dA = \frac{1}{2}\int_{x_1}^{x_2} [y(x)]^2\, dx \tag{2.21a}$$

$$M_y = \int_A x\, dA = \int_{x_1}^{x_2} xy(x)\, dx \tag{2.21b}$$

The coordinates of the centroid $(\bar{x}, \bar{y})$ of the plane ABCD can be obtained by using Equations 2.21a and 2.21b with the area obtained from Equation 2.20

Example 2.11

Determine the location of the centroid of a semicircle with radius a as illustrated in Figure 2.40:

Solution:

The equation describing a circle of radius a is $y^2 + x^2 = a^2$ from analytical geometry, from which we have the functions

$$y(x) = \sqrt{a^2 - x^2} \quad \text{or} \quad x(y) = \sqrt{a^2 - y^2}$$

Figure 2.40 Centroid of a semicircular plate.

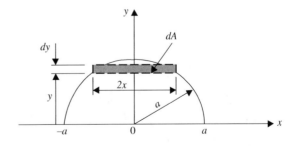

We realize by observation that there is no need to determine the x-coordinate of the centroid because of the symmetry of the geometry about the *y*-axis of the semicircle. This symmetry in geometry leads to $\bar{x} = 0$. However, the determination of the *y*-coordinate of the centroid (i.e., $\bar{y}$) requires the determination of the area moment M_x as indicated in Equation 2.19.

The plane area in Figure 2.40 shows the area of the element $dA = 2x\,dy$ and the area moment is $M_x = y\,dA$ from Equation 2.18a. Thus, the *y*-coordinate of the centroid is determined using Equation 2.21a to be

$$\bar{y} = \frac{\displaystyle\int y\,dA}{A} = \frac{2\displaystyle\int xy\,dy}{2\displaystyle\int x\,dy}$$

$$= \frac{\displaystyle\int_0^a y\sqrt{a^2 - y^2}\,dy}{\displaystyle\int_0^a \sqrt{a^2 - y^2}\,dy}$$

At this stage of the analysis, the engineer may either use an electronic calculator to perform the integrations in the above expression, or use mathematical handbooks such as the *CRC Standard Mathematical Tables and Formulae* (Zwillinger, 2003) to identify the solution. The latter approach will lead to the following for the location of the centroid along the *y*-coordinate:

$$\bar{y} = \frac{-\frac{1}{3}\sqrt{(a^2 - y^2)^3}\,\Big|^a}{\frac{1}{2}\left[y\sqrt{a^2 - y^2} + a^2 \sin^{-1}\frac{y}{|a|}\right]\Big|_0}$$

$$= \frac{\frac{1}{3}a^3}{\frac{1}{2}a^2\left(\frac{\pi}{2}\right)} = \frac{4a}{3\pi}$$

2.3.3.2 Centroid of a Solid with Plane Geometry Defined by Multiple Functions

Cases arise of solids of plane geometry with one edge defined by a multiple functions, such as that illustrated in Figure 2.41. In Figure 2.41, the top edge of the plane is defined by three functions; $y_1(x)$, $y_2(x)$, and $y_3(x)$. One may divide the plane into three regions: A_1, A_2, and A_3. For the corresponding areas of these three subdivisions:

$(\bar{x}_1, \bar{y}_1)$ = the coordinates of the centroid in subdivision A_1
$(\bar{x}_2, \bar{y}_2)$ = the coordinates of the centroid in subdivision A_2
$(\bar{x}_3, \bar{y}_3)$ = the coordinates of the centroid in subdivision A_3

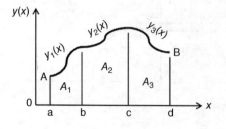

Figure 2.41 Solid of plane geometry with edges defined by multiple functions.

The centroid of the entire plane (ABda), denoted $(\bar{x}, \bar{y})$, may be obtained as the weighted average of those of the three subdivisions as follows:

$$\bar{x} = \frac{\bar{x}_1 A_1 + \bar{x}_2 A_2 + \bar{x}_3 A_3}{A_1 + A_2 + A_3} \tag{2.22a}$$

$$\bar{y} = \frac{\bar{y}_1 A_1 + \bar{y}_2 A_2 + \bar{y}_3 A_3}{A_1 + A_2 + A_3} \tag{2.22b}$$

Example 2.12

Determine the location of the centroid in a coupler made of a triangular plate as illustrated in Figure 2.42a, with the dimensions shown in Figure 2.42b.

Solution:

The plate in Figure 2.42a has three straight edges, so we may choose to have one of the edges BC to coincide with the x-axis as shown in Figure 2.42b. This leaves the two edges AB and AC to define the plate geometry with two distinct functions: $y_1(x)$ for edge AB and $y_2(x)$ for the other edge, in analogy with Figure 2.41.

We first derive the functions $y_1(x)$ and $y_2(x)$ using the coordinates of apexes A, B, and C in Figure 2.42b:

$$y_1(x) = 0.9483x \tag{a}$$

$$y_2(x) = -2.2x + 18.26 \tag{b}$$

We may compute the plane areas for the two divisions in Figure 2.42b, as well as the centroid of these two divisions using Equations 2.13, 2.19, and 2.21 as follows:

$$A_1 = \int_0^{5.8} y_1(x)\,dx = 15.95$$

$$M_{x1} = \frac{1}{2}\int_0^{5.8} [y_1(x)]^2\,dx = 29.24$$

$$M_{y1} = \int_0^{5.8} x y_1(x)\,dx = 61.6743$$

We thus have:

$$\bar{x}_1 = \frac{M_{y1}}{A_1} = 3.8667 \quad \text{and} \quad \bar{y}_1 = \frac{M_{x1}}{A_1} = 1.8332$$

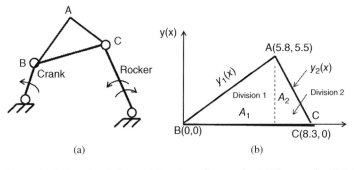

(a)　　　　　　　　　　　　　　　(b)

Figure 2.42 Four-bar linkage with a triangular coupler. (a) The coupler ABC. (b) Dimensions of the coupler.

We may proceed to division 2, following the same procedure as for division 1, and obtain the following:

$$A_2 = \int_{5.8}^{8.3} y_2(x)\,dx = 6.875$$

$$M_{x2} = \frac{1}{2}\int_{5.8}^{8.3} [y_2(x)]^2\,dx = 12.6822$$

$$M_{y2} = \int_{5.8}^{8.3} xy_2(x)\,dx = 45.6042$$

with

$$\bar{x}_2 = \frac{M_{y2}}{A_2} = 6.6333 \quad \text{and} \quad \bar{y}_2 = \frac{M_{x2}}{A_2} = 1.8374$$

The coordinates of the centroid for the entire region can thus be obtained from Equations 2.22a,b to be:

$$\bar{x} = \frac{\bar{x}_1 A_1 + \bar{x}_2 A_2}{A_1 + A_2} = 4.7 \quad \text{and} \quad 4\bar{y} = \frac{\bar{y}_1 A_1 + \bar{y}_2 A_2}{A_1 + A_2} = 1.8345$$

2.3.4 Average Value of Continuous Functions

Often, engineers need to determine the average value of a function that represents a certain physical phenomenon or quantity. It is not a problem for functions with discrete values. For instance, the average load on a beam in Figure 2.5 can be easily determined by adding the four discrete values of the weights of people standing on the beam and then dividing the total weight by a factor of 4. The answer to the same question for the average value of continuous functions such as those illustrated in Figures 2.6 and 2.15, however, cannot be found by calculating the arithmetic average. Integration of the function between the ranges of the variable is the only method for the purpose.

Take, for example, the continuous function $y(x)$ in Figure 2.43a; the average value of this function over the range $x = a$ and $x = b$ can be determined by

$$y_{av} = \frac{\text{Area of rectangle } (A)}{\text{Base of the rectangle}}$$

in which area (A) is the area bounded by the function $y(x)$ in Figure 2.43a between $x = a$ and $x = b$. The same area is equal to the area of the rectangle in Figure 2.43b with the base $(b - a)$.

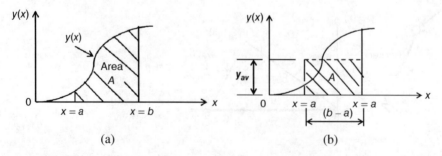

Figure 2.43 Average value of a continuous function. (a) A continuous function. (b) The average value of the function.

We may thus express the above relationship in the form

$$y_{av} = \frac{\int_a^b y(x)\,dx}{b - a} \tag{2.23}$$

Example 2.13

A meteorologist recorded ambient temperature at the San Jose International Airport at the following three time points during a day:

Time of day	Ambient temperature readings (°F)
10:00 a.m.	72
2:00 p.m.	82
6:00 p.m.	65

Estimate the following:

a) The average temperature of the day between 6:00 a.m. and 9:00 p.m.
b) The maximum temperature during that period.

Solution:

We realize that the change of ambient temperature at a specific location is a continuous phenomenon. As such, we need to derive a continuous function that will include the three measured values for the required answers.

We may choose to express the three measured ambient temperatures on a 24-hour clock basis as follows:

Hour of the day (based on 24-hour/day clock)	Measured ambient temperature (°F)
10	72
14	82
18	65

We will use a polynomial function to fit the curve that will include the three measured temperatures and will have the form

$$T(t) = a_0 + a_1 t + a_2 t^2 \tag{a}$$

where $T(t)$ is the continuous function representing the ambient temperature at variable time t, and a_0, a_1, and a_2 are coefficients to be determined by the measured values of temperature in the above table.

Following the procedures presented in Section 2.2.4, we may determine the values of the three coefficients to be

$$a_0 = -71.1, \qquad a_1 = 22.75, \qquad a_2 = -0.844$$

which leads to the function $T(t)$ in the form

$$T(t) = -71.1 + 22.75t - 0.844t^2 \tag{b}$$

The function in Equation (b) allows us to either interpolate or extrapolate temperatures at the specific location that are respectively within or outside the range of the measured temperature.

a) The average temperature of the day between 6:00 a.m. and 9:00 p.m. (or the 21st hour of the day) can thus be estimated using Equation (2.23):

$$T_{av} = \frac{\int_6^{21} T(t)\,dt}{21 - 6} = \frac{\int_6^{21} (-71.1 + 22.75t - 0.844t^2)\,dt}{15} = 66.38°F$$

b) The time and the value of the maximum temperature of the day may be computed in the following way:

Since we have already derived a continuous function for the temperature variation of the day as shown in Equation (b), we may use on Equation (b) the formula from calculus for determining the maximum value of a continuous function $f(x)$.

Thus, we solve for the value of the associate variable by letting x_m be the value at which the function $f(x)$ has a maximum or minimum value by letting $df(x)dx = 0$, and finding a maximum value at x_m if

$$\left. \frac{d^2 f(x)}{dx^2} \right|_{x=x_m} < 0$$

By following the first step, we may determine that a maximum or minimum ambient temperature of 82.2°F occurs at time $t = 13.48$ hour of the day. Evaluation of the second derivative of the temperature function in Equation (b) with $t = 13.48$ shows a value that is less than zero, which indicates that the ambient temperature of 82.2°F indeed is the maximum ambient temperature of the day at the location, and it occurs at 13.48 hour or 1:48 p.m.

2.4   Special Functions for Mathematical Modeling

Engineering students at the junior or senior levels of their undergraduate studies in many engineering schools will already have gained good experience in handling standard functions such as trigonometric and exponential functions in their mathematical education. However, there are engineering analyses that involve situations requiring mathematical formulas to describe physical phenomena that can only be represented by special mathematical functions. These are referred as "special functions" in engineering analysis because they do not appear in many mathematical analyses as commonly as the trigonometric and exponential functions.

There are generally two types of special functions used in engineering analysis:

- **Type 1. Special functions appear in mathematical formulations and solutions.** We will introduce three common special functions that often appear in solutions of engineering analysis:
  - error function and complementary error functions;
  - gamma function;
  - Bessel functions.
- **Type 2. Special functions describing special physical phenomena.** These functions are useful tools in mathematical modeling of two particular physical phenomena that often present in engineering analyses. We will introduce two such special functions:
  - step functions;
  - impulsive functions.

2.4.1 Special Functions in Solutions in Mathematical Modeling

2.4.1.1 The Error Function and Complementary Error Function

This function often appears in the solution of differential equations such as diffusion equations. Diffusion is the net movement of a substance (e.g., an atom, ion, or molecule) from a region of high concentration to a region of low concentration and is a common phenomenon in nature. Oxidation of metals in a moist environment is a typical example. We will look at a special case of diffusion analysis in which the variation of the concentration of a solvent, solvent A, with concentration C_1 diffuses into another solvent, solvent B, with a lower concentration C_2, as illustrated in Figure 2.44.

The concentration of solvent A in the mixed solvent at the depth x and time t after its diffusion into solvent B, expressed as $C(x,t)$, can be obtained using the differential equation already presented in Equation 2.5 (Hsu, 2008):

$$\frac{\partial C(x,t)}{\partial t} = D\frac{\partial^2 C(x,t)}{\partial x^2} \tag{2.5}$$

in which D is the diffusivity of solvent A into solvent B. Diffusivity is a material constant for many substances involved in diffusion processes and the values are available from materials handbooks. The following conditions are specified for the present case of analysis:

$$C(x,0) = 0; \quad C(0,t) = C_s; \quad C(\infty,t) = 0$$

where C_s is the concentration of solvent A at the interface of the two solvents and $C(\infty,t)$ is the concentration of solvent in the mixed solvent far from the interface at time t.

The solution of Equation 2.5 with the specified conditions is

$$C(x,t) = C_s \operatorname{erfc}\left(\frac{x}{2\sqrt{Dt}}\right) \tag{2.24}$$

The function $\operatorname{erfc}(x/2\sqrt{Dt})$ or $\operatorname{erfc}(X)$ with $X = x/2\sqrt{Dt}$ in Equation 2.24 is called the complementary error function, which is related to the error function $\operatorname{erf}(X)$ by the relationship $\operatorname{erfc}(X) = 1 - \operatorname{erf}(X)$. The error function is given by

$$\operatorname{erf}(X) = \frac{2}{\sqrt{\pi}}\int_0^\infty e^{-t^2}\,dt \tag{2.25}$$

Just as for the sine and cosine functions, the value of the error function $\operatorname{erf}(x)$ can be determined from its argument X. Values of the error function with given argument X are available in many textbooks and handbooks (Hsu, 2008; Zwillinger, 2003). It is, however, convenient to recognize that $\operatorname{erf}(0) = 0$ and $\operatorname{erf}(\infty) = 1$.

Figure 2.44 Diffusion of substance A into solvent B.

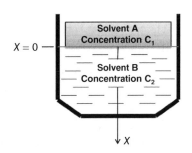

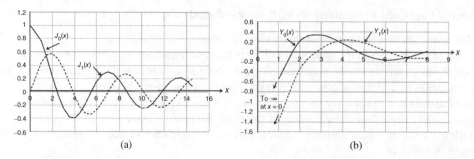

Figure 2.45 Bessel functions. (a) Bessel functions $J_0(x)$ and $J_1(x)$. (b) The Neumann functions $Y_0(x)$ and $Y_1(x)$.

2.4.1.2 The Gamma Function

Like the error function, the gamma function $\Gamma(t)$ also appears in solutions of mathematical modeling. It takes the form

$$\Gamma(t) = \int_0^\infty x^{t-1} e^{-x}\, dx \tag{2.26}$$

The gamma function with integer argument t in Equation (2.26) can be evaluated by the simple formulas

$$\Gamma(t+1) = t\Gamma(t) \quad \text{and} \quad \Gamma(n) = (n-1)!$$

The sign "!" designates the "factorial" value of an integer number. Values of gamma functions are available in mathematics handbooks such as Zwillinger (2003).

2.4.1.3 Bessel Functions

Bessel functions are special functions that often appear in engineering analyses involving problems of "circular," "cylindrical," and "spherical" geometry. They are named after Friedrich Bessel, a German astronomer and mathematician (1784–1846). Bessel developed the Bessel functions in 1824 from his study of the elliptical motion of planets.

Bessel functions behave in a similar way to the periodic sine and cosine functions except that they oscillate with gradual reduction of both the amplitude and frequency.

Bessel Equation and Bessel Functions Bessel functions are solutions of the Bessel equation, which has the form

$$x^2 \frac{d^2 y(x)}{dx^2} + x \frac{dy(x)}{dx} + (\lambda^2 x^2 - n^2) y(x) = 0 \tag{2.27}$$

where $\lambda = $ constant and $n = $ the integers 1, 2, 3, ...

The solution of the Bessel equation (2.27) has the form

$$y(x) = C_1 J_n(\lambda x) + C_2 Y_n(\lambda x) \tag{2.28}$$

in which C_1 and C_2 are arbitrary constants.

The functions $J_n(\lambda x)$ and $Y_n(\lambda x)$ are Bessel functions with specific names:

$J_n(\lambda x)$ is the Bessel function of the first kind of order n.
$Y_n(\lambda x)$ is the Bessel function of the second kind of order n; it is often called a Neumann function.

Figures 2.45a and b illustrate plots of the first two orders of Bessel functions. We will notice that the values of Bessel functions oscillate about the x-axis but with gradual reduction of amplitudes and periods of oscillation.

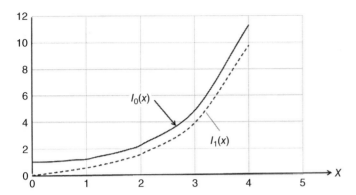

Figure 2.46 Graphical illustration of modified Bessel functions $I_0(x)$ and $I_1(x)$.

A slightly different form of the Bessel equation from Equation 2.27 is

$$x^2\frac{d^2y(x)}{dx^2} + x\frac{dy(x)}{dx} - (\lambda^2 x^2 + n^2)y(x) = 0 \tag{2.29}$$

The solution of the modified Bessel equation in Equation 2.29 is

$$y(x) = C_1 I_n(\lambda x) + C_2 K_n(\lambda x) \tag{2.30}$$

in which $I_n(\lambda x)$ is a modified Bessel function of the first kind of order n, and $K_n(\lambda x)$ is a modified Bessel function of the second kind of order n.

It may be noted from Figure 2.46 that the modified Bessel functions in Equation 2.30 do not oscillate as do the Bessel functions in Equation 2.28.

Example 2.14 Solve the following differential equation:

$$x^2\frac{d^2y(x)}{dx^2} + x\frac{dy(x)}{dx} + (9x^2 - 4)y(x) = 0$$

We recognize that this differential equation is similar in form to Equation 2.27 with $\lambda^2 = 3^2$, and $n^2 = 2^2$. We may thus express the solution as that shown in Equation 2.28 with

$$y(x) = C_1 J_2(3x) + C_2 Y_2(3x)$$

The constants C_1 and C_2 in the above solution may be determined according to specified conditions.

Likewise, the solution of the modified Bessel equation

$$x^2\frac{d^2y(x)}{dx^2} + x\frac{dy(x)}{dx} - (9x^2 + 4)y(x) = 0$$

may be expressed in the form of Equation 2.30:

$$Y(x) = C_1 I_2(3x) + C_2 K_2(3x)$$

Values of Bessel functions of order 0 and 1 with specified arguments are available in mathematical handbooks such as Zwillinger (2003), and as illustrated in Figure 2.45. For instance, $J_0(10.5) = -0.23664$, $J_1(10.5) = -0.07885$, $J_2(10.5) = 0.22162$, and $Y_0(11) = -0.16884$, $Y_1(11) = 0.16370$, $Y_2(11) = 0.19861$. It is useful to be able to recognize the values of Bessel functions of certain special orders and arguments, for instance:

$$J_0(0) = 1, \quad J_n(0) = 0, \quad Y_n(0) \rightarrow \infty$$
$$I_0(0) = 1, \quad I_n(0) = 0, \quad K_n(0) \rightarrow \infty$$

Recurrence Relations of Bessel Functions One will find that many handbooks offer only the values of Bessel functions of order 0 and 1; Bessel functions of higher orders can be obtained by using the following recurrence relations:

$$J_{n-1}(x) + J_{n+1}(x) = \frac{2n}{x} J_n(x) \tag{2.31}$$

$$Y_{n-1}(x) + Y_{n+1}(x) = \frac{2n}{x} Y_n(x) \tag{2.32}$$

Thus, for example, the value of function $J_2(x)$ can be determine by letting $n = 1$ in Equation 2.31 to get

$$J_2(x) = \frac{2}{x} J_1(x) - J_0(x)$$

with the values of $J_0(x)$ and $J_1(x)$ being available in mathematical handbooks (Zwillinger, 2003).

The modified Bessel functions $I_n(x)$ and $K_n(x)$ may be related to the Bessel functions by the formulas:

$$I_n(ix) = i^n J_n(x) \quad \text{and} \quad I_n(x) = i^{-n} J_n(x)$$

where $i = \sqrt{-1}$.

$$K_n(ix) = \frac{\pi i}{2} e^{-(n\pi/2)i} [-J_n(x) + i Y_n(x)]$$

Differentiation and Integration of Bessel Functions The following expressions are used for differentiation:

$$\frac{d}{dx} J_n(ax) = \frac{n}{x} J_n(ax) - a J_{n+1}(ax) \tag{2.33}$$

$$\frac{d}{dx} Y_n(ax) = \frac{n}{x} Y_n(ax) - a Y_{n+1}(ax) \tag{2.34}$$

and for integration:

$$\int x^n J_{n-1}(x)\, dx = x^n J_n(x) \tag{2.35}$$

2.4.2   Special Functions for Particular Physical Phenomena

2.4.2.1   Step Functions

Step functions originated from "Heaviside step functions." They are used to describe physical phenomena that come to existence at a given instant in a "time" domain, or at a position in the "space" domain. These physical phenomena remain in their original state thereafter. Physical phenomena of this kind may be a "weight" or "force" existing in a portion of a solid structure, or for the switching on or off of an electric circuit, resulting in the electric current beginning to flow or ceasing to flow in the circuit thereafter.

The physical situation that a step function represents can be represented diagrammatically as in Figure 2.47. The mathematical expression of the step function in Figure 2.47 is

$$u(x - a) \ \text{ or } \ u_a(x) = 0 \quad \text{for} -\infty < x < a$$
$$= \alpha \quad \text{for } a < x < \infty \tag{2.36a}$$

and

$$u(x - a) = u_a(x) \rightarrow \frac{\alpha}{2} \quad \text{for } x = a \tag{2.36b}$$

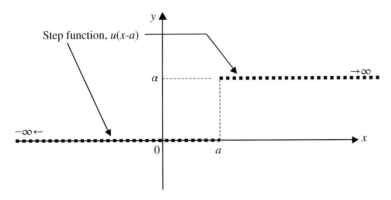

Figure 2.47 Graphical description of a step function.

The function $u_a(x)$ in Equations 2.36a,b denotes a step function with non-zero values, which begins at $x = a$. Step functions can be added or subtracted to represent physical quantities existing over infinite range of variables from $-\infty < x < \infty$ as in Equations 2.36a,b

Example 2.15
Use a step function to describe a physical quantity that varies in the form of "rectangular wave" as illustrated in Figure 2.48.

Solution:
The situation in Figure 2.48 can be described by the superposition of two step functions as illustrated in Figure 2.49. Thus, a physical quantity existing between $x = a$ and $x = b$ in an infinite variable domain such as shown Figure 2.48 can be modeled mathematically by the following expression:

$$f(x) = f_1(x) - f_2(x)$$
$$= Pu(x - a) - Pu(x - b)$$
$$= P[u(x - a) - u(x - b)]$$
$$= Pu_a(x) - Pu_b(x) \tag{2.37}$$

Example 2.16
Use a step function to describe a beam that is subjected to a uniformly distributed load of intensity $w(x)$ over part of its length, as illustrated in Figure 2.50 so that the function describing the load on the beam is valid for the range $-\infty < x < +\infty$.

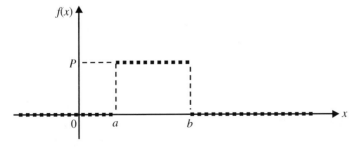

Figure 2.48 General form of a function existing in a finite range in an infinite variable domain.

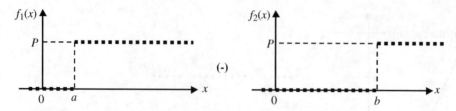

Figure 2.49 Superposition of two step functions.

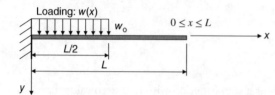

Figure 2.50 A cantilever beam subjected to partial distributed loading.

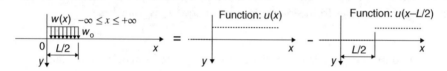

Figure 2.51 Application of a step function to a partially loaded beam.

Solution:

The beam is subjected to a uniformly distributed load $w(x)$ with an intensity w_0, as shown in Figure 2.50, whereas the other part of the beam is load free. The loading function is in the shape of a "step" with the height of the step being w_0 and on the span defined by $0 \leq x \leq L/2$. The loading for the *entire* beam can be described by a step function that is equal to the difference of two step functions with same "height" of the steps, i.e., w_0, but with one function beginning at $x = 0$ and the other with the "step" beginning at $x = L/2$, as illustrated in Figure 2.51.

The loading function for the entire beam length is thus

$$w(x) = w_0 \left[u(x) - u\left(x - \frac{L}{2}\right) \right] \quad \text{with} \quad -\infty \leq x \leq +\infty \tag{2.38}$$

2.4.2.2 Impulsive Functions

Unlike the step function that describes a physical quantity that comes to existence at a specific time or spatial location and remains at its original value thereafter, the *impulsive function* is used to describe physical phenomena that exist only for an *extremely short period of time* or in an extremely small physical extent (i.e., localized physical phenomena). Impulsive functions have other names such as the "delta function" or the "Dirac function." A graphic representation of this function is illustrated in Figure 2.52.

The mathematical expression of the impulsive function is

$$\delta(x) = 0 \quad \text{for} \ x \neq 0$$
$$\rightarrow \infty \quad \text{for} \ x = 0 \tag{2.39}$$

In practice, there is no physical quantity that has an infinite magnitude along with zero pulse width, as illustrated in Figure 2.52. Therefore, for real physical phenomena we deal with finite magnitude and finite but very small pulse width, as illustrated in Figure 2.53. The corresponding function must be modified to be for the following cases.

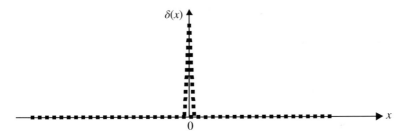

Figure 2.52 Graphical definition of impulsive function.

Figure 2.53 Impulse at origin of a coordinate system.

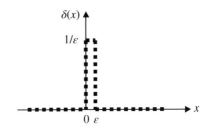

a) The pulse occurs at the origin (Figure 2.53):

$$\delta(x) = 0 \quad x \leq 0$$
$$= 1/\varepsilon \quad 0 < x \leq \varepsilon$$
$$= 0 \quad x > 0 \tag{2.40}$$

b) The pulse occurs at a point $x = a$ away from the origin (Figure 2.54):

$$\delta(x - a) \operatorname{or} \delta_a(x) = 0 \quad x \leq a$$
$$= 1/(\varepsilon - a) \quad a < x < \varepsilon$$
$$= 0 \quad x \geq \varepsilon \tag{2.41}$$

where $\delta_a(x)$ is an impulsive function with the pulse located at $x = a$.

The following are some useful properties of impulsive functions in mathematical modeling:

$$\int_{-\infty}^{\infty} \delta(t)\, dt = 1 \tag{2.42a}$$

$$\int_{0}^{t} \delta(t - a)\, dt = u(t - a) \tag{2.42b}$$

$$\int_{-\infty}^{\infty} f(t)\delta(t - a)\, dt = f(a) \tag{2.42c}$$

$$\frac{du(t)}{dt} = \delta(t) \tag{2.42d}$$

Figure 2.54 Off-origin impulsive function.

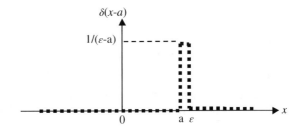

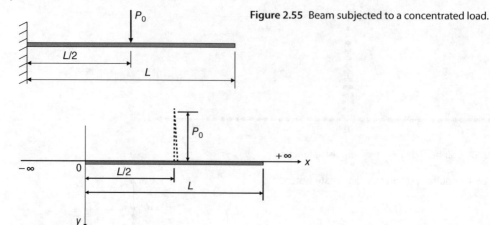

Figure 2.55 Beam subjected to a concentrated load.

Figure 2.56 Impulsive function with a pulse P_0.

Example 2.17

Use an impulsive function to describe the concentrated loading P_0 on a cantilever beam as shown in Figure 2.55.

Solution:

Because the load is a concentrated force that applies at a "spot" located at $x = L/2$, we may use an impulsive function to describe this type of loading as illustrated in Figure 2.56 with a pulse of height P_0 covering the range of $-\infty \leq x \leq +\infty$.

The loading function that covers the variable domain (x) defined by $-\infty \leq x \leq \infty$ may be expressed by an impulsive function as

$$P(x) = P_0\delta(x - L/2) \tag{2.43}$$

2.5 Differential Equations

We began this book with the statement that "Engineering analysis involves the application of scientific principles and approaches ... to reveal the physical state of an engineering system, a machine or device, or structure under study." This properly reminds engineers that the laws of physics should be used as the basis of all engineering analyses. Consequently, all mathematical modeling in engineering analysis must adhere to the laws of physics.

2.5.1 The Laws of Physics for Derivation of Differential Equations

There are three universal laws of physics that all principles and approaches of engineering analysis must follow:

1) Conservation of mass
2) Conservation of energy
3) Conservation of momentum.

Many "application" laws of physics have been derived from these three fundamental laws of physics. For mechanical engineering analyses, the following application laws are often used as bases for the derivation of differential equations used in engineering analysis such as:

- Newton's laws for the mechanics of solids in static, dynamic, and kinematic analyses of machines and rigid bodies.
- Fourier's law for the conduction of heat in solids (it relates the "heat flows" and the induced "temperature gradients," or vice versa, in solids).
- Newton's cooling law for convective heat transfer in fluids.
- Bernoulli's law for fluid dynamics.

The following examples illustrate how other laws of physics can be used to derive differential equations in mathematical modeling.

Example 2.18

Derive an equation to determine the variation of weight of a freely hung solid cone along its length. The cone has root radius R and length L as illustrated in Figure 2.57.

Solution:

We realize that the weight, W, of a solid can be determined by the product of the volume of the solid (V) and its mass density (ρ): or $W = \rho V$. A solution to the gradual weight increase of the cone along its length can be determined by the variation of its volume along its length. Our task is thus to deriving an equation to determine the volumetric variation of the cone along its length.

Referring to Figure 2.57, the volume of the cone can be obtained by summing up all the element volume in the cross-hatched area in the figure. At an arbitrary distance from the fixed end at $y = 0$, the volume of the shaded element equals $\Delta V = \pi [r(y)]^2 \, \Delta y$, in which Δy is the thickness of the volume element.

We may relate the radius $r(y)$ to the variable y by

$$r(y) = R\left(1 - \frac{y}{L}\right)$$

from which we may formulate the volume of the element:

$$\frac{\Delta V}{\Delta y} = \pi R^2 \left(1 - \frac{y}{L}\right)^2$$

We also realize that the volume of the cross-hatched element increases continuously with the increment of y in the last expression. Mathematically, this validates the condition that the

Figure 2.57 Freely hung solid cone.

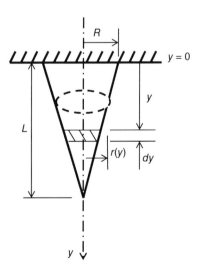

relation holds under the condition that Δy is infinitesimally small, or $\Delta y \to 0$, which leads to the following differential equation:

$$\frac{dV(y)}{dy} - \pi R^2 \left(1 - \frac{y}{L}\right)^2 = 0 \tag{2.44}$$

with the condition $V(0) = 0$.

Example 2.19

Derive the equation of motion of a falling ball of mass m as illustrated in Figure 2.58.

Solution:

We realize that the ball falls from zero initial velocity and falls with a constant gravitational acceleration g toward the ground. Due to this gravitational acceleration, the velocity of the falling ball v increases with time t.

If we let $v(t)$ be the velocity of the mass at time t, the forces that act on the ball at any moment t are as shown in the right-hand part of Figure 2.58. The dynamic equilibrium condition requires that the sum of all these forces be set to zero according to Newton's first law. Mathematically, this is expressed as

$$\sum F_x = -F(t) - R(t) + mg = 0 \tag{a}$$

where $F(t) =$ dynamic (inertial) force, which is in the opposite direction to the acceleration (g in this case) according to Newton's second law; $R(t) =$ the force of air resistance from the air the ball is passing through is proportional to the velocity, or $R(t) = cv(t)$ where c is a proportionality constant; $mg =$ the weight of the ball.

The dynamic force acting on the falling ball can be expressed by Newton's second law as $F(t) = M\, a(t)$, where $a(t) =$ the gravitational acceleration (in this case), which can be expressed as

$$a(t) = \frac{dv(t)}{dt} = \frac{d}{dt}\left(\frac{dx(t)}{dt}\right) = \frac{d^2 x(t)}{dt^2}$$

where $x(t)$ is the instantaneous position of the falling ball.

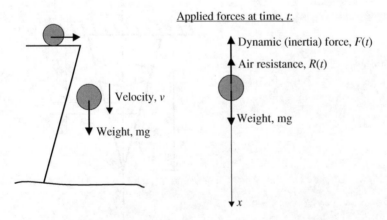

Figure 2.58 Forces on a free fall rigid body.

By substituting the expressions for all the forces into Equation (a), we may derive the following differential equation to be solved for the ball's instantaneous position, $x(t)$, and from which we may obtain the corresponding instantaneous velocity $v(t)$.

$$m\frac{d^2x(t)}{dt^2} + c\frac{dx(t)}{dt} - mg = 0 \tag{2.45}$$

2.6 Problems

2.1 What is the difference between a real number and an imaginary number, and between a constant and a parameter?

2.2 What is mathematical modeling and why does it play a vital role in engineering analysis?

2.3 How would you select an appropriate function with appropriate associated variables that could be used in the solution of the following physical problems?
  a) The time required to drain a swimming pool 20 m × 40 m × 2 m deep.
  b) The time required to transport a silicon wafer by a moving chuck from one end to the other end of a platform of given dimensions.
  c) The time required for a large rotor of an electricity generator made of steel to be heated up to 500°C from room temperature in a furnace in a heat treatment process.
  d) The position of your car determined by a GPS (global positioning system).
  e) The amplitude of vibration of a mass attached to a spring after the application of a small but instantaneous disturbance to the initially motionless mass.
  f) The bending stress in a cantilever beam subjected to a concentrated load (also formulate the function and variable representing the bending stress in a diving board over a swimming pool).

2.4 Give at least one "real-world" situation that can be represented by (a) function of discrete values, and (b) continuous functions.

2.5 Why do we need to use Equation 2.23 to determine the average value of a continuous function?

2.6 Use no more than 25 words to outline the application of curve fitting techniques in engineering analysis.

2.7 Use no more than 25 words to describe the application of the step function and the impulsive function in engineering analysis.

2.8 What are the principal applications of Bessel functions in mathematical modeling of engineering problems?

2.9 Describe geometrically the families of lines represented by the following functions: (a) $y = mx - 3$, and (b) $y = 4x + b$, where m and b are real numbers.

2.10 If $f(x) = x^2 - x$, show that $f(x+1) = f(-x)$.

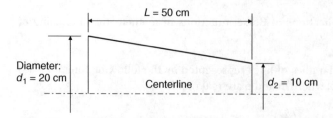

Figure 2.59 Temperature in a nuclear reactor vessel wall.

2.11 Draw the graph of the function
$f(x) = x-1$ if $0 < x < 1$
$f(x) = 2x$ if $x \geq 1$.

2.12 Find the derivatives of each of the following functions:
a) $y = 1/x^2$
b) $y = (1 + 2x)^{1/2}$
c) $y = 1/(2 + x)^{1/2}$

2.13 Derive an expression for the temperature $T(r)$ in a nuclear reactor vessel wall as illustrated in Figure 2.59. The temperature across the reactor wall fits an exponential function with its inner wall maintained at 400°C and the outer wall at 100°C. Constants a and b in Figure 2.59 are the respective inner and outer radii of the wall of the reactor vessel.

2.14 Find the average ordinate of (a) semicircle of radius r; (b) the parabola $= 4 - x^2$ from $x = -2$ to $+2$.

2.15 Temperature is measured in either Fahrenheit or Celsius degrees. Fahrenheit (F) and Celsius (C) temperature are related by the expression $F = aC + b$ in which a and b are constants. The freezing point of water is 0°C or 32°F, and the boiling point of water is 100°C or 212°F. (a) Find the equation relating C and F. (b) At what temperature will the temperatures on both scales be equal in value?

2.16 Derive a function for the variation of the radius along the length of a tapered metal rod with dimensions shown in Figure 2.60 below:

Figure 2.60 Variation of diameter of a rod.

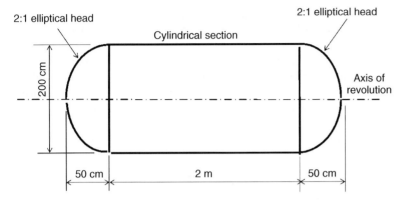

Figure 2.61 A Pressure vessel containing liquid.

2.17 Find the variation of the cross-sectional area, the volume, and the mass of the tapered rod in Figure 2.60 along its length if the mass density of the rod material is ρ g/cm^3.

2.18 A common shape of pressure vessel used in industry involves a hollow cylinder covered at both its ends with welded "2:1 elliptical covers" (or "heads") as illustrated in Figure 2.61. (a 2:1 elliptical head means that the major axis of the ellipse is twice that of the minor axis). Use the integration method to determine the volume content of the pressure vessel with the interior dimensions shown in the figure.

2.19 An engineer designs a circular filling funnel for a water bottling company with a filling volume of 8 fluid ounces (or 236 cm^3). The filling funnel consists of three sections as

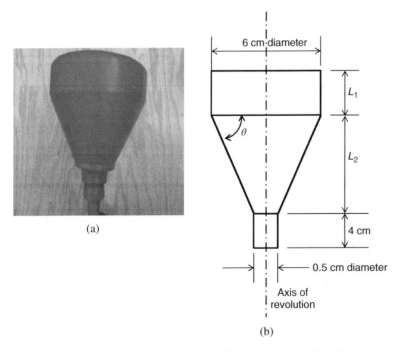

Figure 2.62 Optimal design of a three-section funnel. (a) Image of the three-section funnel. (b) Dimensions of the funnel.

illustrated in Figure 2.62. Determine the approximate tapering angle θ and the lengths L_1 and L_2 that satisfy a design constraint on available space that requires $L_1 + L_2 \leq 10$ cm.

2.20 Use an integration method to determine the volume content of a square tapered chute as illustrated in Figure 2.63 with dimensions $W = 20$ cm, $H = 10.8$ cm, and $h = 1.2$ cm. (Hint: You may check your answer with the formula for the volume of a pyramid from handbooks: Volume of 4-side base pyramids $V = $ (base area)(altitude)/3.)

2.21 A measuring cup shown in Figure 2.64a has overall dimensions given in Figure 2.64b. Determine the following: (a) the overall volume of the cup; (b) the height L_1 for the volume of 200 ml; and (c) the height L_2 for the volume of 100 ml.

2.22 IV bottles such as the one shown in Figure 2.65 are commonly used in hospitals. Use the idealized geometry of the IV bottle shown at the right of the figure to determine
a) The volume of the liquid in the bottle.
b) The diameter of the straight portion of the bottle in Figure 2.65 below if the capacity is designed to be 1200 ml (cm³).

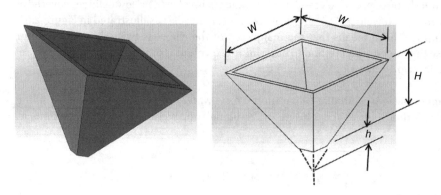

Figure 2.63 Square tapered chute.

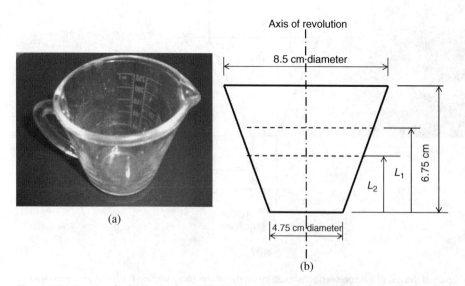

(a)

(b)

Figure 2.64 Design of a measuring jug. (a) Image of the measuring jug. (b) The overall dimensions of the jug.

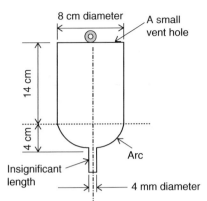

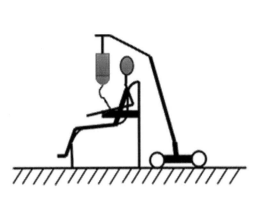

Figure 2.65 IV bottle for hospital use.

Figure 2.66 A partially loaded beam.

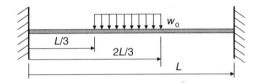

2.23 Use appropriate step functions to describe the loading on the beam illustrated in Figure 2.66.

2.24 Use appropriate step functions to describe the loading on the beam illustrated in Figure 2.67.

2.25 Use the step function to express the loading function $P(x)$ on a partially loaded beam so that the loading function is valid for the region $-\infty < x < +\infty$. The loading condition on the beam is illustrated in Figure 2.68.

2.26 Use an appropriate special function to describe the loading on the beam illustrated in Figure 2.69.

2.27 Derive a mathematical expression for the distributed loads, $P(x)$ on the beam illustrated in Figure 2.70.

2.28 Use the integration method to locate the centroid of a flat plate bounded by three corners A, B, and C with the dimensions shown in Figure 2.71.

Figure 2.67 Nonuniformly loaded beam.

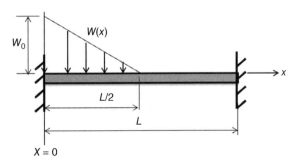

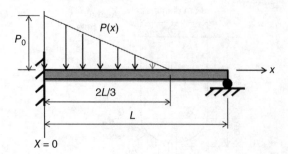

Figure 2.68 Partially loaded beam.

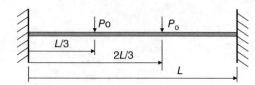

Figure 2.69 A beam subjected to concentrated loads.

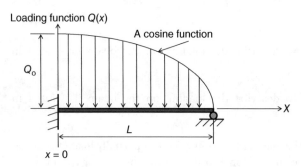

Figure 2.70 A cantilever beam subjected to a distributed load.

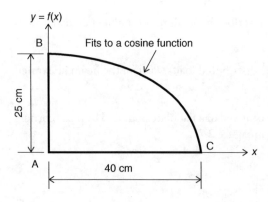

Figure 2.71 A plate with a curved edge.

Figure 2.72 A plate of quarter-circle geometry.

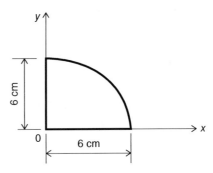

Figure 2.73 A plate of quarter-ellipse geometry.

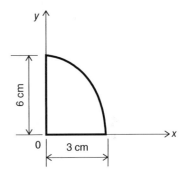

2.29 Use the integration method to locate the centroid of a plate made of a quarter-circle as illustrated in Figure 2.72.

2.30 Use the integration method to locate the centroid of a quarter-ellipse as illustrated in Figure 2.73.

2.31 Use the integration method to locate the centroid of a plate of the geometry and the dimensions shown in Figure 2.74.
Hint: The use of "double integration" technique in determining the enclosed area and area moments of the plane area along both x- and y-coordinates will prove to be a proficient way to reach solutions of this problem

2.32 Use the integration method to determine the area and the location of the centroid of a plate coupler of a four-bar linkage similar to that illustrated in Figure 2.36. This coupler

Figure 2.74 A plate of geometry bounded by a half ellipse and a half circle.

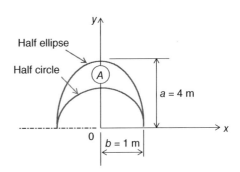

Coordinates: A(0,0), B(2,3), C(6,5), D(8,0)   **Figure 2.75** Plate coupler of a four-bar linkage.

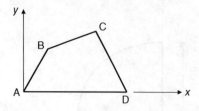

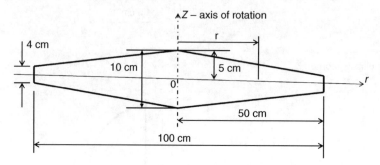

Figure 2.76 Cross-section of a flywheel with a tapered profile.

plate has dimensions defined by the coordinates of the four corners ABCD as shown in Figure 2.75.

2.33 According to a textbook on dynamics (Hibbeler, 2007), the (mass) moment of inertia (I) of a solid machine component is an important quantity that is used as a measure of the resistance of a rigid body to angular acceleration (α) caused by moment $\mathbf{M} = I\alpha$, in which the boldface characters indicate vectorial quantities. The mass moment of inertia of a rigid body can be determined by an integral:

$$I = \int_v r^2 \rho \, dv \tag{a}$$

where r is the perpendicular distance of the moment arm from the axis of rotation, ρ is the mass density of the rigid body, and v is the volume of the rigid body.

Use the above expression to determine the mass moment of inertia of a flywheel with a linearly tapered profile. The flywheel is made of aluminum with a mass density of $2.7 \, \text{g/cm}^3$. The cross-section of the wheel is shown in Figure 2.76. Determine the mass moment of inertia of this flywheel using the integral (Equation a) shown above.

3

Vectors and Vector Calculus

Chapter Learning Objectives

- Recap the distinction between scalar and vector quantities in engineering analysis.
- Learn vector calculus and its applications in engineering analysis.
- Learn to manipulate expressions of vectors and vector functions.
- Refresh vector algebra.
- Learn the dot and cross products of vectors and their physical meanings.
- Learn about derivatives, gradient, divergence, and curl in vector calculus.
- Learn to apply vector calculus in engineering analysis.
- Learn to apply vector calculus in rigid body dynamics in rectilinear and plane curvilinear motion along paths and in both rectangular and cylindrical polar coordinate systems.

3.1   Vector and Scalar Quantities

In Section 2.2.3, we introduced functions that represent physical quantities in engineering analyses, and whose values vary with the values of the associated independent variables in space (x, y, z) in a rectangular coordinate system) and time (t). These quantities are called as *scalar quantities*.

There is another group of physical quantities for which not only the magnitude but also the position and the direction are significant and must be represented.. These are called *vector quantities*. Thus, a *speed* of 80 km/h of a moving car is a *scalar* quantity, but a *velocity* of 80 km/h implies that the car is traveling in specific direction on the road at this speed, so that it is a vector quantity. Engineering analyses involving vectorial quantities will require the use of vector calculus, which will be described later in Section 3.5.

A vector, as stated, is characterized by both its *magnitude* and its *direction*. Examples of vectorial quantities include the velocity of an object traveling either in a (2D) plane or in a (3D) space. As well as the example of the velocity of a moving vehicle, the concept includes related quantities such as acceleration. The force acting on an object is another vectorial quantity that has to be defined in terms of its magnitude and the direction in which the force is acting. Other vectorial quantities include the electric field, current flow, and heat transmission in solids and fluids.

In contrast to vectorial quantities, common physical quantities in engineering analysis such as the temperature, speed, mass of an object, heat and energy, and electric potential are scalar quantities.

A vector may be represented as a directed line segment as shown in Figure 3.1. We will use boldfaced letters and notation to designate the vector quantities throughout this book.

Applied Engineering Analysis, First Edition. Tai-Ran Hsu.
© 2018 John Wiley & Sons Ltd. Published 2018 by John Wiley & Sons Ltd.
Companion Website: www.wiley.com/go/hsu/applied

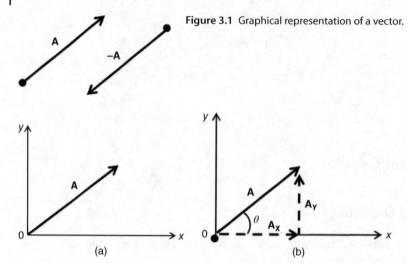

Figure 3.1 Graphical representation of a vector.

Figure 3.2 Rectangular x–y coordinate system for a vector. (a) Vector **A** in x–y coordinates. (b) Decomposition of vector **A**.

In Figure 3.1 the vector **A** is characterized by its magnitude |**A**| and its direction indicated with an arrowhead. Graphical representation of this vector includes an "initial point" shown as a solid circle at one end and a "terminal point" indicated by the arrowhead. A vector with the same magnitude of vector **A** but acting in the opposite direction carries a negative sign (i.e., −**A**) as also shown in Figure 3.1.

Vector quantities are usually defined by a coordinate system in engineering analyses, for example in an x–y coordinate system as shown in Figure 3.2.

Once a vector has been defined in a coordinate system, such as that shown in Figure 3.2a, it can be decomposed into components along those coordinate directions, as illustrated in Figure 3.2b, where the vector **A** is decomposed to component **A**$_x$ with magnitude A_x and the vector component **A**$_y$ with magnitude A_y along the x- and y-axis respectively.

The magnitude of vector **A**, i.e., |**A**| can be obtained as

$$|\mathbf{A}| = \sqrt{|\mathbf{A}_x|^2 + |\mathbf{A}_y|^2} = \sqrt{A_x^2 + A_y^2} \tag{3.1}$$

The **direction** of vector **A** is determined by the angle θ in Figure 3.2b:

$$\tan \theta = \frac{A_y}{A_x} \tag{3.2}$$

Example 3.1

A vector **A** as in Figure 3.2b has its two components along the x- and y-axes with respective magnitudes of 6 units and 4 units. Find the magnitude and direction of the vector **A**.

Solution:

We are given the magnitudes of the vector components along the two coordinates to be $A_x = 6$ and $A_y = 4$; according to Equation (3.1), the magnitude of vector **A** is

$$|\mathbf{A}| = A = \sqrt{6^2 + 4^2} = \sqrt{52} = 7.21 \text{ units}$$

The direction of the vector **A** is determined from Equation (3.2):

$$\tan \theta = \frac{4}{6} = 0.67$$

which gives

$$\theta = \tan^{-1}(0.67) = 33.82°$$

with θ as defined in Figure 3.2b.

3.2 Vectors in Rectangular and Cylindrical Coordinate Systems

Two coordinate systems frequently used in engineering analysis are (1) the rectangular coordinate system with coordinates x, y, and z and (2) the cylindrical coordinate system (r,θ,z), as illustrated in Figure 3.3a.

Figure 3.3b illustrates another way to represent a vector, in this case in three-dimensional space; for instance, the position vector $\mathbf{A}$ with its initial point coinciding with the origin of the coordinate system and its terminal point situated at point P at a position with coordinates (x,y,z) in a rectangular coordinate system. This position vector may be expressed in the following way:

$$\mathbf{A} = x\mathbf{i} + y\mathbf{j} + z\mathbf{k} \tag{3.3}$$

where $\mathbf{i}$, $\mathbf{j}$, and $\mathbf{k}$ are the "unit vectors" along the x-, y-, and z-coordinate directions, respectively, as shown Figure 3.3b. Unit vectors are used to designate the direction of scalar quantities. For example, in Equation 3.3, the term $y\mathbf{j}$ indicates the vector quantity with a magnitude of y but in the direction of the vector $\mathbf{j}$ along the y-coordinate. Each of these three unit vectors in Figure 3.3b has a magnitude of 1.0.

The magnitude of vector $\mathbf{A}$ can be obtained by using the Pythagorean rule:

$$|\mathbf{A}| = A = \sqrt{\left(\sqrt{x^2 + y^2}\right)^2 + z^2} = \sqrt{x^2 + y^2 + z^2} \tag{3.4}$$

3.2.1 Position Vectors

A position vector such as the vector $\mathbf{A}$ of a point located at (x_1, y_1) illustrated in Figure 3.3a. The position of this point may be represented by a vector $\mathbf{r}$ in Figure 3.4a. This vector often is known as the location vector of the point P in space, or the radius vector that represents the position of point P in a space defined by a coordinate system with an arbitrary reference origin

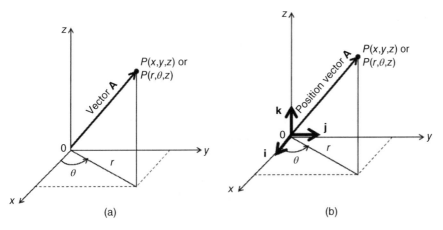

(a) (b)

Figure 3.3 Vector in rectangular and cylindrical coordinate systems. (a) A vector in three-dimensional space. (b) Rectangular unit vectors.

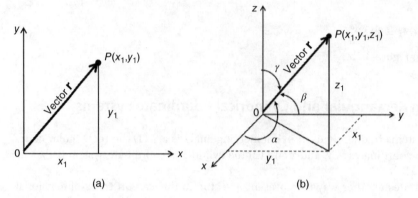

Figure 3.4 Position vectors. (a) In the two-dimensional plane. (b) In three-dimensional space.

O in Figure 3.4b. The position vector is used in describing the motion of "particles" or "rigid bodies" in a planar two-dimensional space or in three-dimensional space.

The position vector $\mathbf{r}$ in Figure 3.4b can be decomposed into three components as $\mathbf{r} = \mathbf{r}_x + \mathbf{r}_y + \mathbf{r}_z$, with $\mathbf{r}_x$, $\mathbf{r}_y$, and $\mathbf{r}_z$ being the components of the position vector $\mathbf{r}$ along the x-, y-, and z-coordinates, respectively. These components may be expressed using the unit vectors $\mathbf{i}$, $\mathbf{j}$, and $\mathbf{k}$ as defined in Figure 3.3b, with $\mathbf{r}_x = x_1\mathbf{i}$, $\mathbf{r}_y = y_1\mathbf{j}$, and $\mathbf{r}_z = z_1\mathbf{k}$, and the vector's magnitude being

$$|\mathbf{r}| = r = \sqrt{|\mathbf{r}_x|^2 + |\mathbf{r}_y|^2 + |\mathbf{r}_z|^2} = \sqrt{x_1^2 + y_1^2 + z_1^2} \tag{3.5}$$

The angles α, β, and γ between the position vector and the respective x-, y- and z-coordinate directions can be computed using the expressions

$$\cos\alpha = \frac{x_1}{|\mathbf{r}|}, \quad \cos\beta = \frac{y_1}{|\mathbf{r}|}, \quad \cos\gamma = \frac{z_1}{|\mathbf{r}|} \tag{3.6}$$

Example 3.2

Express the vector $\mathbf{A}$ in Figure 3.2 in terms of unit vectors as defined in Figure 3.3b. This vector $\mathbf{A}$ has two components with magnitudes and directions given in Example 3.1. Show the magnitudes of the components in both rectangular and cylindrical coordinate systems.

Solution:

The magnitudes of the two components of vector $\mathbf{A}$ with

$$|\mathbf{A}_x| = 6 \quad \text{and} \quad |\mathbf{A}_y| = 4$$

as given in Example 3.1. Following Equation 3.3, we can express this vector in the form of the two corresponding unit vectors $\mathbf{i}$ and $\mathbf{j}$ as follows:

$$\mathbf{A} = 6\mathbf{i} + 4\mathbf{j}$$

The unit vectors in the cylindrical coordinates (r,θ,z) in Figure 3.3a can be derived from the relationship $x = r\cos\theta$, $y = r\sin\theta$, and $z = z$, from which we may express the position vector $\mathbf{A}$ in Figure 3.3b in terms of unit vectors in the following way:

$$\mathbf{A} = r\mathbf{i} + \theta\mathbf{j} + z\mathbf{k} \tag{3.7}$$

in which the magnitudes r, θ, and z in the cylindrical coordinate system can be related to the coordinates x, y, and z in the rectangular coordinate system by the following matrix relation:

$$\begin{Bmatrix} r \\ \theta \\ z \end{Bmatrix} = \begin{bmatrix} \cos\theta & \sin\theta & 0 \\ -\sin\theta & \cos\theta & 0 \\ 0 & 0 & 1 \end{bmatrix} \begin{Bmatrix} x \\ y \\ z \end{Bmatrix} \tag{3.8}$$

Example 3.3
Given vector $\mathbf{A} = 2\mathbf{i} + 4\mathbf{k}$ and vector $\mathbf{B} = 5\mathbf{j} + 6\mathbf{k}$, determine (a) in what planes do these two vectors exist, and (b) their respective magnitudes.

Solution:

a) Vector $\mathbf{A}$ may be expressed as $\mathbf{A} = 2\mathbf{i} + 0\mathbf{j} + 4\mathbf{k}$, so it is positioned in the x–z plane in Figure 3.3. Vector $\mathbf{B}$ may be expressed as $\mathbf{B} = 0\mathbf{i} + 5\mathbf{j} + 6\mathbf{k}$, with no value along the x-coordinate. It is therefore positioned in the y–z plane in a rectangular coordinate system.
b) The magnitude of vector $\mathbf{A}$ is

$$|\mathbf{A}| = A = \sqrt{2^2 + 4^2} = \sqrt{20} = 4.47$$

and the magnitude of vector $\mathbf{B}$ is

$$|\mathbf{B}| = B = \sqrt{5^2 + 6^2} = \sqrt{61} = 7.81$$

Example 3.4
Given two vectors: $\mathbf{A} = 3\mathbf{i} + 2\mathbf{j}$ and $\mathbf{B} = 4\mathbf{i} - 5\mathbf{j}$, plot graphs of (a) $\mathbf{A} + \mathbf{B}$ and (b) $\mathbf{A} - \mathbf{B}$ in the x–y plane.

Solution:

a) $\mathbf{A} + \mathbf{B} = (3 + 4)\mathbf{i} + (2 - 5)\mathbf{j} = 7\mathbf{i} - 3\mathbf{j}$
b) $\mathbf{A} - \mathbf{B} = (3 - 4)\mathbf{i} + [2 - (-5)]\mathbf{j} = -\mathbf{i} + 7\mathbf{j}$

Graphical results for both (a) and (b) are shown in Figure 3.5.

Example 3.5
Determine the angle of a position vector $\mathbf{r} = 6\mathbf{i} + 4\mathbf{j}$ in Figure 3.4a in the x–y plane.

Solution:
Following the definition of unit vectors $\mathbf{i}$ and $\mathbf{j}$ in Figure 3.3b, we have the magnitude of the x-component of the vector $\mathbf{r}$ as $x_1 = 6$ units along the x-coordinate and that of the y-component as $y_1 = 4$ units along the y-coordinate; and the magnitude of the vector $\mathbf{r}$ is $|\mathbf{r}| = \sqrt{6^2 + 4^2} = 7.21$. Using the expressions shown in Equation 3.6, we determine the angles α and β to be

$$\cos\alpha = \frac{6}{7.21} = 0.832, \quad \text{or} \quad \alpha = 33.68°$$

Figure 3.5 Summation and subtraction of the two vectors in Example 3.4.

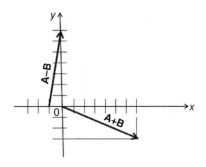

and

$$\cos \beta = \frac{4}{7.21} = 0.555 \quad \text{or} \quad \beta = 56.3°$$

A check of the above computation is that $\alpha + \beta = 33.68 + 56.3 = 89.984° \approx 90°$, as it should be.

3.3 Vectors in 2D Planes and 3D Spaces

In reality, vectors may operate in both two-dimensional planes and three-dimensional space. Figure 3.6a shows a vehicle being pushed uphill in a two-dimensional plane.

In Figure 3.6b, we have all the force vectors acting on the vehicle represented by $\mathbf{F}$ = the force vector that pushes the vehicle uphill against a slope defined by the angle θ, $\mathbf{W}$ = the weight of the vehicle, and the friction force vectors $\mathbf{F}_f$ at the contact of the wheels of the vehicle and the road surface. All these force vectors act in the plane defined by the x–y coordinates, as illustrated in Figure 3.7.

Vector quantities will often act in 3D space. A common case of space vectors is illustrated in Figure 3.8a, with the position of the supports of the three structural components specified in a 3D space defined by the (x,y,z) coordinates. Figure 3.8b shows the corresponding force vectors that are induced in these members by the application of a weight at point C in Figure 3.8a. The applied force, the $\mathbf{F}$ vector, may be expressed as $\mathbf{F} = -W_z\mathbf{k}$ in which the magnitude of the weight is W_z, or in a general form as $\mathbf{F} = F_x\mathbf{i} + F_y\mathbf{j} + F_z\mathbf{k}$, with F_x, F_y and F_z being the magnitudes of the three components of the force vector along the x-, y-, and z-coordinate directions, respectively. The unit vectors $\mathbf{i}$, $\mathbf{j}$, and $\mathbf{k}$ are defined in Figure 3.3b. The induced force vectors in the structural members AC, BC, and OC are expressed as $\mathbf{F}_{AC} = d\mathbf{i} - e\mathbf{j} + b\mathbf{k}$, $\mathbf{F}_{BC} = d\mathbf{i} + f\mathbf{j} + c\mathbf{k}$, and $\mathbf{F}_{OC} = d\mathbf{i}$, as shown in Figure 3.8b.

(a)   (b)

Figure 3.6 Force vectors in a 2D plane. (a) Force acting in a plane. (b) Multiple forces acting in a plane.

Figure 3.7 Force vectors in the x–y plane.

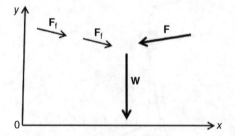

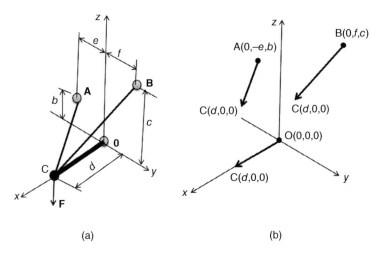

(a)   (b)

Figure 3.8 Vectors in 3D space. (a) A space structure. (b) Force vectors in the structural members.

3.4 Vector Algebra

3.4.1 Addition of Vectors

The resultant vector from summing of vectors can be obtained by joining the initial point of a vector onto the terminal point of the other vector as illustrated in Figure 3.9 for the summation of two vectors.

In Figure 3.9b, vector **C** is the resultant vector of the summation of vectors **A** and **B** in Figure 3.9a. Figure 3.9c shows the summation of two vectors using the parallelogram law.

Summation of more than two vectors is illustrated graphically in Figure 3.10. Here the summation is of the four vectors **A**, **B**, **C**, and **D** shown in Figure 3.10a with the resultant vector **R** = **A** + **B** + **C** + **D** shown in Figure 3.10b.

3.4.2 Subtraction of Vectors

Subtraction of two vectors can be handled by summing the first vector with the second vector in reversed direction. Figure 3.11a shows two vectors **A** and **B** in the same plane. The difference of these two vectors follows the rule of addition of two vectors as illustrated in Figure 3.9b but with vector **B** in opposite direction to that presented in Figure 3.11a. Figure 3.11b shows the

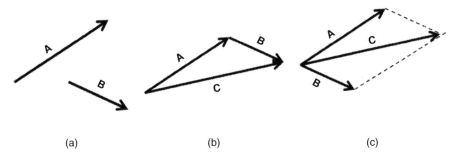

(a)   (b)   (c)

Figure 3.9 Summations of two plane vectors. (a) Two free vectors. (b) Summation of two vectors. (c) An alternative representation of vector summation.

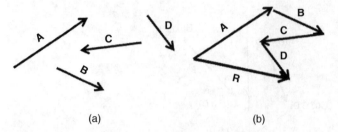

(a)   (b)

Figure 3.10 Summation of multiple plane vectors. (a) Four free vectors. (b) Summation of four vectors.

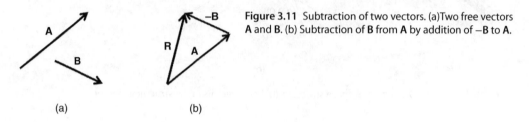

Figure 3.11 Subtraction of two vectors. (a)Two free vectors A and B. (b) Subtraction of **B** from **A** by addition of −**B** to **A**.

(a)   (b)

addition of vector **A** and vector −**B**, giving the difference of these two vectors as the vector **R** where $\mathbf{R} = \mathbf{A} + (-\mathbf{B}) = \mathbf{A} - \mathbf{B}$.

3.4.3 Addition and Subtraction of Vectors Using Unit Vectors in Rectangular Coordinate Systems

A convenient way of adding and subtracting vectors is to express the vectors to be summed or subtracted in terms of rectangular unit vectors as shown in Equation 3.3. The resultant vector is obtained by summing the magnitudes of the components of the vectors associated with the unit vectors **i**, **j**, and **k** along the respective x-, y-, and z-coordinate directions.

Figure 3.12 illustrates how the two vectors $\mathbf{A} = A_x\mathbf{i} + A_y\mathbf{j}$ and $\mathbf{B} = B_x\mathbf{i} + B_y\mathbf{j}$ are summed to give the resultant vector $\mathbf{C} = C_x\mathbf{i} + C_y\mathbf{j}$, in which $C_x = A_x + B_x$ and $C_y = A_y + B_y$.

In Figure 3.12, we have the magnitudes of the components of vector **A** as $A_x = 3$ units and $A_y = 6$ units, and those of vector **B** as $B_x = 5$ units and $B_y = 3$ units; we may determine the magnitudes of the two components $C_x = 3 + 5 = 8$ units and $C_y = 6 + 3 = 9$ units for the resultant vector **C**, as illustrated in Figure 3.12. The resultant vector **C** shown in the figure thus can be expressed as $\mathbf{C} = 8\mathbf{i} + 9\mathbf{j}$.

Figure 3.12 Summation of two vectors in the same plane.

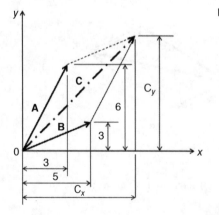

Figure 3.13 Navigation routes of a cruise ship.

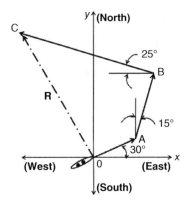

The subtraction of vectors may be handled in a similar way to that demonstrated in Section 3.4.2.

Example 3.6

A cruise ship begins its journey from port O to its destination of port C with intermediate stops over two ports at A and B as shown in Figure 3.13. The ship sails 100 km in the direction 30° to northeast to port A. From port A, the ship sails 180 km in a direction 15° northeast from port A to port B. The last leg of the cruise is from port B to port C in a direction 25° northwest to the north of Port C. Find the total distance the ship travels from port O to port C.

Solution:

Let us assign the position vector **A** to represent the change of position of the ship from port O to port A, vector **B** to represent the ship's change of position from port A to port B, and vector **C** for the change of position from port B to port C. Expressions for the three position vectors in terms of the unit vectors **i** and **j** along the respective coordinates x and y in Figure 3.13 are

$$\mathbf{A} = 100 \left(\cos 30° \right) \mathbf{i} + 100 \left(\sin 30° \right) \mathbf{j} = 86.6\mathbf{i} + 50\mathbf{j}$$
$$\mathbf{B} = 180 \left[\cos \left(30 + 15 \right)° \right] \mathbf{i} + 180 \left[\sin \left(30 + 15 \right)° \right] \mathbf{j} = 127.28\mathbf{i} + 127.28\mathbf{j}$$
$$\mathbf{C} = 350 \left[\cos \left(90 + 25 \right)° \right] \mathbf{i} + 350 \left[\sin \left(90 + 25 \right)° \right] \mathbf{j} = -147.92\mathbf{i} + 317.21\mathbf{j}$$

The resultant vector **R** is the sum of these three position vectors:

$$\mathbf{R} = \mathbf{A} + \mathbf{B} + \mathbf{C} = (86.6 + 127.28 + -147.92)\mathbf{i} + (50 + 127.28 + 317.21)\mathbf{j}$$
$$= 65.96\mathbf{i} + 494.5\mathbf{j}$$

The magnitude of vector **R** is the distance between port O and port C and is

$$|\mathbf{R}| = R = \sqrt{(65.96)^2 + (494.5)^2} = \sqrt{248\,881} = 498.88 \text{ km}$$

3.4.4 Multiplication of Vectors

There are three types of multiplications of vectors: (1) scalar product, (2) dot product, and (3) cross product.

3.4.4.1 Scalar Multiplier

The scalar product is the product of a scalar m with a vector **A**. Mathematically, it is expressed as

$$\mathbf{R} = m(\mathbf{A}) = m\mathbf{A} \tag{3.9}$$

Thus for vector $\mathbf{A} = A_x\mathbf{i} + A_y\mathbf{j} + A_z\mathbf{k}$, where A_x, A_y, and A_z are the magnitudes of the components of vector $\mathbf{A}$ along the x-, y-, and z-coordinate directions, respectively. The resultant vector $\mathbf{R}$ is expressed as

$$\mathbf{R} = mA_x\mathbf{i} + mA_y\mathbf{j} + mA_z\mathbf{k} \tag{3.10}$$

3.4.4.2 Dot Product

The dot product of two vectors results not in another vector but in a scalar quantity that has magnitude but no direction. Physically, the dot product of two vectors is a scalar. It is the projection of one vector onto the other one which acts at an angle θ with it such as that illustrated in Figure 3.14. Mathematically, the dot product of vector $\mathbf{A}$ and vector $\mathbf{B}$ is

$$\mathbf{A} \cdot \mathbf{B} = |\mathbf{A}||\mathbf{B}|\cos\theta = \text{a scalar} \tag{3.11}$$

where θ is the angle between vector $\mathbf{A}$ and vector $\mathbf{B}$, as illustrated in Figure 3.14.

The physical meaning of the dot product of two vectors may be illustrated as in Figure 3.15. It involves the displacement of a vehicle by a displacement vector $\mathbf{d}$ of magnitude d. This displacement is produced by a force vector $\mathbf{F}$ of magnitude F. The work done by the force vector $\mathbf{F}$ for producing a displacement of d of the vehicle is defined as

$$\text{work} = \text{force} \times \text{displacement}$$

Mathematically, the work done on the vehicle is expressed as

$$W = \mathbf{F} \cdot \mathbf{d} = Fd\cos\theta$$

Here the force vector $\mathbf{F}$ and displacement vector $\mathbf{d}$ are in the same direction, as shown in Figure 3.15; thus the angle between these two vectors $\theta = 0$, or $\cos\theta = 1$. We will thus have the work $W = Fd$, which is a scalar quantity in this case.

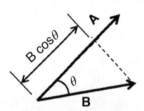

We may also demonstrate that the dot product of the same vectors shown in Equation 3.11 is same with order of multiplication reversed, that is

$$\mathbf{B} \cdot \mathbf{A} = |\mathbf{B}||\mathbf{A}|\cos\theta = \text{the same scalar as in Equation 3.11}$$

The algebraic definition of dot product of vectors can be written as

$$\mathbf{A} \cdot \mathbf{B} = A_xB_x + A_yB_y + A_zB_z \tag{3.12}$$

where A_x, A_y, and A_z are the magnitudes of the components of vector $\mathbf{A}$ along the x-, y-, and z-coordinate directions, respectively, and

Figure 3.14 Dot product of two vectors.

B_x, B_y, and B_z are the magnitudes of the components of vector $\mathbf{B}$ along the same rectangular coordinate directions.

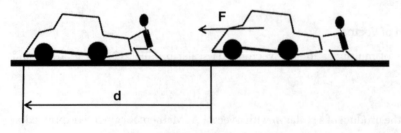

Figure 3.15 Work done in a displacing a solid.

One may thus derive that

$$\mathbf{A} \cdot \mathbf{B} = \mathbf{B} \cdot \mathbf{A} \tag{3.13}$$

The verification of the relation in Equation 3.13 is obvious. For instance, if the magnitudes of the two vectors are $|\mathbf{A}| = 6$ units for vector $\mathbf{A}$ and $|\mathbf{B}| = 4$ units for vector $\mathbf{B}$ with the angle between the two vectors being $\theta = 60^\circ$, the dot product of these two vectors according to Equation (3.11) will be

$$\mathbf{A} \cdot \mathbf{B} = 6 \times 4 \times \cos 60^\circ = 12 \text{ units}$$

and the dot product of vector $\mathbf{B}$ and $\mathbf{A}$ will be the same:

$$\mathbf{B} \cdot \mathbf{A} = 4 \times 6 \cos 60^\circ = 12 \text{ units}$$

The definition of the dot product of vectors also leads to the following useful relations:

$$\mathbf{i} \cdot \mathbf{i} = \mathbf{j} \cdot \mathbf{j} = \mathbf{k} \cdot \mathbf{k} = 1$$
$$\mathbf{i} \cdot \mathbf{j} = \mathbf{j} \cdot \mathbf{i} = \mathbf{i} \cdot \mathbf{k} = \mathbf{k} \cdot \mathbf{i} = \mathbf{j} \cdot \mathbf{k} = \mathbf{k} \cdot \mathbf{j} = 0 \tag{3.14}$$

where $\mathbf{i}$, $\mathbf{j}$, and $\mathbf{k}$ are the unit vectors in a rectangular coordinate systems as illustrated in Figure 3.3.

An algebraic definition with an expression similar to the one in Equation (3.12) for the dot product of two vectors $\mathbf{A} \cdot \mathbf{B}$ may be derived as shown in Equation (3.15). The latter equation is derived by expressing both these vectors with the unit vectors $\mathbf{i}$, $\mathbf{j}$, and $\mathbf{k}$ in rectangular coordinates for $\mathbf{A} = A_x\mathbf{i} + A_y\mathbf{j} + A_z\mathbf{k}$, and $\mathbf{B} = B_x\mathbf{i} + B_y\mathbf{j} + B_z\mathbf{k}$ and using the relationships given in Equation (3.14):

$$\mathbf{A} \cdot \mathbf{B} = A_xB_x + A_yB_y + A_zB_z \tag{3.15}$$

where A_x, A_y, and A_z are the magnitudes of the components of vector $\mathbf{A}$ along the x-, y-, and z-coordinate directions, respectively, and B_x, B_y, and B_z are the magnitudes of the components of vector $\mathbf{B}$ along the same rectangular coordinate directions. The unit vectors $\mathbf{i}$, $\mathbf{j}$, and $\mathbf{k}$ are defined in Figure 3.3.

A useful rearrangement of the expression for the dot product of two vectors gives the angle between two vector such as the vectors $\mathbf{A}$ and $\mathbf{B}$ in Equation (3.15). The angle between these two vectors θ may be obtained as:

$$\cos\theta = \frac{A_xB_x + A_yB_y + A_zB_z}{AB} \tag{3.16}$$

Example 3.7

Determine (a) the dot product of the two vectors: $\mathbf{A} = 2\mathbf{i} + 7\mathbf{j} + 15\mathbf{k}$ and $\mathbf{B} = 21\mathbf{i} + 31\mathbf{j} + 41\mathbf{k}$, and (b) the angle between these two vectors.

Solution:

a) Using Equation 3.15, we get the result of the dot product of vectors $\mathbf{A}$ and $\mathbf{B}$ to be

$$\mathbf{A} \cdot \mathbf{B} = 2 \times 21 + 7 \times 31 + 15 \times 41 = 874$$

b) Since the magnitudes of vectors $\mathbf{A}$ and $\mathbf{B}$ are

$$|\mathbf{A}| = A = \sqrt{2^2 + 7^2 + 15^2} = 16.67 \quad \text{and} \quad |\mathbf{B}| = B = \sqrt{21^2 + 31^2 + 41^2} = 55.52$$

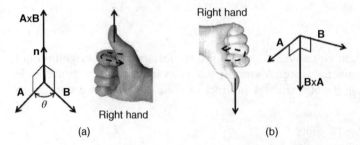

Figure 3.16 Cross product of two vectors **A** and **B** in a right-handed coordinate system. (a) Cross product **A** × **B**. (b) Cross product **B** × **A**.

and the numerator of Equation 3.16 for this case was computed in (a) to be 874, we have

$$\cos\theta = \frac{874}{16.67 \times 55.52} = 0.9443$$

giving the angle between the two vectors as $\theta = \cos^{-1}(0.9443) = 19.21°$

3.4.4.3 Cross Product

The cross product of two vectors **A** and **B** results in a vector **C** that has direction perpendicular to the plane on which both vectors **A** and **B** exist, and has magnitude that is equal to the magnitudes of **A** and **B** and the sine of the angle θ between them. Figure 3.16 shows the cross product of two vectors **A** and **B** in a right-handed coordinate system.

Thus the cross product of two vectors is a vector **R** = **A** × **B** (reads "A cross B"). The magnitude of the resultant vector **R** is the product of the magnitudes of **A** and **B** and the sine of the angle between them (the angle θ in Figure 3.16a). The direction of vector **R** is perpendicular to the plane of **A** and **B** as shown in Figure 3.16. Mathematically, we have

$$\mathbf{R} = \mathbf{A} \times \mathbf{B} = |\mathbf{A}||\mathbf{B}|(\sin\theta)\mathbf{n} = AB(\sin\theta)\mathbf{n} \quad \text{with} \quad 0 \le \theta \le \pi \tag{3.17}$$

where the vector **n** is a unit vector in the normal direction of the cross product of vectors **A** and **B**: that is, **A** × **B**, as shown in Figure 3.16a. If **A** = **B** or if **A** is parallel to **B**, then $\sin\theta = 0$, and **A** × **B** = 0.

Figure 3.17 illustrates the principle of the cross product of two vectors. This case is related to the tightening of a threaded pipe to a faucet using a plumber's wrench. The applied force vector **F** is normal to the axis y, along which is the moment arm vector **d**. One may easily envision that the resultant moment vector $\mathbf{M}_z$ is along the z-axis, which is perpendicular to the plane on which vectors **F** and **d** exist. Mathematically, the resultant moment $\mathbf{M}_z$ may be obtained by $\mathbf{M}_z = \mathbf{F} \times \mathbf{d}$.

The magnitude of the resultant vector **R** in Equation 3.17 may be expressed by a determinant for vector $\mathbf{A} = A_x\mathbf{i} + A_y\mathbf{j} + A_z\mathbf{k}$ and vector $\mathbf{B} = B_x\mathbf{i} + B_y\mathbf{j} + B_z\mathbf{k}$ in a rectangular coordinate system, where A_x, A_y, A_z, and B_x, B_y, B_z are the magnitudes of the components of vectors **A** and **B**, respectively, along the x-, y-, and z-coordinates, and **i**, **j**, and **k** are the unit vectors defined in Figure 3.3b. The resultant vector of the cross product of vectors **A** and **B** may be obtained from the following determinant:

$$\mathbf{R} = \mathbf{A} \times \mathbf{B} = \begin{vmatrix} \mathbf{i} & \mathbf{j} & \mathbf{k} \\ A_x & A_y & A_z \\ B_x & B_y & B_z \end{vmatrix} \tag{3.18}$$

in which the elements between two vertical straight lines represent a *determinant*.

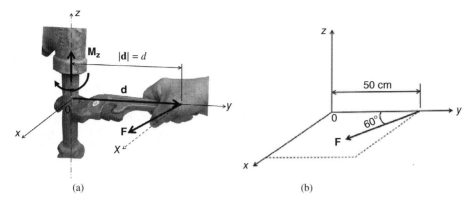

Figure 3.17 Cross product of two vectors giving the torque applied to a pipe. (a) Physical interpretation. (b) Numerical situation. *See color section.*

Example 3.8

Determine the torque applied to the pipe in Figure 3.17b by a force $F = 45\,\text{N}$ with an angle $\theta = 60°$ to the y-axis at a distance $d = 50\,\text{cm}$ from the centerline of the pipe.

Solution:

We may express the force vector

$$\mathbf{F} = (F\sin\theta)\mathbf{i} + (F\cos\theta)\mathbf{j} = (45\sin 60°)\mathbf{i} + (45\cos 60°)\mathbf{j}$$

or

$$\mathbf{F} = 38.97\mathbf{i} + 22.5\mathbf{j}$$

The moment arm vector $\mathbf{d}$ may be expressed as $\mathbf{d} = d\mathbf{j} = 50\mathbf{j}$. The resultant vector $\mathbf{M}_z = \mathbf{F} \times \mathbf{d}$ can thus be computed using Equation 3.18 to be

$$\mathbf{M}_z = \begin{vmatrix} \mathbf{i} & \mathbf{j} & \mathbf{k} \\ 38.97 & 22.5 & 0 \\ 0 & 50 & 0 \end{vmatrix}$$

$$= (22.5 \times 0 - 0 \times 50)\mathbf{i} - (38.97 \times 0 - 0 \times 0)\mathbf{j} + (38.97 \times 50 - 22.5 \times 0)\mathbf{k}$$

$$= 1948.5\mathbf{k}$$

The resultant torque on the pipe thus has a magnitude of $M_z = 1948.5\,\text{N cm}$ in the direction along the z-axis.

Example 3.9

Prove that $\mathbf{A} \times \mathbf{B} = -\mathbf{B} \times \mathbf{A}$.

Solution:

By the definition of cross product of two vectors given in Equation 3.17, we have $\mathbf{A} \times \mathbf{B} = \mathbf{C}$ with the magnitude of vector $\mathbf{C}$ being $C = AB\sin\theta$ and in the direction such that $\mathbf{A}$, $\mathbf{B}$, and $\mathbf{C}$ form a right-handed system as illustrated in Figure 3.16a. The cross product of $\mathbf{B} \times \mathbf{A} = \mathbf{D}$ has the same magnitude, but the vector $\mathbf{D}$ is in the opposite direction as shown in Figure 3.16b in the negative direction. Thus we have proved that $\mathbf{A} \times \mathbf{B} = -\mathbf{B} \times \mathbf{A}$.

Example 3.10

If vectors $\mathbf{A} = \mathbf{i} - \mathbf{j} + 2\mathbf{k}$ and $\mathbf{B} = 2\mathbf{i} + 3\mathbf{j} + 4\mathbf{k}$, determine $\mathbf{A} \times \mathbf{B}$.

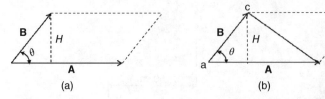

Figure 3.18 Plane areas by cross product of two vectors. (a) Area of a parallelogram. (b) Area of a triangle.

Solution:

We may use Equation 3.18 for the solution:

$$\mathbf{A} \times \mathbf{B} = \begin{vmatrix} \mathbf{i} & \mathbf{j} & \mathbf{k} \\ 1 & -1 & 2 \\ 2 & 3 & 4 \end{vmatrix} = \mathbf{C}$$

from which the vector is

$$\mathbf{C} = [(-1 \times 4) - (2 \times 3)]\mathbf{i} + [(1 \times 4) - (2 \times 2)]\mathbf{j} + [(1 \times 3) - (-1 \times 2)]\mathbf{k}$$
$$= -10\mathbf{i} + 5\mathbf{k}$$

3.4.4.4 Cross Product of Vectors for Plane Areas

The cross product of two vectors may be used to determine the area of a plane parallelogram with vectors **A** and **B** forming the two adjacent sides, as illustrated in Figure 3.18.

Figure 3.18 shows two plane areas: one a parallelogram and the other a triangle. The area of a parallelogram is equal to the base, represented by the magnitude of vector **A**, times the height H, with $H = |\mathbf{B}| \sin \theta$. Consequently, the area of the parallelogram in Figure 3.18a is

$$\text{Area of parallelogram} = \text{Base} |\mathbf{A}| \times \text{Height} \, H = |\mathbf{A}||\mathbf{B}| \sin \theta$$

The above expression is identical to the magnitude of the cross product of vectors A and B as indicated in Equation 3.17. We may thus express the area of a plane parallelogram to be equal to the cross product of vectors **A** and **B**, or

$$\text{Area of parallelogram} = |\mathbf{A} \times \mathbf{B}| \tag{3.19}$$

with vectors **A** and **B** being the two sides of the parallelogram.

A similar expression for the area of the triangle in Figure 3.18b may be derived with the relationship of Area of triangle $= \frac{1}{2}(\text{Base} \times \text{Height})$, with the height of the triangle being $H = |\mathbf{B}| \sin \theta$, the area of the triangle abc can be obtained with the expression

$$\text{Area of triangle abc} = \frac{1}{2}[\text{Base} |\mathbf{A}| \times \text{Height} \, H] = \frac{1}{2}|\mathbf{A}||\mathbf{B}| \sin \theta = \frac{1}{2}\mathbf{A} \times \mathbf{B}$$

3.4.4.5 Triple product

There are times when an engineering analysis requires the multiplication of three vectors **A**, **B**, and **C**. Mathematical manipulations of triple products will obviously be more complicated than the case of dot and cross products of two vectors.

The following rules apply for triple products of vectors **A**, **B**, and **C**:

$$(\mathbf{A} \cdot \mathbf{B})\mathbf{C} \neq \mathbf{A}(\mathbf{B} \cdot \mathbf{C}) \tag{3.20a}$$

$$\mathbf{A} \cdot (\mathbf{B} \times \mathbf{C}) = \mathbf{B} \cdot (\mathbf{C} \times \mathbf{A}) = \mathbf{C} \cdot (\mathbf{A} \times \mathbf{B}) \tag{3.20b}$$

$$\mathbf{A} \times (\mathbf{B} \times \mathbf{C}) \neq (\mathbf{A} \times \mathbf{B}) \times \mathbf{C} \tag{3.20c}$$

$$\mathbf{A} \times (\mathbf{B} \times \mathbf{C}) = (\mathbf{A} \cdot \mathbf{C})\mathbf{B} - (\mathbf{A} \cdot \mathbf{B})\mathbf{C} \tag{3.20d}$$

$$(\mathbf{A} \times \mathbf{B}) \times \mathbf{C} = (\mathbf{A} \cdot \mathbf{C})\mathbf{B} - (\mathbf{B} \cdot \mathbf{C})\mathbf{A} \tag{3.20e}$$

Example 3.11

Expand $\mathbf{A} \cdot (\mathbf{B} \times \mathbf{C})$ in Equation 3.20b if vectors $\mathbf{A}$, $\mathbf{B}$, and $\mathbf{C}$ can be expressed as follows:

$$\mathbf{A} = A_x\mathbf{i} + A_y\mathbf{j} + A_z\mathbf{k}$$
$$\mathbf{B} = B_x\mathbf{i} + B_y\mathbf{j} + B_z\mathbf{k}$$
$$\mathbf{C} = C_x\mathbf{i} + C_y\mathbf{j} + C_z\mathbf{k}$$

in which $\mathbf{i}$, $\mathbf{j}$, and $\mathbf{k}$ are unit vectors in a rectangular coordinate system and the coefficients associated with these unit vectors are the magnitudes of the components of vectors $\mathbf{A}$, $\mathbf{B}$, and $\mathbf{C}$, respectively.

Solution:

The product $\mathbf{A} \cdot (\mathbf{B} \times \mathbf{C})$ can be shown to equal

$$\mathbf{A} \cdot (\mathbf{B} \times \mathbf{C}) = (A_x\mathbf{i} + A_y\mathbf{j} + A_z\mathbf{k}) \begin{vmatrix} \mathbf{i} & \mathbf{j} & \mathbf{k} \\ B_x & B_y & B_z \\ C_x & C_y & C_z \end{vmatrix} = \begin{vmatrix} A_x & A_y & A_z \\ B_x & B_y & B_z \\ C_x & C_y & C_z \end{vmatrix} \tag{3.21}$$

Note: The relations for dot products of unit vectors in Equation 3.14 were used in reaching this solution.

3.4.4.6 Additional Laws of Vector Algebra

A summary of the laws of vector algebra will be useful in engineering analysis (in which m and n are real number constants and vectors $\mathbf{A} = A_x\mathbf{i} + A_y\mathbf{j} + A_z\mathbf{k}$ and $\mathbf{B} = B_x\mathbf{i} + B_y\mathbf{j} + B_z\mathbf{k}$):

Commutative law for addition:	$\mathbf{A} + \mathbf{B} = \mathbf{B} + \mathbf{A}$
Associative law for addition:	$\mathbf{A} + (\mathbf{B} + \mathbf{C}) = (\mathbf{A} + \mathbf{B}) + \mathbf{C}$
Associative law for multiplication:	$m(n\mathbf{A}) = (mn)\mathbf{A} = n(m\mathbf{A})$
Distributive law:	$(m + n)\mathbf{A} = m\mathbf{A} + n\mathbf{A}$
Distributive law:	$m(\mathbf{A} + \mathbf{B}) = m\mathbf{A} + m\mathbf{B}$
Commutative law for dot product:	$\mathbf{A} \cdot \mathbf{B} = \mathbf{B} \cdot \mathbf{A}$
Distributive law for dot product:	$\mathbf{A} \cdot (\mathbf{B} + \mathbf{C}) = \mathbf{A} \cdot \mathbf{B} + \mathbf{A} \cdot \mathbf{C}$
Scalar multiplier for dot product:	$m(\mathbf{A} \cdot \mathbf{B}) = (m\mathbf{A}) \cdot \mathbf{B} = \mathbf{A} \cdot (m\mathbf{B}) = (\mathbf{A} \cdot \mathbf{B})m$
Dot product of unit vectors:	$\mathbf{i} \cdot \mathbf{j} = \mathbf{j} \cdot \mathbf{k} = \mathbf{k} \cdot \mathbf{i} = 0$
Scalar resultant of dot product:	$\mathbf{A} \cdot \mathbf{B} = A_xB_x + A_yB_y + A_zB_z$
	$\mathbf{A} \cdot \mathbf{A} = A^2 = A_x^2 + A_y^2 + A_z^2$
	$\mathbf{B} \cdot \mathbf{B} = B^2 = B_x^2 + B_y^2 + B_z^2$

3.4.4.7 Use of Triple Product of Vectors for Solid Volume

It is well known that the volume of a parallelepiped solid such as shown in Figure 3.19 can be obtained from the expression:

Volume of a parallelepiped solid = Height × Area of the parallelogram (PARA).

We have already derived the expression for the area PARA as Area $= \mathbf{B} \times \mathbf{C}$ such as shown in Equation 3.19. The height of the parallelepiped solid is $|\mathbf{A}|\sin\theta$, in which θ is the angle between vectors $\mathbf{A}$ and $\mathbf{B}$.

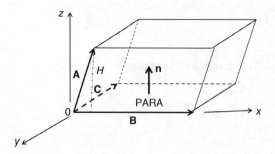

Figure 3.19 A parallelepiped solid.

We thus have that the volume of the parallelepiped solid in Figure 3.19 is

$$V = |\mathbf{A}| \sin\theta (\mathbf{B} \times \mathbf{C})$$
$$= |\mathbf{A}|(\mathbf{B} \times \mathbf{C}) \sin\theta$$
$$= \mathbf{A} \cdot (\mathbf{B} \times \mathbf{C})$$
$$= \mathbf{A} \cdot \begin{bmatrix} \mathbf{i} & \mathbf{j} & \mathbf{k} \\ B_x & B_y & B_z \\ C_x & C_y & C_z \end{bmatrix} = \begin{bmatrix} A_x & A_y & A_z \\ B_x & B_y & B_z \\ C_x & C_y & C_z \end{bmatrix}$$

where A_x, A_y, A_z, B_x, B_y, etc. are the magnitudes of the components of vectors $\mathbf{A}$, $\mathbf{B}$, and $\mathbf{C}$ along the x-, y- and z-coordinates, respectively.

3.5 Vector Calculus

Vector calculus deals with problems involving continuous variations of the magnitudes and directions of vector quantities. Unlike most single-valued functions described in Section 2.2.3 and the formulation of derivatives of these functions in Section 2.2.5, we need to keep track of these continuous changes of *both* the magnitudes and directions of vector quantities in the analysis. The mathematical formulation for doing this is called vector calculus.

The magnitude of a vector as well as it direction may change with any scalar quantity that may relate to its positions defined by (x,y,z) in a rectangular coordinate system and/or time t, as may be illustrated by the following situation.

Continuous variations of both magnitude and direction of physical quantities in position and time are common in reality, such as in Example 3.6 in which a cruise ship travels to several ports at various distances apart and in various directions, as illustrated in Figure 3.13. Another example is the velocity of vehicles cruising on narrow and winding streets in urban areas. Figure 3.20a shows such a scene with vehicles traveling along a narrow and winding Lombard Street in the City of San Francisco, California. This 180-meter-long paved crooked block involves eight sharp turns on a steep down slope at 27%, which is much too steep by any standard for urban streets. It is a major tourist attraction of the city (https://en.wikipedia.org/wiki/Lombard_Street_(San_Francisco)). Figure 3.20b shows that drivers on this stretch of the street need to vary the velocity of their cars in both direction and magnitude at all times in order to avoid accidents by driving off the street and hitting the cars ahead or being hit by the cars behind.

3.5.1 Vector Functions

In Section 2.2.3 we defined a function to be a physical quantity with its value determined by variables in a space defined by (x, y, z) in rectangular coordinates and/or time t. As stressed at the

(a) (b)

Figure 3.20 Lombard Street in San Francisco. (a) A narrow and winding street. (b) Breath-taking driving. (*Source*: (a) © Gaurav1146 Wikimedia Commons.). *See color section.*

beginning of this chapter, the definition of vector quantities requires not only the magnitudes but also their directions. Consequently, we may express a vector function that involves both its magnitude and/or direction defined by scalar quantities with position and/or time variables.

We may thus express a vector function of vector **A** in the form **A**(*u*), in which *u* is the scalar that defines the vector **A** (Spiegel, 1963). As already discussed, the scalar *u* in the expression **A**(*u*) can be space designated by (*x*,*y*,*z*) in a rectangular coordinate system and/or time (*t*).

For vectors in rectangular coordinate systems, we may express the vector **A**(*u*) in the following form with **i**, **j**, and **k** being the unit vectors along the *x*-, *y*-, and *z*-coordinate directions, respectively:

$$\mathbf{A}(u) = \mathbf{A}_x(u) + \mathbf{A}_y(u) + \mathbf{A}_z(u) \tag{3.22a}$$

or

$$\mathbf{A}(u) = A_x(u)\mathbf{i} + A_y(u)\mathbf{j} + A_z(u)\,\mathbf{k} \tag{3.22b}$$

where $\mathbf{A}_x$, $\mathbf{A}_y$, and $\mathbf{A}_z$ in Equation 3.22a denote the components of vector **A**(*u*) along the *x*-, *y*-, and *z*-coordinate directions, respectively, whereas A_x, A_y, and A_z in Equation 3.22b are the magnitudes of the components of vector **A**(*u*) along the same coordinates.

3.5.2 Derivatives of Vector Functions

The rate of change of a continuous vector function **A**(*u*) as expressed in Equation 3.22a with respect to the scalar variable *u* often arises in engineering analyses; its applications will be demonstrated in later sections of this chapter. Derivatives that describe the rate of change of vector functions may be expressed in a similar way to those for the scalar functions in Equation 2.9 in Chapter 2. The mathematical formulation of the derivative of continuous vector function **A**(*u*) is

$$\frac{d\mathbf{A}(u)}{du} = \lim_{\delta u \to 0} \frac{\mathbf{A}(u + \Delta u) - \mathbf{A}(u)}{\Delta u} \tag{3.23a}$$

For the vector function **A**(*u*) in Equation (3.22b) in a rectangular coordinate system, the derivative of this function becomes

$$\frac{d\mathbf{A}(u)}{du} = \frac{dA_x(u)}{du}\mathbf{i} + \frac{dA_y(u)}{du}\mathbf{j} + \frac{dA_z(u)}{du}\mathbf{k} \tag{3.23b}$$

When the vector function **A**(*x,y,z*) is written as $\mathbf{A}(x, y, z) = A_x(x, y, z)\mathbf{i} + A_y(x, y, z)\mathbf{j} + A_z(x, y, z)\mathbf{k}$, we will get the differential of this vector function **A**(x,y,z) in the form

$$d\mathbf{A} = \frac{\partial \mathbf{A}}{\partial x}dx + \frac{\partial \mathbf{A}}{\partial y}dy + \frac{\partial \mathbf{A}}{\partial z}dz \tag{3.23c}$$

Example 3.12

A position vector $\mathbf{r}$ in a rectangular coordinate system as illustrated in Figure 3.4a has both its magnitude and direction varying with time t, and its two components $\mathbf{r}_x$ and $\mathbf{r}_y$ vary with time according to $r_x = 1 - t^2$ and $r_y = 1 + 2t$, respectively. Determine the rate of variation of the position vector with respect to time variable t.

Solution:

We may express the position vector $\mathbf{r}$ in the form

$$\mathbf{r}(t) = \mathbf{r}_x(t) + \mathbf{r}_y(t) = r_x(t)\mathbf{i} + r_y(t)\mathbf{j}$$

in which $\mathbf{i}$ and $\mathbf{j}$ are the unit vectors along the x- and y-coordinate directions, respectively.

The rate of change of the position vector $\mathbf{r}(t)$ with respect to variable t may be obtained using Equation 3.23b as shown below:

$$\frac{d\mathbf{r}(t)}{dt} = \frac{dr_x(t)}{dt}\mathbf{i} + \frac{dr_y(t)}{dt}\mathbf{j} = \left[\frac{d}{dt}(1 - t^2)\right]\mathbf{i} + \left[\frac{d}{dt}(1 + 2t)\right]\mathbf{j} = (-2t)\mathbf{i} + 2\mathbf{j}$$

The following rules on the differentiation of products of vector functions are used often in engineering analysis for vector functions $\mathbf{A}(x, y, z) = A_x(x, y, z)\mathbf{i} + A_y(x, y, z)\mathbf{j} + A_z(x, y, z)\mathbf{k}$ and $\mathbf{B}(x, y, z) = B_x(x, y, z)\mathbf{i} + B_y(x, y, z)\mathbf{j} + B_z(x, y, z)\mathbf{k}$ in rectangular coordinate systems, where $\mathbf{i}$, $\mathbf{j}$, and $\mathbf{k}$ are the respective unit vectors along the x-, y-, and z-coordinate directions:

$$\frac{\partial}{\partial x}(\mathbf{A} \cdot \mathbf{B}) = \mathbf{A} \cdot \frac{\partial \mathbf{B}}{\partial x} + \frac{\partial \mathbf{A}}{\partial x} \cdot \mathbf{B} \tag{3.24a}$$

$$\frac{\partial}{\partial y}(\mathbf{A} \cdot \mathbf{B}) = \mathbf{A} \cdot \frac{\partial \mathbf{B}}{\partial y} + \frac{\partial \mathbf{A}}{\partial y} \cdot \mathbf{B} \tag{3.24b}$$

$$\frac{\partial}{\partial z}(\mathbf{A} \cdot \mathbf{B}) = \mathbf{A} \cdot \frac{\partial \mathbf{B}}{\partial z} + \frac{\partial \mathbf{A}}{\partial z} \cdot \mathbf{B} \tag{3.24c}$$

$$\frac{\partial}{\partial x}(\mathbf{A} \times \mathbf{B}) = \mathbf{A} \times \frac{\partial \mathbf{B}}{\partial x} + \frac{\partial \mathbf{A}}{\partial x} \times \mathbf{B} \tag{3.25a}$$

$$\frac{\partial}{\partial y}(\mathbf{A} \times \mathbf{B}) = \mathbf{A} \times \frac{\partial \mathbf{B}}{\partial y} + \frac{\partial \mathbf{A}}{\partial y} \times \mathbf{B} \tag{3.25b}$$

$$\frac{\partial}{\partial z}(\mathbf{A} \times \mathbf{B}) = \mathbf{A} \cdot \frac{\partial \mathbf{B}}{\partial z} + \frac{\partial \mathbf{A}}{\partial z} \times \mathbf{B} \tag{3.25c}$$

Example 3.13

Determine $d\mathbf{A}$ if the vector function $\mathbf{A}(x, y, z) = (x^2 \sin y)\mathbf{i} + (z^2 \cos y)\mathbf{j} - (xy^2)\mathbf{k}$.

Solution:

We may use Equation 3.23c for the solution as follows:

$$d\mathbf{A} = \frac{\partial \mathbf{A}}{\partial x}dx + \frac{\partial \mathbf{A}}{\partial y}dy + \frac{\partial \mathbf{A}}{\partial z}dz$$

$$= \left[(\sin y)\mathbf{i}\frac{d}{dx}(x^2) - (y^2)\mathbf{k}\frac{dx}{dx}\right]dx + \left[x^2\mathbf{i}\frac{d}{dy}(\sin y) + z^2\mathbf{j}\frac{d}{dy}(\cos y)\right]dy$$

$$+ \left[(\cos y)\mathbf{j}\frac{d}{dz}(z^2)\right]dz$$

$$= \left[(2x \sin y)\mathbf{i} - y^2\mathbf{k}\right]dx + \left[(x^2 \cos y)\mathbf{i} - (z^2 \sin y)\mathbf{j} - 2xy\mathbf{k}\right]dy + \left[(2z \cos y)\mathbf{j}\right]dz$$

$$= (2x \sin y\, dx + x^2 \cos y\, dy)\mathbf{i} + (2z \cos y\, dz - z^2 \sin y\ dy)\mathbf{j} - (y^2\, dx + 2xy\, dy)\mathbf{k}$$

3.5.3 Gradient, Divergence, and Curl

Gradient, divergence, and curl are frequently used when dealing with variations of vectors using a vector operator designated by ∇ (pronounced "del") defined as follows:

$$\nabla = \mathbf{i}\frac{\partial}{\partial x} + \mathbf{j}\frac{\partial}{\partial y} + \mathbf{k}\frac{\partial}{\partial z} \tag{3.26}$$

in a rectangular coordinate systems (x,y,z).

3.5.3.1 Gradient

The gradient relates to the variation of the magnitudes of vector quantities with a scalar quantity φ, and is defined as

$$\operatorname{grad}\varphi = \nabla\varphi = \left(\mathbf{i}\frac{\partial}{\partial x} + \mathbf{j}\frac{\partial}{\partial y} + \mathbf{k}\frac{\partial}{\partial z}\right)\varphi = \frac{\partial\varphi}{\partial x}\mathbf{i} + \frac{\partial\varphi}{\partial y}\mathbf{j} + \frac{\partial\varphi}{\partial z}\mathbf{k} \tag{3.27}$$

3.5.3.2 Divergence

The divergence of vector function $\mathbf{A}(x,y,z)$ implies the "growth" or "contraction" of this vector function in its components along the coordinates. The divergence of the vector function $\mathbf{A}(x,y,z)$ is defined as

$$\operatorname{div}\mathbf{A} = \nabla \cdot \mathbf{A} = \left(\mathbf{i}\frac{\partial}{\partial x} + \mathbf{j}\frac{\partial}{\partial y} + \mathbf{k}\frac{\partial}{\partial z}\right) \cdot (A_x\mathbf{i} + A_y\mathbf{j} + A_z\mathbf{k})$$

$$= \frac{\partial A_x}{\partial x} + \frac{\partial A_y}{\partial y} + \frac{\partial A_z}{\partial z} \tag{3.28}$$

where A_x, A_y, and A_z are the magnitudes of the components of vector $\mathbf{A}$ along the x-, y-, and z-coordinate directions, respectively.

3.5.3.3 Curl

The curl of a vector function $\mathbf{A}$ is related to the "rotation" of this vector. It is defined as

$$\operatorname{curl}\mathbf{A} = \nabla \times \mathbf{A} = \left(\mathbf{i}\frac{\partial}{\partial x} + \mathbf{j}\frac{\partial}{\partial y} + \mathbf{k}\frac{\partial}{\partial z}\right) \times (A_x\mathbf{i} + A_y\mathbf{j} + A_z\mathbf{k})$$

$$= \begin{vmatrix} \mathbf{i} & \mathbf{j} & \mathbf{k} \\ \dfrac{\partial}{\partial x} & \dfrac{\partial}{\partial y} & \dfrac{\partial}{\partial z} \\ A_x & A_y & A_z \end{vmatrix}$$

$$= \mathbf{i}\begin{vmatrix} \dfrac{\partial}{\partial y} & \dfrac{\partial}{\partial z} \\ A_y & A_z \end{vmatrix} - \mathbf{j}\begin{vmatrix} \dfrac{\partial}{\partial x} & \dfrac{\partial}{\partial z} \\ A_x & A_z \end{vmatrix} + \mathbf{k}\begin{vmatrix} \dfrac{\partial}{\partial x} & \dfrac{\partial}{\partial y} \\ A_x & A_y \end{vmatrix}$$

$$= \left(\frac{\partial A_z}{\partial y} - \frac{\partial A_y}{\partial z}\right)\mathbf{i} - \left(\frac{\partial A_z}{\partial x} - \frac{\partial A_x}{\partial z}\right)\mathbf{j} + \left(\frac{\partial A_y}{\partial x} - \frac{\partial A_x}{\partial y}\right)\mathbf{k} \tag{3.29}$$

Example 3.14

If $\varphi = xy^2z^3$ and vector $\mathbf{A} = A_x\mathbf{i} + A_y\mathbf{j} + A_z\mathbf{k}$, use Equations 3.27, 3.28, and 3.29 to determine (a) grad φ; (b) div(φA); (c) curl(φA).

Solution:

a)

$$\operatorname{grad}\varphi = \nabla\varphi = \left(\mathbf{i}\frac{\partial}{\partial x} + \mathbf{j}\frac{\partial}{\partial y} + \mathbf{k}\frac{\partial}{\partial z}\right)\varphi$$

$$= \frac{\partial\varphi}{\partial x}\mathbf{i} + \frac{\partial\varphi}{\partial y}\mathbf{j} + \frac{\partial\varphi}{\partial z}\mathbf{k}$$

$$= \frac{\partial}{\partial x}\left(xy^2z^3\right)\mathbf{i} + \frac{\partial}{\partial y}\left(xy^2z^3\right)\mathbf{j} + \frac{\partial}{\partial z}\left(xy^2z^3\right)\mathbf{k}$$

$$= y^2z^3\mathbf{i} + 2xyz^3\mathbf{j} + 3xy^2z^2\mathbf{k}$$

b)

$$\operatorname{div}(\varphi A) = \nabla\cdot(\varphi A) = \nabla\cdot\left[\left(xy^2z^3\right)A_x\mathbf{i} + \left(xy^2z^3\right)A_y\mathbf{j} + \left(xy^2z^3\right)A_z\mathbf{k}\right]$$

$$= \frac{\partial}{\partial x}\left[\left(xy^2z^3\right)A_x\right] + \frac{\partial}{\partial y}\left[\left(xy^2z^3\right)A_y\right] + \frac{\partial}{\partial z}\left[\left(xy^2z^3\right)A_z\right]$$

c) Likewise, we may use Equation 3.29 to show that

$$\operatorname{curl}(\varphi A) = \nabla\times(\phi A) = \left(\mathbf{i}\frac{\partial}{\partial x} + \mathbf{j}\frac{\partial}{\partial y} + \mathbf{k}\frac{\partial}{\partial z}\right)\times(\phi A_x\mathbf{i} + \phi A_y\mathbf{j} + \phi A_z\mathbf{k})$$

$$= \begin{vmatrix} \mathbf{i} & \mathbf{j} & \mathbf{k} \\ \dfrac{\partial}{\partial x} & \dfrac{\partial}{\partial y} & \dfrac{\partial}{\partial z} \\ \phi A_x & \phi A_y & \phi A_z \end{vmatrix}$$

$$= \mathbf{i}\begin{vmatrix} \dfrac{\partial}{\partial y} & \dfrac{\partial}{\partial z} \\ \varphi A_y & \varphi A_z \end{vmatrix} - \mathbf{j}\begin{vmatrix} \dfrac{\partial}{\partial x} & \dfrac{\partial}{\partial z} \\ \varphi A_x & \varphi A_z \end{vmatrix} + \mathbf{k}\begin{vmatrix} \dfrac{\partial}{\partial x} & \dfrac{\partial}{\partial y} \\ \varphi A_x & \varphi A_y \end{vmatrix}$$

$$= \left(\frac{\partial\varphi A_z}{\partial y} - \frac{\partial\varphi A_y}{\partial z}\right)\mathbf{i} - \left(\frac{\partial\varphi A_z}{\partial x} - \frac{\partial\varphi A_x}{\partial z}\right)\mathbf{j} + \left(\frac{\partial\varphi A_y}{\partial x} - \frac{\partial\varphi A_x}{\partial y}\right)\mathbf{k}$$

The following expressions relating to gradient, divergence, and curl of vector **A** are useful in vector calculus (Spiegel, 1963):

$$\nabla\cdot(\phi\mathbf{A}) = (\nabla\phi)\cdot\mathbf{A} + \phi(\nabla\cdot\mathbf{A}) \tag{3.30a}$$

$$\nabla^2\phi = \text{Laplacian operator of } \phi = \nabla\cdot\nabla\phi = \frac{\partial^2\phi}{\partial x^2} + \frac{\partial^2\phi}{\partial y^2} + \frac{\partial^2\phi}{\partial z^2} \tag{3.30b}$$

$$\operatorname{div}\operatorname{curl}\mathbf{A} = 0 \tag{3.30c}$$

3.6   Applications of Vector Calculus in Engineering Analysis

Vectors and vector calculus as presented in Section 3.5 are much used as effective mathematical tools in deriving mathematical models, mainly in terms of differential equations, in numerous disciplines in physics and engineering such as the governing equations of electrodynamics (Maxwell's equations) and fluid dynamics (Navier–Stokes equations), and in heat conduction in solids, etc. These equations involve the Laplacian, curl, and divergence terms as presented in the previous section. Fundamentally, all the phenomena involving electricity, magnetism, and fluid flow are related to grad/div/curl/Laplacian operators on vector functions and their derivatives. Differentiation operators with vector functions are presented in many different forms of Hamiltonian and Schrödinger (wave) equations. Thus, fundamentally, vector calculus applies to everything down to the quantum level.

In more practical and specific cases, such as in antenna/scattering problems, there is a need to find the vector and scalar potentials due to a current distribution and to use them to characterize the electromagnetic fields of an object. Determining the fields from the potentials requires the use of grad/div/curl operators as presented in Section 3.5. Many problems involving surface or volume integrals can be transformed to "lower-dimensional" integrals by the use of vector calculus.

The following examples are a few specific applications of gradient, divergence, and curl of vectors in engineering analysis ("Notes on applications of vector calculus" by W.L. Kath, http://people.esam.northwestern.edu/~kath/courses.html).

3.6.1 In Heat Transfer

We start with the Fourier law of heat conduction in solids:

$$\mathbf{q} = -k\nabla T \tag{3.31a}$$

where $\mathbf{q}$ is the heat flux vector (defined as heat flow in a specific direction per unit area and time) at a point in the solid, k is the thermal conductivity of the solid, and T is the temperature at a point inside the solid. Derivation of this law will be presented in Chapter 7.

The following heat conduction equation is derived using the principle of conservation of energy and the Fourier law in Equation 3.31a:

$$\frac{\partial T}{\partial t} = \alpha \nabla^2 T \tag{3.31b}$$

in which $\alpha = k/(\rho c)$ is the thermal diffusivity of the solid, where ρ is the mass density and c is the specific heat of the solid.

Equation 3.31b leads to the steady-state heat conduction equation in solids in the following form:

$$\nabla^2 T = 0 \tag{3.31c}$$

3.6.2 In Fluid Mechanics

Fluid flow in space follows the law of continuity, expressed by the following equation in conjunction with the principle of conservation of mass:

$$\frac{\partial \rho}{\partial t} + \mathbf{v} \cdot \nabla \rho + \rho \nabla \cdot \mathbf{v} = 0 \tag{3.32a}$$

where the vector function $\mathbf{v}$ is the local velocity of the fluid.

Equation (3.32a) leads to the following simpler equation for a noncompressible fluid, in which the mass density ρ does not vary with time and space:

$$\nabla \cdot \mathbf{v} = 0 \tag{3.32b}$$

For fluid moving in conduits or open channels, the Bernoulli equation that will be presented in Section 7.3.2 is often used to assess the relationship between the velocity vectors of the moving fluid flow ($\mathbf{v}$) and the applied pressure (p) and the potential energy (φ). This equation is derived from the principle of conservation of energy with absence of rotation al flow of the fluid

$$\frac{\partial \varphi}{\partial t} + \frac{1}{2}|\mathbf{v}|^2 + \frac{p}{\rho} + \mathbf{U} = 0 \tag{3.32c}$$

where **v** is the velocity vector of the fluid in motion and **U** is the body force vector. Both these vectors may vary with position and direction in fluid flow.

3.6.3 In Electromagnetism with Maxwell's Equations

One of the great discoveries of modern-day physics in the nineteenth century was electromagnetism. It arose from the study of the electromagnetic forces generated in electrical conductors situated in magnetic fields, resulting in motion of the conductor—a situation that is exploited today in many electromagnetic devices such as electric motors and generators as well as electromagnetic actuators and devices with broad applications in various disciplines of engineering. The three principal quantities involved in this physical situation are (1) the magnetic field, with magnetic flux density **B**; (2) the electric current flow which is in the direction of the electric field **E**; and (3) the force acting on or generated by the conductor and its motion with a velocity **v**. These three principal quantities, **B**, **E**, **v**, are vectors that vary with the spatial arrangement and with time. The commonly used Faraday's right-hand rule, illustrated in Figure 3.21, describes the relationship of the these principal vectorial quantities: the thumb points the direction of the velocity of the moving conductor (**v**), the index finger points the direction of the magnetic flux density (**B**), and the middle finger points the direction of electric current flow or the electric field (**E**).

The design of devices that involve electromagnetism requires the use of mathematical models in the forms of differential equations. The following four Maxwell's equations are applicable for this purpose.

Maxwell's first equation

$$\nabla \cdot (\varepsilon \mathbf{E}) = \rho \tag{3.33a}$$

where ε = the permittivity or dielectric constant of the medium between the conductor and the magnetic field, and ρ = the charge density in the conductor.

Maxwell's second equation

$$\nabla \cdot \mathbf{B} = 0 \tag{3.33b}$$

Maxwell's third equation

$$\nabla \times \mathbf{H} = \mathbf{J} + \varepsilon \frac{\partial \mathbf{E}}{\partial t} \tag{3.33c}$$

where **H** = magnetic field intensity and **J** = electric current flow.

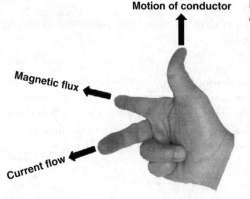

Motion of conductor

Figure 3.21 Faraday's right-hand rule in electromagnetism.

Magnetic flux

Current flow

Maxwell's fourth equation

$$\nabla \times \mathbf{E} = -\frac{\partial B}{\partial t} \tag{3.33d}$$

where $B =$ the magnitude of vector **B**, referred to as the magnetic flux density.

These above four Maxwell's equations allow the derivation of the equation that is widely used in determining the electric field **E**:

$$\nabla^2 \mathbf{E} = \frac{1}{c^2} \frac{\partial^2 \mathbf{E}}{\partial t^2} \tag{3.34}$$

in which the constant c is the speed of light.

3.7 Application of Vector Calculus in Rigid Body Dynamics

We will focus this section on the use of vector calculus in an elementary level of rigid body dynamic analysis. Dynamic analysis is a special branch of solid mechanics and is an important part of the design of any moving machine or structure, regardless of their size from giant space stations and jumbo jet airplanes to small components of sensors and actuators at the minute scale of micrometers (Hsu, 2002, 2008).

Dynamics involves both kinematics and kinetics of moving solids (Beer et al., 2004). *Kinematics* is the study of the geometry of motion. It relates displacement, velocity, and acceleration of moving solids at given times. However, it does not deal with the cause of the motion. *Kinetics* relates the forces acting on a moving rigid body, the mass of the body, and the motion of the body. It is also used to predict the motion caused by given forces or to determine the forces required to produce a given motion.

This section will focus on the kinematics of rigid bodies in motion and the coverage will be confined to planar motions. As indicated, kinematic analysis of moving rigid bodies involves the determination of the instantaneous positions, velocities, and accelerations of those moving rigid bodies. Accelerations (or decelerations) of moving bodies are the sources of dynamic forces according to Newton's second law. Often vector calculus is used in determining acceleration in the subsequent kinetic analysis of moving bodies.

Since the following presentation is intended to illustrate the application of vector calculus in dynamics analysis of solids, we will make no distinction between the terminologies of "particles" and "rigid bodies." In this sense, the mass of the rigid bodies in the subsequent analyses is neglected as would be done in the case of "particle dynamics."

Most of the mathematical formulations are available in two major references: Beer et al. (2004) and Hibbler (2007).

3.7.1 Rigid Body in Rectilinear Motion

In general, a rigid body in motion is characterized by its instantaneous position denoted by a position vector **r**, the velocity vector **v**, and the acceleration vector **a**.

Figure 3.22 illustrates a rigid body in rectilinear motion. This motion takes place along a straight line as shown in the figure.

The rigid body is originally located at point 0 in the coordinate system shown in Figure 3.22. It travels to a new position at point a in time t. We may thus define its new position by the position vector function $\mathbf{r}(t)$, which can be written

$$\mathbf{r}(t) = S(t)\mathbf{i} \quad \text{or} \quad \mathbf{r}(t) = x(t)\mathbf{i} \tag{3.35a}$$

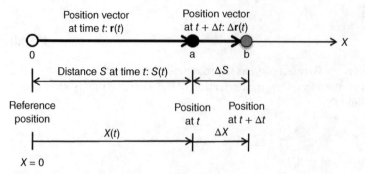

Figure 3.22 Rectilinear motion of a rigid body.

in which **i** is the unit vector along the x-coordinate direction, and $S(t)$ is the distance that this body has traveled from point 0 to point a in time t.

Figure 3.22 further indicates that the average velocity vector function of the rigid body $\mathbf{v}(t)$ is

$$\mathbf{v}(t) = \frac{d\mathbf{r}(t)}{dt} = \frac{d}{dt}[S(t)\mathbf{i}]$$

$$= \left(\frac{d}{dt}[S(t)]\right)\mathbf{i} = v(t)\mathbf{i} \tag{3.35b}$$

where $v(t)$ is the magnitude of the velocity vector $\mathbf{v}(t)$ at time t. The acceleration vector function $\mathbf{a}(t)$ for the moving body may be expressed as

$$\mathbf{a}(t) = \frac{d\mathbf{v}(t)}{dt} = \frac{d}{dt}\left(\frac{d}{dt}\mathbf{r}(t)\right)$$

$$= \frac{d}{dt}\left[\frac{d}{dt}S(t)\mathbf{i}\right] = \frac{d^2}{dt^2}[S(t)\mathbf{i}]$$

$$= \left[\frac{d^2 S(t)}{dt^2}\right]\mathbf{i} = a(t)\mathbf{i} \tag{3.35c}$$

in which $a(t)$ is the magnitude of the acceleration vector $\mathbf{a}(t)$.

Example 3.15

A rigid body is traveling along the x-axis as shown in Figure 3.22. Assume that the instantaneous position of the body may be represented by a function $x(t) = 11t^2 - 2t^3$, x in meters and time t in seconds. Assume the mass of the body is negligible; determine the vector functions of the velocity and acceleration of the moving rigid body.

Solution:

With the magnitude of the position vector $x(t) = 11t^2 - 2t^3$, we may use Equation 3.35b to determine the magnitude of the velocity vector function $\mathbf{v}(t)$ and Equation 3.35c for the magnitude of the acceleration vector function $\mathbf{a}(t)$ as shown in Equation (a):

$$\mathbf{v}(t) = \frac{d\mathbf{x}(t)}{dt} = \frac{d}{dt}\left(11t^2 - 2t^3\right) = 22t - 6t^2$$

$$\mathbf{a}(t) = \frac{d\mathbf{v}(t)}{dt} = \frac{d}{dt}\left(22t - 6t^2\right) = 22 - 12t \tag{a}$$

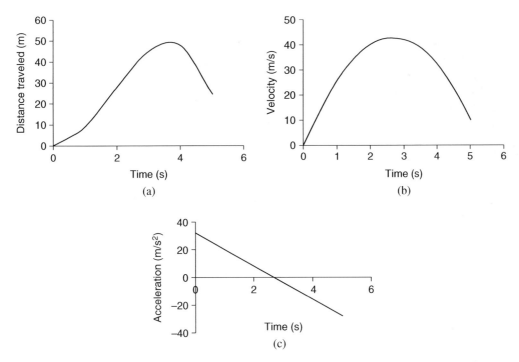

Figure 3.23 Vector functions of instantaneous position, velocity, and acceleration in rectilinear motion of a rigid body. (a) Instantaneous position. (b) Instantaneous velocity.(c) Instantaneous acceleration.

The vector functions of the velocity and acceleration of the moving rigid body thus can be expressed in Equations (b) and (c), respectively:

$$\mathbf{v}(t) = (22t - 6t^2)\mathbf{i} \tag{b}$$

$$\mathbf{a}(t) = (22 - 12t)\mathbf{i} \tag{c}$$

Figure 3.23 shows the variation of the magnitudes of the instantaneous position vector $\mathbf{x}(t)$, the velocity vector $\mathbf{v}(t)$, and the acceleration vector $\mathbf{a}(t)$ using the functions derived in Equations (a), (b), and (c).

The magnitude of the position vector $\mathbf{x}(t)$ in Figure.3.23a increases to the maximum value of 49 m at time $t = 3.67$ seconds, and decreases thereafter. A similar trend appears in the velocity along the same direction as shown in Figure 3.23b. The acceleration $a(t)$, however, decreases continuously from the beginning of the motion and in a dictated linear fashion, as expressed in Equation (c) and illustrated in Figure 3.23c.

Another noteworthy feature is that all the three principal quantities—the position vector function $\mathbf{r}(t)$, the velocity vector function $\mathbf{v}(t)$, and the acceleration vector function $\mathbf{a}(t)$—lie in the same line (or on the same path) at all times in rectilinear motions of rigid bodies.

3.7.2 Plane Curvilinear Motion in Rectangular Coordinates

The analysis of rigid body dynamics along curvilinear paths is much more complicated than that of rectilinear motion as presented in Section 3.7.1 and Example 3.15. Figure 3.24a illustrates a rigid body traveling along a curved path with its initial position at point $P_0(u)$ where u represents a scalar variable, either in coordinates represented by x, y, z or/and time t. The corresponding

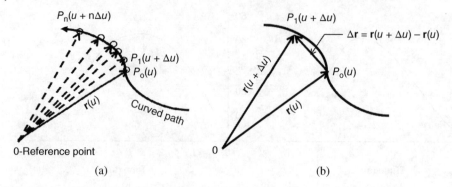

Figure 3.24 Rigid body moving along a curved path. (a) Motion on a curved path. (b) Change of a position vector with variable *u*.

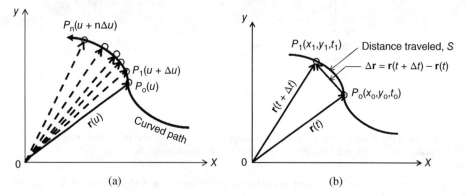

Figure 3.25 Rigid body moving along a curved path in the *x–y* plane. (a) Traveling on a curved path. (b) Change of position with time *t*

position vector of its initial location is $\mathbf{r}(u)$ according to the definition in Figure 3.24 with an arbitrarily chosen reference point 0. The sequential positions of the body along the curved path are given by the corresponding position vectors defined by the increment in variable u as shown in Figure 3.24a. The shift of the position vector between two adjacent variables u and $(u + \Delta u)$ is illustrated in Figure 3.24b, with the curved path situated in the *x–y* plane.

Figure 3.25a illustrates a rigid body traveling along a curved path in the *x–y* plane, with Figure 3.25b showing the variation of the position vectors with time t. From this we may derive the position vector function $\mathbf{r}(t)$ representing the "continuous" variation of the position of the moving body in Figure 3.25a with respect to time. The rate of the change of this position vector in Figure 3.25b can be mathematically expressed as follows:

$$\frac{d\mathbf{r}(t)}{dt} = \lim_{\Delta t \to 0} \frac{\mathbf{r}(t + \Delta t) - \mathbf{r}(t)}{\Delta t} \tag{3.36}$$

in which the position vector has the form

$$\mathbf{r}(t) = x(t)\mathbf{i} + y(t)\mathbf{j} \tag{3.37a}$$

in a rectangular coordinate system defined by *x*- and *y*-coordinates.

The velocity vector function can be expressed using Equation 3.35b as

$$\mathbf{v}(t) = \frac{d\mathbf{r}(t)}{dt} = \left[\frac{dx(t)}{dt}\right]\mathbf{i} + \left[\frac{dy(t)}{dt}\right]\mathbf{j} \tag{3.37b}$$

Figure 3.26 Position, velocity, and acceleration vectors of a rigid body moving on a curved path.

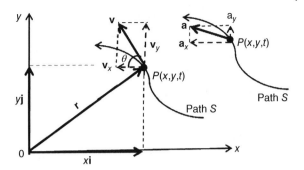

and the acceleration vector function using Equation 3.35c as

$$\mathbf{a}(t) = \frac{d\mathbf{v}(t)}{dt} = \left[\frac{d^2x(t)}{dt^2}\right]\mathbf{i} + \left[\frac{d^2y(t)}{dt^2}\right]\mathbf{j} \qquad (3.37c)$$

A graphical representation of Equations 3.37a,b,c is shown in Figure 3.26, in which the point P moves along a path S with time t and the position vector $\mathbf{r}(t)$ is as given in Equation 3.37a. The velocity vector $\mathbf{v}(t)$ at position P is along the tangent of the curved path at P, and is perpendicular to the position vector $\mathbf{r}$. The curved path that is made up by the tips of the arrowheads of velocity vector $\mathbf{v}$ is called the "hodograph." Since the acceleration vector $\mathbf{a}$ is a derivative of the velocity vector $\mathbf{v}$, as shown in Equation 3.37c, we may observe that the acceleration vector $\mathbf{a}$ is along the tangent of the hodograph but not the path on which the solid moves. Consequently, it is important for engineers to recognize that the "hodograph" does not coincide with the curved path on which the rigid body travels, as illustrated in the insert of Figure 3.26

Example 3.16
A moving rigid body with a position vector $\mathbf{r}(t) = t\mathbf{i} + 2t^3\mathbf{j}$ in a plane defined by rectangular coordinates x and y and with $\mathbf{i}$ and $\mathbf{j}$ being the unit vectors along the respective x- and y-coordinate directions. If $\mathbf{r}(t)$ has units of meters and the time t in seconds, find the velocity vector $\mathbf{v}$ and acceleration vector $\mathbf{a}$, and their respective magnitudes at $t = 3$.

Solution:
The given position vector indicates the coordinates at which the rigid body is situated:

$$x(t) = t \quad \text{and} \quad y(t) = 2t^3$$

We may find the corresponding velocity vector and acceleration vector from Equations 3.37b and 3.37c, respectively:

$$\mathbf{v}(t) = \left[\frac{d(t)}{dt}\right]\mathbf{i} + \left[\frac{d(2t^3)}{dt}\right]\mathbf{j} = \mathbf{i} + 6t^2\mathbf{j}$$

$$\mathbf{a}(t) = \frac{d\mathbf{v}(t)}{dt} = \left[\frac{d(1)}{dt}\right]\mathbf{i} + \left[\frac{d(6t^2)}{dt}\right]\mathbf{j} = 0\mathbf{i} + 12t\mathbf{j} = 12t\mathbf{j}$$

The magnitude of the velocity vector is

$$v(t) = |\mathbf{v}(t)| = \sqrt{1^2 + (6t)^2} = \sqrt{1 + 36t^2} \quad \text{m/s}$$

and the magnitude of the acceleration is

$$a(t) = |\mathbf{a}(t)| = \sqrt{(12t)^2} = 12t \quad \text{m/s}^2$$

which give the magnitudes of the velocity and acceleration at $t = 3$ as

$$v(3) = \sqrt{1^2 + 36(3)^2} = \sqrt{325} = 18.03 \text{ m/s}$$

and

$$a(3) = 12(3) = 36 \text{ m/s}^2$$

3.7.3 Application of Vector Calculus in the Kinematics of Projectiles

The kinematics of projectiles is a unique class of problems in which a rigid body (the projectile) is launched upward into the atmosphere with an initial velocity and at a positive launch angle. The kinematic analysis of the motion of the projectile will establish how high and how far it will travel with negligible air resistance but subject to gravitational acceleration (g) at all times.

Figure 3.27 illustrates the flight path of a projectile, in the course of which its velocity vector **v** will change not only in magnitude but also in direction along the flight path (shown with a dashed line) under the influence of gravitational acceleration g expressed by the acceleration vector $\mathbf{a} = -g\mathbf{j}$, which acts in a vertically downward direction.

We may also observe from Figure 3.27 that the projectile is launched into motion with the initial velocity v_0, which has components along the x- and the y-coordinate directions in the x-y plane. This velocity vector may be expressed as

$$\mathbf{v}_0 = (v_x)_0 \mathbf{i} + (v_y)_0 \mathbf{j}$$

where **i** and **j** are the unit vectors along the x- and y-coordinate directions, respectively. We may readily express the magnitudes of these components of the initial velocity as

$$(v_x)_0 = v_0 \cos\theta \quad \text{and} \quad (v_y)_0 = v_0 \sin\theta$$

where θ is the angle at the launching of the body. It is often referred to as the "jump slope angle." Figure 3.27 shows the continuous variation of the position of the projectile along the flight path. We write the position vector function $\mathbf{r}(t)$ of the projectile at an arbitrary position $P(x,y)$ at time t in the form

$$\mathbf{r}(t) = x(t)\mathbf{i} + y(t)\mathbf{j} \tag{3.38}$$

in which $x(t)$ and $y(t)$ determine the distance the projectile travels and also the maximum height it will reach. We realize that the variation of function $x(t)$ is independent of that of $y(t)$ and vice versa.

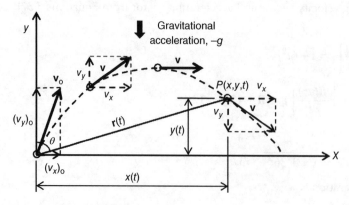

Figure 3.27 Flight path of a projectile.

We will thus handle our kinematic analysis of the motion of the projectile in separate directions along the x- and y-coordinate directions with reference to the coordinate system shown in Figure 3.27 in the following analysis.

As mentioned at the beginning of this subsection, kinematic analysis deals with finding the trajectory (i.e., flight path), the maximum height that the projectile will reach, and the range with the given initial velocity of the projectile $\mathbf{v}_0$ and the "jump slope angle θ." The following equations may be used in such an analysis for a projectile flying in a plane defined by an x–y coordinate system as illustrated in Figure 3.27.

One may begin the analysis with the known acceleration of the projectile—that is, the gravitational acceleration $\mathbf{a}(t) = -g\mathbf{j}$, where $g = 32.2$ ft/s^2 in the traditional system ("English" units), or 9.81 m/s^2 in the SI or metric system of units. We may obtain the velocity vector function using the following expression relating to Equation 3.37c:

$$\mathbf{v}(t) = \int \mathbf{a}(t)\,dt + c_1 = \int (-g\mathbf{j})\,dt + c_1 \tag{3.39}$$

where c_1 is the integration constant that can be determined by the available initial velocity vector function $\mathbf{v}(0)$.

Likewise, the trajectory of the projectile represented by the position vector can be determined using

$$\mathbf{r}(t) = \int \mathbf{v}(t)\,dt + c_2 \tag{3.40}$$

in which the integration constant c_2 can be determined by the initial position of the projectile at $\mathbf{r}(0)$.

The expression for $\mathbf{r}(t)$ in Equation 3.40 will lead to the solution for the instantaneous location of the projectile, given by $x(t)$ and $y(t)$ in Equation 3.38, thereby allowing determination of the maximum height of the projection y_{max} at time t_m from the following equations:

$$\left.\frac{dy(t)}{dt}\right|_{t=t_m} = 0 \quad \text{with} \quad \left.\frac{d^2y(t)}{dt^2}\right|_{t=t_m} < 0$$

The time at which the projectile returns to the ground (t_e) can be obtained by solution of the equation $x(t_e) = 0$. This time t_e is also used to determine the range of the projectile as $x(t)$ evaluated at t_e.

Example 3.17

A projectile is launched at the origin of a rectangular coordinate system (x–y) as shown in Figure 3.28 with an initial velocity $V_0 = 200$ m/s at a jump slope angle $\theta = 30°$. Determine the following:

a) The instantaneous position vector function of the projectile $\mathbf{r}(t)$.
b) The maximum height the projectile attains y_m.
c) The range R.
d) The impact velocity v_e.

Solution:

We realize that in addition to the given initial velocity and the jump slope angle, we also know the acceleration vector $\mathbf{a}(t) = -g\mathbf{j}$, or $\mathbf{a}(t) = -9.81\mathbf{j}$ m/s^2 of the projectile. We may express the initial velocity vector using these specified conditions as

$$\mathbf{v}_0 = \mathbf{v}(t)|_{t=0} = (\mathbf{v}_x)_0 + (\mathbf{v}_y)_0 = (200\cos 30°)\mathbf{i} + (200\sin 30°)\mathbf{j} = 173.21\mathbf{i} + 100\mathbf{j} \tag{a}$$

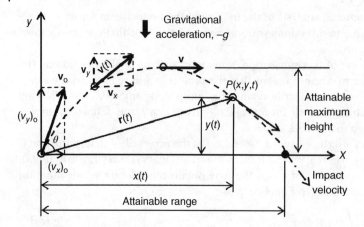

Figure 3.28 The path of a projectile in a rectangular coordinate system.

The velocity vector function $\mathbf{v}(t)$ of the projectile may be obtained using Equation 3.39:

$$\mathbf{v}(t) = \int \mathbf{a}(t)\,dt + c_1 = \int (-9.81\mathbf{j})\,dt + c_1 = (-9.81\mathbf{j})t + c_1 \tag{b}$$

The constant c_1 in Equation (b) may be determined using the initial velocity as expressed in Equation (a), or $c_1 = 173.21\mathbf{i} + 100\mathbf{j}$. Consequently, the velocity vector function of the projectile can be expressed as

$$\mathbf{v}(t) = 173.21\mathbf{i} + (100 - 9.81t)\mathbf{j} \tag{c}$$

a) **Determine the instantaneous position of the projectile, or the position vector function $r(t)$.**

We may use Equation (3.40) to derive the expression of the position vector function $\mathbf{r}(t)$ as follows:

$$\mathbf{r}(t) = \int \mathbf{v}(t)\,dt + c_2 = \int [173.21\mathbf{i} + (100 - 9.81t)\mathbf{j}]\,dt + c_2$$

$$= (173.21t)\mathbf{i} + (100t)\mathbf{j} - \left(\frac{9.81}{2}t^2\right)\mathbf{j} + c_2$$

$$= (173.21t)\mathbf{i} + (100t - 4.905t^2)\mathbf{j} + c_2 \tag{d}$$

The constant c_2 in Equation (d) may be determined by using the initial condition that $\mathbf{r}(0) = 0$, leading to $c_2 = 0$. Consequently, we have

$$\mathbf{r}(t) = (173.21t)\mathbf{i} + (100t - 4.905t^2)\mathbf{j} \tag{e}$$

The general expression of the position vector function $\mathbf{r}(t) = x(t)\mathbf{i} + y(t)\mathbf{j}$ as shown in Equation 3.38 leads to the following two expressions for the magnitudes of the components $x(t)$ and $y(t)$ of the position vector of the projectile from Equation (e):

$$x(t) = 173.21t \tag{f}$$
$$y(t) = 100t - 4.905t^2 \tag{g}$$

b) **Determine the maximum height of the moving projectile.**

We may first use the function $y(t)$ in Equation (g) to determine the time t_m required to gain the maximum attainable height of the projectile, that is, the maximum value, $y(t_m)$ or y_m. This requires the calculus techniques for finding the maximum or minimum value of a function. In the present case, we need to solve the following equation:

$$\frac{dy(t)}{dt}\bigg|_{t=t_m} = \frac{d}{dt}(100t - 4.905t^2)\bigg|_{t=t_m} = 0$$

leading to the equation $100 - 9.81t_m = 0$, from which we solve for t_m to get $t_m = 5.0968$ s. To ensure that this t_m would result in $y(t_m)$ to be the maximum of the function $y(t)$, we will need to show that

$$\frac{d^2y(t)}{dt^2}\bigg|_{t=t_m} = -9.81 < 0$$

Thus, we have obtained the maximum attainable height of the projectile:

$$y_m = y(t_m) = 100(5.0968) - 4.95(5.0968)^2 = 382.26\,\text{m}$$

c) *The attainable range (R)*

We will need first to find the time that the projectile requires to touch down on the ground again. Mathematically, this means that $y(t_e) = 0$, where t_e is the required time that can be obtained by solving the following equation:

$$y(t_e) = 100t_e - 4.902t_e^2 = 0$$

The solution of the above equation is $t_e = 20.3874$ s (the other solution $t_e = 0$ is not realistic). The range of the projection can thus be obtained from the expression for $x(t_e)$ in Equation (f) as

$$R = x(20.3874) = 173.21(20.3874) = 3531.3\,\text{m}$$

d) *The impact velocity* (v_e)

Now that we have determined the time of the impact ($t_e = 20.3874$ s), we may determine the impact velocity vector by evaluating the velocity vector function in Equation (c) at this instant as

$$\mathbf{v}_e = \mathbf{v}(t_e) = \mathbf{v}(20.3874) = 173.21\mathbf{i} + (100 - 9.81 \times 20.3874)\mathbf{j} = 173.21\mathbf{i} - 100\mathbf{j}$$

The magnitude of the impact velocity is

$$|\mathbf{v}_e| = \sqrt{(173.21)^2 + (-100)^2} = 200 \quad \text{m/s}$$

The direction of the impact velocity is $\theta = \tan^{-1}(100/173.21) = -18.43°$ from the x-coordinate.

3.7.4 Plane Curvilinear Motion in Cylindrical Coordinates

In Section 3.2, we presented a position vector in both rectangular coordinates defined by (x,y,z) and a cylindrical coordinate system involving (r,θ,z) coordinates in Figures 3.3a and 3.4b. In this section, we formulate the kinematics of a rigid body traveling in a plane defined by the r–θ coordinates, as illustrated in Figure 3.29a.

Instead of using the unit vectors $\mathbf{i}$, $\mathbf{j}$, and $\mathbf{k}$ in the rectangular coordinate systems as illustrated in Figure 3.3b, we define the unit vectors in an r–θ coordinate system as

$\mathbf{u_r}$ = the unit vector along the r-coordinate

$\mathbf{u_\theta}$ = the unit vector along the coordinate that follows the trend of the positive θ-coordinate (i.e., in the counterclockwise direction) in the direction that is perpendicular to the r-coordinate.

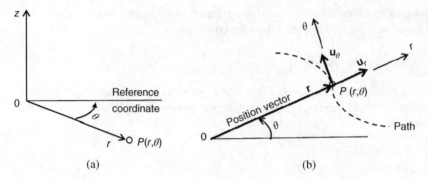

Figure 3.29 Vectors in plane cylindrical coordinate system. (a) A plane defined by r–θ coordinates. (b) A position vector in the r–θ plane.

The directions of both these unit vectors are illustrated in Figure 3.29b. One should note that the directions of these unit vectors vary with change of the position vector with respect to time t.

Thus, by following the definition of unit vectors as stipulated in Equation (3.3), the current position vector of the rigid body $\mathbf{r}(r,\theta)$ can be expressed as

$$\mathbf{r}(r, \theta) = r\mathbf{u}_r + \theta\mathbf{u}_\theta \tag{3.41}$$

where r is the radial distance of the curved path along which the rigid body travels.

The position vector is a function of time t, or $\mathbf{r}(t)$, and the corresponding time-varying unit vector functions are $\mathbf{u}_r(t)$ and $\mathbf{u}_\theta(t)$, leading to the following position vector functions such as

$$\mathbf{r}(t) = r(t)\mathbf{u}_r(t) + \theta\mathbf{u}_\theta(t) \tag{3.42a}$$

for the time-varying position vector $\mathbf{r}(t)$, where $r(t)$ is the magnitude of the vector $\mathbf{r}(t)$ at time t.

We may also derive expressions for the velocity vector function by using Equation 3.37b:

$$\begin{aligned}
\mathbf{v}(t) &= \frac{d\mathbf{r}(t)}{dt} \\
&= \frac{d}{dt}[\mathbf{r}(t)\mathbf{u}_r(t)] \\
&= r(t)\frac{d[\mathbf{u}_r(t)]}{dt} + \mathbf{u}_r(t)\frac{d[r(t)]}{dt} \\
&= r(t)\dot{\mathbf{u}}_r(t) + \dot{r}(t)\mathbf{u}_r(t)
\end{aligned} \tag{3.42b}$$

The rate of change of the radial unit vector in a cylindrical coordinate system (r,θ) with respect to time (t) i.e., $\dot{\mathbf{u}}_r(t)$, in Equation 3.42b is derived from the situation in which the rigid body moves from position P to P', together with the movement of the unit vector function $\mathbf{u}_r(t)$ as illustrated in Figure 3.30a. The same movement results in a variation of the angular movement with an increment of $\Delta\theta$ from θ to θ'.

In Figure 3.30b, we observe that the position vector function of the body at $P'(t)$ may be related to its previous position vector function by the following relationship:

$$\mathbf{u}_r'(t) = \mathbf{u}_r(t) + \Delta\mathbf{u}_r(t)$$

where $\Delta\mathbf{u}_r(t)$ is the difference between the unit vector $\mathbf{u}_r'$ at the current location P' and $\mathbf{u}_r$ at the previous location P.

But on the assumption of a small displacement of the body in a continuous variation process, the corresponding distance that the body has traveled $\hat{S}$ is also very small, giving the arc $\hat{S} = \rho\Delta\theta$, where ρ is the radius of the curved path and $\Delta\theta$ is the angle corresponding to the

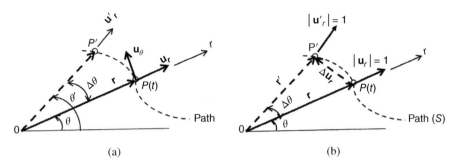

Figure 3.30 Variation of unit vector functions in plane cylindrical coordinates. (a) Displacement of a body from P to P'. (b) The net increase of unit vector $\mathbf{u}_r(t)$.

arc. Consequently, since $\Delta\mathbf{u}_r(t)$ is very small in its magnitude, we may legitimately relate the increment of the unit vector function $\mathbf{u}_r(t)$ to the unit vector function $\mathbf{u}_\theta(t)$ as illustrated in Figure 3.30a as

$$\Delta\mathbf{u}_r(t) = (\Delta\theta)\ \mathbf{u}_\theta(t) \tag{3.43}$$

where $\Delta\theta$ is the corresponding variation of the θ-coordinate associated with the shift of the position vector from position P to P' as shown in Figure 3.30b.

Equation 3.43 leads to the following formulation for the derivatives of the unit vector functions:

$$\dot{\mathbf{u}} = \lim_{\Delta t \to 0} \frac{\Delta\mathbf{u}_r(t)}{\Delta t} = \lim_{\Delta t \to 0} \frac{(\Delta\theta)\mathbf{u}_\theta}{\Delta t} = \lim_{\Delta t \to 0} \left(\frac{\Delta\theta}{\Delta t}\right)\mathbf{u}_\theta$$

from which we may derive a useful expression that relates the two unit vectors in a cylindrical coordinate system:

$$\dot{\mathbf{u}}_r(t) = \dot{\theta}(t)\mathbf{u}_\theta \tag{3.44}$$

By substituting the relationship in Equation 3.44 into Equation 3.43, we may obtain the velocity vector function in the r–θ coordinate system, corresponding to Equation 3.42b in the form

$$\mathbf{v}(t) = r(t)\dot{\theta}(t)\mathbf{u}_\theta + \dot{r}(t)\mathbf{u}_r \tag{3.45}$$

We thus have the magnitude of the following two velocity components as

$$v_r(t) = \dot{r} \tag{3.46a}$$

$$v_\theta(t) = r(t)\dot{\theta}(t) \tag{3.46b}$$

where $v_r(t)$ and $v_\theta(t)$ are the magnitudes of the velocity vector $\mathbf{v}(t)$ along the respective radial and tangential directions in a cylindrical coordinate system.

The magnitude of the velocity vector at time t is thus

$$v(t) + \sqrt{[r(t)\dot{\theta}(t)]^2 + [\dot{r}(t)]^2} \tag{3.47}$$

The derivation of the acceleration vector function may begin with the relationship of the acceleration vector function $\mathbf{a}(t) = d\mathbf{v}(t)/dt$ and the velocity vector function $\mathbf{v}(t)$ as expressed in Equation 3.45. Differentiating $\mathbf{v}(t)$:

$$\mathbf{a}(t) = \frac{d}{dt}\mathbf{v}(t) = \frac{d}{dt}\left[r(t)\dot{\theta}(t)\mathbf{u}_\theta + \dot{r}(t)\mathbf{u}_r\right]$$
$$= \left[\dot{r}(t)\dot{\theta}(t)\mathbf{u}_\theta + r(t)\ddot{\theta}(t)\mathbf{u}_\theta + r(t)\dot{\theta}(t)\dot{\mathbf{u}}_\theta\right] + \left[\ddot{r}(t)\mathbf{u}_r + \dot{r}(t)\dot{\mathbf{u}}_r\right] \tag{3.48}$$

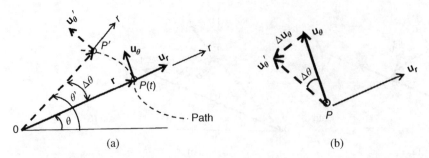

Figure 3.31 Variation of the transverse unit vector function. (a) Displacement of unit vector $\mathbf{u}_\theta$. (b) Corresponding variation of unit vector $\mathbf{u}_\theta$.

Again, we will need to formulate the derivative of $\dot{\mathbf{u}}_\theta$ in Equation 3.48. This can be done similarly to what we did for the expression of $\dot{\mathbf{u}}_r$ in Equation 3.44.

Figure 3.31a illustrates the displacement of the transverse unit vector $\mathbf{u}_\theta$ associated with the travel of the particle from P to P' along the path S. We note that although the magnitude of vector $\mathbf{u}_\theta$ remain unchanged, its direction has varied from θ to θ' by an amount $\Delta\theta$, as indicated in Figure 3.31b with a shift that is equal to $\Delta\mathbf{u}_\theta = \mathbf{u}'_\theta - \mathbf{u}_\theta$. We may use a similar argument on small displacement of the body moving along the path S; the variation of the magnitude of the unit vector $\mathbf{u}_\theta$ with a change of the angle θ is

$$\Delta\mathbf{u}_\theta \approx (\Delta\theta)\mathbf{u}_\theta$$

or in terms of the magnitudes of the unit vectors:

$$|\Delta\mathbf{u}_\theta| = \Delta u_\theta \approx (\Delta\theta)|\mathbf{u}_\theta| = (\Delta\theta)(1) = (\Delta\theta)$$

In view of the fact that the magnitudes of both unit vectors $\mathbf{u}_\theta$ and $\mathbf{u}_r$ are by definition unity—that is, $|\mathbf{u}_\theta| = |\mathbf{u}_r| = 1.0$ —the above expression may be written

$$|\Delta\mathbf{u}_\theta| = \Delta u_\theta \approx (\Delta\theta)|\mathbf{u}_\theta| = (\Delta\theta)|\mathbf{u}_r|$$

which leads to the following equality:

$$\Delta\mathbf{u}_\theta = -(\Delta\theta)\mathbf{u}_r \tag{3.49}$$

One will note that the negative sign added to the right-hand side of Equation 3.49 makes the vector $\Delta\mathbf{u}_\theta$ in opposite direction to the positive direction of vector $\mathbf{u}_r$ as indicated in Figure 3.31b.

Recognizing that $\dot{\mathbf{u}}_\theta$ in Equation 3.48 is the derivative of the transverse unit vector $\mathbf{u}_\theta$ with respect to the time variable t, we may express it using in Equation 3.49 as

$$\dot{\mathbf{u}}_\theta = \lim_{\Delta t \to 0} \frac{\Delta\mathbf{u}_\theta}{\Delta t} = \left(-\lim_{\Delta t \to 0} \frac{\Delta\theta(t)}{\Delta t} \right) \mathbf{u}_r = -\dot{\theta}(t)\mathbf{u}_r \tag{3.50}$$

Thus, by substituting the term $\dot{\mathbf{u}}_\theta$ in Equation 3.48 with $-\dot{\theta}(t)\mathbf{u}_r$ in Equation 3.50, we obtain the following expression for the acceleration vector function:

$$\mathbf{a}(t) = a_r\mathbf{u}_r + a_\theta\mathbf{u}_\theta \tag{3.51}$$

in which the magnitudes of the acceleration in the respective radial and tangential directions a_r and a_θ are

$$a = \ddot{r}(t) - r[\dot{\theta}(t)]^2 \tag{3.52a}$$

$$a_\theta = r[\ddot{\theta}(t)] + 2[\dot{r}(t)\,\dot{\theta}(t)] \tag{3.52b}$$

Figure 3.32 Acceleration vector in a cylindrical coordinate system.

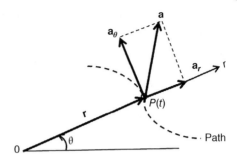

The magnitude of the acceleration vector in Equation 3.51 can thus be obtained using Equation 3.53:

$$a = |\mathbf{a}| = \sqrt{a_r^2 + a_\theta^2}$$

$$= \sqrt{\left\{ \ddot{r}(t) - r\left[\dot{\theta}(t)\right]^2 \right\}^2 + \left\{ r\left[\ddot{\theta}(t)\right] + 2\left[\dot{r}(t)\,\dot{\theta}(t)\right] \right\}^2} \tag{3.53}$$

Equation 3.52a indicates that a_r is the coefficient of the unit vector $\mathbf{u}_r$ along the r-coordinate, as expressed in Equation 3.51. It also represents the magnitude of the component of the acceleration vector $\mathbf{a}$ along the radial direction in a cylindrical coordinate system. The vector $\mathbf{a}_r$ at time t is shown in Figure 3.32. Likewise, a_θ in Equation 3.52b is the magnitude of the vector $\mathbf{a}_\theta$, which is the component of the acceleration vector $\mathbf{a}$ in the transverse, or tangential direction in the same cylindrical coordinate as shown Figure 3.32. The term $\ddot{\theta}$ in Equation (3.52b) is referred to as the angular acceleration in analysis.

Example 3.18
An automobile is traveling along a section of road of circular curvature with radius $r = 25$ m at the location and instant as shown in Figure 3.33. The rate of change of its angular displacement is $\dot{\theta} = 0.5\,\text{rad/s}$, with an angular acceleration of $\ddot{\theta} = 0.15\,\text{rad/s}^2$. Determine the magnitude of the automobile's velocity and acceleration at this instant.

Solution:
The position vector of the vehicle at time t is

$$\mathbf{r}(t) = r\mathbf{u}_r + \theta\mathbf{u}_\theta = 25\mathbf{u}_r + \theta\mathbf{u}_\theta$$

in which r and θ are the components of the position vector $\mathbf{r}(t)$ in the radial direction and the angle from the reference shown line in Figure 3.33, with the angular velocity being $\dot{\theta} = 0.5\,\text{rad/s}$ and the angular acceleration $\ddot{\theta} = 0.15\,\text{rad/s}^2$.

Figure 3.33 A vehicle traveling along an arc.

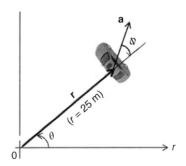

We will use Equations 3.46a and 3.46b to determine the radial and tangential velocity components as follows:

$$v_r(t) = \dot{r} = \frac{dr(t)}{dt}$$

$$= \frac{d(25)}{dt} = 0$$

$$v_\theta(t) = r\dot{\theta}(t)$$

$$= 25 \times 0.5 = 12.5 \, \text{m/s}$$

Thus, the magnitude of the velocity vector of the moving automobile at the time t is

$$v = \sqrt{v_r^2 + v_\theta^2}$$

$$= \sqrt{0^2 + 12.5^2} = 12.5 \, \text{m/s}$$

The radial and tangential components of the acceleration vector of the vehicle may be computed using Equations 3.52a and 3.52b:

$$a_r = \ddot{r} - r(\dot{\theta})^2$$

$$= 0 - 25(0.5)^2 = -6.25 \, \text{m/s}^2$$

The negative sign attached to value of a_r in the above expression indicates that this component of the acceleration is in the opposite direction to the positive r-coordinate. The tangential component according to Equation 3.52b is

$$a_\theta = r\ddot{\theta} - 2\dot{r}\dot{\theta}$$

$$= 25 \times 0.15 - 2 \times 0 \times 0.5 = 3.75 \, \text{m/s}^2$$

The magnitude of the acceleration vector is thus equal to:

$$a = \sqrt{(-6.25)^2 + (3.75)^2} = 7.29 \, m/s^2$$

The angle φ that the direction of the acceleration vector **a** makes with the radial direction r in Figure 3.33 is

$$\varphi = \tan^{-1}\frac{a_\theta}{a_r} = \tan^{-1}\frac{3.75}{(-6.25)} = -31°$$

Note that the acceleration of the vehicle is not in the same direction as the velocity of the vehicle, as illustrated in Figure 3.34, with its radial component pointing toward the center of the arc.

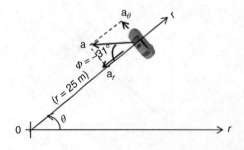

Figure 3.34 Magnitudes of the components of the acceleration vector of a vehicle traveling along a circular road.

3.7.5 Plane Curvilinear Motion with Normal and Tangential Components

We have demonstrated in the foregoing sections that of the three principal physical properties involving a rigid body in motion along a curvilinear path, both the position vector **r** and velocity vector **v** moving in trajectories along the tangent of the path S, but the acceleration vector **a** is in the direction of tangent to a hodograph that does not coincide with the path S, as illustrated in Figure 3.26 and Figure 3.32.

It is often desirable to express the acceleration vectors with their components in both the radial and tangential directions in a kinematic analysis of moving rigid bodies. We may derive expressions for such components from the acceleration vector function in general curvilinear motion in cylindrical coordinates as presented in Equations 3.48 and 3.52a, b.

Figure 3.35a shows the cylindrical coordinates (r,θ) and the two unit vectors along the linear coordinate r and the angular coordinate θ. We will derive expressions for the magnitudes of acceleration vectors along the linear radial direction r, and the tangential component normal to the r-coordinate as illustrated in Figure 3.35b.

We begin the derivation of expressions for the two components $\mathbf{a}_r$ and $\mathbf{a}_\theta$ of the vector **a** in Figure 3.35b by combining Equations 3.48 and 3.52a,b.

Equation 3.51 gives the acceleration vector function in the form

$$\mathbf{a}(t) = a_r(t)\mathbf{u}_r + a_\theta(t)\mathbf{u}_\theta$$

in which the magnitude of the component along the r-coordinate at time t has the form $a_r = \ddot{r} - r\dot{\theta}^2$ and that of the component along the θ-coordinate has the form $a_\theta = r\ddot{\theta} + 2r\dot{\theta}$, as shown in Equations 3.52a and 3.52b, respectively.

Now, since the velocity of the traveling rigid body is along the tangential direction of the given curved path as shown in Figure 3.35b, we may write the velocity vector as $\mathbf{v}(t) = v(t)\mathbf{u}_\theta(t)$, in which $v(t)$ is the magnitude of the velocity vector $\mathbf{v}(t)$ and $\mathbf{u}_\theta(t)$ is the unit vector in the θ-coordinate direction.

The corresponding acceleration vector function $\mathbf{a}(t)$ can thus be obtained as the following derivative of the velocity vector with respect to time:

$$\mathbf{a}(t) = \frac{d\mathbf{v}(t)}{dt} = \frac{d[v(t)\mathbf{u}_\theta(t)]}{dt}$$

$$= \frac{dv(t)}{dt}\mathbf{u}_\theta(t) + v(t)\frac{d\mathbf{u}_\theta(t)}{dt}$$

$$= \frac{dv(t)}{dt}\mathbf{u}_\theta(t) + v(t)\dot{\mathbf{u}}_\theta(t)$$

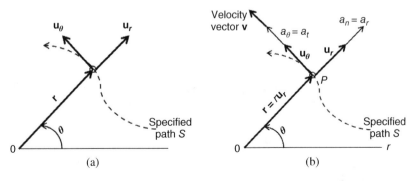

Figure 3.35 Plane curvilinear motion in a cylindrical coordinate system. (a) General curvilinear motion. (b) Showing the radial and tangential components

We may substitute the proven relationship $\dot{\mathbf{u}}_\theta(t) = -\dot{\theta}\mathbf{u}_r(t)$ from Equation (3.50) into the above expression and obtain

$$\mathbf{a}(t) = \frac{dv(t)}{dt}\mathbf{u}_\theta(t) + v(t)[-\dot{\theta}\mathbf{u}_r(t)]$$

$$= \dot{v}\mathbf{u}_\theta(t) - v(t)(\dot{\theta})\mathbf{u}_r(t)$$

$$= a_r\mathbf{u}_r(t) + \dot{v}\mathbf{u}_\theta(t) \tag{3.54}$$

The acceleration vector function $\mathbf{a}(t)$ expressed in Equation (3.54) may be written in the following typical vector form:

$$\mathbf{a} = a_r\mathbf{u}_r + a_t\mathbf{u}_\theta$$

in which $a_r = a_n =$ the magnitude of normal component of the acceleration, and $a_\theta = a_t =$ the magnitude of tangential component of the acceleration vector. Comparing the above expressions with those in Equation 3.54 with $a_n = -v\dot{\theta}$ and $a_t = \dot{v}$, and from Equation 3.46b, we have the magnitude of velocity component in the tangential of the paths as $v_\theta = r(t)\dot{\theta}$, from which we obtain

$$\dot{\theta} = \frac{v_\theta}{r} = \frac{v}{r} \quad \text{at time } t$$

Here we equate $v_\theta = v$ because the rigid body travels with a velocity $\mathbf{v}(t)$ along the tangent of the path.

We may thus express the magnitudes of the normal and tangential components of the acceleration in the following forms:

$$a_t = \dot{v} = \frac{dv(t)}{dt} \tag{3.55a}$$

$$a_n = -\frac{v^2}{r} \tag{3.55b}$$

The negative sign in Equation 3.55b indicates that the normal acceleration component has a direction toward the center 0 in Figure 3.35b.

The magnitude of the normal acceleration component Equation 3.55b results in the same value of $a_n = a_r = 6.25$ m/s^2 with $v = 12.5$ m/s as in Example 3.18.

Example 3.19

An automobile travels on a circular track with a radius of 25 m as shown in Figure 3.36. If the magnitude of the velocity vector function of the vehicle is $v(t) = 1 + 2t^2$ m/s, and the starting location of the vehicle is as shown in the figure, determine the following:

a) The magnitude of the acceleration of the vehicle at 4 seconds from standstill location.
b) The distance and the number of laps that the car has traveled in 10 seconds.

Solution:

We may begin our solution with the given magnitude of velocity vector function $\mathbf{v}(t)$, which is $v(t) = 1 + 2t^2$ m/s, from which we may obtain the magnitude of the tangential component of the acceleration vector function:

$$a_t(t) = |\mathbf{a}_t(t)| = \frac{dv(t)}{dt}$$

$$= \frac{d(1 + 2t^2)}{dt} = 4t \tag{a}$$

Figure 3.36 A vehicle traveling on a circular track.

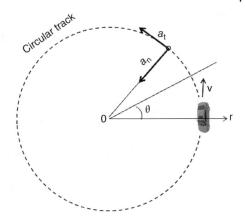

a) ***The acceleration at time t = 4 seconds***

The tangential component $a_t = 4t = 4 \times 4 = 16$ m/s^2. The magnitude of the radial (or normal) component of the acceleration at time $t = 4$ s is obtained using the expression in Equation (3.55b), as

$$a_n = -\left.\frac{v(t)^2}{r}\right|_{t=4} = -\frac{\left[1 + 2(4)^2\right]^2}{25} = -43.56 \, \text{m/s}^2 \tag{b}$$

or 43.56 m/s^2 toward the center of the circular track.

The magnitude of the acceleration vector of the vehicle at $t = 4$ s can thus be computed from the two components in Equations (a) and (b) as

$$a(4) = \sqrt{a_t^2 + a_n^2} = \sqrt{16^2 + 43.56^2} = 46 \, \text{m/s}^2$$

b) ***The distance traveled by the vehicle at time t = 10 seconds***

The distance $S(t)$ that the vehicle has traveled after time t can be computed from the relationship $v(t) = dS(t)/dt$, which leads to

$$\int_0^S dS(t) = \int_0^t v(t) \, dt$$

with the given velocity vector function $\mathbf{v}(t)$.

We may thus compute the distance traveled by the vehicle after 10 seconds to be

$$S(10) = \int_0^{10} (1 + 2t^2) \, dt = \left.\left(t + \frac{2}{3}t^3\right)\right|_0^{10} = 676.67 \, \text{m}$$

The equivalent number of laps that the vehicle has traveled is

$$N = \frac{S(10)}{2\pi r} = \frac{676.67}{2 \times 3.14 \times 25} = 4.31 \text{ laps}$$

Example 3.20

Figure 3.37a illustrates a baggage conveyor in a major airport in Germany. We assume that a box, shown in gray in Figure 3.37b, is being transported from the exit of the collection station by the moving conveyor. Approximate dimensions and the motion of the conveyor are illustrated in Figure 3.37b. We further assume that the conveyor is designed to move the baggage from zero initial velocity at entry location 0 with acceleration $a(t) = 0.001t$ m/s^2 between location 0 and point c, but that it moves the baggage at a constant velocity with no acceleration thereafter. Determine the following:

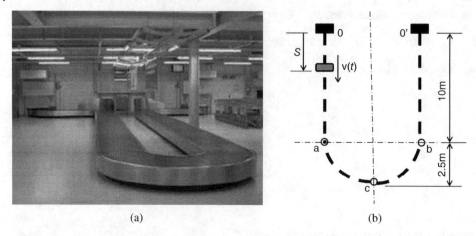

Figure 3.37 Baggage transportation conveyor. (a) An airport baggage transportation conveyor in Frankfurt, Germany. (b) Dimensions and travel of a piece of baggage [here a box]. ((a) Photograph courtesy of Frankfurt-Hahn Airport.)

a) The velocity of the box baggage at locations a and c.
b) The acceleration and its direction of the baggage at the same locations.
c) The time required for the baggage to reach point c.
d) The time required for the baggage to return to the collection station at $0'$ if it is not picked up by any passenger.
e) The time for a complete excursion.

Solution:
We need to find the velocity vector function of the baggage that will lead to the computation of the distance the box has traveled from location 0 to point c. The magnitude of the tangential velocity vector function $\mathbf{v}_t(t)$ can be found from the following expressions:

We are given the magnitude of the acceleration vector function $a(t) = a_t(t) = 0.001t$, but this acceleration component is related to the rate of change of the velocity $v_t(t)$ as given in Equation 3.37c with

$$\mathbf{a}_t(t) = \frac{d\mathbf{v}_t(t)}{dt}$$

which yields an expression for the magnitude of the velocity vector function $\mathbf{v}_t(t)$ by integration:

$$\int_0^t d\mathbf{v}_t(t)\, dt = \int_0^t \mathbf{a}_t(t)\, dt = \int_0^t (0.001t)\, dt$$

From this the velocity function is

$$v_t(t) = 0.0005\, t^2 \tag{a}$$

If we let $S(t)$ be the distance that the box has traveled from the starting point 0 after time t, the relation $v_t(t) = dS(t)/dt$ will let us calculate $S(t)$:

$$\int_0^t dS(t) = \int_0^t v_t(t)\, dt = \int_0^t (0.0005t^2)\, dt$$

which results in

$$S(t) = 1.67 \times 10^{-4}\, t^3 \tag{b}$$

We are now ready to deal with the specific problems in the example.

a) **Determination of velocity of the box at locations a and c and the time required for the box to reach these two locations**

We may use the relationship shown in Equation (b) to compute the time for the box to reach locations a and c by the distance S traveled.

If we let the time required for the box to reach location a to be t_a with a distance of travel $S(t_a) = 10$ m, we can compute $t_a = [10/(1.67 \times 10^{-4})]^{1/3} = 39.12$ seconds. Likewise, the time for the box to reach location c from the entry location 0 is t_c, with $S(t_c) = 10 + 2\pi r/4 = 10 + (2 \times 3.14 \times 2.5)/4 = 13.925$ m, giving $t_c = [13.925/(1.67 \times 10^{-4})]^{1/3} = 46.69$ seconds.

With $t_a = 39.12$ seconds and $t_c = 46.69$ seconds, we may compute the velocity at locations a and c using Equation (a) as follows.

At location a:

$$v_t(t_a) = 0.0005(39.12)^2 = 0.765 \, \text{m/s}$$

and at location c:

$$v_t(t_c) = 0.0005(46.69)^2 = 1.09 \, \text{m/s}$$

b) **The acceleration of the box at locations a and c**

The acceleration can be determined from the given acceleration $a_t(t) = 0.01t$ with $t_a = 39.12$ s and $t_c = 46.69$ s as follows.

At location a:

$$a_t(t_a) = 0.001 \times 39.12 = 0.04 \, \text{m/s}^2$$

and at location c:

$$a_t(t_c) = a(t_c)$$
$$= 0.001 \times 46.69 = 0.0467 \, \text{m/s}^2$$

However, location c is on a curvilinear path with a radius of curvature $r = 2.5$ m. Thus, using Equation 3.55b, we may compute the magnitude of the normal component of the acceleration:

$$a_n(t_c) = \frac{[v_t(t_c)]^2}{r^2}$$
$$= (1.09)^2/2.5 = 0.4752 \, \text{m/s}^2$$

The magnitude of the acceleration at location c is thus

$$a_c = \sqrt{a_t^2 + a_n^2}$$
$$= \sqrt{(0.0467)^2 + (0.4752)^2} = 0.4776 \, \text{m/s}^2$$

c) **The time required for the baggage to reach point c.**

The time required to pass location c is $t_c = 46.69$ seconds.

d) **The time required for uncollected baggage to return to the collection station at 0′**

We realize that the conveyor moves the baggage with no acceleration beyond location c—that is, the box moves at constant speed from location c to the end location 0′. The speed for the remaining portion of the movement is the speed at location c, or $v = 1.09$ m/s. The time required to move the box from location c to 0′ can be computed with $t_{c-0'} = S_{c-0'}/v = (10 + 2\pi r/4)/1.09 = 12.78$ seconds.

e) *The time for a complete excursion*

The time for the entire excursion of the box movement by this particular conveyor is thus equal to $t = t_c + t_{c-0'} = 46.69 + 12.78 = 59.47$ seconds ≈ 1 minute.

3.8 Problems

3.1 Given two vectors $\mathbf{A} = 2\mathbf{i} + 3\mathbf{j} + 4\mathbf{k}$ and $\mathbf{B} = \mathbf{i} + 2\mathbf{j} + 3\mathbf{k}$, compute the following: (a) $\mathbf{A} + \mathbf{B}$; (b) $\mathbf{A} - \mathbf{B}$; (c) $\mathbf{B} - \mathbf{A}$; (d) $\mathbf{A} \cdot \mathbf{A}$; (e) $\mathbf{A} \cdot \mathbf{B}$; (f) $\mathbf{A} \times \mathbf{B}$. Compute also the magnitudes of each of resultant vectors in each case.

3.2 Express the position vector $\mathbf{r}$ for a point $P(x,y,z)$ in terms of the unit vectors $\mathbf{i}$, $\mathbf{j}$, and $\mathbf{k}$ in a space defined by a rectangular coordinate system with coordinates x, y, and z. Also, compute the magnitude of this vector.

3.3 Illustrate the position vectors related to point $P_1(3,5)$ in an x–y rectangular coordinate system, and $P_2(3,5,8)$ in a rectangular coordinate system employing x-, y-, and z-coordinates.

3.4 For position vectors $\mathbf{r}_1$ for point $P_1(3,5,4)$ and $\mathbf{r}_2$ for point $P_2(1,0,3)$ in a rectangular coordinate systems employing x-, y-, and z-coordinates, show graphically (a) $\mathbf{r}_1 + \mathbf{r}_2$; (b) $\mathbf{r}_1 - \mathbf{r}_2$; (c) the magnitudes of the resultant vectors in each case.

3.5 Calculate the dot product of two position vectors, $\mathbf{r}_1$ for $P_1(3,5,4)$ and $\mathbf{r}_2$ for $P_2(1,0,3)$, and determine the angle between these two vectors.

3.6 Find the magnitude of a position vector $\mathbf{r}$ with coordinates $(2,3,5)$ in a 3D space as shown in Figure 3.3a. Express the position vector in terms of the unit vectors $\mathbf{i}$, $\mathbf{j}$, and $\mathbf{k}$, and determine the angles α, β, and γ defined in Figure 3.4b.

3.7 Find the magnitudes of the two vectors $\mathbf{A}$ and $\mathbf{B}$ in Problem 3.1 using unit vectors $\mathbf{i}$, $\mathbf{j}$, and $\mathbf{k}$ defined in Figure 3.3b, and the angles α, β and γ between these two vectors as defined in Figure 3.4b.

3.8 Given a vector $\mathbf{C} = 3\mathbf{i} - 2\mathbf{j} + 4\mathbf{k}$, determine $\mathbf{A} \cdot (\mathbf{B} \times \mathbf{C})$, with vectors $\mathbf{A}$ and $\mathbf{B}$ as defined in Problem 3.1.

3.9 Prove the following: (a) $\mathbf{A} \cdot \mathbf{B} = \mathbf{B} \cdot \mathbf{A}$; (b) $\mathbf{A} \times \mathbf{B} \neq \mathbf{B} \times \mathbf{A}$; and (c) $\mathbf{i} \cdot \mathbf{j} = \mathbf{j} \cdot \mathbf{k} = \mathbf{k} \cdot \mathbf{i} = 0$; (d) $\mathbf{i} \cdot \mathbf{i} = \mathbf{j} \cdot \mathbf{j} = \mathbf{k} \cdot \mathbf{k} = 1$.

3.10 Use vector representations to determine the volume of the parallelepiped solid shown in Figure 3.38 with lengths of the sides $A = 5$ units, $B = 8$ units, and $C = 4$ units.

3.11 A rigid body moves along both the x- and y-coordinate directions according to the functions $x(t) = \cos(5\pi t)$ and $y(t) = 3 \sin(\pi t)$. Find the magnitude and direction of the velocity and acceleration vectors at $t = 0.25$.

3.12 If vector $\mathbf{A} = (x^2 \sin y)\mathbf{i} + (z^2 \cos y)\mathbf{j} - (xy^2)\mathbf{k}$, determine the following:

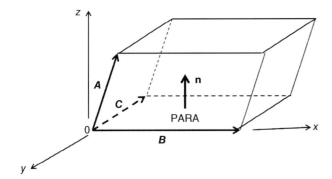

Figure 3.38 Volume of a parallelepiped solid.

a) $\dfrac{\partial A}{\partial x}$,

b) $\dfrac{\partial A}{\partial y}$,

c) $\dfrac{\partial A}{\partial z}$,

d) dA

3.13 If vector $\mathbf{A} = xz\mathbf{i} - y^2\mathbf{j} + 2x^2 y\mathbf{k}$ and $\varphi = x^2 yz^2$, determine the following:

a) $\nabla \phi$,

b) $\nabla \cdot \mathbf{A}$,

c) $\nabla \times \mathbf{A}$,

d) $\operatorname{div}(\varphi\mathbf{A})$,

e) $\operatorname{curl}(\varphi\mathbf{A})$.

3.14 If $\varphi = 2x^2 y - xz^3$, find:

a) $\nabla\varphi = \dfrac{\partial\varphi}{\partial x}\mathbf{i} + \dfrac{\partial\varphi}{\partial y}\mathbf{j} + \dfrac{\partial\varphi}{\partial z}\mathbf{k}$

b) $\nabla^2\varphi = \nabla \cdot \nabla\varphi$

3.15 Consider a projectile traveling in a space defined by a rectangular coordinate system with x-, y-, and z-coordinates. Unit vectors $\mathbf{i}$, $\mathbf{j}$, and $\mathbf{k}$ along the respective x-, y-, and z-coordinate directions are used in the analysis. The position of the projectile may be expressed by the position vector $\mathbf{r} = (25 \cos t)\mathbf{i} + (12 \sin t)\mathbf{j} + (9t)\mathbf{k}$, in which t is time in seconds and the unit of the magnitude of this position vector is meters. Determine the following:

a) The magnitude of the velocity vector $\mathbf{v}$.

b) The expression for the distance (S) the projectile has traveled.

3.16 An airplane starts from an airport located at the origin 0 in Figure 3.39 and follows the pattern as illustrated for its flight. It begins the flight of 150 miles in the direction 20° northeast to A. From A, the airplane then flies 200 miles in direction 23° northwest to B, and from B it flies 240 miles in the direction 10° southwest to C. Determine the distance between location C and its original location 0 in terms of a position vector $\mathbf{r}$ as shown in Figure 3.39.

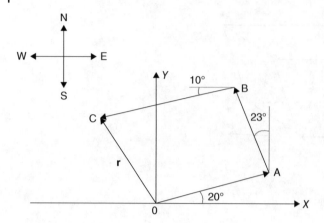

Figure 3.39 Flight path of an airplane.

3.17 A rigid body is moving along a curved path $y = x - (x^2/100)$ in the plane defined by an x–y rectangular coordinate system, in which x and y in the above function are in units of meters. The magnitude of the velocity component along the x-coordinate is $v_x = 0.9$ m/s and remains constant. Determine the velocity and acceleration vectors when $x = 8$ m.

3.18 A jet fighter airplane is dropping a bomb to hit the target on the ground as illustrated in Figure 3.40. The airplane flies horizontally over the target area at an altitude H at a constant velocity V. Use the vector calculus method with $H = 320$ m and $V = 600$ km/h to determine:
a) The horizontal distance the dropped bomb travels relative to the point from which it was dropped.
b) The line-of-sight angle θ at which the bomb should be released in order to hit the target indicated in Figure 3.40?
Assume that the aerodynamic drag on the falling bomb is negligible.

3.19 A jet fighter airplane travels along a vertical parabolic path as shown in Figure 3.41. The airplane travels at a speed of 180 m/s with an acceleration of 0.5 m/s². Compute the magnitude and the direction of the acceleration of the airplane at point A in the figure.
Hint: The radius of curvature r of a curve in the x–y plane may be obtained with the expression

$$r = \frac{d^2y(x)/dx^2}{\left[1 + (dy(x)/dx)^2\right]^{3/2}}$$

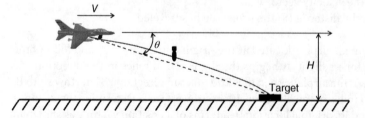

Figure 3.40 A jet airplane bombing a target on the ground.

Figure 3.41 Flight path of a fighter jet.

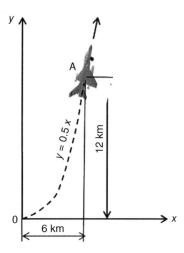

3.20 In Example 3.20, the maximum velocity of the baggage on a transportation conveyor was 1.09 m/s at the farthest location, point C in Figure 3.37b. What value for the acceleration of the conveyor would reduce the maximum velocity to 0.5 m/s at the same location C?

4

Linear Algebra and Matrices

Chapter Learning Objectives

- Linear algebra and its applications
- Forms of linear functions and linear equations
- Expression of simultaneous linear equations in matrix forms
- Distinction between matrices and determinates
- Different forms of matrices for different applications
- Transposition of matrices
- Addition, subtraction and multiplication of matrices
- Inversion of matrices
- Solution of simultaneous equations using matrix inversion method
- Solution of large numbers of simultaneous equations using Gaussian elimination method
- Eigenvalues and eigenfunctions in engineering analysis

4.1   Introduction to Linear Algebra and Matrices

Linear algebra is concerned mainly with systems of linear equations, matrices, vector space, linear transformations, eigenvalues, and eigenvectors. It is widely applied to other areas of mathematics and to other various disciplines of engineering and science (Hogben, 1987).

Linear functions are often used to represent straight lines. For example, the function $y(x) = mx + b$ is used to represent a straight line in the x–y plane, in which m and b are constants, with m representing the "slope" of the straight line and b the intercept of the line with the y-axis as illustrated in Figure 4.1. Equations involving linear functions are linear equations.

Engineers need to deal with various types of equations; some are linear and others are non-linear. For instance, Equation 4.1 is linear:

$$-4x_1 + 3x_2 - 2x_3 + x_4 = 0 \tag{4.1}$$

with x_1, x_2, x_3, and x_4 being the unknown quantities.

At least one unknown being of a higher power, such as in Equations 4.2a, results in this type of equations being nonlinear:

$$-4x_1 + 3x_2 - 2x_3^2 + x_4^3 = 0 \tag{4.2a}$$

Other nonlinear equations may have the forms:

$$x^2 + y^2 = 1 \tag{4.2b}$$

$$xy = 1 \tag{4.2c}$$

$$\sin x = y \tag{4.2d}$$

Applied Engineering Analysis, First Edition. Tai-Ran Hsu.
© 2018 John Wiley & Sons Ltd. Published 2018 by John Wiley & Sons Ltd.
Companion Website: www.wiley.com/go/hsu/applied

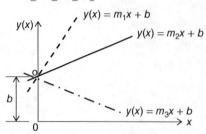

$m_1 > m_2 > m_3 (m_3$ is negative):

Figure 4.1 Linear functions.

There are occasions in which engineers need to solve simultaneous linear equations with unknowns x_1, x_2, and x_3, such as

$$8x_1 + 4x_2 + x_3 + 12 \tag{4.3a}$$

$$2x_1 + 6x_2 - x_3 = 3 \tag{4.3b}$$

$$x_1 - 2x_2 + x_3 = 2 \tag{4.3c}$$

One may readily solve the simultaneous equations in Equations 4.3 for the roots $x_1 = 1$, $x_2 = 0.5$, $x_3 = 2$ using Cramer's rule.

Engineers are often required to solve large number of simultaneous linear equations, such as shown in Equation 4.4:

$$
\begin{aligned}
a_{11}x_1 + a_{12}x_2 + a_{13}x_3 + \cdots + a_{1n}x_n &= r_1 \\
a_{21}x_1 + a_{22}x + a_{23}x_3 + \cdots + a_{2n}x_n &= r_2 \\
a_{31}x_1 + a_{32}x_2 + a_{33}x_3 + \cdots + a_{3n}x &= r_3 \\
&\vdots \\
a_{n1}x + a_{n2}x + a_{n3}x_3 + \cdots + a_{nn}x_n &= r_n
\end{aligned}
\tag{4.4}
$$

In Equation 4.4, a_{ij} with subscripts $i = 1, 2, 3, \ldots, m$ and $j = 1, 2, 3, \ldots, n$, are constant coefficients, x_j with $j = 1, 2, 3, \ldots, n$ are the unknowns to be solved for from the equations, and r_j with $j = 1, 2, 3, \ldots, n$ are the given constants in each individual equation.

Engineers may be expected to solve large numbers of simultaneous linear equations, exceeding 100 000 for problems relating to complex geometry, loading, and boundary conditions, such as when assessing the stresses near sharp crack tips in solid structures using the finite-element method (Hsu, 1986). In reality, almost all modern numerical analytical methods commonly used in engineering analysis, whether the finite-element method (FEM) such as presented in Chapter 11, finite-difference method (FDM) presented in Chapter 10, or boundary element method (BEM) require effective and efficient ways of handling large numbers of data associated with a large number of simultaneous equations. Modern digital computers, with their super-fast speed in performing arithmetic operations and enormous data storage capabilities, are the only practical equipment for handling these analyses. However, digital computers with the current state-of-the-arts are not intelligent enough to handle functions, even the simple exponential or trigonometric functions, without having them converted into simple arithmetic operations. The same process is required to convert simultaneous linear or differential equations for solutions. The means of such conversions—the use of algorithms—is necessary in virtually all engineering analyses; matrix techniques are a viable tool in developing algorithms for analytical problems involving large number of data in the analyses.

4.2   Determinants and Matrices

Determinants and matrices are the logical and convenient choice for representation of large quantities of real numbers, variables, and vectors involved in engineering analyses. These entities are arranged in arrays in rows and columns such as shown below:

$$
\begin{matrix}
a_{11} & a_{12} & a_{13} & \bullet & \bullet & \bullet & \bullet & a_{1n} \\
a_{21} & a_{22} & a_{23} & \bullet & \bullet & \bullet & \bullet & a_{2n} \\
a_{31} & a_{32} & a_{33} & \bullet & \bullet & \bullet & \bullet & a_{3n} \\
\bullet & \bullet & \bullet & \bullet & \bullet & \bullet & \bullet & \bullet \\
\vdots & & & & & & & \\
\bullet & \bullet & \bullet & \bullet & \bullet & \bullet & \bullet & \\
a_{m1} & a_{m2} & a_{m3} & \bullet & \bullet & \bullet & \bullet & a_{mn}
\end{matrix}
$$

in which $a_{11}, a_{12}, \ldots, a_{mn}$ represent groups of data that may include real numbers, variables, and vector quantities expressed in a number of rows (m) and columns (n). The location of individual data in these arrays is defined by the subscripts i and j—for instance, a_{ij}, with i and j denoting the respective row and column numbers at which this particular data is located in the array.

Consequently, we may express the above data set in the following forms with an equal number of rows and columns:

$$
|A| = |a_{ij}| = \begin{vmatrix}
a_{11} & a_{12} & a_{13} & \bullet & \bullet & \bullet & \bullet & a_{1n} \\
a_{21} & a_{22} & a_{23} & \bullet & \bullet & \bullet & \bullet & a_{2n} \\
\bullet & \bullet & \bullet & \bullet & \bullet & \bullet & \bullet & \bullet \\
\bullet & \bullet & \bullet & \bullet & \bullet & \bullet & \bullet & \bullet \\
a_{m1} & a_{m2} & a_{m3} & \bullet & \bullet & \bullet & \bullet & a_{mn}
\end{vmatrix}
\tag{4.5}
$$

or

$$
[A] = [a_{ij}] = \begin{bmatrix}
a_{11} & a_{12} & a_{13} & \bullet & \bullet & \bullet & \bullet & a_{1n} \\
a_{21} & a_{22} & a_{23} & \bullet & \bullet & \bullet & \bullet & a_{2n} \\
\bullet & \bullet & \bullet & \bullet & \bullet & \bullet & \bullet & \bullet \\
\bullet & \bullet & \bullet & \bullet & \bullet & \bullet & \bullet & \bullet \\
a_{m1} & a_{m2} & a_{m3} & \bullet & \bullet & \bullet & \bullet & a_{mn}
\end{bmatrix}
\tag{4.6}
$$

The form of an array of data enclosed within two vertical lines in Equation 4.5 is referred to as the *determinant* of the data set, whereas Equation 4.6 with the data set enclosed in square brackets is the *matrix* of the same data set. Both determinant and matrix involve data sets designated by subscripts $i = 1, 2, 3, \ldots, m$ and $j = 1,2,3, .., n$, in which m denotes the row numbers and n the column numbers.

The difference between the determinant and the matrix of the same data set is that a determinant can be evaluated to a single number, or a scalar quantity, but this is not the case for matrices—matrices cannot be evaluated to single numbers or variables. Matrices represent arrays of data and they remain so in mathematical operations throughout engineering analyses.

4.2.1   Evaluation of Determinants

The evaluation of determinants begins with the application of a simple rule for the simplest form of 2×2 determinants of the type:

$$
|A| = \begin{vmatrix}
a_{11} & a_{12} \\
a_{21} & a_{22}
\end{vmatrix}
\tag{4.7}
$$

The data in this determinant has two rows and two columns—that is, $m=2$ and $n=2$. Its numerical value may be obtained by summing the cross products of the elements in the determinant according to the following rule:

$$|A| = a_{11}a_{22} - a_{12}a_{21} \tag{4.8}$$

Example 4.1

Determine the value of the determinant

$$|A| = \begin{vmatrix} 4 & 1 \\ 1 & -2 \end{vmatrix}$$

Solution:

The value of the determinant may be calculated using Equation 4.8 as

$$|A| = \begin{vmatrix} 4 & 1 \\ 1 & -2 \end{vmatrix} = 4 \times (-2) - 1 \times 1 = -9$$

For determinants that have more than 2×2 data, the process of evaluating the determinants will be to reduce the full determinant to a summation of subdeterminants of size 2×2, and apply the aforementioned rule to obtain the value of the determinant in the same way as illustrated in Example 4.1.

Example 4.2

Evaluate the following 3×3 determinant:

$$|A| = \begin{vmatrix} 1 & 2 & 3 \\ 0 & -1 & 4 \\ -2 & 5 & -3 \end{vmatrix}$$

Solution:

We observe that the above 3×3 determinant has the elements $a_{11} = 1$, $a_{12} = 2$, $a_{13} = 3$; $a_{21} = 0, a_{22} = -1, a_{23} = 4; a_{31} = -2, a_{32} = 5, a_{33} = -3$. We will attempt to reduce it to a group of 2×2 determinants, and thereafter apply the simple rule of cross products for the solution.

A general rule for the size reduction process is to use the formula

$$|C^n| = |c_{ij}^n| = \sum_{n=1}^{s} (-1)^{i+j} |c_{ij}^n| \tag{4.9a}$$

where the superscript n denotes the reduction step number. Determinant $|C^n|$ is the determinant of $|A|$ after the n-step reduction in size. The elements c_{ij}^n in these matrices are in the determinants that exclude the elements in the ith row and jth column in the previous matrices. The symbol s in Equation 4.9a denotes the row or column number of the determinant $|A|$.

We further observe that the determinant to be evaluated in this example is 3×3, which requires only one step to reduce the sizes of the determinants $|C^n|$ to a group of subdeterminants of size 2×2:

$$|A| = \begin{vmatrix} 1 & 2 & 3 \\ 0 & -1 & 4 \\ -2 & 5 & -3 \end{vmatrix}$$

$$= (1)(-1)^{1+1} \begin{vmatrix} -1 & 4 \\ 5 & -3 \end{vmatrix} + (2)^{1+2} \begin{vmatrix} 0 & 4 \\ -2 & -3 \end{vmatrix} + (3)^{1+3} \begin{vmatrix} 0 & -1 \\ -2 & -3 \end{vmatrix}$$

$$= -39$$

An alternative way to evaluate a determinant (Zwillinger, 2003) is to sum over all permutations $i_1 \neq i_2 \neq i_3 \neq i_4 \cdots \neq i_n$, with δ denoting the number of transpositions necessary to bring the sequence $(i_1, i_2, i_3, \ldots, i_n)$ back to the natural order $(1, 2, 3, \ldots, n)$ as shown below:

$$|A| = \sum (-1)^{\delta} a_{1i1} a_{2i2} a_{3i3} \cdots a_{nin} \tag{4.9b}$$

The following is an example of evaluating a 3×3 determinant using the alternative method given by Equation 4.9b.

$$|A| = \begin{vmatrix} a_{11} & a_{12} & a_{13} \\ a_{21} & a_{22} & a_{23} \\ a_{31} & a_{32} & a_{33} \end{vmatrix}$$

$$= a_{11} a_{22} a_{33} + a_{12} a_{23} a_{31} + a_{13} a_{21} a_{32} - a_{13} a_{22} a_{31} - a_{11} a_{23} a_{32} - a_{12} a_{21} a_{33} \tag{4.10}$$

Example 4.3
Use Equations 4.9 and 4.10 to evaluate the determinant in Example 4.2.

Solution:
With the elements of the determinant identified in Example 4.2, using the expression provided in Equation 4.10 we may evaluate the value of the determinant as follows:

$$|A| = (1)(-1)(-3) + (2)(4)(-2) + (3)(0)(5) - (3)(-1)(-2) - (1)(4)(5) - (2)(0)(-3)$$
$$= -39$$

We obtain the same answer as in Example 4.2.

4.2.2 Matrices in Engineering Analysis

As mentioned in Section 4.2, matrices cannot be evaluated to a single value. Rather they will always be in the form of matrices. However, matrices can added, subtracted, and multiplied like vectors presented in Section 3.4 in Chapter 3.

4.3 Different Forms of Matrices

There are several different forms of matrix that engineers can use in expressing sets of data in logical sequences. The following are eight commonly used matrices in engineering analysis.

4.3.1 Rectangular Matrices

A rectangular matrix consists of arrays of numbers or variables, termed "elements," denoted by a_{ij}, with the first subscript i indicating the row number and the second subscript j indicating the column number. A general form of this type of matrix is shown below:

$$[A] = [a_{ij}]$$

$$= \begin{bmatrix} a_{11} & a_{12} & a_{13} & \bullet & \bullet & \bullet & \bullet & a_{1n} \\ a_{21} & a_{22} & a_{23} & \bullet & \bullet & \bullet & \bullet & a_{2n} \\ \bullet & \bullet & \bullet & \bullet & \bullet & \bullet & \bullet & \bullet \\ \bullet & \bullet & \bullet & \bullet & \bullet & \bullet & \bullet & \bullet \\ a_{m1} & a_{m2} & a_{m3} & \bullet & \bullet & \bullet & \bullet & a_{mn} \end{bmatrix} \tag{4.11}$$

in which the subscript $i = 1, 2, \ldots, m$, and the subscript $j = 1, 2, \ldots, n$, where m is the total row number and n is the total column number in matrix $[A]$.

4.3.2 Square Matrices

Square matrices are a special case of the rectangular matrices shown Equation 4.11 where the total number of rows m equals the total number of columns n. For instance, a 3×3 matrix can be expressed as

$$[A] = \begin{bmatrix} a_{11} & a_{12} & a_{13} \\ a_{21} & a_{22} & a_{23} \\ a_{31} & a_{32} & a_{33} \end{bmatrix} \tag{4.12}$$

Square matrices commonly appear in engineering analyses.

4.3.3 Row Matrices

In a row matrix, the total number of row $m = 1$, with the total number of columns being n:

$$\{A\} = \begin{Bmatrix} a_{11} & a_{12} & a_{13} & \bullet & \bullet & \bullet & \bullet & a_{1n} \end{Bmatrix} \tag{4.13}$$

4.3.4 Column Matrices

This type of matrix is similar to the row matrix but with transposed numbers of rows and columns. In this case, the total number of rows is m and there is only one column, or $n = 1$:

$$\{A\} = \begin{Bmatrix} a_{11} \\ a_{21} \\ a_{31} \\ \bullet \\ \bullet \\ \bullet \\ \bullet \\ a_{m1} \end{Bmatrix} \tag{4.14}$$

Column matrices, represented generally in Equation 4.14, often are used to express a vector quantity, with the elements of this matrix representing the components of that vector quantity. For instance, a force vector in a three-dimensional space defined by (x, y, z) coordinate systems may be written

$$\{F\} = \begin{Bmatrix} F_x \\ F_y \\ F_z \end{Bmatrix}$$

in which F_x, F_y, and F_z are the components of the force vector along the x-, y-, and z-coordinate directions, respectively.

4.3.5 Upper Triangular Matrices

A close look at the square matrix in Equation 4.12 will reveal that a diagonal line may be drawn through elements in a square matrix, as in Figure 4.2.

In an *upper triangular* or *upper diagonal matrix*, all the elements below the diagonal line are zero, as illustrated for a 3×3 square matrix in Equation 4.15:

$$[A] = \begin{bmatrix} a_{11} & a_{12} & a_{13} \\ 0 & a_{22} & a_{23} \\ 0 & 0 & a_{33} \end{bmatrix} \tag{4.15}$$

Figure 4.2 Diagonal of a square matrix.

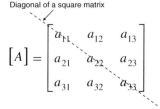

Diagonal of a square matrix

$$[A] = \begin{bmatrix} a_{11} & a_{12} & a_{13} \\ a_{21} & a_{22} & a_{23} \\ a_{31} & a_{32} & a_{33} \end{bmatrix}$$

4.3.6 Lower Triangular Matrices

The lower triangular matrix is the opposite case to the upper triangular matrix; here all the elements above the diagonal line are zero, as shown for a 3×3 square matrix in Equation 4.16:

$$[A] = \begin{bmatrix} a_{11} & 0 & 0 \\ a_{21} & a_{22} & 0 \\ a_{31} & a_{32} & a_{33} \end{bmatrix} \tag{4.16}$$

4.3.7 Diagonal Matrices

In diagonal matrices, the only nonzero elements are those on the diagonals. Equation 4.17 shows an example of a diagonal matrix of size of 4×4:

$$[A] = \begin{bmatrix} a_{11} & 0 & 0 & 0 \\ 0 & a_{22} & 0 & 0 \\ 0 & 0 & a_{33} & 0 \\ 0 & 0 & 0 & a_{44} \end{bmatrix} \tag{4.17}$$

4.3.8 Unit Matrices

This type of matrix is a diagonal matrix with all the nonzero elements on the diagonal lines having a value of unity (i.e. 1.0). Equation 4.18 shows a unit matrix of size 4×4:

$$[I] = \begin{bmatrix} 1.0 & 0 & 0 & 0 \\ 0 & 1.0 & 0 & 0 \\ 0 & 0 & 1.0 & 0 \\ 0 & 0 & 0 & 1.0 \end{bmatrix} \tag{4.18}$$

Unit matrices are usually represented by [I].

Unit matrices have the properties indicated in Equations 4.19a and 4.19b:

$$\alpha[I] = \alpha \begin{bmatrix} 1 & 0 & 0 \\ 0 & 1 & 0 \\ 0 & 0 & 1 \end{bmatrix} = \begin{bmatrix} \alpha & 0 & 0 \\ 0 & \alpha & 0 \\ 0 & 0 & \alpha \end{bmatrix} \quad \text{(a diagonal matrix)} \tag{4.19a}$$

where α is a scalar or a real number.

$$[A][I] = [I][A] \tag{4.19b}$$

4.4 Transposition of Matrices

Transposition of a matrix [A] is often required in engineering analysis. The transpose of [A] is denoted by $[A]^T$. Transposition of matrix [A] is carried out by interchanging the subscripts that define the locations of the elements in matrix [A]. The mathematical operation

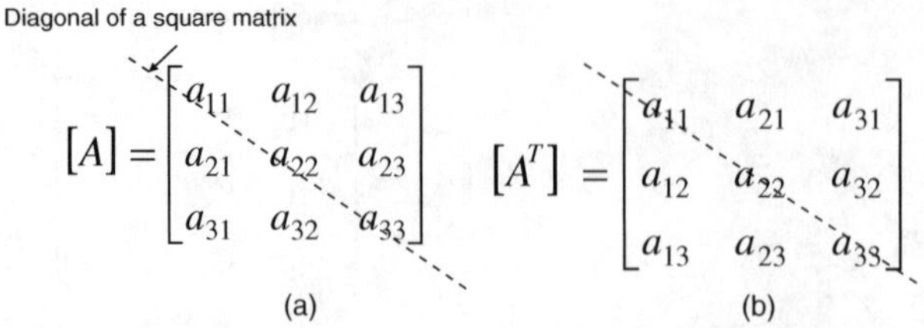

Diagonal of a square matrix

$$[A] = \begin{bmatrix} a_{11} & a_{12} & a_{13} \\ a_{21} & a_{22} & a_{23} \\ a_{31} & a_{32} & a_{33} \end{bmatrix} \qquad [A^T] = \begin{bmatrix} a_{11} & a_{21} & a_{31} \\ a_{12} & a_{22} & a_{32} \\ a_{13} & a_{23} & a_{33} \end{bmatrix}$$

(a) (b)

Figure 4.3 Transposition of a square matrix.

of matrix transposition will achieved by letting $[a_{ij}]^T = [a_{ji}]$. For instance, a column matrix in Equation 4.14 becomes a row matrix after transposition, as shown below:

$$\{A\}^T = \left\{ \begin{array}{c} a_{11} \\ a_{21} \\ a_{31} \\ \bullet \\ \bullet \\ \bullet \\ a_{m1} \end{array} \right\}^T = \left\{ a_{11} \ a_{21} \ a_{31} \ \bullet \bullet \bullet \bullet \ a_{m1} \right\}$$

and a rectangular matrix can be transposed into another rectangular matrix:

$$\begin{bmatrix} a_{11} & a_{12} & a_{13} \\ a_{21} & a_{22} & a_{23} \end{bmatrix}^T = \begin{bmatrix} a_{11} & a_{21} \\ a_{12} & a_{22} \\ a_{13} & a_{23} \end{bmatrix}$$

Likewise, the transposition of a square matrix will result in another square matrix of the same size as illustrated in Figure 4.3.

4.5 Matrix Algebra

We saw in Section 4.1 that a matrix represents an array of real numbers or variables, or vector quantities. Arithmetic operations of matrices, such as addition, subtraction, multiplication, and division cannot be undertaken as for single numbers. The following rules apply when performing these arithmetic operations between matrices.

4.5.1 Addition and Subtraction of Matrices

Addition or subtraction of two matrices requires that both matrices have the same size; that is, that they have equal number of rows and columns. For example, the addition or subtraction of matrices [A] and [B] with respective elements a_{ij} and b_{ij} can be carried out as

$$[A] \pm [B] = [C] \quad \text{with} \quad c_{ij} = a_{ij} \pm b_{ij} \tag{4.20}$$

In which a_{ij}, b_{ij}, and c_{ij} are the elements of the matrices [A], [B] and [C] respectively.

4.5.2 Multiplication of a Matrix by a Scalar Quantity α

For multiplication of a matrix by a scalar quantity α, the following expression applies:

$$\alpha[C] = [\alpha c_{ij}] \tag{4.21}$$

where c_{ij} are the elements of matrix [C].

4.5.3 Multiplication of Two Matrices

For the multiplication of two matrices they must satisfy the condition that

The total number of columns in the first matrix

= the total number of rows in the second matrix

We translate this statement into a mathematical expression in Equation 4.22:

$$\underset{(m\times p)}{[C]} = \underset{(m\times n)}{[A]} \times \underset{(n\times p)}{[B]} \tag{4.22}$$

where the notations shown in the parentheses below the matrices denote the number of rows and columns in each of these matrices.

The following recurrence relationship may be used to determine the element in the product matrix [C]:

$$c_{ij} = a_{i1}b_{1j} + a_{i2}b_{2j} + \cdots + a_{in}b_{nj} \tag{4.23}$$

for $i = 1, 2,\ldots, m$, and $j = 1, 2,\ldots, n$.

Example 4.4

Demonstrate the multiplication of two matrices [A] and [B], both of size 3×3:

$$[A] = \begin{bmatrix} a_{11} & a_{12} & a_{13} \\ a_{21} & a_{22} & a_{23} \\ a_{31} & a_{32} & a_{33} \end{bmatrix} \quad \text{and} \quad [B] = \begin{bmatrix} b_{11} & b_{12} & b_{13} \\ b_{21} & b_{22} & b_{23} \\ b_{31} & b_{32} & b_{33} \end{bmatrix}$$

Solution:

We first confirm that multiplication of these two matrices is possible by checking that the number of columns of the first matrix [A] equals the number of rows of the second matrix [B]. This is the case, with three columns and three rows in each case. We may then proceed to multiply these two matrices according to the recurrence formula provided in Equation 4.23. The matrix [C] that is the product of matrices [A] and [B] thus takes the form

$$[C] = [A][B]$$

$$= \begin{bmatrix} a_{11} & a_{12} & a_{13} \\ a_{21} & a_{22} & a_{23} \\ a_{31} & a_{32} & a_{33} \end{bmatrix} \begin{bmatrix} b_{11} & b_{12} & b_{13} \\ b_{21} & b_{22} & b_{23} \\ b_{31} & b_{32} & b_{33} \end{bmatrix}$$

$$= \begin{bmatrix} a_{11}b_{11} + a_{12}b_{21} + a_{13}b_{31} & a_{11}b_{12} + a_{12}b_{22} + a_{13}b_{32} & a_{11}b_{13} + a_{12}b_{23} + a_{13}b_{33} \\ a_{21}b_{11} + a_{22}b_{21} + a_{23}b_{31} & \bullet & \bullet \\ \bullet & \bullet & \bullet \end{bmatrix}$$

Example 4.5

Carry out the multiplication of a rectangular matrix [C] with 2 rows and 3 columns and a second matrix {x} with 1 column and 3 rows.

Solution:

$$\{y\} = [C]\{x\}$$

$$= \begin{bmatrix} c_{11} & c_{12} & c_{13} \\ c_{21} & c_{22} & c_{23} \end{bmatrix} \begin{Bmatrix} x_1 \\ x_2 \\ x_3 \end{Bmatrix}$$

$$= \begin{Bmatrix} c_{11}x_1 + c_{12}x_2 + c_{13}x_3 \\ c_{21}x_1 + c_{22}x_2 + c_{23}x_3 \end{Bmatrix}$$

$$= \begin{Bmatrix} y_1 \\ y_2 \end{Bmatrix}$$

Example 4.6 Demonstrating the importance of the *order* of multiplication.

a) Multiplication of a row matrix with 1 row and 3 columns matrix and column matrix with 3 rows and 1 column:

$$\{a_{11} \quad a_{12} \quad a_{13}\} \begin{Bmatrix} b_{11} \\ b_{21} \\ b_{31} \end{Bmatrix} = a_{11}b_{11} + a_{12}b_{21} + a_{13}b_{31}$$

(The product is a single number, or a scalar.)

b) Multiplication of a column matrix with 3 rows and 1 column and a row matrix with 1 row and 3 columns:

$$\begin{Bmatrix} a_{11} \\ a_{21} \\ a_{31} \end{Bmatrix} \{b_{11} \quad b_{12} \quad b_{13}\} = \begin{bmatrix} a_{11}b_{11} & a_{11}b_{12} & a_{11}b_{13} \\ a_{21}b_{11} & a_{21}b_{12} & a_{21}b_{13} \\ a_{31}b_{11} & a_{31}b_{12} & a_{31}b_{13} \end{bmatrix}$$

(The product is a square matrix.)

Example 4.7 Multiplication of a square matrix with 3 rows and 3 columns and a column matrix of 3 rows and one column:

$$\begin{bmatrix} a_{11} & a_{12} & a_{13} \\ a_{21} & a_{22} & a_{23} \\ a_{31} & a_{32} & a_{33} \end{bmatrix} \begin{Bmatrix} x \\ y \\ z \end{Bmatrix} = \begin{Bmatrix} a_{11}x + a_{12}y + a_{13}z \\ a_{21}x + a_{22}y + a_{23}z \\ a_{31}x + a_{32}y + a_{33}z \end{Bmatrix}$$

(The product is a column matrix.)

4.5.4 Matrix Representation of Simultaneous Linear Equations

The general rule for multiplication of matrices and the result in Example 4.7 provide a powerful tool for expressing simultaneous linear equations in matrix form; for example,

Equations 4.3a,b,c can be expressed in the following form:

$$\begin{bmatrix} 8 & 4 & 1 \\ 2 & 6 & -1 \\ 1 & -2 & 1 \end{bmatrix} \begin{Bmatrix} x_1 \\ x_2 \\ x_3 \end{Bmatrix} = \begin{Bmatrix} 12 \\ 3 \\ 2 \end{Bmatrix} \tag{4.24a}$$

In general, simultaneous linear equations can be expressed in the matrix form:

$$\begin{bmatrix} a_{11} & a_{12} & a_{13} & \bullet & \bullet & \bullet & \bullet & a_{1n} \\ a_{21} & a_{22} & a_{23} & \bullet & \bullet & \bullet & \bullet & a_{2n} \\ \bullet & \bullet & \bullet & \bullet & \bullet & \bullet & \bullet & \bullet \\ \bullet & \bullet & \bullet & \bullet & \bullet & \bullet & \bullet & \bullet \\ a_{n1} & a_{n2} & a_{n3} & \bullet & \bullet & \bullet & \bullet & a_{nn} \end{bmatrix} \begin{Bmatrix} x_1 \\ x_2 \\ \bullet \\ \bullet \\ x_n \end{Bmatrix} = \begin{Bmatrix} r_1 \\ r_2 \\ \bullet \\ \bullet \\ r_n \end{Bmatrix} \tag{4.24b}$$

We may conveniently express the simultaneous linear equations in Equations 4.24a and 4.24b in the following simplified form:

$$[A] \ \{x\} = \{r\} \tag{4.25}$$

where matrix $[A]$ is referred to as the "coefficient matrix," $\{x\}$ is the "unknown matrix," and $\{r\}$ is the "resultant" matrix. The expression in Equation 4.25 is commonly used for solutions of large number of simultaneous linear equations as will be presented in Section 4.7.3.

4.5.5 Additional Rules for Multiplication of Matrices

From the basic rule for multiplication of matrices in Section 4.5.3 we learned that not all pairs of matrices can be multiplied together. The following rules regarding the multiplication of matrices need to be followed in engineering analyses:

a) Distributive law: $\qquad\qquad\qquad [A]([B] + [C]) = [A][B] + [A][C]$
b) Associative law: $\qquad\qquad\qquad [A]([B][C]) = [A][B]([C])$
c) Noncommutativity: $\qquad\qquad\qquad\qquad [A][B] \neq [B][A]$
d) The product of two transposed matrices: $\quad ([A][B])^T = [B]^T[A]^T$

4.6 Matrix Inversion, [A]⁻¹

There is no such operation as the division of one matrix by another. The closest thing to this operation in matrix algebra is achieved by *matrix inversion*. We define the inverse of matrix $[A]$, which is denoted $[A]^{-1}$, by

$$[A][A]^{-1} = [A]^{-1}[A] = [I] \tag{4.26}$$

where $[I]$ is a unit matrix as defined in Equation 4.18.

Note that inversion of a matrix $[A]$ is possible only if the equivalent determinant of $[A]$ is not zero: that is, $|A| \neq 0$. The matrix $[A]$ is called "singular matrix" if $|A| = 0$.

Inverting a matrix $[A]$ involves the following general steps:

Step 1: Evaluate the equivalent determinant of the matrix $[A]$, and check that $|A| \neq 0$.

Step 2: If the elements of matrix $[A]$ are a_{ij}, we may determine the elements of the co-factor matrix $[C]$ as

$$c_{ij} = (-1)^{i+j} |A'| \tag{4.27}$$

where $|A'|$ is the equivalent determinant of a matrix $[A']$ that has all the elements of $[A]$ except for those in the *i*th row and *j*th column.

Step 3: Transpose the co-factor matrix from [C] to [C]^T.

Step 4: The inverse matrix [A]$^{-1}$ of matrix [A] may be formulated as

$$[A]^{-1} = \frac{1}{|A|}[C]^T \tag{4.28}$$

Example 4.8

Find the inverse of matrix [A] in the following form:

$$[A] = \begin{bmatrix} 1 & 2 & 3 \\ 0 & -1 & 4 \\ -2 & 5 & -3 \end{bmatrix}$$

Solution:

Step 1: Evaluate the equivalent determinant of [A]:

$$|A| = \begin{vmatrix} 1 & 2 & 3 \\ 0 & -1 & 4 \\ -2 & 5 & -3 \end{vmatrix}$$

$$= 1\begin{vmatrix} -1 & 4 \\ 5 & -3 \end{vmatrix} - 2\begin{vmatrix} 0 & 4 \\ -2 & -3 \end{vmatrix} + 3\begin{vmatrix} 0 & -1 \\ -2 & -3 \end{vmatrix}$$

$$= -39 \quad (\neq 0)$$

Step 2: Use Equation 4.27 to find the elements of the co-factor matrix [C]:

$$c_{11} = (-1)^{1+1}[(-1)(-3) - (4)(5)] = -17$$
$$c_{12} = (-1)^{1+2}[(0)(-3) - (4)(-2)] = -8$$
$$c_{13} = (-1)^{1+3}[(0)(5) - (-1)(-2)] = -2$$
$$c_{21} = (-1)^{2+1}[(2)(-3) - (3)(5)] = 21$$
$$c_{22} = (-1)^{2+2}[(1)(-3) - (3)(-2)] = 3$$
$$c_{23} = (-1)^{2+3}[(1)(5) - (2)(-2)] = -9$$
$$c_{31} = (-1)^{3+1}[(2)(4) - (3)(-1)] = 11$$
$$c_{32} = (-1)^{3+2}[(1)(4) - (3)(0)] = -4$$
$$c_{33} = (-1)^{3+3}[(1)(-1) - (2)(0)] = -1$$

We thus have the co-factor matrix, [C] in the form

$$[C] = \begin{bmatrix} -17 & -8 & -2 \\ 21 & 3 & -9 \\ 11 & -4 & -1 \end{bmatrix}$$

Step 3: Transpose the matrix [C] according to the rule given in Section 4.4:

$$[C]^T = \begin{bmatrix} -17 & 21 & 11 \\ -8 & 3 & -4 \\ -2 & -9 & -1 \end{bmatrix}$$

Step 4: Determine the inverse matrix [A]$^{-1}$ using Equation 4.28:

$$[A]^{-1} = \frac{[C]^T}{|A|} = \frac{1}{-39}\begin{bmatrix} -17 & 21 & 11 \\ -8 & 3 & -4 \\ -2 & -9 & -1 \end{bmatrix} = \frac{1}{39}\begin{bmatrix} 17 & -21 & -11 \\ 8 & -3 & 4 \\ 2 & 9 & 1 \end{bmatrix}$$

4.7 Solution of Simultaneous Linear Equations

4.7.1 The Need for Solving Large Numbers of Simultaneous Linear Equations

Linear algebra is a powerful mathematical tool for solving the very large numbers of simultaneous equations required by a some commercial analytical codes using the finite-element method (FEM) or the finite-difference method (FDM). These codes are used extensively by industry in solving a variety of engineering problems that involve solids and fluids of complicated geometry subjected to realistic boundary and loading conditions. The following example illustrates the requirement to solve a large number of simultaneous linear simultaneous equations. This case relates to the computation of stress distribution in a perforated rectangular plate like the one illustrated in Figure 4.4. The rectangular plate containing a small hole with radius d is subjected to tensile force F acting along longer dimension.

The stress analysis of the perforated plate in Figure 4.4a is complicated by the presence of the hole in the plate. The induced stresses in a solid plate subjected to the applied force F can readily be computed to be $\sigma = F/(DW)$, where D and W are the respective width and thickness of the plate. For a perforated plate, however, the stress in the plate near the hole is much more complex because of the geometric change in the perforated region. According to Volterra and Gaines (1971), the stress field near the hole may be computed from the following equations derived from the theory of linear elasticity:

$$\sigma_r(r, \theta) = \frac{\sigma}{2}\left(1 - \frac{a^2}{r^2}\right) + \frac{\sigma}{2}\left(1 + \frac{3a^4}{r^4} - \frac{4a^2}{r^2}\right)\cos 2\theta \tag{4.29a}$$

$$\sigma_\theta(r, \theta) = \frac{\sigma}{2}\left(1 + \frac{a^2}{r^2}\right) - \frac{\sigma}{2}\left(1 + \frac{3a^4}{r^4}\right)\cos 2\theta \tag{4.29b}$$

$$\sigma_{r\theta}(r, \theta) = -\frac{\sigma}{2}\left(1 - \frac{3a^4}{r^4} + \frac{2a^2}{r^2}\right)\sin 2\theta \tag{4.29c}$$

where stresses σ_r and σ_θ in Equations 4.29a and 4.29b are stress components along the r- and θ-directions respectively in the (r,θ) coordinate system shown in Figure 4.4c. The stress component $\sigma_{r\theta}$ in Equation 4.29c is the shearing stress. Also appearing in Equations 4.29a,b,c is σ, which is the nominal stress along the x-coordinate away from the hole with $\sigma = F/(DW)$. The notations a and r in Equations 4.29 are respectively radius of the hole and the linear distance from the origin of the coordinate systems in Figure 4.4c.

Equations 4.29a,b,c are used to evaluate the stresses in the perforated plate. These equations show that the distribution of stresses is by no means uniform in the plate. For instance, we may compute the maximum stress in the plate to a $(\sigma_\theta)_{max} = 3\sigma$ at the "top and bottom" of the rim of the hole in Figure 4.4c with respective values of $\theta = \pi/2$ and $3\pi/2$ in Equation 4.29b. This value of the maximum stress leads to a well-known "stress concentration factor of 3" for a perforated plates subject to unilaterally applied force F.

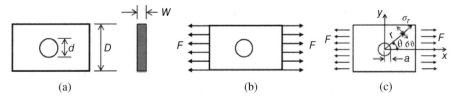

Figure 4.4 A perforated plate subjected to lateral stretching forces. (a) The perforated plate. (b) Application of force F. (c) The stress field.

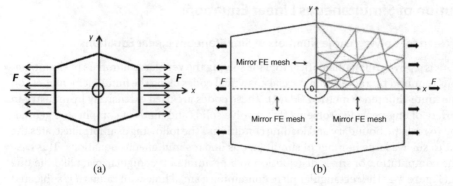

Figure 4.5 Discretization of a perforated plate with tapered edges. (a) Loading of the tapered plate. (b) Discretized model of the plate.

Stress concentrations exist in all load-bearing machine and building structures wherever there is abrupt change of geometry. The magnitude of concentration factors may vary gently with gradual change of geometry in structures such as tapered bars and rods to reach huge values in structures at the vicinity of shark crack tips, in which case the stress normal to the crack face may reach "infinity" in theory. Solution of such structural analyses is beyond the reach of classical methods. One should further realize that the classical solution of stress distributions in a perforated plate given in Equation 4.29 may cease to be valid should the geometry of the plate deviate from rectangular to one with tapered edges as shown in Figure 4.5a. Alternative method is needed for the solution of stress distributions in a plate with this new geometry. Numerical methods such as the "finite-element method" (FEM) or the "finite-difference method" (FDM) are two viable alternative techniques for engineering analyses of this kind.

As will be presented in Chapters 10 and 11, the essence of both FEM and FDM is to discretize a solid structure or a volume of fluid of complex geometry and shape into an assembly of a "finite" number of subdivisions of specific but simpler geometry (called "elements") interconnected at the apexes and/or other designated points (called "nodes"). Figure 4.5b illustrates how the perforated plate with tapered edges in Figure 4.5a can be discretized for computing the induced stress distributions when the plate is subjected to the applied force F. After the discretization the structure is no longer a continuous and homogeneous plate. It is an assembly of elements of solid plates with elements of three or four sides interconnected at the nodes. One may conceive that the FEM offers an "approximation" of the sought solutions, with the solution being closer to the "real" solution the more elements are used in the discretized model.

It is common practice in finite-element stress analysis for the primary unknowns required in the solution to be the displacements of the discretized substance. In the case of the stress analysis of the perforated plate shown in Figure 4.5, there are 31 hybrid triangular and quadrilateral (four-sided) elements interconnected at 26 nodes. The primary unknown quantities in this analysis are the displacements at the nodes reacting to the applied forces F. The in-plane deformation of the plate results in all nodes deforming in the x–y plane. We realize from the physical situation that each node should have two degrees of freedom—that is, will deform in both the x- and y-directions. However, the loading conditions and the symmetry of the geometry of the plate allow all the nodes on the straight boundaries of the modeled area to deform with only one component, but not in both directions. For example, the nodes on the bottom straight edge of the modeled area may deform along the x-direction and those on the two straight vertical edges may deform along the y-direction due to the Poisson effect. We may thus count the total degrees of- freedom of nodal deformation as follows:

- Number of interior nodes plus the one node at the top-right corner and the one on the rim of the hole with two degree of freedom = 17.
- Number of nodes along two straight vertical edges with one degree of freedom = 4.
- Number of nodes along the horizontal straight edge with one degree-of-freedom = 5.
- Total number of degrees of freedom in the discretized model thus equals $17 \times 2 + 4 + 5 = 43$.

Thus the number of simultaneous linear equations that have to solved for the solution of the displacement of the 26 nodes in the model in Figure 4.5b is 43.

The finite-element model with meshes as depicted in Figure 4.5b is by no means an optimal arrangement for this particular analysis. An accurate model would have placed many smaller elements near the rim of the hole where there is steep variation of stress. The model in Figure 4.5b is used only to illustrate the principle of discretization of solid structures of complex geometry.

This case illustration of the stress analysis of a perforated plate with tapered edges may lead the reader to an over-simplified impression that the solution of the simultaneous linear equations required is not a major effort with the use of advanced digital computers. The reality, however, is very different from this illustration; problems involving more than 10,000 degrees of freedom, and the corresponding need to solve this many number of simultaneous equations, are most likely in typical finite-element analysis of most engineering problems, and the matrix solution technique that will be presented in the next section appears to be the only practical way of achieving the solutions.

4.7.2 Solution of Large Numbers of Simultaneous Linear Equations Using the Inverse Matrix Technique

The inverse of a matrix as defined in Equation 4.26 in Section 4.6 is often used to solve n simultaneous equations. A typical set of simultaneous equations is shown in Equation 4.30:

$$
\begin{aligned}
a_{11}x_1 + a_{12}x_2 + a_{13}x_3 + \cdots + a_{1n}x_n &= r_1 \\
a_{21}x_1 + a_{22}x_2 + a_{23}x_3 + \cdots + a_{2n}x_n &= r_2 \\
a_{31}x_1 + a_{32}x_2 + a_{33}x_3 + \cdots + a_{3n}x_n &= r_3 \\
\vdots \\
\vdots \\
a_{n1}x_1 + a_{n2}x_2 + a_{n3}x_3 + \cdots + a_{nn}x_n &= r_n
\end{aligned}
\tag{4.30}
$$

This set of n simultaneous equations may be expressed in matrix form as in Equation 4.31:

$$
\begin{bmatrix}
a_{11} & a_{12} & a_{13} & \bullet & \bullet & \bullet & \bullet & a_{1n} \\
a_{21} & a_{22} & a_{23} & \bullet & \bullet & \bullet & \bullet & a_{2n} \\
\bullet & \bullet & \bullet & \bullet & \bullet & \bullet & \bullet & \bullet \\
\bullet & \bullet & \bullet & \bullet & \bullet & \bullet & \bullet & \bullet \\
a_{n1} & a_{n2} & a_{n3} & \bullet & \bullet & \bullet & \bullet & a_{nn}
\end{bmatrix}
\begin{Bmatrix} x_1 \\ x_2 \\ \bullet \\ \bullet \\ x_n \end{Bmatrix}
=
\begin{Bmatrix} r_1 \\ r_2 \\ \bullet \\ \bullet \\ r_n \end{Bmatrix}
\tag{4.31}
$$

or in a compact version form as

$$
[A]\{x\} = \{r\}
\tag{4.32}
$$

where [A] is usually called the coefficient matrix, {x} is the unknowns matrix, and {r} is the resultant matrix. The elements in both [A] and {r} matrices are given constants in the simultaneous equations in Equation 4.31.

The unknowns matrix {x} in Equation 4.32 may be solved by multiplication of an inverse matrix of [A] on both sides of the equation as follows:

$$[A]^{-1}[A]\{x\} = [A]^{-1}\{r\}$$

which will lead to

$$[I]\{x\} = [A]^{-1}\{r\}$$

or

$$\{x\} = [A]^{-1}\{r\} \tag{4.33}$$

for the solutions of the unknowns.

Example 4.9

Solve the following simultaneous equations using the inverse matrix method:

$$4x_1 + x_2 = 24 \tag{a}$$

$$x_1 - 2x_2 = -21 \tag{b}$$

Solution:

We may express the above simultaneous equations in the matrix form $[A]\{x\} = \{r\}$ similar to that in Equation 4.32, where

$$[A] = \begin{bmatrix} 4 & 1 \\ 1 & -2 \end{bmatrix}; \quad \{x\} = \begin{Bmatrix} x_1 \\ x_2 \end{Bmatrix}, \quad \text{and} \quad \{r\} = \begin{Bmatrix} 24 \\ -21 \end{Bmatrix}$$

We need first to evaluate the equivalent determinant of matrix [A] to ensure that inversion of this matrix is possible:

$$|A| = \begin{vmatrix} 4 & 1 \\ 1 & -2 \end{vmatrix}$$

$$= -8 - 1$$

$$= -9 \quad (|A| \neq 0)$$

We may thus proceed to find the matrix of co-factors [C]. We derive the elements of the co-factor matrix to be $c_{11} = -2$, $c_{12} = -1$, $c_{21} = -1$, and $c_{22} = 4$, from which the co-factor matrix [C] takes the form

$$[C] = \begin{bmatrix} -2 & -1 \\ -1 & 4 \end{bmatrix}$$

and its transpose is

$$[C]^T = \begin{bmatrix} -2 & -1 \\ -1 & 4 \end{bmatrix}$$

The inverse of matrix [A] can thus be determined according to Equation 4.28:

$$[A]^{-1} = \frac{[C]^T}{|A|}$$

$$= -\frac{1}{9} \begin{bmatrix} -2 & -1 \\ -1 & 4 \end{bmatrix}$$

$$= \frac{1}{9} \begin{bmatrix} 2 & 1 \\ 1 & -4 \end{bmatrix}$$

Using Equation 4.33, we obtain the solution of the simultaneous equations as

$$\{x\} = \begin{Bmatrix} x_1 \\ x_2 \end{Bmatrix} = [A]^{-1}\{r\}$$

$$= \frac{1}{9} \begin{bmatrix} 2 & 1 \\ 1 & -4 \end{bmatrix} \begin{Bmatrix} 24 \\ -21 \end{Bmatrix}$$

$$= \frac{1}{9} \begin{Bmatrix} 2 \times 24 - 1 \times 21 = 27 \\ 1 \times 24 + (-4) \times (-21) = 108 \end{Bmatrix}$$

$$= \begin{Bmatrix} 3 \\ 12 \end{Bmatrix}$$

from which we have the two unknowns in Equations (a) and (b) as $x_1 = 3$ and $x_2 = 12$.

4.7.3   Solution of Simultaneous Equations Using the Gaussian Elimination Method

As was indicated at the beginning of Section 4.7, numerical analyses—whether they are based on the finite-element method or the finite-difference method for solving industrial-scale problems—require the solution of large numbers of simultaneous linear equations. When the popular finite-element method is used, the number of these equations is related to the number of degrees of freedom (the unknown quantities) involved in the problem and often is in the thousands or more. In such a case, the familiar Cramer's rule that is taught in secondary schools for solving simultaneous equations can no longer be used for such solutions, nor can the matrix inversion technique presented in Section 4.7.2 be used effectively. The Gaussian elimination method and its several derivatives appears to be the only practical solution techniques for solving large number of simultaneous equations using digital computers.

Here, we will present only the basic formulation of the Gaussian elimination method with recurrence relations that involve only simple arithmetic operations of real numbers that can be readily programmed in the language of digital computers, which feature has made this method most effective with the use of digital computers. Several derivatives of this popular method are available in textbooks, such as the Gauss–Jordan elimination method (Wylie and Barrett, 1995), and the Gauss–Seidel method (Jeffrey, 2002).

The essence of the Gaussian elimination method is to convert the coefficient matrix, such as matrix [A] in Equation 4.32, into the form of an upper triangular matrix as in Equation 4.15 using an "elimination process." Once this is done, the last unknown quantity associated with the last row of the converted upper triangular matrix in the simultaneous equations becomes immediately available. The second-last unknown quantity in the unknown matrix {x} may be obtained by substituting the newly found numerical value of the last unknown quantity into the second-last row in the converted upper triangular coefficient matrix [A]. The remaining unknown quantities may be obtained by a similar procedure, called "back substitution." We will illustrate the procedure of the Gaussian elimination method by considering a typical set of three simultaneous equations as shown in Equations 4.34a,b,c.

$$a_{11}x_1 + a_{12}x_2 + a_{13}x_3 = r_1 \tag{4.34a}$$

$$a_{21}x_1 + a_{22}x_2 + a_{23}x_3 = r_2 \tag{4.34b}$$

$$a_{31}x_1 + a_{32}x_2 + a_{33}x_3 = r_3 \tag{4.34c}$$

where $a_{11}, a_{12},\dots, a_{33}$ are specified coefficients; x_1, x_1, x_2, x_3 are the unknown quantities to be solved; and r_1, r_2, r_3 are the specified resultant constants.

These simultaneous equations may be expressed in a matrix form as shown in Equation 4.35:

$$\begin{bmatrix} a_{11} & a_{12} & a_{13} \\ a_{21} & a_{22} & a_{23} \\ a_{31} & a_{32} & a_{33} \end{bmatrix} \begin{Bmatrix} x_1 \\ x_2 \\ x_3 \end{Bmatrix} = \begin{Bmatrix} r_1 \\ r_2 \\ r_3 \end{Bmatrix} \tag{4.35}$$

or in the form $[A]\{x\} = \{r\}$.

Let us first express the unknown quantity x_1 in the first equation of Equation 4.35 in terms of the other two unknown quantities using Equation (4.34a) with

$$x_1 = \frac{r_1}{a_{11}} - \frac{a_{12}}{a_{11}} x_2 - \frac{a_{13}}{a_{11}} x_3$$

Substituting that expression into Equations 4.34b and 4.34c will result in the following new forms in Equations 4.36:

$$a_{11} x_1 + a_{12} x_2 + a_{13} x_3 = r_1 \tag{4.36a}$$

$$0 + \left(a_{22} - a_{21} \frac{a_{12}}{a_{11}} \right) x_2 + \left(a_{23} - a_{21} \frac{a_{13}}{a_{11}} \right) x_3 = r_2 - \frac{a_{21}}{a_{11}} r_1 \tag{4.36b}$$

$$0 + \left(a_{32} - a_{31} \frac{a_{12}}{a_{11}} \right) x_2 + \left(a_{33} - a_{31} \frac{a_{13}}{a_{11}} \right) x_3 = r_3 - \frac{a_{31}}{a_{11}} r_1 \tag{4.36c}$$

We notice that the terms $a_{21} x_1$ and $a_{31} x_1$ are "eliminated" from Equations 4.34b and 4.34c respectively with the above substitutions. Equations 4.36a,b,c can be expressed in matrix form as

$$\begin{bmatrix} a_{11} & a_{12} & a_{13} \\ 0 & a_{22}^1 & a_{23}^1 \\ 0 & a_{32}^1 & a_{33}^1 \end{bmatrix} \begin{Bmatrix} x_1 \\ x_2 \\ x_3 \end{Bmatrix} = \begin{Bmatrix} r_1 \\ r_2^1 \\ r_3^1 \end{Bmatrix} \tag{4.37}$$

where

$$a_{22}^1 = a_{22} - a_{21} \frac{a_{12}}{a_{11}}$$

$$a_{23}^1 = a_{23} - a_{21} \frac{a_{13}}{a_{11}}$$

$$a_{32}^1 = a_{32} - a_{31} \frac{a_{12}}{a_{11}}$$

$$a_{33}^1 = a_{33} - a_{31} \frac{a_{13}}{a_{11}}$$

and the required corresponding modifications of r_2 and r_3 in the resultant matrix $\{r\}$ are

$$r_2^1 = r_2 - \frac{a_{21}}{a_{11}} r_1$$

$$r_3^1 = r_3 - \frac{a_{31}}{a_{11}} r_1$$

with the condition that $a_{11} \neq 0$. The superscript "1" attached to the elements in Equation 4.37 designates the "elimination step number"—step 1 in this case.

We have thus "eliminated" x_1 in both Equation 4.34b and Equation 4.34c as shown in Equations 4.36 and 4.37. The next required step is to eliminate x_2 in Equation 4.34c using Equations 4.36c and 4.37, resulting in a partial completion of inversion the coefficient matrix in

Equation 4.35 into an upper triangular matrix. Ultimately, we will have the following converted form for the simultaneous equations in Equation 4.38:

$$
\begin{bmatrix} a_{11} & a_{12} & a_{13} \\ 0 & a_{22}^2 & a_{23}^2 \\ 0 & 0 & a_{33}^2 \end{bmatrix} \begin{Bmatrix} x_1 \\ x_2 \\ x_3 \end{Bmatrix} = \begin{Bmatrix} r_1^2 \\ r_2^2 \\ r_3^2 \end{Bmatrix}
\tag{4.38}
$$

in which the superscript "2" in Equation 4.38 designates the second elimination step. For a 3×3 coefficient matrix, two steps are all that are required for inverting it into an upper triangular form.

The last unknown quantity, x_3 in Equation 4.34, may be readily determined from the last equation in Equation 4.38 to give

$$
a_{33}^2 x_3 = r_3^2
$$

or

$$
x_3 = \frac{r_3^2}{a_{33}^2}
$$

We may obtain the next unknown quantity x_2 from the second-last equation in Equation 4.38 with the solution of x_3:

$$
a_{22}^2 x_2 + a_{23}^2 x_3 = r_2^2
$$

or

$$
x_2 = \frac{r_2^2 - a_{23}^2 x_3}{a_{22}^2} = \frac{r_2^2 - a_{23}^2 \dfrac{r_3^2}{a_{33}^2}}{a_{22}^2}
$$

with x_3 obtained from the previous step.

We may determine the value of the remaining unknown quantity, x_1 in Equation 4.35, by substituting the values of x_2 and x_3 into the first equation in Equation 4.38 following a procedure similar to that outlined above.

Two useful recurrence relationships for the elimination process can be shown as follows:

$$
a_{ij}^n = a_{ij}^{n-1} - a_{in}^{n-1} \frac{a_{nj}^{n-1}}{a_{nn}^{n-1}}
\tag{4.39a}
$$

$$
r_i^n = r_i^{n-1} - a_{in}^{n-1} \frac{r_n^{n-1}}{a_{nn}^{n-1}}
\tag{4.39b}
$$

where the superscript n denotes the elimination step number. Numbers with superscript zero (i.e., $n = 0$ in Equation 4.39) are those with the value of the original elements a_{ij}, with $a_{11}^0 = a_{11}, a_{12}^0 = a_{12}, \dots, a_{33}^0 = a_{33}$ in Equation 4.35.

One must bear in mind that the subscripts i and j used in the elimination process must satisfy the conditions that $i > n$ and $j > n$.

The recurrence relationship for "back substitutions" is

$$
x_i = \frac{r_i - \displaystyle\sum_{j=i+1}^{n} a_{ij} x_j}{a_{ii}} \quad \text{with} \quad i = n-1, n-2, \dots, 1
\tag{4.40}
$$

in which a_{ij}, r_i, and x_j are the elements in the final matrices at the conclusion of the elimination process.

Example 4.10

Solve the following simultaneous equations by the Gaussian elimination method using the recurrence relationships in Equation 4.39a and 4.39b.

$$4x_1 + x_2 = 24 \tag{a}$$

$$x_1 - 2x_2 = -21 \tag{b}$$

Solution:

We may express Equations (a) and (b) in the following matrix form:

$$\begin{bmatrix} 4 & 1 \\ 1 & -2 \end{bmatrix} \begin{Bmatrix} x_1 \\ x_2 \end{Bmatrix} = \begin{Bmatrix} 24 \\ -21 \end{Bmatrix} \tag{c}$$

Recognize that

$$a_{11}^0 = a_{11} = 4, a_{12}^0 = a_{12} = 1, a_{21}^0 = a_{21} = 1, a_{22}^0 = a_{22} = -2, r_1^0 = r_1 = 24, r_2^0 = r_2 = -21$$

We are now ready to use the recurrence relationships shown in Equations 4.39 a,b for the Gaussian elimination procedure. We recognize that in this example only one step is required to convert the coefficient matrix in Equation (c) for two simultaneous equations in Equations (a) and (b).

Step 1 with $n = 1$, $i > n = 2$, and $j > n = 2$: We obtain the elements for the coefficient and resultant matrices after step 1 elimination as

$$a_{22}^1 = a_{22}^0 - a_{21}^0 \frac{a_{12}^0}{a_{11}^0}$$

$$= -2 - 1 \times \frac{1}{4} = -2 - \frac{1}{4}$$

$$= -\frac{9}{4}$$

and

$$r_2^1 = r_2^0 - a_{21}^0 \frac{r_1^0}{a_{11}^0}$$

$$= -21 - 1 \times \frac{24}{4} = -21 - 6$$

$$= -27$$

The coefficient matrix in Equation (c) after step 1 elimination becomes

$$\begin{bmatrix} 4 & 1 \\ 0 & -\frac{9}{4} \end{bmatrix} \begin{Bmatrix} x_1 \\ x_2 \end{Bmatrix} = \begin{Bmatrix} 24 \\ -27 \end{Bmatrix} \tag{d}$$

Since the original coefficient matrix is of size 2×2, one step of elimination is all that is required for the solution. Thus, from Equation (c), we have the solution for x_2 as

$$-\frac{9}{4}x_2 = -27 \quad \rightarrow \quad x_2 = 12$$

Use back substitution in Equation 4.39b to get the solution of other unknown, x_1, with $n = 2$ as follows:

$$x_1 = \frac{r_1 - \displaystyle\sum_{j=2}^{2} a_{1j}x_j}{a_{11}^0} = \frac{r_1 - a_{12}^0 x_2}{a_{11}^0} = \frac{24 - 1 \times 12}{4} = 3$$

The solutions to Equations (a) and (b) are thus $x_1 = 3$ and $x_2 = 12$. (Note that the same results were obtained using the matrix inversion method as presented in Example 4.8.)

Example 4.11

Solve the following three simultaneous equations using the Gaussian elimination method:

$$8x_1 + 4x_2 + x_3 = 12 \tag{a}$$

$$2x_1 + 6x_2 - x_3 = 3 \tag{b}$$

$$x_1 - 2x_2 + x_3 = 2 \tag{c}$$

Solution:

Let us express Equations (a), (b), and (c) in matrix form:

$$\begin{bmatrix} 8 & 4 & 1 \\ 2 & 6 & -1 \\ 1 & -2 & 1 \end{bmatrix} \begin{Bmatrix} x_1 \\ x_2 \\ x_3 \end{Bmatrix} = \begin{Bmatrix} 12 \\ 3 \\ 2 \end{Bmatrix} \tag{d}$$

We will require two steps (i.e., $n = 2$) in the elimination process for three simultaneous equations. We will proceed to apply the Gaussian elimination process to Equation (d) using Equations 4.39a and 4.39b.

Step 1 with $n = 1$, $i > n = 2$, and $j > n = 2$:

$$a_{22}^1 = a_{22}^0 - a_{21}^0 \frac{a_{12}^0}{a_{11}^0} = a_{22} - a_{21} \frac{a_{12}}{a_{11}}$$

$$= 6 - 2 \times \frac{4}{8}$$

$$= 5$$

with $i = 2, j = 2$:

$$a_{23}^1 = a_{23}^0 - a_{21}^0 \frac{a_{13}^0}{a_{11}^0} = a_{23} - a_{21} \frac{a_{13}}{a_{11}}$$

$$= -1 - 2 \times \frac{1}{8}$$

$$= -1.25$$

with $i = 2, j = 3$:

$$r_2^1 = r_2^0 - a_{21}^0 \frac{r_1^0}{a_{11}^0} = r_2 - a_{21} \frac{r_1}{a_{11}}$$

$$= 3 - 2 \times \frac{12}{8}$$

$$= 0$$

with $i = 3$ and $j = 2$:

$$a_{32}^1 = a_{32}^0 - a_{31}^0 \frac{a_{12}^0}{a_{11}^0} = a_{32} - a_{31} \frac{a_{12}}{a_{11}}$$

$$= -2 - 1 \times \frac{4}{8}$$

$$= -2.5$$

$$a_{33}^1 = a_{33}^0 - a_{31}^0 \frac{a_{13}^0}{a_{11}^0} = a_{33} - a_{31} \frac{a_{13}}{a_{11}}$$

$$= 1 - 1 \times \frac{1}{8}$$

$$= 0.875$$

$$r_3^1 = r_3^0 - a_{31}^0 \frac{r_1^0}{a_{11}^0} = r_3 - a_{31} \frac{r_1}{a_{11}}$$

$$= 2 - 1 \times \frac{12}{8}$$

$$= 0.5$$

We may thus express Equation (d) after step 1 elimination in the form

$$\begin{bmatrix} 8 & 4 & 1 \\ 0 & 5 & -1.25 \\ 0 & -2.5 & 0.875 \end{bmatrix} \begin{Bmatrix} x_1 \\ x_2 \\ x_3 \end{Bmatrix} = \begin{Bmatrix} 12 \\ 0 \\ 0.5 \end{Bmatrix} \tag{e}$$

We now proceed to step 2 in the elimination process to convert the coefficient matrix in Equation (e) into an upper triangular matrix.

Step 2 with $n = 2$, $i > n = 3$ and $j > n = 3$: We recognize that $a_{21}^2 = a_{31}^2 = a_{32}^2 = 0$ because the subscripts i and j of these matrix elements are less than $n = 2$.

$$a_{33}^2 = a_{33}^1 - a_{32}^1 \frac{a_{23}^1}{a_{22}^1}$$

$$= 0.875 - (-2.5) \times \frac{(-1.25)}{5}$$

$$= 0.25$$

$$r_3^2 = r_3^1 - a_{32}^1 \frac{r_2^1}{a_{22}^1}$$

$$= 0.5 - (-2.5) \times \frac{0}{5}$$

$$= 0.5$$

We have completed the conversion of the matrix equation in Equation (e) to a new form of upper triangular coefficient matrix [A] with modified resultant matrix {r} in Equation (f) after step 2 elimination as

$$\begin{bmatrix} 8 & 4 & 1 \\ 0 & 5 & -1.25 \\ 0 & 0 & 0.25 \end{bmatrix} \begin{Bmatrix} x_1 \\ x_2 \\ x_3 \end{Bmatrix} = \begin{Bmatrix} 12 \\ 0 \\ 0.5 \end{Bmatrix} \tag{f}$$

One may readily see from the last line in Equation (f) that the solution for x_3 is

$$x_3 = 0.5/0.25 = 2$$

The values of the remaining two unknowns, x_2 and x_1, may be obtained using the recurrence relation of back substitution as given in Equation 4.39b as follows. We begin with $n = 3$ in Equation 4.39b:

$$x_i = \frac{r_i - \displaystyle\sum_{j=i+1}^{3} a_{ij} x_j}{a_{ii}} \quad \text{with} \quad i = 2,1$$

Hence, to determine x_2 with $i = 2$:

$$x_2 = \frac{r_2 - \sum_{j=3}^{3} a_{2j}x_j}{a_{22}} = \frac{r_2 - a_{23}x_3}{a_{22}}$$

$$= \frac{0 - (-1.25) \times 2}{5} = 0.5$$

and to determine x_1 with $i = 1$:

$$x_1 = \frac{r_1 - \sum_{j=2}^{3} a_{1j}x_j}{a_{11}} = \frac{r_1 - (a_{12}x_2 + a_{13}x_3)}{a_{11}}$$

$$= \frac{12 - (4 \times 0.5 + 1 \times 2)}{8} = 1$$

The solution of the simultaneous equations in Equation (a) is thus $x_1 = 1$, $x_2 = 0.5$, and $x_3 = 2$.

4.8 Eigenvalues and Eigenfunctions

The term "eigenvalue" was derived from the German word *Eigenwert*, which means "proper" or "characteristic" value; similarly, the term "eigenfunction" was derived from another German word *Eigenfunktion* meaning "proper" or "characteristic" function.

Eigenvalues and eigenfunctions appear in engineering analyses involving linear transformation of functions in vector spaces (Hogben, 1987). They also appear in the characteristic equations associated with solutions in engineering analyses of mechanical vibration, heat conduction, and electromagnetism, as will be presented in Chapter 9. The roots of these characteristic equations are known as eigenvalues, which provides nontrivial solutions to the differential equations that satisfy the prescribed boundary conditions. Nontrivial solutions of equations are solutions other than those that result in legitimate zero in the solution of the equation. Functions that provide multiple eigenvalues are called eigenfunctions.

"Characteristic equations" often appear in engineering analyses. For example, natural frequencies in modal analyses of structures, in which natural frequencies of the structures are denoted by ω_n with mode numbers $n = 1, 2, 3, \ldots$, are important design parameters of structures' vibrational behavior due to periodic excitation with frequencies ω. Uncontrollable, and often devastating, vibration called "resonant vibration" of a structure can occur when the frequency ω of an excitation force matches any of the natural frequencies of the structure. The governing differential equations used to determine the natural frequencies of structures are homogeneous differential equations, as will be presented in Chapters 8 and 9. The characteristic equation associated with the solution for the amplitude of vibration $y(x)$ from these equations will have a general form

$$y(x) = C \sin \beta x$$

Imposition of a condition $y(L) = 0$, with L being the length of the structure, will lead to the expression in Equation 4.41:

$$y(L) = 0 = C \sin \beta L \tag{4.41}$$

We recognize that there are two options to satisfy Equation 4.41: option 1, by letting the constant coefficient $C = 0$; and option 2, by letting $\sin \beta L = 0$.

We further see that option 1 will lead to a trivial solution of the problem, meaning $y(x) = 0$ at all times with all values of x, which does not meet the requirements. With option 2, on the other hand, with the function $\sin \beta L = 0$, the characteristic equation (Equation 4.41) will have multiple roots that satisfy the condition with $\beta = 0$, π/L, $2\pi/L$, $3\pi/L$, ..., or $\beta = n\pi/L$ with $n = 1, 2, 3, 4, \ldots$.

A close look at the above roots of the characteristic equation $\sin \beta L = 0$ reveals that although the root $\beta = 0$ leads to a trivial solution of $y(x)$, the other roots, $\beta = \pi/L$, $2\pi/L$, $3\pi/L$, ..., $n\pi/L$ do not give trivial solutions of Equation 4.41. We thus classify $\beta = \pi/L$, $2\pi/L$, $3\pi/L$, ... as the eigenvalues of the characteristic equation $\sin \beta L = 0$.

4.8.1 Eigenvalues and Eigenvectors of Matrices

In this subsection, we will describe the method for determining the eigenvalues from characteristic equations involved in linear transformation of vector quantities from one space to another. Most of these transformations involve vectors expressed as square matrices, which is a common way of expressing vectors as was illustrated in Chapter 3. We will thus focus our attention on determining the eigenvalues in this particular type of applications.

We follow the derivation of solutions of eigenvalues and eigenvectors as given in two reference works (Hogben, 1987; Malek-Madani, 1998). Certain terminology is commonly used in the literature in relation to eigenvalue problems:

- *Eigenfunctions:* These are functions that are derived in the course of solving homogeneous differential equations involving homogeneous boundary conditions.
- *Characteristic equations:* These are the equations that involve eigenfunctions, similarly to Equation 4.41.
- *Eigenvalues:* The roots of the characteristic equations, such as the roots β described for Equation 4.41.
- *Eigenvectors:* These are the vectors involved in linear transformation, either in a two-dimensional plane or in a three-dimensional space.

For cases involving linear transformation of vectors we may represent the characteristic equations and the eigenvalues by the mathematical expressions discussed next.

4.8.2 Mathematical Expressions of Eigenvalues and Eigenvectors of Square Matrices

Let

$$\{x\} = \begin{Bmatrix} x_1 \\ x_2 \\ x_3 \\ \bullet \\ \bullet \\ x_n \end{Bmatrix}$$

be an eigenvector and [A] be a square matrix with real-number elements, such that

$$[A]\{x\} = \{0\} \tag{4.42}$$

Further let

$$[A]\{x\} - \lambda\{x\} = \{0\} \tag{4.43}$$

from which we get

$$[A]\{x\} = \lambda\{x\} \tag{4.44}$$

The real numbers λ that satisfy Equations 4.43 and 4.44 are defined as the eigenvalues of the eigenvector $\{x\}$.

Equation 4.43 may be expressed in another form:

$$([A] - \lambda[I])\{x\} = \{0\} \qquad (4.45)$$

where $[I]$ is the unit matrix defined in Equation 4.18.

The eigenvalues λ in Equations 4.43 and 4.44 may be obtained by solving the following polynomial equation (Equation 4.46):

$$\det|[A] - \lambda[I]| = 0 \qquad (4.46)$$

Example 4.12

Show that the vector

$$\{\mathbf{x}\} = \begin{Bmatrix} 1 \\ 1 \\ 1 \end{Bmatrix}$$

is an eigenvector corresponding to the eigenvalue 2 for the matrix

$$[A] = \begin{bmatrix} 4 & -1 & -1 \\ 0 & -1 & 3 \\ 1 & 3 & -2 \end{bmatrix}$$

Solution:

Using Equation 4.23 for multiplication of two matrices, we obtain the following relationship for $[A]\{x\}$:

$$[A]\{\mathbf{x}\} = \begin{bmatrix} 4 & -1 & -1 \\ 0 & -1 & 3 \\ 1 & 3 & -2 \end{bmatrix} \begin{Bmatrix} 1 \\ 1 \\ 1 \end{Bmatrix}$$

$$= \begin{Bmatrix} 2 \\ 2 \\ 2 \end{Bmatrix} = 2 \begin{Bmatrix} 1 \\ 1 \\ 1 \end{Bmatrix} = 2\{\mathbf{x}\}$$

$$= (\text{eigenvalue } 2) \times (\text{eigenvector}\{\mathbf{x}\})$$

By comparing the above expression with Equation 4.44, we have the eigenvalue $\lambda = 2$.

Example 4.13

Find eigenvalues and eigenvectors of the matrix

$$[A] = \begin{bmatrix} -1 & 2 \\ -7 & 8 \end{bmatrix}$$

Solution:

We first derive the polynomial characteristic equation with eigenvalue λ using Equation 4.46:

$$|[A] - \lambda[I]| = \begin{vmatrix} -1 - \lambda & 2 \\ -7 & 8 - \lambda \end{vmatrix} = 0$$

from which we solve for the eigenvalues $\lambda_1 = 1$ and $\lambda_2 = 6$.

Next, we determine the eigenvectors corresponding to these two eigenvalues.

For eigenvalue $\lambda_1 = 1$ We use Equation 4.45 to find the associated eigenvectors:

$$\left(\begin{bmatrix} -1 & 2 \\ -7 & 8 \end{bmatrix} - 1 \begin{bmatrix} 1 & 0 \\ 0 & 1 \end{bmatrix} \right) \begin{Bmatrix} x_1 \\ x_2 \end{Bmatrix} = \begin{bmatrix} -1-1 & 2 \\ -7 & 8-1 \end{bmatrix} \begin{Bmatrix} x_1 \\ x_2 \end{Bmatrix}$$

$$= \begin{Bmatrix} 0 \\ 0 \end{Bmatrix}$$

which leads to the following simultaneous equations:

$$-2x_1 + 2x_2 = 0 \tag{a}$$

$$-7x_1 + 7x_2 = 0 \tag{b}$$

We will find that if we let $x_2 = p$, a nonzero real number, we will have the same answer for x_1, that is, $x_1 = p$. Consequently, we may establish that vector $\{x\}$ is

$$\begin{Bmatrix} v_1 \\ v_2 \end{Bmatrix} = \begin{Bmatrix} p \\ p \end{Bmatrix} = p \begin{Bmatrix} 1 \\ 1 \end{Bmatrix}$$

In fact, p could be chosen as any real nonzero number. In other words, a nonzero constant such as p as a multiple of any eigenvector is also an eigenvector. Thus, we conclude that the vector

$$\begin{Bmatrix} 1 \\ 1 \end{Bmatrix}$$

is the eigenvector corresponding to eigenvalue $\lambda_1 = 1$.

For eigenvalue $\lambda_2 = 6$ We follow the same procedure as for λ_1, with the following equation in matrix form:

$$\begin{bmatrix} -1-6 & 2 \\ -7 & 8-6 \end{bmatrix} \begin{Bmatrix} x_1 \\ x_2 \end{Bmatrix} = \begin{Bmatrix} 0 \\ 0 \end{Bmatrix}$$

leading to the following simultaneous equations:

$$-7x_1 + 2x_2 = 0 \tag{c}$$

$$-7x_1 + 2x_2 = 0 \tag{d}$$

Again, if we assume $x_2 = p$ in Equation (c), this will lead to x_1 in Equation (d) as $x_1 = 2/7$. We then obtain the eigenvector:

$$\begin{Bmatrix} x_1 \\ x_2 \end{Bmatrix} = \begin{Bmatrix} \frac{2}{7}p \\ p \end{Bmatrix} = 7p \begin{Bmatrix} 2 \\ 7 \end{Bmatrix}$$

We thus conclude that the eigenvector corresponding to eigenvalue $\lambda_2 = 6$ is

$$\begin{Bmatrix} 2 \\ 7 \end{Bmatrix}$$

The matrices that relate to linear transform of vectors in three-dimensional space are usually of size 3×3, in which case the finding of eigenvalues and eigenvectors is a little more complicated, as will be demonstrated in the following example.

Example 4.14

Find the eigenvalues and eigenvectors of the following 3×3 matrix:

$$[A] = \begin{bmatrix} 5 & -1 & 0 \\ 0 & -5 & 9 \\ 5 & -1 & 0 \end{bmatrix} \tag{a}$$

Solution:

Following the established procedure for finding the eigenvalues of the matrix {v} by solving the characteristic equation as shown in Equation 4.46 (or$|([A] - \lambda[I])| = 0$), we will obtain the following expression using the matrix [A] in Equation (a):

$$\begin{vmatrix} 5 - \lambda & -1 & 0 \\ 0 & -(5 + \lambda) & 9 \\ 5 & -1 & -\lambda \end{vmatrix} = 0 \tag{b}$$

We may solve Equation (b) for the three roots $\lambda_1 = 0$, $\lambda_2 = 4$, $\lambda_3 = -4$.

We are now ready to find the corresponding eigenvectors corresponding to these eigenvalues.

For the case of $\lambda_1 = 0$ We solve the following equation by substituting $\lambda = \lambda_1 = 0$ into Equation (b), or, in matrix format:

$$\begin{bmatrix} 5 & -1 & 0 \\ 0 & -5 & 9 \\ 5 & -1 & 0 \end{bmatrix} \begin{Bmatrix} x_1 \\ x_2 \\ x_3 \end{Bmatrix} = \begin{Bmatrix} 0 \\ 0 \\ 0 \end{Bmatrix} \tag{c}$$

There are several ways one may obtain the eigenvectors from Equation (c); one may assign various real constants to the unknown vectors in Equation (c), similarly to what was done in Example 4.13; or one may assign a nonzero real number to only one of the three unknown vectors in the same equation and relate the other two vectors to the same assigned real number. The latter would require the use of the Gaussian elimination method such as was described in Section 4.7.3.

In this example, we use the Gaussian elimination method to convert the coefficient matrix in Equation (c) into an upper triangular matrix as

$$\begin{bmatrix} 5 & -1 & 0 \\ 0 & -5 & 9 \\ 0 & 0 & 25 \end{bmatrix} \begin{Bmatrix} x_1 \\ x_2 \\ x_3 \end{Bmatrix} = \{0\} \tag{d}$$

We assume $x_3 = p$ where p is a nonzero real constant. The second equation in Equation (d) will give $x_2 = 9x_3/5 = 9p/5$. The value of x_1 in Equation (d) may be obtained by expansion of the first equation derived from Equation (d): $x_1 = x_2/25 = 9p/25$.

We may thus find that

$$\{x\} = \begin{Bmatrix} x_1 \\ x_2 \\ x_3 \end{Bmatrix} = 25 \begin{Bmatrix} 9 \\ 45 \\ 25 \end{Bmatrix}$$

with the eigenvector

$$\begin{Bmatrix} 9 \\ 45 \\ 25 \end{Bmatrix}$$

We may also obtain the other two eigenvectors:

$$\begin{Bmatrix} 1 \\ 1 \\ 1 \end{Bmatrix} \text{ with } \lambda_2 = 4; \quad \begin{Bmatrix} 1 \\ 9 \\ 1 \end{Bmatrix} \text{ with } \lambda_3 = -4$$

Finding eigenvalues and corresponding eigenvectors for matrices higher than sizes 3×3 becomes very tedious; the use of numerical techniques and of commercial software packages such as MatLAB (Malek-Madani, 1998) appears to be a viable way to obtain the solutions.

4.8.3 Application of Eigenvalues and Eigenfunctions in Engineering Analysis

Eigenvalues and eigenfunctions are used engineering analyses raging from control theory, through vibration analysis, electric circuits, and rigid body dynamics to quantum mechanics. We will use the following case to illustrate the finding of natural frequencies of a multimode mass–spring system (Chapra, 2012).

Figure 4.6 illustrates a physical situation in which two masses (each m) are attached to three springs with identical spring constants (k), in a multiple-degrees-of freedom vibrational system. Slight instantaneous disturbances were simultaneously applied to both the upper and lower masses m, resulting in free vibration of these masses with instantaneous displacements $y_1(t)$ and $y_2(t)$, as illustrated on the right-hand side of Figure 4.6, where t is the time after the inception of vibration of the masses.

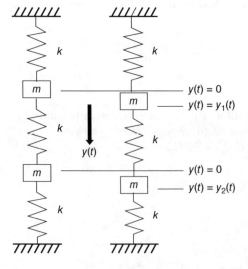

Figure 4.6 Free vibration of multiple mass–spring system.

We may derive the following two simultaneous differential equations for the amplitudes of the deviation of the two masses, $y_1(t)$ and $y_2(t)$, from their initial equilibrium conditions:

$$m\frac{d^2y_1(t)}{dt^2} = -ky_1(t) + k[y_2(t) - y_1(t)] \tag{4.47a}$$

$$m\frac{d^2y_2(t)}{dt^2} = -k[y_2(t) - y_1(t)] - ky_2(t) \tag{4.47b}$$

The solution for $y_1(t)$ and $y_2(t)$ in Figure 4.6 requires the solution of the simultaneous differential equations in Equations 4.47a and 4.47b because of the "coupling" effect of the multiple degrees of freedom of the vibration of the masses in the system. However, the theory of vibration of masses in mass–spring systems, as will be presented in Chapter 8, indicates that the solution $y_1(t)$ and $y_2(t)$ will be in the form

$$y_i(t) = Y_i \sin(\omega t) \tag{4.48}$$

the subscript $i = 1,2$ in Equation 4.48 denotes the amplitudes of vibration of the upper and lower masses, $y_1(t)$ and $y_2(t)$ respectively. y_i with $i = 1$ and 2 are the respective maximum magnitudes of the displacement of the two masses. The natural frequency of the vibrating mass is denoted ω, with the unit rad/s.

Upon substitution of the typical solutions for the vibrating masses in Equation 4.48 into Equations 4.47a and 4.47b, we obtain the expressions in Equations 4.49a and 4.49b:

$$\left(\frac{2k}{m} - \omega^2\right)Y_1 - \frac{k}{m}Y_2 = 0 \tag{4.49a}$$

$$-\frac{k}{m}Y_1 + \left(\frac{2k}{m} - \omega^2\right)Y_2 = 0 \tag{4.49b}$$

These may be written in matrix form as

$$\begin{bmatrix} \left(\dfrac{2k}{m} - \omega^2\right) & -\dfrac{k}{m} \\ -\dfrac{k}{m} & \left(\dfrac{2k}{m} - \omega^2\right) \end{bmatrix} \begin{Bmatrix} Y_1 \\ Y_2 \end{Bmatrix} = \begin{Bmatrix} 0 \\ 0 \end{Bmatrix} \tag{4.50a}$$

Equation 4.50a may be expressed differently as

$$\left[\begin{bmatrix} \dfrac{2k}{m} & -\dfrac{k}{m} \\ -\dfrac{k}{m} & \dfrac{2k}{m} \end{bmatrix} - \omega^2 \begin{bmatrix} 1 & 0 \\ 0 & 1 \end{bmatrix} \right] \begin{Bmatrix} Y_1 \\ Y_2 \end{Bmatrix} = \begin{Bmatrix} 0 \\ 0 \end{Bmatrix} \tag{4.50b}$$

Comparing Equation 4.50b with Equation 4.45, we obtain the following equivalences:

$$[A] = \begin{bmatrix} \dfrac{2k}{m} & -\dfrac{k}{m} \\ -\dfrac{k}{m} & \dfrac{2k}{m} \end{bmatrix}, \quad \{x\} = \begin{Bmatrix} Y_1 \\ Y_2 \end{Bmatrix}, \quad \text{and} \quad \lambda = \omega^2 \tag{4.51}$$

or in a simple matrix form, $[A]\{Y\} = \{0\}$ with $[A]$ given in the first expression of Equation 4.51. The last expression in Equation 4.51 as

$$\omega = \sqrt{\lambda} \tag{4.52}$$

in which λ is the eigenvalue of the amplitudes of the mass vibrations is of particular significance for evaluating the natural frequencies of mass–spring vibrational systems.

Example 4.15
We will derive numerical solutions for the system illustrated in Figure 4.6 by letting $m = 10\,\text{kg}$ and spring constant $k = 200\,\text{N/m}$. Determine the natural frequencies of the system in Figure 4.6.

Solution:
Substituting $m = 10\,\text{kg}$ and $k = 200\,\text{N/m}$ into the first expression of Equation 4.51 will result in

$$[A] = \begin{bmatrix} 40 & -20 \\ -20 & 40 \end{bmatrix}$$

Use Equation 4.46 to solve for the eigenvalues λ:

$$\begin{aligned} |[A] - \lambda[I]| &= \left| \begin{bmatrix} 40 & -20 \\ -20 & 40 \end{bmatrix} - \lambda \begin{bmatrix} 1 & 0 \\ 0 & 1 \end{bmatrix} \right| \\ &= \begin{vmatrix} 40 - \lambda & -20 \\ -20 & 40 - \lambda \end{vmatrix} = 0 \end{aligned}$$

from which we solve for $\lambda = \lambda_1 = 20$ and $\lambda = \lambda_2 = 60$.

We may then use Equation 4.52 to determine the natural frequencies of the two modes to be $\omega_1 = 4.47\,\text{rad/s}$ and $\omega_2 = 7.75\,\text{rad/s}$ for the mass–spring system shown in Figure 4.6.

4.9 Problems

4.1 Add two matrices $[A] + [B]$ where

$$[A] = \begin{bmatrix} 1 & 4 & 7 \\ 2 & 5 & 8 \\ 3 & 6 & 9 \end{bmatrix} \quad \text{and} \quad [B] = \begin{bmatrix} 3 & 6 & 9 \\ 2 & 5 & 8 \\ 1 & 4 & 7 \end{bmatrix}$$

4.2 Multiply two matrices as follows:
 a) $[A][B]$
 b) $[A]\{C\}$
 c) $[B][A]$
 d) $[B][A]^T$
 e) $[A]^T[B]$
 f) $\{C\}\{D\}$
 g) $\{C\}^T\{D\}$
 h) $\{C\}^T\{D\}^T$

with [A] and [B] being the same as in Problem 4.1 and

$$\{C\} = \begin{Bmatrix} 1 \\ 4 \\ 7 \end{Bmatrix} \quad \text{and} \quad \{D\} = \begin{Bmatrix} 2 \\ 5 \\ 8 \end{Bmatrix}$$

4.3 Find the co-factor matrix [C] and the inverse matrix $[A]^{-1}$ (if it exists) with the matrix [A] in the form of

1. (a) $[A] = \begin{bmatrix} 1 & -1 \\ 1 & 1 \end{bmatrix}$ (b) $[A] = \begin{bmatrix} 3 & 1 & 2 \\ -1 & 2 & 1 \\ 0 & 1 & 1 \end{bmatrix}$

4.4 Find the eigenvalues (λ) and eigenvector of the matrix

$$[A] = \begin{bmatrix} -1 & 2 \\ -5 & 1 \end{bmatrix}$$

4.5 Find the eigenvalues (λ) and the corresponding eigenvectors of the matrix

$$[A] = \begin{bmatrix} 3 & 0 & 0 \\ 0 & 2 & 0 \\ 4 & 0 & 1 \end{bmatrix}$$

4.6 Use (a) the matrix inversion method and (b) the Gaussian elimination method to solve the following simultaneous equations:

$$3x + 2y - 6z = 0$$
$$x - y + z = 4$$
$$y + z = 3$$

4.7 Use both the matrix inversion and the Gaussian elimination methods to solve the following simultaneous equations:

$$2x - 3y = 32$$
$$x + 4y = -6$$

Show every step of your computation for the solutions.

4.8 Determine the natural frequencies of the mass–spring system illustrated in Figure 4.6 and described in Example 4.15 but with only two springs supporting masses m_1 and m_2.

5

Overview of Fourier Series

Chapter Learning Objectives

- Appreciate that Fourier series are the mathematical form for periodic physical phenomena.
- Learn to use Fourier series to represent periodic physical phenomena in engineering analysis.
- Learn the required conditions for deriving Fourier series.
- Appreciate the principle of using Fourier series derived from the function for one period to apply the same Fourier series for other periods.
- Derive the mathematical expressions of Fourier series representing common physical phenomena.
- Understand the convergence of Fourier series of continuous periodic functions.
- Understand the convergence of Fourier series of piecewise continuous functions.
- Understand the convergence of Fourier series at discontinuities

5.1   Introduction

We learned in Chapter 2 that functions are used to represent physical phenomena in engineering analysis. In Section 2.4.2, we learned that step functions can be used to represent physical phenomena that exist at the inception of a specific time or location in spectra, and that impulsive functions describe phenomena of extremely short duration in time or extremely small extent in space. In this chapter, we will present another useful function that represents physical phenomena of periodic nature. Being periodical in their existence, these functions start and end at specific instants in time or location in space. They are expressed in the form of infinite series that is called the *Fourier series*. The functions that describe the specific physical quantities by the Fourier series can be used to represent the same *periodic* physical quantities over the *entire spectrum* that the variable of these functions covers.

Periodic phenomena are common in our daily lives; Figure 5.1a shows a merry-go-round in an amusement park. The large revolving platform consists of a number of mechanisms driving model horses, cars, carousels, and the like, on which people (especially, children) ride round and round, sometimes with up-and-down motions as well. Figure 5.1b shows a close-up view of a model pony on the revolving platform; this has periodic up-and-down motions activated by a supporting column below the mechanical pony. The forces required to lift up and pull down the pony, and other carousels revolving on the platform, are also of periodic nature. Both the motions and the forces required to activate the individual mechanisms are periodic, and they can be represented by Fourier series in dynamics and stress analyses.

Another example of periodic motion that is common in domestic situations is provided by the sewing machine, such as that shown in Figure 5.2a. This machine stitches fabrics with the

Applied Engineering Analysis, First Edition. Tai-Ran Hsu.

(a)   (b)

Figure 5.1 A mechanism with periodic motions. (a) Riding horse on a merry-go-round. (b) Up-and-down motion of a mechanical pony. *See color section.*

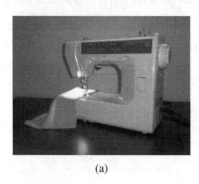
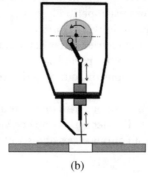

(a)   (b)

Figure 5.2 Periodic motion of the needle of a sewing machines. (a) A typical sewing machine. (b) The repetitive cyclical motion of the needle.

up-and-down motion of a threaded needle. The mechanism that produces such motion of the needle is illustrated in Figure 5.2b. The forces that are applied to the needle in its piercing of the fabric will also be of periodic nature. In engineering analyses, Fourier series can thus be used to represent both the periodic motion of the needle and also the induced forces that produces the motions of the needle.

Many machine tools involve periodic motions. An examples is given relating to the production of metal cups and dishes using a sheet metal stamping machine, the working principle of which is illustrated in Figure 5.3. A die in the desired shape of the finished product is lies beneath a flat metal sheet; the pressure of the plunger/punch on the sheet metal deforms the metal into the shape of the die surface. The force on the plunger that stamps metal into the required shape (which may be a shallow dish or deep cup) by means of the impact forces during the stamping process can be represented by a Fourier series because of its periodic nature.

5.2 Representing Periodic Functions by Fourier Series

The Fourier series, which was derived to represent periodic phenomena, is a useful tool in engineering analysis because it permits mathematical representation of periodic phenomena that are commonly encountered, as described in Section 5.1. The principal advantage of using the

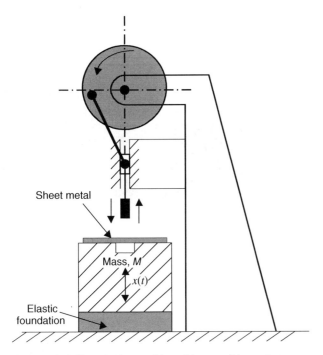

Figure 5.3 A die stamping machine with a repetitive action.

Fourier series derived from the periodic function for one period is that it can be used for all other periods.

The functions that the Fourier series represents may be either continuous or piecewise continuous within each period. The periods of these functions may cover $(-\pi, \pi)$ or $(-L, L)$, as illustrated in Figure 5.4. Note also the two different types of periodic functions represented by these Fourier series: a "continuous" function is illustrated in Figure 5.4a, and a "piecewise continuous functions" in Figure 5.4b.

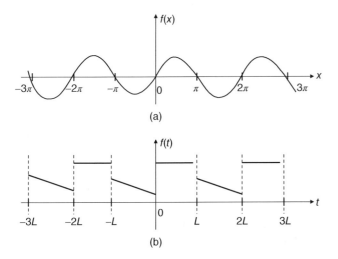

Figure 5.4 Periodic functions in typical Fourier series. (a) A periodic function with period $(-\pi, \pi)$. (b) A periodic function with period $(-L, L)$.

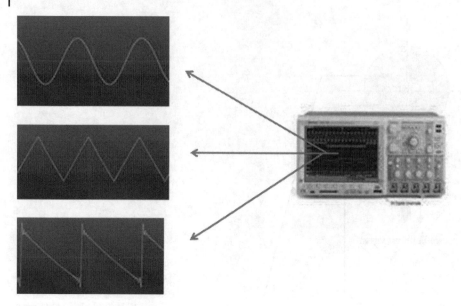

Figure 5.5 Measured signals representing periodic physical phenomena. *See color section.*

The periodic functions illustrated in Figure 5.4 appear often in measurements of various physical phenomena and are frequently visualized on oscilloscope screens as shown in Figure 5.5. Three different forms of signal are shown: (1) a continuous "sinusoidal function" at the top; (2) a piecewise continuous function in the shapes of "saw teeth" in the middle; and (3) a "discontinuous linear function" at the bottom left. Fourier series can be derived to represent all of these types of periodic functions.

5.3 Mathematical Expression of Fourier Series

Mathematical forms of Fourier series for periodic phenomena may be derived subject to the following conditions:

1) The mathematical expression of the function (e.g., $f(x)$) over one full period must be available.
2) If the period of the function is denoted by $2L$, then

$$f(x) = f(x \pm 2L) = f(x \pm 4L) = \cdots = f(x \pm nL)$$

where n is an even number.
3) The function within one period of $2L$ is defined for an interval $c < x < c + 2L$, where $c = 0$ or any arbitrarily chosen real number within the specified period. In practice, this allows the engineer to choose any starting point for his or her definition of the periodic behavior of the function.
4) The function $f(x)$ and its first-order derivative $f'(x)$ must be continuous or piecewise continuous in the period of $c < x < c+2L$.

If the above conditions are satisfied, expression of the function in any chosen period in the following infinite series form applies also for *all other periods*:

$$f(x) = \frac{a_0}{2} + \sum_{n=1}^{\infty} \left(a_n \cos \frac{n\pi x}{L} + b_n \sin \frac{n\pi x}{L} \right) = f(x \pm 2L) = f(x \pm 4L) = \cdots \tag{5.1}$$

where a_0, a_n, and b_n are Fourier series coefficients to be determined by the following integrals:

$$a_n = \frac{1}{L} \int_c^{c+2L} f(x) \cos \frac{n\pi x}{L} dx \quad n = 0, 1, 2, 3, \ldots \tag{5.2a}$$

$$b_n = \frac{1}{L} \int_c^{c+2L} f(x) \sin \frac{n\pi x}{L} dx \quad n = 1, 2, 3, \ldots \tag{5.2b}$$

Occasionally the coefficient a_0, as a special case of a_n with $n = 0$ in Equation 5.2a, needs to be determined separately as the following integral:

$$a_0 = \frac{1}{L} \int_c^{c+2L} f(x) dx \tag{5.2c}$$

Example 5.1
Derive a Fourier series for a periodic function, $f(x)$ with a period of $(-\pi, \pi)$.

Solution:
We let $c = -\pi$ and the period $2L = \pi - (-\pi) = 2\pi$, which leads to $L = \pi$. We thus have the Fourier series using Equations 5.1 and 5.2 as

$$f(x) = \frac{a_0}{2} + \sum_{n=1}^{\infty} (a_n \cos(nx) + b_n \sin(nx)) \tag{5.3}$$

with

$$a_n = \frac{1}{\pi} \int_{-\pi}^{\pi} f(x) \cos(nx) dx \quad n = 0, 1, 2, 3, \ldots \tag{5.4a}$$

and

$$b_n = \frac{1}{\pi} \int_{-\pi}^{\pi} f(x) \sin(nx) dx \quad n = 1, 2, 3, \ldots \tag{5.4b}$$

Example 5.2
Derive a Fourier series for a periodic function $f(x)$ with a period of $(-l, l)$.

Solution:
We have $c = -l$ and the period, $2L = l - (-l) = 2l$, and hence the half-period $L = l$.
Substituting this into Equations 5.1 and 5.2 will result in the following Fourier series:

$$f(x) = \frac{a_0}{2} + \sum_{n=1}^{\infty} \left(a_n \cos \frac{n\pi x}{l} + b_n \sin \frac{n\pi x}{l} \right) \tag{5.5}$$

where the coefficients may be obtained with the following integrals:

$$a_n = \frac{1}{l} \int_{-l}^{l} f(x) \cos \frac{n\pi x}{l} dx \quad n = 0, 1, 2, 3, \ldots \tag{5.6a}$$

$$b_n = \frac{1}{l} \int_{-l}^{l} f(x) \sin \frac{n\pi x}{l} dx \quad n = 1, 2, 3, \ldots \tag{5.6b}$$

Example 5.3
Derive a Fourier series for the function $f(x)$ with a period of $(0, 2L)$.

Solution:

As in the previous examples, we let $c = 0$ and the period $2L = (2L - 0) = 2L$, giving $L = L$.

The following Fourier series with appropriate coefficients is derived by using Equations 5.1 and 5.2:

$$f(x) = \frac{a_0}{2} + \sum_{n=1}^{\infty} \left(a_n \cos \frac{n\pi x}{L} + b_n \sin \frac{n\pi x}{L} \right) \tag{5.7}$$

with

$$a_0 = \frac{1}{L} \int_0^{2L} f(x)\, dx \tag{5.8a}$$

$$a_n = \frac{1}{L} \int_0^{2L} f(x) \cos \frac{n\pi x}{L}\, dx \quad n = 1, 2, 3, \dots \tag{5.8b}$$

$$b_n = \frac{1}{L} \int_0^{2L} f(x) \sin \frac{n\pi x}{L}\, dx \quad n = 1, 2, 3, \dots \tag{5.8c}$$

The different forms of Fourier series presented in Equations 5.3, 5.5, and 5.7 are also available in various engineering mathematical textbooks. The expression in Equation 5.7 with coefficients as in Equation 5.8 is a particularly useful form in many engineering analyses.

Example 5.4

Find the Fourier series for the periodic sine function in Figure 5.6a with the numerical scale shown in Figure 5.6b.

Solution:

We express the period of the sine function in Figure 5.6b by letting the period $2L = 8$ ($L = 4$); then the sine function over one period shown in Figure 5.6b has the form

$$f(t) = 3 \sin \frac{\pi}{4} t \quad \text{for} \quad 0 \le t \le 8 \tag{a}$$

The Fourier series for the sine function in Equation (a) can be expressed according to Equation 5.7 with $L = 4$ as

$$f(t) = \frac{a_0}{2} + \sum_{n=1}^{\infty} \left(a_n \cos \frac{n\pi t}{4} + b_n \sin \frac{n\pi t}{4} \right) \tag{5.7}$$

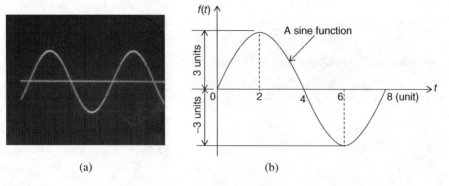

(a)　　　　　　　　　(b)

Figure 5.6 Periodic sine function. (a) Signal on oscilloscope screen. (b) Quantitative details of the signal over one period.

We have the coefficient a_0 determined by Equation (5.8a) as

$$a_0 = \frac{1}{4}\int_0^8 f(t)\,dt = \frac{1}{4}\int_0^8 \left(3\sin\frac{\pi}{4}t\right) dt = 0 \tag{b}$$

and the other coefficients a_n and b_n determined by Equations 5.8b and 5.8c respectively as

$$a_n = \frac{1}{4}\int_0^8 \left(3\sin\frac{\pi t}{4}\right)\left(\cos\frac{n\pi t}{4}\right) dt = 0 \tag{c}$$

$$b_n = \frac{1}{4}\int_0^8 \left(3\sin\frac{\pi t}{4}\right)\left(\sin\frac{n\pi t}{4}\right) dt = \frac{3n}{\pi(1-n^2)} \quad n \neq 1 \tag{d}$$

The fact that $n \neq 1$ in the coefficients b_n in Equation (d) requires the determination of coefficient b_1 using the integral in Equation 5.2b with $n = 1$:

$$b_1 = \frac{1}{4}\int_0^8 \left(3\sin\frac{\pi t}{4}\right)\left(\sin\frac{\pi t}{4}\right) dt = 3 \tag{e}$$

The Fourier series for the signal shown in Figure 5.6a can thus be expressed by substituting the coefficients in Equations (b), (c), (d), (e) into Equation (5.7), resulting in

$$f(t) = 3 + \frac{3}{\pi}\sum_{n=2}^{\infty} \left(\frac{n}{1-n^2}\right)\sin\frac{n\pi}{4}t \tag{f}$$

The function $f(t)$ in Equation (f) will enable engineers to determine the values of the signal depicted in Figure 5.6b beyond the period of $0 \leq t \leq 8$. For instance, engineers will be able to find the value of the function at $t = 12$ units from $f(12-8) = f(4) = 0$ in Equation (f) in accordance with the definition of Fourier series given at the beginning of Section 5.3, or the value of the function at $t = 36$ units from $f(36-(4\times 8)) = f(4) = 0$ from Equation (f).

Example 5.5
Find the Fourier series for the periodic "sawtooth" function in Figure 5.5 with the numerical scale shown in Figure 5.7.

Solution:
We see that this function has a period $2L = 8$ units (or $L = 4$) as in Example 5.4. The function has to be defined in three distinct subperiods between $t = 0$ and $t = 2$, between $t = 2$ and $t = 6$,

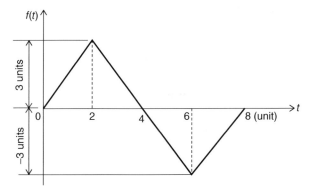

Figure 5.7 A sawtooth signal waveform, such as the signal from an oscilloscope screen (Figure 5.5).

and between $t = 6$ and $t = 8$, as shown in Figure 5.7. We thus have the function over one period defined in the following way:

$$f(t) = 1.5t \quad 0 \le t < 2$$
$$= -1.5t + 6 \quad 2 \le t < 6 \tag{a}$$
$$= 1.5t - 12 \quad 6 \le t < 8$$

We will use Equation 5.7 for the Fourier series, and Equations 5.8a, 5.8b, and 5.8c to determine the coefficients using the function $f(t)$ in Equation (a) as shown below:

$$a_0 = \frac{1}{4} \left[\int_0^2 (1.5t)\,dt + \int_2^6 (-1.5t + 6)\,dt + \int_6^8 (1.5t - 12)\,dt \right]$$
$$= 0 \tag{b}$$

$$a_n = \frac{1}{4} \left[\int_0^2 (1.5t) \cos \frac{n\pi t}{4}\,dt + \int_2^6 (-1.5t + 6) \cos \frac{n\pi t}{4}\,dt + \int_6^8 (1.5t - 12) \cos \frac{n\pi t}{4}\,dt \right]$$
$$= \frac{6}{n\pi} \sin \frac{n\pi}{2} \tag{c}$$

$$b_n = \frac{1}{4} \left[\int_0^2 (1.5t) \sin \frac{n\pi t}{4}\,dt + \int_2^6 (-1.5t + 6) \sin \frac{n\pi t}{4}\,dt + \int_6^8 (1.5t - 12) \sin \frac{n\pi t}{4}\,dt \right]$$
$$= \frac{6}{n\pi} \left(\cos \frac{3n\pi}{2} - \frac{12}{n\pi} \sin \frac{3n\pi}{2} \right) \tag{d}$$

The Fourier series for the sawtooth function depicted in Figure 5.7 can thus be expressed by substituting the coefficients in Equations (b), (c), and (d) into Equation 5.7:

$$f(t) = \frac{6}{\pi} \sum_{n=1}^{\infty} \frac{1}{n} \left[\left(\sin \frac{n\pi}{2} \right) \cos \frac{n\pi t}{4} + \left(\cos \frac{3n\pi}{2} - \frac{12}{n\pi} \sin \frac{3n\pi}{2} \right) \sin \frac{n\pi t}{4} \right] \tag{e}$$

The Fourier series in Equation (e) will enable engineers to determine the values of the function at values of the variable t preceding or after the period $(0, 8)$, as described in Example 5.4. However, the function has discontinuities at $t = 2$ and $t = 6$. The value of the function in Fourier series at these values of the variable will be described in Section 5.5.

Example 5.6
Find the Fourier series for the periodic piecewise continuous linear signal shown on the screen of an oscilloscope in Figure 5.5 with the numerical scale shown in Figure 5.8.

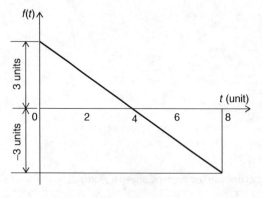

Figure 5.8 A piecewise continuous linear signal waveform, such as the signal from an oscilloscope screen (Figure 5.5).

Solution:
As in the two previous examples, the period of the function $f(t)$ is $2L = 8$ units, which gives $L = 4$ units. The function $f(t)$ for one period is

$$f(t) = -\frac{3}{4}t + 3 \quad \text{for } 0 < t < 8 \tag{a}$$

The Fourier series for the periodic function $f(t)$ may be expressed using Equation 5.7, with the coefficients determined from Equations 5.8a,b,c:

$$f(t) = \frac{a_0}{2} + \sum_{n=1}^{\infty} \left(a_n \cos \frac{n\pi t}{4} + b_n \sin \frac{n\pi t}{4} \right) \tag{b}$$

and with the coefficients:

$$a_0 = \frac{1}{4} \int_0^8 \left(-\frac{3}{4}t + 3 \right) dt = 0 \tag{c}$$

$$a_n = \frac{1}{4} \int_0^8 \left(-\frac{3}{4}t + 3 \right) \cos \frac{n\pi t}{4} \, dt = -\frac{6}{n^2 \pi^2} \tag{d}$$

$$b_n = \frac{1}{4} \int_0^8 \left(-\frac{3}{4}t + 3 \right) \sin \frac{n\pi t}{4} \, dt = \frac{9}{n\pi} \tag{e}$$

We thus obtain the Fourier series of the function in Figure 5.8 by substituting the coefficients in Equations (c), (d), and (e) into (b), to obtain

$$f(t) = \sum_{n=1}^{\infty} \left[\left(-\frac{6}{n^2 \pi^2} \right) \cos \frac{n\pi t}{4} + \frac{9}{n\pi} \sin \frac{n\pi t}{4} \right]$$

Example 5.7
Derive the Fourier series for a piecewise continuous periodic function $f(t)$ of period $(-\pi, \pi)$ with $f(t) = 0$ in the subperiod $(-\pi < t < 0)$ and $f(t) = t$ in the subperiod $(0 < t < \pi)$. The function is shown graphically in Figure 5.9.

Solution:
Since this function has a period defined in $(-\pi, \pi)$, we may use the Fourier series expressed in Equations 5.3 and 5.4 for the Fourier series and the corresponding coefficients. The Fourier series has the form of Equation 5.3:

$$f(x) = \frac{a_0}{2} + \sum_{n=1}^{\infty} (a_n \cos(nx) + b_n \sin(nx)) \tag{a}$$

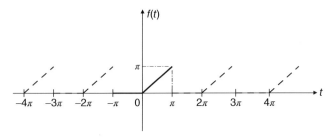

Figure 5.9 A periodic function with period defined in $(-\pi, \pi)$.

Coefficient a_0 is given by

$$a_0 = \frac{1}{\pi} \int_{-\pi}^{\pi} f(t)\,dt = \frac{1}{\pi} \int_{0}^{\pi} t\,dt = \frac{\pi}{2} \tag{b}$$

We may evaluate the other coefficients a_n and b_n following Equations 5.4a,b:

$$
\begin{aligned}
a_n &= \frac{1}{\pi} \int_{-\pi}^{\pi} f(t) \cos nt\,dt \\
&= \frac{1}{\pi} \left[\int_{-\pi}^{0} 0 \cdot \cos nt\,dt + \int_{0}^{\pi} t \cos nt\,dt \right] \\
&= \frac{1}{\pi} \left[\frac{t}{n} \sin nt + \frac{1}{n^2} \cos nt \right]\Big|_{0}^{\pi} \\
&= \frac{1}{n^2 \pi} (\cos n\pi - 1) \qquad n = 1,2,\dots
\end{aligned}
\tag{c}
$$

$$
\begin{aligned}
b_n &= \frac{1}{\pi} \int_{-\pi}^{\pi} f(t) \sin nt\,dt \\
&= \frac{1}{\pi} \left[\int_{-\pi}^{0} 0 \cdot \sin nt\,dt + \int_{0}^{\pi} t \sin nt\,dt \right] \\
&= \frac{1}{\pi} \left[\frac{t}{n} \cos nt + \frac{1}{n^2} \sin nt \right]\Big|_{0}^{\pi} \\
&= \frac{1}{n} \cos n\pi \qquad n = 1,2,\dots
\end{aligned}
\tag{d}
$$

Substituting Equations (b), (c) and (d) into Equation (a), we have the Fourier series in the form

$$f(t) = \frac{\pi}{4} + \sum_{n=1}^{\infty} \left[\frac{1}{n^2 \pi} (\cos n\pi - 1) \cos nt - \frac{1}{n} \cos n\pi \sin nt \right]$$

After replacing $\cos(n\pi) = (-1)^n$ with $n = 1, 2, 3, \dots$, in Equations (c) and (d), the Fourier series in Equation (a) can be expressed as

$$f(t) = \frac{\pi}{4} + \sum_{n=1}^{\infty} \left[\frac{(-1)^n - 1}{n^2 \pi} \cos nt - \frac{(-1)^n}{n} \sin nt \right]$$

Example 5.8

Derive a Fourier series for a piecewise continuous periodic function defined over one period as $f(x) = 0$ in the subperiod $(-5 < x < 0)$ and $f(x) = 3$ in the subperiod $(0 < x < 5)$. The function is shown graphically in Figure 5.10.

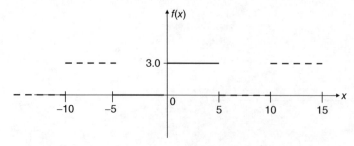

Figure 5.10 A periodic function with period defined in $(-5, 5)$.

Solution:

We have the period $2L = 5 - (-5) = 10$, and $L = 5$. We may choose the starting point $c = -5$.

The Fourier series in this case fits that shown in Equations 5.5 and 5.6a,b:

$$f(x) = \frac{a_0}{2} + \sum_{n=1}^{\infty} \left(a_n \cos \frac{n\pi x}{5} + b_n \sin \frac{n\pi x}{5} \right) \tag{a}$$

$$a_n = \frac{1}{5} \int_{-5}^{5} f(x) \cos \frac{n\pi x}{5} dx$$

$$= \frac{1}{5} \left[\int_{-5}^{0} (0) \cos \frac{n\pi x}{5} dx + \int_{0}^{5} (3) \cos \frac{n\pi x}{5} dx \right]$$

$$= \frac{3}{5} \int_{0}^{5} \cos \frac{n\pi x}{5} dx$$

We may readily integrate the cosine function in the above integral to get

$$a_n = \frac{3}{5} \left(\frac{5}{n\pi} \sin \frac{n\pi x}{5} \right) \Big|_{0}^{5} = 0 \quad \text{if} \quad n \neq 0 \tag{b}$$

For the case with $n = 0$, we have

$$a_0 = \frac{3}{5} \int_{0}^{5} \cos \frac{(0)\pi x}{5} dx = \frac{3}{5} \int_{0}^{5} dx = 3 \tag{c}$$

The other coefficients can be obtained as

$$b_n = \frac{1}{5} \int_{-5}^{5} f(x) \sin \frac{n\pi x}{5} dx$$

$$= \frac{1}{5} \left[\int_{-5}^{0} (0) \sin \frac{n\pi x}{5} dx + \int_{0}^{5} (3) \sin \frac{n\pi x}{5} dx \right]$$

$$= \frac{3}{5} \int_{0}^{5} \sin \frac{n\pi x}{5} dx$$

Evaluating the above integral gives

$$b_n = \frac{3}{5} \left(-\frac{5}{n\pi} \cos \frac{n\pi x}{5} \right) \Big|_{0}^{5} = \frac{3(1 - \cos n\pi)}{n\pi} = \frac{3[1 - (-1)^n]}{n\pi} \tag{d}$$

Substituting the expressions given in Equation (b), (c), and (d) into Equation (a) will result in the Fourier series in the form

$$f(x) = \frac{3}{2} + \sum_{n=1}^{\infty} \frac{3(1 - \cos n\pi)}{n\pi} \sin \frac{n\pi x}{5} = \frac{3}{2} + \sum_{n=1}^{\infty} \frac{3[1 - (-1)^n]}{n\pi} \sin \frac{n\pi x}{5}$$

5.4 Convergence of Fourier Series

We note that the Fourier series expressed in Equation 5.1 and those in the previous examples all consist of an *infinite* number of terms in the Fourier series. In other words, summing an infinite number of terms is required to get the Fourier series to be equal to the relevant function for one period as shown on the left-hand side of the equation. In reality, of course, it is not practical to sum an infinite number of terms. An obvious question is "How many terms are required?"; that is, what is the number of terms (the number n) in a Fourier series such as that in Equation 5.1 that should one include in determining the value of the function for a given value of the variable x?" Conversely, how will a Fourier series converge to the function's value with inclusion of a *reasonable number of terms* in the calculation? While there is no clear answer to this question in general, we may get a sense of such convergence from the following example.

The issue of the convergence of Fourier series is related to the behavior of the function it represents. However, Fourier series usually converge well with only a handful terms for many continuous functions in the specified periods, as will be demonstrated by the following case study (Wylie and Barret, 1995). This case involves the derivation of a Fourier series for the following function over one period:

$$f(t) = \begin{cases} 0 & -\pi \leq t \leq 0 \\ \sin t & 0 \leq t \leq \pi \end{cases} \tag{5.9}$$

The function is depicted graphically in Figure 5.11.

Since the period of the function is $(-\pi, \pi)$, the Fourier series in Equation 5.3 is used:

$$f(x) = \frac{a_0}{2} + \sum_{n=1}^{\infty} [a_n \cos(nx) + b_n \sin(nx)] \tag{a}$$

The coefficients in Equation (a) can be evaluated using Equation 5.4a:

$$\begin{aligned} a_n &= \frac{1}{\pi} \int_{-\pi}^{\pi} f(t) \cos(nt) \, dt \\ &= \frac{1}{\pi} \int_{-\pi}^{0} (0) \cos nt \, dt + \frac{1}{\pi} \int_{0}^{\pi} \sin t \cos nt \, dt \\ &= \frac{1 + \cos n\pi}{(1 - n^2)\pi} \end{aligned} \tag{b}$$

We will have to work out the special case of a_n with $n = 1$ because the expression for a_n in Equation (b) is not valid with $n = 1$:

$$a_1 - \frac{1}{\pi} \int_{0}^{\pi} \sin t \cos t \, dt = \left. \frac{\sin^2 t}{2\pi} \right|_{0}^{\pi} = 0 \tag{c}$$

The coefficients b_n are evaluated with Equation 5.4b:

$$\begin{aligned} b_n &= \frac{1}{\pi} \int_{-\pi}^{\pi} f(t) \sin(nt) \, dt \\ &= \frac{1}{\pi} \int_{-\pi}^{0} (0) \sin nt \, dt + \frac{1}{\pi} \int_{0}^{\pi} \sin t \sin nt \, dt \qquad n = 1, 2, 3, \ldots \end{aligned}$$

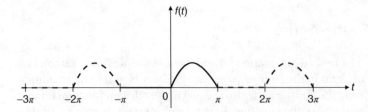

Figure 5.11 A periodic function with period of $(-\pi, \pi)$.

We may obtain the coefficients b_n from the above integral as

$$b_n = \frac{1}{\pi} \left\{ \frac{1}{2} \left[\frac{\sin(1-n)t}{1-n} - \frac{\sin(1+n)t}{1+n} \right] \right\} \Bigg|_0^\pi = 0 \quad \text{for} \quad n \neq 1 \tag{d}$$

The special case of coefficient b_1 is

$$b_1 = \frac{1}{\pi} \int_0^\pi \sin t \sin t \, dt = \frac{1}{2} \tag{e}$$

We may substitute the coefficients from Equations (b), (c), (d), and (e) into Equation (a) to obtain the Fourier series in the form

$$f(t) = \frac{1}{\pi} + \frac{\sin t}{2} + \sum_{n=2}^{\infty} (a_n \cos nt + b_n \sin nt) \tag{5.10}$$

The function $f(t)$ in Equation 5.10 can be evaluated by expanding the series to give

$$f(t) = \frac{1}{\pi} + \frac{\sin t}{2} - \frac{2}{\pi} \left(\frac{\cos 2t}{3} + \frac{\cos 4t}{15} + \frac{\cos 6t}{35} + \frac{\cos 8t}{63} + \cdots \right) \tag{5.11}$$

Let us examine the accuracy of the Fourier series in Equation 5.11 according to the number of terms included in the computation. The following cases will demonstrate the correlation.

a) **Use only one term of the Fourier series in Equation 5.11.** The correlation of the Fourier series with only one and the first term, $f_1 = 1/\pi$, and the full function is depicted in Figure 5.12.

b) **Use two terms of the Fourier series in Equation 5.11.** In this case the Fourier series has the form $f_2(t) = 1/\pi + (\sin t)/2$. The correlation of this series and the full function is depicted in Figure 5.13.

c) **Use three terms of the Fourier series in Equation 5.11.** In this case we have the Fourier series:

$$f_3(t) = \frac{1}{\pi} + \frac{\sin t}{2} - \frac{2 \cos 2t}{3\pi}$$

Correlation of this truncated series and the full function is depicted in Figure 5.14.

One may readily observe from the three graphical illustrations of correlations of the periodic function $f(t)$ and its Fourier series representation that the correlation increases dramatically with the increase of the number of terms of the Fourier series that are included.

Figure 5.12 Correlation of a function with the Fourier series with one term.

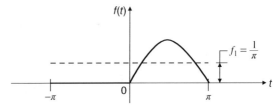

Figure 5.13 Correlation of a function with the Fourier series with two terms.

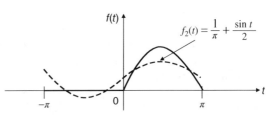

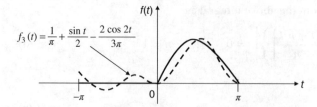

$$f_3(t) = \frac{1}{\pi} + \frac{\sin t}{2} - \frac{2\cos 2t}{3\pi}$$

Figure 5.14 Correlation of a function with the Fourier series with three terms.

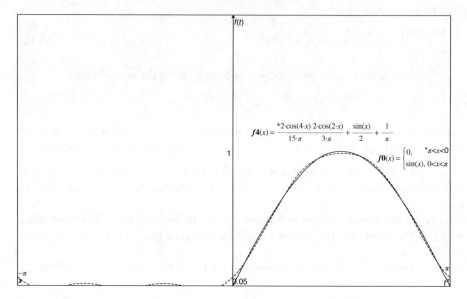

$$f4(x) = \frac{*2\cdot\cos(4\cdot x)}{15\cdot\pi} - \frac{2\cdot\cos(2\cdot x)}{3\cdot\pi} + \frac{\sin(x)}{2} + \frac{1}{\pi}$$

$$f0(x) = \begin{cases} 0, & *\pi < x < 0 \\ \sin(x), & 0 < x < \pi \end{cases}$$

Figure 5.15 Correlation of a function with the Fourier series with four terms.

Figure 5.15 shows the correlation of the Fourier series with four terms from Equation 5.11 with the full function. Clearly, a much closer correlation is obtained using only four terms of the Fourier series, as demonstrated in Figure 5.15.

5.5 Convergence of Fourier Series at Discontinuities

For periodic functions of continuous nature such as illustrated in Figure 5.4a and the uppermost of the three cases shown in Figure 5.5, good convergence of the Fourier series to the full function is obtainable with the inclusion of only a handful of terms of the derived Fourier series, as illustrated in Section 5.4. We have demonstrated that as few as four terms ($n = 4$) of the derived Fourier series in Equation 5.11 resulted in a good enough representation of the given function that represents the periodic phenomenon illustrated in Figure 5.15. In contrast, for functions that are not continuous within the periods (see Figure 5.4b, for example), the Fourier series will not converge to the value at the discontinuity regardless of how many terms of the Fourier series are included in the evaluation.

Let us examine the convergence of Fourier series at the discontinuity points x_0, x_1, and x_2 of the discontinuous function shown in Figure 5.16.

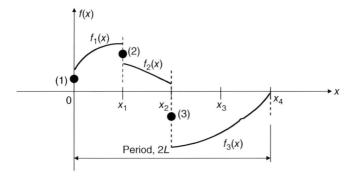

Figure 5.16 Fourier series at discontinuities.

In Figure 5.16, we have the function $f(x)$ containing the following piecewise continuous functions defined in a period of $2L$ as

$f_1(x)$ for $0 < x < x_1$
$f_2(x)$ for $x_1 < x < x_2$
$f_3(x)$ for $x_2 < x < x_4$

The Fourier series for this periodic function $f(x)$ has a similar form to that shown in Equation 5.1:

$$f(x) = f(x \pm 2L) = f(x \pm 4L) = \cdots = f(x \pm nL)$$

$$= \frac{a_0}{2} + \sum_{n=1}^{\infty} \left(a_n \cos \frac{n\pi x}{L} + b_n \sin \frac{n\pi x}{L} \right) \tag{5.12}$$

where a_0, a_n, and b_n are Fourier series coefficients.

One will observe from Figure 5.16 that the function $f(x)$ over one period contains discontinuities at $x = 0$, x_1 and x_2. We will find that the values of the function represented by the Fourier series in Equation 5.12 will converge at these discontinuities at the values obtained from the expressions shown below regardless of how many terms are included in the Fourier series in Equation 5.12. Instead, engineers will find that the Fourier series will converge to the values at the filled circles in Figure 5.16 at the variables at which the discontinuities of the function occur. Mathematically, the converged values of the function at these discontinuities can be obtained by the following expressions:

$f(0) = \frac{1}{2} f(0)$ at Point (1)

$f(x_1) = \frac{1}{2}[f_1(x_1) + f_2(x_1)]$ at Point (2)

$f(x_2) = \frac{1}{2}[f_2(x_2) + f_3(x_2)]$ at Point (3)

$f(x_4) = f_3(x_4) = \frac{1}{2} f_1(0)$ same as Point (1)

where Points (1), (2) and (3) are as indicated in Figure 5.16.

Example 5.9
Determine the value of Fourier series representing the periodic function in Example 5.8 at $x = 0$, 5, 10, 12.5, −5, −7.5 and −10.

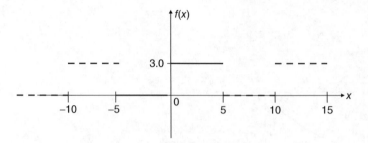

Figure 5.17 A piecewise continuous periodic function.

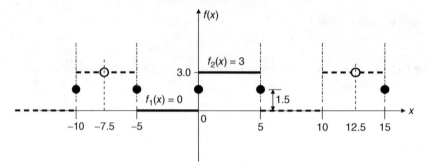

Figure 5.18 Convergence of the Fourier series for a piecewise continuous periodic function.

Solution:

The function in this example is represented graphically in Figure 5.17. The Fourier series for the periodic function was derived in Example 5.8 in the form

$$f(x) = \frac{3}{2} + \sum_{n=1}^{\infty} \frac{3(1 - \cos n\pi)}{n\pi} \sin \frac{n\pi x}{5}$$

$$= \frac{3}{2} + \sum_{n=1}^{\infty} \frac{3[1 - (-1)^n]}{n\pi} \sin \frac{n\pi x}{5}$$

(a)

Using the convergence criterion outlined earlier in this section for discontinuous functions, we find that the values of the Fourier series in Equation (a) with a large number of terms converge at both the points shown in open circles and those shown in filled circles in Figure 5.18.

The filled circles in Figure 5.18 depict the converged values of the Fourier series at the discontinuities. They are obtained from the following relations:

$$f(x)|_{x=0,5,10,-5,-10} = \frac{1}{2}[f_1(x) + f_2(x)]|_{x=0,5,10,-5,-10}$$

$$= \frac{1}{2}(0 + 3) = 1.5$$

The open circles in Figure 5.18 are in the continuous portion of the function, and their converged values are obtained from the function that is represented by the Fourier series.

Example 5.10   Derive the Fourier series for the piecewise continuous periodic function illustrated in Figure 5.19, with the expression for one period being

$$f(t) = \begin{cases} 3t & (0 < t < 1) \\ 1 & (1 < t < 4) \end{cases}$$

(5.14)

Figure 5.19 Fourier series of a piecewise continuous periodic function.

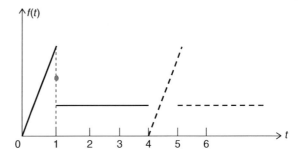

We will use Equation 5.1 to derive the Fourier series of the function specified in Equation 5.14:

$$f(t) = \frac{a_0}{2} + \sum_{n=1}^{\infty} \left(a_n \cos \frac{n\pi t}{L} + b_n \sin \frac{n\pi t}{L} \right) \tag{a}$$

with a period of $2L = 4$.

The Fourier series coefficients for the function illustrated in Figure 5.19 are derived from Equations 5.2a and 5.2b in the forms:

$$a_0 = \frac{15}{4} \tag{b}$$

$$a_n = \frac{6}{n^2\pi^2} \left(\cos \frac{n\pi}{2} - 1 \right) + \frac{1}{n^3\pi^3}(12 - n^2\pi^2) \sin \frac{n\pi}{2} \quad \text{with} \quad n = 1, 2, 3, 4, 5, \ldots \tag{c}$$

$$b_n = \frac{6}{n^2\pi^2} \sin \frac{n\pi}{2} - \frac{1}{n\pi} \left(4\cos \frac{n\pi}{2} + \cos 2n\pi \right) \quad \text{with} \quad n = 1, 2, 3, 4, 5, \ldots \tag{d}$$

Graphical representations of the Fourier series using the terms to $n = 3$, 15, and 80 are illustrated in Figure 5.20, Figure 5.21, and Figure 5.22, respectively.

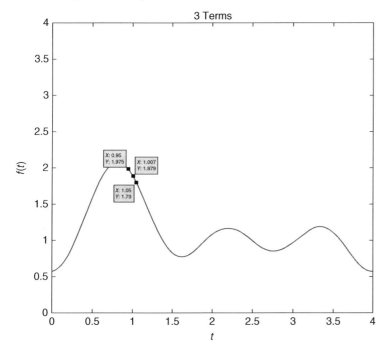

Figure 5.20 Fourier series with 3 terms ($n = 3$).

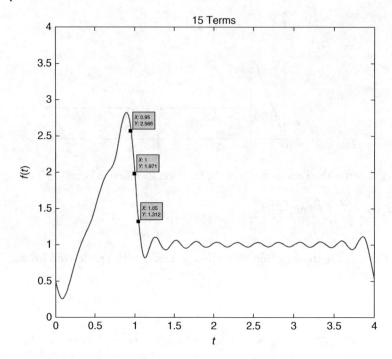

Figure 5.21 Fourier series with 15 terms ($n = 15$).

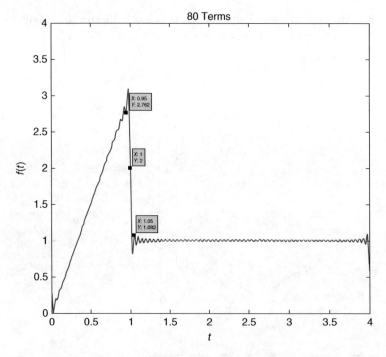

Figure 5.22 Fourier series with 80 terms ($n = 80$).

We may observe from these figures that good convergence of the derived Fourier series in this case is achieved with close to 80 terms (i.e., $n = 80$), and also find the converged value at $t = 1$ to be $f(1) = 2$, as expected.

5.6 Problems

5.1 Derive the Fourier series with a graphical representation of three periods for each of the following periodic functions. One period of the function is defined:
a) $f(t) = -t$ for $-\pi < t < 0$; and $f(t) = t$ for $0 < t < \pi$.
b) $f(t) = t^2$ for $-\pi < t < \pi$.
c) $f(x) = \cos(x/2)$ with $-\pi < x < \pi$.

5.2 Derive Fourier series for the following periodic physical phenomena with one period graphically illustrated in Figure 5.23, Figure 5.24, and Figure 5.25.

5.3 Derive the Fourier series and then evaluate the functions representing the following periodic physical phenomena in Figure 5.26 and Figure 5.27, at specific values of the variable:
a) The parabolic function in Figure 5.26 at $t = -4, -2, 2, 4, 8$.
b) The step function in Figure 5.27 at $t = 1, 1.5, 2, 2.5, 4$.

5.4 Determine the values of the function in Example 5.6 at $t = -6, -14, -16, 10, 16, 18, 24$.

5.5 Determine the values of the Fourier series in Example 5.7 at $t = -3\pi, -2.5\pi, 4.5\pi, 5\pi$.

Figure 5.23 A step function.

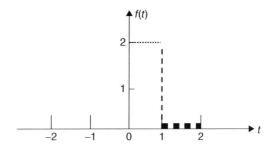

Figure 5.24 A piecewise continuous function.

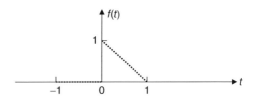

Figure 5.25 A parabolic function.

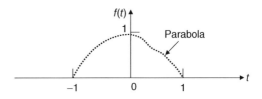

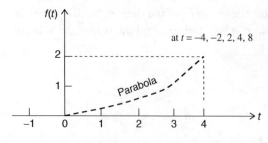

Figure 5.26 Evaluation of the Fourier series for a parabolic function.

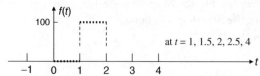

Figure 5.27 Evaluation of the Fourier series for a step function.

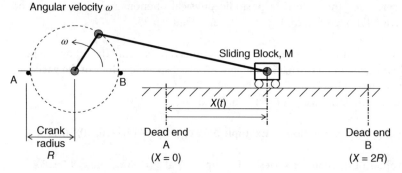

Figure 5.28 Crank–slider linkage.

5.6 Derive a mathematical expression using a Fourier series to describe the periodic motion of the plunger/punch in a stamping machine such as shown in Figure 5.3. The "crank–slider" linkage is frequently used in mechanism design.

This problem requires the derivation of a function describing the position of the sliding block M during one period of the crank–slider mechanism motion, as illustrated in Figure 5.28 if the crank rotates at a constant velocity of 5 rpm. Illustrate the periodic function over three periods, and derive the Fourier series describing the position of the sliding block, $x(t)$, where t is the time in minutes.

6

Introduction to the Laplace Transform and Applications

Chapter Learning Objectives

- Learn the application of Laplace transform in engineering analysis.
- Learn the required conditions for transforming variable or variables in functions by the Laplace transform.
- Learn the use of available Laplace transform tables for transformation of functions and the inverse transformation.
- Learn to use partial fraction and convolution methods in inverse Laplace transforms.
- Learn the Laplace transform for ordinary derivatives and partial derivatives of different orders.
- Learn how to use Laplace transform methods to solve ordinary and partial differential equations.
- Learn the use of special functions in solving indeterminate beam bending problems using Laplace transform methods.

6.1　Introduction

The Laplace transform is named after Pierre-Simon Laplace (1749–1829), a renowned French mathematician and astronomer. It is a mathematical operation that is used to "transform" a *variable* from a variable domain to a parametric domain. In layman's terms, the Laplace transform can be used to "convert" a variable of a function into a parameter. After the transformation, that variable is no longer a variable but it can be treated as a "parameter," which—again in layman's terms—is a "constant under specific conditions." Transformation of variables into parameters can significantly simplify the mathematical analysis of many physical problems. A word of caution, however, is that *the Laplace transform can only be used to transform variables that cover a range from zero (0) to infinity (∞). Any variable that does not vary within this range cannot be transformed by the Laplace transform.* Consequently, the Laplace transform is often used to transform the time variable (t) of a function, because in much of engineering analysis this variable usually covers the range $(0,\infty)$.

The Laplace transform is a useful tool in converting variables to parameters in functions—a practice that is often used in automatic control of the motion of machine components, in which time t is often a variable in the analysis. It is also an effective tool for solving differential equations in mathematical analyses of physical problems. In the latter cases, major effort is required in the inversion of the transformed expressions with Laplace transform parameters into the original functions involving the corresponding variables for the solutions.

Applied Engineering Analysis, First Edition. Tai-Ran Hsu.
© 2018 John Wiley & Sons Ltd. Published 2018 by John Wiley & Sons Ltd.
Companion Website: www.wiley.com/go/hsu/applied

6.2 Mathematical Operator of Laplace Transform

The mathematical operator used to transform a function of variable u is $L[f(u)]$ where the variable u covers the range $(0,\infty)$. The mathematical expression of the Laplace transform of function $f(t)$ with $0 < t < \infty$ has the form

$$L[f(t)] = \int_0^\infty f(t)e^{-st}\,dt = F(s) \qquad (6.1)$$

where s is a parameter of the Laplace transform corresponding to the variable t in Equation (6.1), and $F(s)$ is the Laplace transform of function $f(t)$ with $0 < t < \infty$. It is not a function of s. It is merely an expression that involves the Laplace transform parameter s.

The "inverse Laplace transform" operates in the reverse sense; that is, it inverts the transformation expression $F(s)$ in Equation 6.1 back to its original function $f(t)$. Mathematically, it has the form

$$L^{-1}[F(s)] = f(t) \qquad (6.2)$$

where the operator L^{-1} denotes the inverse Laplace transform.

Laplace transformation is useful in "eliminating" variables in the course of many mathematical manipulations.

Example 6.1

Find the Laplace transform of a function $f(t) = t^2$ with $0 < t < \infty$.

Solution:

We may use the definition of Laplace transform in Equation 6.1 to find the Laplace transform of the function $f(t) = t^2$ with $0 < t < \infty$ as follows:

$$L[f(t)] = \int_0^\infty e^{-st}(t^2)\,dt = F(s) \qquad (a)$$

We may use either integration tables from mathematical handbooks or electronic calculators/computers to find the integral in Equation (a) as

$$L[f(t)] = \int_0^\infty e^{-st}(t^2)\,dt = e^{-st}\left[-\frac{2t^2}{2s} - \frac{2t}{s^2} - \frac{2}{s^3}\right]\Big|_0^\infty = \frac{2}{s^3} = F(s)$$

Laplace transforms of many relatively simple functions are available in "Laplace Transform Tables," such as that presented in Appendix 1. The answer to the above example is identical to Case (3) with $n = 3$ in the table of Appendix 1.

We emphasize that the functions that can be transformed using Laplace transform must have their variables covering the range from zero (0) to infinity (∞). However, one should not misinterpret this statement by the fact that there can be more than one function involved in this overall spectrum, and the Laplace transform can be performed over these functions within this overall spectrum $(0, \infty)$. For example, the physical phenomena described by special functions in Section 2.4.2 are eligible for Laplace transformation. The following is an example of transforming a "ramp function" that is commonly used in mathematical modeling of automatic motion control of machines, and many production processes can be transformed using the Laplace transform.

Figure 6.1 A ramp function.

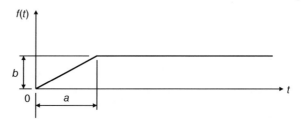

Example 6.2
Perform the Laplace transform on the ramp function illustrated in Figure 6.1.

Solution:
We may define the ramp function in Figure 6.1 with two functions in the overall range $(0,\infty)$:

$$f(t) = \frac{b}{a}t \quad 0 \leq t < a$$
$$= b \quad a < t < \infty \tag{a}$$

The Laplace transform may be carried out on both these functions in Equation (a) using the integral in Equation 6.1 as follows:

$$L[f(t)] = \int_0^\infty f(t)e^{-st}\,dt = F(s)$$
$$= \int_0^a \frac{b}{a}te^{-st}\,dt + \int_a^\infty be^{-st}\,dt \tag{b}$$

We may find the integrals in Equation (b), either from mathematical handbooks or using electronic calculators, as

$$F(s) = \frac{b}{a}\frac{e^{-st}}{(-s)^2}(-st-1)\Big|_0^a + \frac{b}{(-s)}e^{-st}\Big|_a^\infty$$
$$= -\frac{b}{as^2}(as+1)e^{-as} + \frac{b}{as^2} + \frac{b}{s}e^{-as}$$

Laplace transformation is also used to transform the step functions described in Section 2.4.2. The following example will illustrate such an application.

Example 6.3
Apply Laplace transforms to (a) the step function $u_0(x)$ in Figure 6.2, and (b) the step function $u_a(x)$ in Figure 6.3.

Figure 6.2 A unit step function.

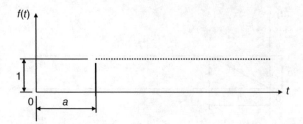

Figure 6.3 A unit step function at $t = a$.

Solution:

a) The function $u_0(x)$ is illustrated in Figure 6.2. The mathematical expression of the step function is given by Equation 2.36a with $\alpha = 1$:

$$f(t) = u_0(t) = 1 \quad \text{with } 0 \leq t < \infty$$

Thus, the Laplace transform of this function is

$$L[u_0(t)] = \int_0^\infty (1)e^{-st}\, dt = -\frac{1}{s}e^{-st}\Big|_0^\infty = \frac{1}{s} \tag{6.3}$$

as in Case 1 in Appendix 1.

b) The Laplace transform is applied to the step function $u_a(x - a)$ illustrated in Figure 6.3. The mathematical expression of the function $f(t)$ in this case is obtained from Equation 2.36a with $\alpha = 1$:

$$f(t) = u_a(t) = \begin{cases} 0 & 0 \leq t < a \\ 1 & a < t < \infty \end{cases}$$

The corresponding Laplace transform is

$$L[u_a(t)] = \int_0^a (0)e^{-st}\, dt + \int_a^\infty (1)e^{-st}\, dt = -\frac{1}{s}e^{-st}\Big|_a^\infty = \frac{1}{s}e^{-as} \tag{6.4}$$

We find that the result for the step function $u_a(t)$ is identical to that shown in Case 15 in the Laplace transform table in Appendix 1.

6.3 Properties of the Laplace Transform

Properties of Laplace transform can be used in conjunction with the Laplace transform tables in Appendix 1, as well as for mathematical manipulations. The following are a few such properties.

6.3.1 Linear Operator Property

$$L[af(t) + bg(t)] = aL[f(t)] + bL[g(t)] \tag{6.5}$$

where $f(t)$ and $g(t)$ are two functions sharing the same variable that cover the range $(0, \infty)$.

Example 6.4

Find the Laplace transform of a function $f(t) = 4t^2 - 3\cos t + 5e^{-t}$.

Solution:

Using the linear operator in Equation 6.5, we may decompose the transform into three groups:

$$L(4t^2 - 3\cos t + 5e^{-t}) = 4L[t^2] - 3L[\cos t] + 5L[e^{-t}] = F(s)$$

From the Laplace transform table in Appendix 1 we will find the individual components of the Laplace transform in the above expression to be

$$F(s) = \frac{8}{s^3} - \frac{3s}{s^2 + 1} + \frac{5}{s + 1}$$

6.3.2 Shifting Property

If the Laplace transform of a function $f(t)$ is $L[f(t)] = F(s)$ as shown in Equation 6.1, then we will have the following relationship to transfer function $f(t)$ with a multiple of e^{at}. In such a case, one needs to replace the parameter s by $(s - a)$ in $F(s)$, or

$$L[e^{at}f(t)] = F(s - a) \tag{6.6}$$

in which a is a constant and is a real number.

Example 6.5   If $f(t) = \sin at$, we have

$$L[f(t)] = L[\sin at] = \frac{a}{s^2 + a^2}$$

from the Laplace transform table.

The shifting property will facilitate finding the Laplace transform of function $F(t) = e^{2t} \sin at$ as

$$L[F(t)] = L[e^{2t} \sin at] = \frac{a}{(s - 2)^2 + a^2}$$

6.3.3 Change of Scale Property

If $L[f(t)] = F(s)$, we will have

$$L[f(at)] = \frac{1}{a} F\left(\frac{s}{a}\right) \tag{6.7}$$

where a is a constant and is a real number.

Example 6.6   If $f(t) = \sin t$, and

$$L[f(t)] = L[\sin t] = \frac{1}{s^2 + 1} = F(s)$$

from the Laplace transform table in Appendix 1, we will have the Laplace transform of function $f(t) = \sin 3t$ as

$$L[\sin 3t] = \frac{1}{3} \frac{1}{\left(\dfrac{s}{3}\right)^2 + 1} = \frac{3}{s^2 + 9}$$

6.4 Inverse Laplace Transform

In contrast to the Laplace transform of functions defined in Equation 6.1, there is no explicit mathematical formula available for inverting a Laplace-transformed expression $F(s)$ back to its original function as expressed in Equation 6.2. However, there are four approaches that generally one may follow to invert a Laplace-transformed expression $F(s)$.

1) Use the Laplace transform table in a reverse manner. This approach involves searching for the function $f(t)$ in the right-hand column of the table in Appendix 1 that matches the $F(s)$ in the left column in the same table.
2) Use the partial fraction method.
3) Use the convolution theorem.
4) Use the Bromwich contour integration.

Approach (1) rarely works for most practical cases. Approaches (2) and (3) will be elaborated in the Sections 6.4.2 and 6.4.3. Method in (4) involves the use of complex variable theory, which is beyond the scope of this book. We will thus focus on the first three approaches.

6.4.1 Using the Laplace Transform Tables in Reverse

This method involves matching the $F(s)$ for the corresponding function $f(t)$ with those shown in the Laplace transform table in Appendix 1, or those from other handbooks to find the corresponding original function $f(t)$. For example, one may find the following inverse Laplace transforms corresponding to the indicated cases in the table in Appendix 1:

$$L^{-1}\left[\frac{1}{s-a}\right] = e^{at} \tag{Case 7}$$

$$L^{-1}\left[\frac{a-b}{(s-a)(s-b)}\right] = e^{at} - e^{bt} \quad a \neq b \tag{Case 11}$$

$$L^{-1}\left[\frac{s}{s^2+\omega^2}\right] = \cos \omega t \tag{Case 18}$$

The likelihood of finding the desired inverse Laplace transform of functions by this method is not always good; other methods have thus been developed for this purpose.

6.4.2 The Partial Fraction Method

This method applies to cases where the expression $F(s)$ is a *rational function*. The partial fraction method is used to express these rational functions of $F(s)$ in a number of simpler fractions that can be inverted to the corresponding original functions using Laplace transform tables. The general procedure of this method may be outlined in the following two steps:

Step 1 Express the $F(s)$ in partial fractions:

$$F(s) = \frac{P(s)}{Q(s)}$$

$$= \frac{A_1}{s-a_1} + \frac{A_2}{s-a_2} + \cdots + \frac{A_n}{s-a_n} = \frac{\text{polynomial of order } (n-1)}{\text{higher-order polynomial of order } n} \tag{6.8}$$

where $A_1, A_2, \ldots, A_n$, and $a_1, a_2, \ldots, a_n$ are constants.

Step 2 Invert the simpler fractions on the right-hand-side of Equation 6.8 using Laplace transform tables.

Example 6.7

Invert the following Laplace transform expression $F(s)$ using the partial fraction method:

$$F(s) = \frac{3s + 7}{s^2 - 2s - 3}$$

Solution:

We may express

$$F(s) = \frac{3s + 7}{s^2 - 2s - 3} = \frac{3s + 7}{(s - 3)(s + 1)}$$

$$= \frac{A}{s - 3} + \frac{B}{s + 1} \tag{a}$$

where A and B are constants.

The two constants A and B in Equation (a) may be determined by multiplying both sides of Equation (a) by $(s - 3)(s + 1)$, giving

$$3s + 7 = A(s + 1) + B(s - 3)$$

$$= (A + B)s + (A - 3B) \tag{b}$$

By comparing coefficients of terms on both sides of Equation (b), we may solve for $A = 4$ and $B = -1$, which leads to the following solution with use of Laplace transform table in Appendix 1:

$$L^{-1}\left[\frac{3s + 7}{(s - 3)(s + 1)}\right] = 4L^{-1}\left(\frac{1}{s - 3}\right) - L^{-1}\left(\frac{1}{s + 1}\right)$$

$$= 4e^{3t} - e^{-t}$$

Example 6.8

Find the following inverse of a Laplace transform using the partial fraction method:

$$L^{-1}[F(s)] = L^{-1}\left[\frac{3s + 1}{s^3 - s^2 + s - 1}\right]$$

Solution:

Since $F(s)$ in this problem is a rational function, we may use the partial fraction method for the inversion. Let $F(s)$ be written in the form

$$\frac{3s + 1}{(s - 1)(s^2 + 1)} = \frac{A}{s - 1} + \frac{Bs + C}{s^2 + 1} \tag{a}$$

Note that we have placed a first-order polynomial of the parameter s in the numerator of the second fraction in the right-hand-side of Equation (a) because the denominator of this fraction is of second order in the parameter s. Following the procedure outlined in Example 6.7, we may solve for $A = 2$, $B = -2$, and $C = 1$, which gives

$$\frac{3s + 1}{(s - 1)(s^2 + 1)} = \frac{2}{s - 1} - \frac{2s}{s^2 + 1} + \frac{1}{s^2 + 1} \tag{c}$$

We obtain the inversion of the Laplace-transformed function $f(t)$ by using the Laplace transform table in Appendix 1:

$$f(t) = L^{-1}[F(s)]$$

$$= L^{-1}\left[\frac{3s + 1}{s^3 - s^2 + s - 1}\right]$$

$$= L^{-1}\left(\frac{2}{s - 1}\right) - L^{-1}\left(\frac{2s}{s^2 + 1}\right) + L^{-1}\left(\frac{1}{s^2 + 1}\right)$$

$$= 2e^t - 2\cos t + \sin t$$

6.4.3 The Convolution Theorem

Use of the convolution theorem appears to be the only way that to invert an expression $F(s)$ involving integrals and is thus a useful method for inverting Laplace transforms.

If we are given the inverted Laplace transforms $L^{-1}[F(s)] = f(t)$ and $L^{-1}[G(s)] = g[t]$ from the Laplace transform table in Appendix 1, in which

$$F(s) = \int_0^\infty e^{-st} f(t)\,dt \text{ and } G(s) = \int_0^\infty e^{-st} g(t)\,dt$$

we may find the inverse Laplace transform of the expression $Q(s) = F(s)G(s)$ with the following integrals:

$$L^{-1}[Q(s)] = L^{-1}[F(s)G(s)] = \int_0^t f(\tau)g(t-\tau)\,d\tau \tag{6.9a}$$

or in another form:

$$L^{-1}[Q(s)] = L^{-1}[F(s)G(s)] = \int_0^t f(t-\tau)g(\tau)\,d\tau \tag{6.9b}$$

The reader is reminded that the notation τ in Equations 6.9a and 6.9b is a dummy integration variable. Consequently, any term associated with variable t may be "factored" out of the integrals in the process of integration.

Example 6.9

Use the convolution theorem to find the inverse Laplace transform with

$$Q(s) = \frac{s}{(s^2 + a^2)^2}$$

in which a is a constant.

Solution:

We can express $Q(s)$ in the form of the product of two rational functions:

$$Q(s) = \frac{s}{(s^2 + a^2)^2} = \frac{s}{s^2 + a^2} \cdot \frac{1}{s^2 + a^2} \tag{a}$$

From the Laplace transform table in Appendix 1 we find the following inverse transforms:

$$L^{-1}\left[\frac{s}{s^2 + a^2}\right] = \cos at = f(t)$$

and

$$L^{-1}\left[\frac{1}{s^2 + a^2}\right] = \frac{\sin at}{a} = g(t)$$

We may use the integral in Equation 6.9a to find the inverse Laplace transform of $Q(s)$:

$$L^{-1}\left[\frac{s}{(s^2 + a^2)^2}\right] = \int_0^t \cos a\tau \frac{\sin a(t-\tau)}{a}\,d\tau$$

$$= \frac{t \sin at}{2a}$$

One will get the same result using the partial fraction method described in Section 6.4.2 or the convolution integral in Equation 6.9b.

Hint: The integrand in the above integral was first expanded using a trigonometric operation according to $\sin(\alpha \pm \beta) = \sin\alpha\cos\beta \pm \sin\beta\cos\alpha$, followed by factoring out the terms associated with variable t from the integrals. The integration will be performed on the dummy variable τ only.

Example 6.10

Use the convolution theorem to find the inverse Laplace transform of the function

$$f(t) = L^{-1}[F(s)] = L^{-1}\left[\frac{1}{s^2(s+1)^2}\right]$$

Solution:

Since

$$\left[\frac{1}{s^2(s+1)^2}\right] = \frac{1}{s^2} \cdot \frac{1}{(s+1)^2}$$

and from Laplace transform table

$$L^{-1}\left[\frac{1}{s}\right] = t \quad \text{and} \quad L^{-1}\left[\frac{1}{(s+1)^2}\right] = te^{-t}$$

we may find the inverse of $F(s)$ from Equation 6.9b as

$$f(t) = L^{-1}\left[\frac{1}{s^2(s+1)^2}\right] = \int_0^t (\tau e^{-\tau})(t-\tau) \, d\tau$$

$$= \int_0^t (\tau t - \tau^2)e^{-\tau} \, d\tau$$

$$= te^{-t} + 2e^{-t} + t - 2$$

Example 6.11

Use (a) the partial fraction method, and (b) the convolution theory method to find the corresponding function $f(t)$ of the following expression of the Laplace transform parameter s:

$$F(s) = \frac{1}{(s+1)(s^2+4)}$$

Solution:

a) *Using the partial fraction method:* We may express $F(s)$ in terms of the following partial fractions:

$$F(s) = \frac{1}{(s+1)(s^2+4)} = \frac{A}{s+1} + \frac{Bs+C}{s^2+4}$$

We can determine the constants A, B, and C in the above to be

$$A = \frac{1}{5}, \quad B = -\frac{1}{5}, \quad C = \frac{1}{5}$$

We thus have $F(s)$ expressed in the following partial fractions:

$$F(s) = \frac{1}{(s+1)(s^2+4)}$$

$$= \frac{1}{5}\frac{1}{s+1} - \frac{1}{5}\frac{s}{s^2+4} + \frac{1}{5}\frac{1}{s^2+4}$$

We may find the inverse Laplace transform of each individual item in the above expression of $F(s)$ from the Laplace transform table in Appendix 1 as follows:

$$F(s) = \frac{1}{5}e^{-t} - \frac{1}{5}\cos 2t + \frac{1}{10}\sin 2t$$

b) *Using the convolution theory:* We first decompose the expression of $F(s)$ into the form of the product of two functions:

$$F(s) = \frac{1}{(s+1)(s^2+4)} = \frac{1}{s+1} \cdot \frac{1}{s^2+4}$$

From the Laplace transform table we find the following inverse Laplace transforms:

$$F(s) = \frac{1}{s+1} = e^{-t} \text{ and } G(s) = \frac{1}{s^2+4} = \frac{1}{2}\sin 2t$$

Then using Equation 6.9b, we have the inverse of $F(s)$ as

$$f(t) = \int_0^t F(t-\tau)G(\tau)\,d\tau = \int_0^t e^{-(t-\tau)}\left(\frac{1}{2}\sin 2\tau\right)d\tau$$

$$= \frac{1}{2}e^{-t}\int_0^t e^{\tau}\sin 2\tau\,d\tau$$

We may obtain the above integral either from integration tables in a mathematical handbook or by using an electronic calculator as

$$f(t) = \frac{1}{2}e^{-t}\left[\frac{e^{\tau}(\sin 2\tau - 2\cos 2\tau)}{1+2^2}\right]\Bigg|_0^t$$

$$= \frac{1}{10}\sin 2t - \frac{1}{5}\cos 2t + \frac{1}{5}e^{-t}$$

We have thus shown that both the partial fraction method and convolution theorem have led to the same result in inverting the Laplace transformed function.

6.5 Laplace Transform of Derivatives

We have learned how to perform a Laplace transform of functions involving variables that cover a range of (0 to ∞). Since derivatives are functions too, we should also be able to transform the variables in derivatives to parameter s using the Laplace transform.

6.5.1 Laplace Transform of Ordinary Derivatives

Let us define a first- order derivative of a function $f(t)$ by $f'(t) = df(t)/dt$. The Laplace transform of $f'(t)$ may be accomplished using the Laplace transform definition in Equation 6.1:

$$L[f'(t)] = \int_0^\infty e^{-st}f'(t)\,dt = \int_0^\infty e^{-st}\left[\frac{df(t)}{dt}\right]dt \tag{6.10}$$

The variable to be transformed in Equation 6.10 should also satisfy the condition $0 < t < \infty$. Integration of the integrand in Equation 6.10 requires the use of a special technique called "integration by parts," using the integral $I = \int u\,dv = uv - \int v\,du$ where u and v are parts of the function to be integrated.

Letting $u = e^{-st}$ and $dv = [df(t)/dt]\,dt$, which leads to $du = -se^{-st}\,dt$ and $v = f(t)$, followed by substituting the above relationships into Equation 6.10, results in the following:

$$L[f'(t)] = \int_0^\infty e^{-st}f'(t)\,dt = e^{-st}f(t)\big|_0^\infty - \int_0^\infty f(t)(-se^{-st})\,dt \tag{6.11}$$

The integral in Equation 6.11 will lead to the following expression for the Laplace transform for the first-order derivative of the function $f(t)$:

$$L[f'(t)] = \int_0^\infty e^{-st}f'(t)\,dt$$

$$= e^{-st}f(t)\big|_0^\infty - \int_0^\infty f(t)(-se^{-st})\,dt$$

$$= -f(0) + s \int_0^\infty e^{-st} f(t)\, dt$$

$$= -f(0) + s L[f(t)]$$

We may thus write the Laplace transform of the first-order derivative of the function $f(t)$ as

$$L[f'(t)] = s L[f(t)] - f(0) \tag{6.12}$$

Following a similar procedure, we may derive the Laplace transform of the second-order derivative of the function $f(t)$ as

$$L[f'(t)] = s^2 L[f(t)] - sf(0) - f'(0) \tag{6.13}$$

A recurrence relation for the Laplace transforms of nth-order of derivatives of the function $f(t)$ can thus be formulated:

$$L[f^n(t)] = s^n L[f(t)] - s^{n-1} f(0) - s^{n-2} f'(0) - s^{n-3} f''(0) - f^{n-1}(0) \tag{6.14}$$

Example 6.12
Perform the Laplace transform on $f''(t)$ with $f(t) = t \sin t$.

Solution:
The Laplace transform of the second order derivative of function $f(t)$ (that is, $f''(t)$) may be obtained either using Equation 6.13 or from the general expression in Equation 6.14 with $n = 2$. To do this, we will first find that $f'(t) = t \cos t + \sin t$.
  Thus, from Equation 6.13), we get

$$L[f''(t)] = s^2 L[f(t)] - sf(0) - f'(0)$$
$$= s^2 L[t \sin t] - s(t \sin t)|_{t=0} - (t \cos t + \sin t)|_{t=0}$$
$$= s^2 L[t \sin t]$$

6.5.2 Laplace Transform of Partial Derivatives

In Section 2.2.5 we saw that partial derivatives involving more than one independent variable appear frequently in engineering analyses. Engineers have to perform Laplace transform of these functions and their derivatives.

Typical functions involving more than one independent variable have such forms as $f(x,t)$, $f(x,y,t)$, or $f(x,y,z,t)$, in which x, y, z, and t are independent variables with (x,y,z) typically being space variables and t representing the time as described in Section 2.2.2. The rate of change of this type of function may be expressed by "partial derivatives" with respective to each of the independent variable involved in the function. Partial derivatives for the function $f(x,t)$ may be expressed as $\partial f(x,t)/\partial x$ for the rate of change of function $f(x,t)$ with respective to variable x, and $\partial f(x,t)/\partial t$ for the rate of change of the function $f(x,t)$ with respective to the variable t.

The function $f(x,t)$ may have higher-order derivatives, as described in Section 2.2.5 for functions with only one variable. For instance, $\partial^2 f(x,t)/\partial x^2$ and $\partial^2 f(x,t)/\partial t^2$ are the second-order partial derivatives of the function $f(x,t)$ with variables x and t, respectively. We may express other higher-order partial derivatives of the function $f(x,t)$ analogously.

The function $f(x,t)$ and its partial derivatives of any order can be transformed using the Laplace transform as defined in Equation 6.1 if the variable in the transformation covers the range $(0,\infty)$ and the function $f(x,t)$ is continuous within the specified range of the transformed variable.

Let us denote the Laplace transforms of a function with multiple variables x and t by

$$L_x[f(x,t)] = \int_0^\infty e^{-sx} f(x,t)\,dx = F^*(s,t) \quad \text{with} \quad 0 < x < \infty \tag{6.15}$$

for the Laplace transform with respect to variable x, and

$$L_t[f(x,t)] = \int_0^\infty e^{-st} f(x,t)\,dt = F^*(x,s) \quad \text{with} \quad 0 < t < \infty \tag{6.16}$$

for the Laplace transform with respect to the variable t.

The subscripts attached to the Laplace transform operator (L) in Equations 6.15 and 6.16 denote the variable to be transformed by the Laplace transform.

The expressions for the Laplace transform of partial derivatives of functions may be derived similarly to the way we used for functions with single variables in Section 6.5.1. Thus, we may derive the Laplace transform of function $f(x,t)$ as shown next.

We realize that there can be two separate Laplace transforms for the two independent variables involved in function $f(x,t)$, with one transform on variable x and the other on variable t. For instance, if variable t in function $f(x,t)$ is being transformed according to Equation 6.16, the Laplace transform of the derivative of the function with respect to the other independent variable, x—that is, $\partial f(x,t)/\partial x$—may be expressed by the following integral that defines the Laplace transform as in Equation 6.1:

$$L_t\left[\frac{\partial f(x,t)}{\partial x}\right] = \int_0^\infty e^{-st}\left[\frac{\partial f(x,t)}{\partial x}\right] dt$$

We may observe from this expression that since the integration variable t is independent of the other variable x, in the above integral we may legitimately "factor out" the differentiation $\partial/\partial x$ from this integral; we may thus express the Laplace transform of the partial derivative of function $f(x,t)$ with transformation variable t as

$$\begin{aligned}
L_t\left[\frac{\partial f(x,t)}{\partial x}\right] &= \int_0^\infty e^{-st}\left[\frac{\partial f(x,t)}{\partial x}\right] dt \\
&= \frac{\partial}{\partial x}\int_0^\infty e^{-st} f(x,t)\,dt \\
&= \frac{\partial F^*(x,s)}{\partial x}
\end{aligned} \tag{6.17}$$

where $F^*(x,s)$ is as defined in Equation 6.16.

The same Laplace transform of the partial derivative of the function with respect to variable t will be derived by following a similar procedure to that used with the Laplace transform of ordinary derivatives for single-variable functions in Section 6.5.1.

Thus, letting the integral for the Laplace transform of the partial derivative with respect to the same variable as that to be transformed:

$$\begin{aligned}
I &= \int_0^\infty e^{-st}\left[\frac{\partial f(x,t)}{\partial t}\right] dt \\
&= \int_0^\infty u\,dv = uv\big|_0^\infty - \int_0^\infty v\,du
\end{aligned} \tag{6.18}$$

in which u and v are parts of the integral I. If we let $u = e^{-st}$ we have $du = -se^{-st}\,dt$, and $dv = [\partial f(x,t)/\partial t]$, leading to $v = f(x,t)$.

On substituting these expressions into Equation 6.18, we will have the Laplace transform of a partial derivative of a function with respective to the same variable as the transformed variable:

$$L_t \left[\frac{\partial f(x,t)}{\partial t} \right] = I = e^{-st} f(x,t)|_{t=0}^{\infty} + \int_0^{\infty} s e^{-st} f(x,t) \, dt$$

$$= -f(x,0) + s \int_0^{\infty} e^{-st} f(x,t) \, dt$$

$$= -f(x,0) + sF^*(x,s) \tag{6.19}$$

We may thus express the Laplace transform of partial derivatives by

$$L_t \left[\frac{\partial f(x,t)}{\partial t} \right] = sF^*(x,s) - f(x,0) \tag{6.20}$$

One will observe that the Laplace transform of partial derivatives such as shown in Equation 6.20 is similar to that for ordinary derivatives shown in Equation 6.12.

Likewise, we may show that the Laplace transform of second-order partial derivatives as follows:

$$L_t \left[\frac{\partial^2 f(x,t)}{\partial t^2} \right] = s^2 F^*(x,s) - sF^*(x,s) - \left. \frac{\partial f(x,t)}{\partial t} \right|_{t=0} \tag{6.21a}$$

$$L_x \left[\frac{\partial^2 f(x,t)}{\partial x^2} \right] = \frac{\partial^2 F^*(x,s)}{\partial x^2} \tag{6.21b}$$

Example 6.13

If the Laplace transform of function $\theta(x,t) = xe^{-t}$ is defined as

$$L_t[\theta(x,t)] = \theta^*(x,s) = \int_0^{\infty} e^{-st} \theta(x,t) \, dt \tag{a}$$

find the following Laplace transforms of the derivatives of the function $\theta(x,t)$:

a) $L_t \left[\dfrac{\partial \theta(x,t)}{\partial x} \right]$

b) $L_t \left[\dfrac{\partial^2 \theta(x,t)}{\partial x^2} \right]$

c) $L_t \left[\dfrac{\partial \theta(x,t)}{\partial t} \right]$

d) $L_t \left[\dfrac{\partial^2 \theta(x,t)}{\partial t^2} \right]$

Solution:

We first establish the Laplace transform of the multiple-variable function $\theta(x,t) = xe^{-t}$ using Equation (a) and obtain

$$L_t[\theta(x,t)] = x L_t[e^{-t}] = \frac{x}{s+1} = \theta^*(x,s)$$

We then proceed to determine the four required Laplace transforms of the derivatives of the function $\theta(x,t)$ using Equations (6.21a,b). The following expressions are obtained:

a) $L_t\left[\dfrac{\partial\theta(x,t)}{\partial x}\right] = \dfrac{\partial\theta^*(x,s)}{\partial x} = \dfrac{\partial}{\partial x}\left(\dfrac{x}{s+1}\right) = \dfrac{1}{s+1}$

b) $L_t\left[\dfrac{\partial^2\theta(x,t)}{\partial x^2}\right] = \dfrac{\partial^2\theta^*(x,s)}{\partial x^2} = \dfrac{\partial}{\partial x}\left(\dfrac{1}{s+1}\right) = 0$

c) $L_t\left[\dfrac{\partial\theta(x,t)}{\partial t}\right] = s\theta^*(x,s) - \theta(x,0) = \dfrac{sx}{s+1} - x = -\dfrac{x}{s+1}$

$L_t\left[\dfrac{\partial^2\theta(x,t)}{\partial t^2}\right] = s^2\theta^*(x,t) - s\theta(x,0) - \left.\dfrac{\partial\theta(x,t)}{\partial t}\right|_{t=0}$

d)
$$= \dfrac{s^2 x}{s+1} - sx - (-x)$$
$$= \dfrac{s^2 x}{s+1} - sx + x = \dfrac{x}{s+1}$$

6.6 Solution of Ordinary Differential Equations Using Laplace Transforms

A common application of the Laplace transform in engineering analysis is solving differential equations, both ordinary differential equations (ODEs) and partial differential equations (PDEs).

The principle is that for an ODE involving functions with single variables covering spectra from (0 to ∞), performing a Laplace transform on the ODE will "eliminate" the only variable in the equation. Consequently, the ODE is converted into an algebraic equation $F(s)$ with no "variable" of any function in the resulting expressions after the transformation. The representation of $F(s)$ obtained from that algebraic equation is then inverted to obtain the corresponding function, which is the sought solution of the ODE. The application of Laplace transformation in solving PDEs works on a similar principle, in which the multiple variables involved in the functions may be transformed one variable at the time, followed by sequential transformations of other variables, resulting an algebraic expression for the Laplace transform parameter s in the expression for $F(s)$.

6.6.1 Laplace Transform for Solving Nonhomogeneous Differential Equations

For the use of the Laplace transform for solving either ordinary or partial differential equations, the following two requirements must be met:

1) The variable(s) in the function in the equations must cover the range from (0 to ∞).
2) All appropriate conditions for the differential equations must be specified.

The procedure for using Laplace transformation to solve ODEs will be demonstrated in Example 6.14.

Example 6.14
Solve the following ODE using the Laplace transform technique:

$$\frac{d^2y(t)}{dt^2} + 2\frac{dy(t)}{dt} + 5y(t) = e^{-t}\sin t \qquad 0 \le t \le \infty \tag{a}$$

with the conditions

$$y(0) = 0 \quad \text{and} \quad y'(0) = 1 \tag{b}$$

Solution:
If we define the Laplace transform of the function in Equation (a) or Equation 6.1 to be

$$L[y(t)] = \int_0^\infty y(t)e^{-st}\,dt = Y(s) \tag{c}$$

We first apply the Laplace transform defined in Equation (c) to each term in the ODE of Equation (a):

$$L\left[\frac{d^2y(t)}{dt^2}\right] + 2L\left[\frac{dy(t)}{dt}\right] + 5L[y(t)] = L[e^{-t}\sin t]$$

The Laplace transform of the derivatives in the above expression can be obtained using the recurrence relation shown in Equation 6.14, and using the Laplace transform table in Appendix 1. We will thus have the following expression for the Laplace transform $Y(s)$ of the function $y(t)$ in Equation (a):

$$[s^2 Y(s) - sy(0) - y'(0)] + 2[s Y(s) - y(0)] + 5Y(s)$$

$$= \frac{1}{(s+1)^2 + 1} = \frac{1}{s^2 + 2s + 2}$$

After applying the conditions specified in Equation (b) in the above expression and rearranging and combining terms, we arrive at the following expression for $Y(s)$:

$$Y(s) = \frac{s^2 + 2s + 3}{(s^2 + 2s + 2)(s^2 + 2s + 5)} \tag{d}$$

We need to invert the expression $Y(s)$ in Equation (d) to get the function $y(t)$, the solution of Equation (a). Since $Y(s)$ in Equation (d) is a rational function involving the parameter s, we use the partial fraction method for the inversion. Thus, letting

$$Y(s) = \frac{s^2 + 2s + 3}{(s^2 + 2s + 2)(s^2 + 2s + 5)} = \frac{As + B}{s^2 + 2s + 2} + \frac{Cs + D}{s^2 + 2s + 5}$$

where A, B, C, and D are constants.

We solve for $A = 0$, $B = 1/3$, $C = 0$, and $D = 2/3$. We have thus expressed $Y(s)$ in two simple fractions as

$$Y(s) = \frac{\dfrac{1}{3}}{s^2 + 2s + 2} + \frac{\dfrac{2}{3}}{s^2 + 2s + 5}$$

$$= \frac{1}{3} \cdot \frac{1}{s^2 + 2s + 2} + \frac{2}{3} \cdot \frac{1}{s^2 + 2s + 5}$$

The result of inversion of the $Y(s)$ and thus the solution of the ODE, $y(t)$ is

$$y(t) = L^{-1}[Y(s)]$$

$$= \frac{1}{3}L^{-1}\left[\frac{1}{s^2 + 2s + 2}\right] + \frac{2}{3}L^{-1}\left[\frac{1}{s^2 + 2s + 5}\right]$$

$$= \frac{1}{3}L^{-1}\left[\frac{1}{(s+1)^2 + 1}\right] + \frac{2}{3}L^{-1}\left[\frac{1}{(s+1)^2 + 4}\right]$$

We may find the function $y(t)$ by using Laplace transform tables and exploiting the "shifting property":

$$y(t) = L^{-1}[Y(s)]$$
$$= \frac{1}{3}e^{-t}\sin t + \frac{2}{3}\left(\frac{1}{2}e^{-t}\sin 2t\right)$$
$$= \frac{1}{3}e^{-t}(\sin t + \sin 2t)$$

Laplace transformation may be used in engineering analysis in conjunction with the special functions presented in Section 2.4.2. The Laplace transform is a powerful tool in solving both statically determinate and indeterminate problems in the bending of beams, as will be illustrated in the following three examples.

6.6.2 Differential Equation for the Bending of Beams

Figure 6.4 illustrates a beam subjected to a distributed bending load $W(x)$ per unit length of the beam. The induced deflection in the beam $y(x)$ at location x can be obtained by solving the Euler–Bernoulli equation in the following differential form (Volterra and Gaines, 1971):

$$\frac{d^4y(x)}{dx^4} = \frac{W(x)}{EI} \tag{6.22}$$

where E is the Young's modulus of the beam material and I is the section moment of inertia of the beam cross-section. The product of EI is referred to as the flexural rigidity of the beam structure.

The bending stress normal to the beam cross-section induced by the applied load $W(x)$ can be computed from the local bending moment $M(x)$ obtainable from the local deflection $y(x)$ from the solution of Equation 6.22. The maximum normal bending stress is a critical design criterion for many beam designs, such as illustrated in Examples 1.1 and 1.2 in Chapter 1.

The deflection $y(x)$ of the beam by the applied distribute load $W(x)$ in Equation 6.22 can be related to other relevant quantities in beam bending such as those presented in Equations 2.10a to 2.10c. We will use the Laplace transform to solve the differential equation in Equation 6.22 for beam bending problems that would otherwise be difficult to solve by conventional methods presented in many mechanics of materials books.

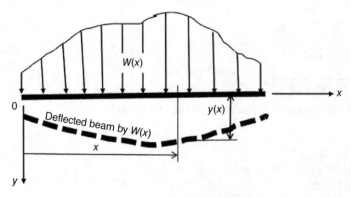

Figure 6.4 A beam deflected under a distributed load.

Example 6.15

Use the Laplace transform method to find the induced deflection function $y(x)$ of a cantilever beam subjected to a uniform distributed load with intensity w_0 on half of the beam span, as illustrated in Figure 6.5.

Solution:

We may find the induced deflection of the beam $y(x)$ by solving the differential equation in Equation 6.22:

$$\frac{d^4y(x)}{dx^4} = \frac{W(x)}{EI} \tag{a1}$$

The applied loading function $W(x)$ in Equation (a) for the present case can be expressed as

$$W(x) = \begin{cases} W_0 & \text{for } 0 \le x \le L/2 \\ 0 & \text{for } L/2 \le x \le L \end{cases}$$

We thus need to solve the following fourth-order differential equation for the deflection of the beam with $y(x)$ in Figure 6.5:

$$\frac{d^4y(x)}{dx^4} = \begin{cases} \dfrac{W_0}{EI} & \text{with } 0 \le x \le L/2 \\ 0 & \text{with } L/2 \le x \le L \end{cases} \tag{a2}$$

with the following end (or boundary) conditions using Equations 2.10a to 2.10c in the following expressions for the present example:

$$y(x)|_{x=0} = y(0) = 0 \tag{b1}$$

$$\frac{dy(x)}{dx}\bigg|_{x=0} = y'(0) = 0 \tag{b2}$$

$$\frac{d^2y(x)}{dx^2}\bigg|_{x=L} = y''(L) = 0 \tag{b3}$$

$$\frac{d^3y(x)}{dx^3}\bigg|_{x=L} = y'''(L) = 0 \tag{b4}$$

The last two end conditions in Equations (b3) and (b4) represent the conditions with zero bending moment and shear forces at the free-hung end of the cantilever beam in Figure 6.5, as described in Equations 2.10b,c in Section 2.2.5.

The loading function $W(x)$ in Figure 6.5 and Equation (a) is a uniformly distributed load covering $0 \le x \le L/2$, and E and I are respectively the Young's modulus of the beam material and the section moment of inertia of the beam.

Legitimate use of the Laplace transform method in solving the differential equation requires the variable x in the function $y(x)$ to cover the range $(0,\infty)$. We thus need to derive the form of Equation (a2) from the variable range $0 \le x \le L$ to the domain $0 \le x < \infty$ in order to use the

Figure 6.5 A cantilever beam subjected to uniform distributed load.

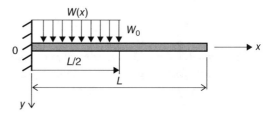

Laplace transform method to solve Equation (a2). A first step that one may take for such conversion is expressing it as the difference between two step functions as presented in Example 2.16 in Chapter 2.

Using the definition of step functions in Section 2.4.2, we may extend the current loading function $W(x)$ from the range $(0,L)$ to $(0,\infty)$ by taking the difference between two step functions as shown in Equation (c):

$$W(x) = W_0[u(x) - u(x - L/2)] \quad \text{with} \quad 0 \le x < \infty \tag{c}$$

where $u(x)$ is the unit step function as defined in Section 2.4.2.

The equation for the deflection function $y(x)$ in Equation (a1) will thus become

$$\frac{d^4 y(x)}{dx^4} = \frac{W_0}{EI}\left[u(x) - u\left(x - \frac{L}{2}\right)\right] \tag{a3}$$

in which we will have $0 \le x < \infty$, and the end conditions in Equations (b1) to (b4) remain unchanged.

We may thus use the Laplace transform method to solve the deflection function $y(x)$ in Equation (a3) with the end conditions in Equations (b1) to (b4).

Let $Y(s)$ be the function $y(x)$ after the Laplace transform, following the definition of the transform as

$$L[f(x)] = \int_0^\infty e^{-sx} f(x)\, dx = Y(s) \tag{d}$$

The following expression (e) for $Y(s)$ will be obtained by using the recurrence relation in Equation 6.14 with the Laplace transform of step functions given in Case 15 in Appendix 1:

$$s^4 Y(s) - s^3 y(0) - s^2 y'(0) - s y''(0) - y'''(0) = \frac{W_0}{RI}\left(\frac{1 - e^{sL/2}}{s}\right) \tag{e}$$

By applying the end conditions specified in Equations (b1) and (b2) into Equation (e) and letting $y''(0) = c_1$ and $y'''(0) = c_2$, with c_1 and c_2 being two arbitrary constants to be determined later, we obtain the expression for $Y(s)$ as follows:

$$Y(s) = \frac{c_1}{s^3} + \frac{c_2}{s^4} + \frac{W_0}{EIs^5}(1 - e^{sL/2}) \tag{f}$$

The induced beam deflection function $y(x)$ may be obtained by inverting the expression $Y(s)$ in Equation (f), resulting in

$$y(x) = L^{-1}[Y(s)] = \frac{c_1 x^2}{2!} + \frac{c_2 x^3}{3!} + \frac{W_0}{EI}\frac{x^4}{4!} - \frac{W_0}{EI}\frac{(x - L/2)^4}{4!} u\left(x - \frac{L}{2}\right) \tag{g}$$

The solution in Equation (g) is valid for the range $(0,\infty)$. The unit step function that appears in the last part of Equation (g) will result in the following solution in the two separate portions in the following expressions:

$$y(x) = \frac{c_1 x^2}{2} + \frac{c_2 x^3}{6} + \frac{W_0 x^4}{24EI} \quad \text{for} \quad 0 \le x \le L/2 \tag{h1}$$

and

$$y(x) = \frac{c_1 x^2}{2} + \frac{c_2 x^3}{6} + \frac{W_0 x^4}{24EI} - \frac{W_0}{24EI}\left(x - \frac{L}{2}\right)^4 \quad \text{for} \quad x > L/2 \tag{h2}$$

We may determine the two arbitrary constants c_1 and c_2 by applying the two remaining end conditions $y''(L) = 0$ and $y'''(L) = 0$ to Equation (h2) to obtain

$$c_1 = \frac{W_0 L^2}{8EI} \quad \text{and} \quad c_2 = -\frac{W_0 L}{2EI}$$

Example 6.16

Use the Laplace transform method to solve Equation 6.22 and find the induced deflection function $y(x)$ of the cantilever beam due to the application of a concentrated force acting at its free end as illustrated in Figure 6.6.

Solution:

There are two issues involved in solving this problem. (1) The loading function in Equation 6.22 is for distributed loads as shown in Figure 6.4 but the current problem is for the beam subjected to a concentrated force P. An distributed loading function $W(x)$ equivalent to a concentrated force needs to be derived. (2) The beam has a finite length L, meaning that the variable x covers a range $(0,L)$, not $(0,\infty)$ as required for the Laplace transform. A conversion is required of the range of coverage of the variable x in the function $y(x)$ from the current range of $(0,L)$ to one of $(0,\infty)$. We will demonstrate that both the aforementioned requirements can be met by using the special "impulsive function" as described in Section 2.4.2.

We first derive the concentrated force P for the current problem equivalent to the distributed load $W(x)$ in the general beam bending situation in Figure 6.4, using the impulsive function described in Section 2.4.2 and letting

$$W(X) = P(x) = -P\delta(x - L) \tag{6.23}$$

where $\delta(x-L)$ is an impulsive function defined in Section 2.4.2. The content of Equation 6.23 is similar to that of Example 2.17.

The loading function $P(x)$ has x valid for the range $(-\infty,+\infty)$ by the definition of the impulsive functions. This range includes the range $(0,\infty)$ and hence legitimizes us in using the Laplace transform method for the solution of the converted differential equation.

We thus have the differential equation for deflection function $y(x)$ in Equation 6.22 expressed in the following equivalent form that is valid for Laplace transformation:

$$\frac{d^4 y(x)}{dx^4} = -\frac{P\,\delta(x - L)}{EI} \quad \text{with} \quad 0 < x < \infty \tag{a}$$

in which E and I are respectively the Young's modulus of the beam material and the section moment of inertia of the beam. The following end conditions need to be satisfied at the fixed end in Equation (b1) and at the free-end in Equation (b2):

$$y(x)|_{x=0} = 0$$
$$\left.\frac{dy(x)}{dx}\right|_{x=0} = y'(0) = 0 \tag{b1}$$

and

$$\left.\frac{d^2 y(x)}{dx^2}\right|_{x=L} = y''(L) = 0$$
$$\left.\frac{d^3 y(x)}{dx^3}\right|_{x=L} = y'''(L) = 0 \tag{b2}$$

Figure 6.6 A cantilever beam subjected to a concentrate force P.

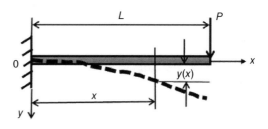

Following the definition of the Laplace transform of the function $y(x)$ in Equation (d) in Example 6.15 and applying the Laplace transform to each term in Equation (a) will result in

$$s^4 Y(s) - s^3 y(0) - s^2 y'(0) - s y''(0) - y'''(0) = -\frac{P}{EI} e^{-Ls} \qquad \text{(c)}$$

We apply the specified condition for the equation with $y(0) = y'(0) = 0$ in Equation (b1) to expression (c). However, we do not have values for $y''(0)$ and $y'''(0)$ for the expression in Equation (c), but we may let $y''(0) = c_1$ and $y'''(0) = c_2$, in which c_1 and c_2 are constants to be determined later. Substituting these values into Equation (c) results in the following expression:

$$Y(s) = \frac{c_1}{s^3} + \frac{c_2}{s^4} - \frac{P}{EI} \frac{e^{-Ls}}{s^4} \qquad \text{(d)}$$

The solution of Equation (a) for the deflection function $y(x)$ of the beam is obtained by inverting the Laplace-transformed $Y(s)$ in Equation (d) to yield

$$y(x) = \frac{c_1 x^2}{2!} + \frac{c_2 x^3}{3!} - \frac{P}{EI} \frac{(x-L)^3}{3!} u(x-L) \qquad \text{(e)}$$

By virtue of the definition of the step function $u(x-L)$ appearing in Equation (e), we may decompose the solution $y(x)$ into two sections:

$$y(x) = \frac{c_1 x^2}{2} + \frac{c_2 x^3}{6} \qquad \text{for} \quad 0 \le x < L \qquad \text{(f1)}$$

$$y(x) = \frac{c_1 x^2}{2} + \frac{c_2 x^3}{6} - \frac{P}{6EI}(x-L)^3 \quad \text{for} \quad x = L \qquad \text{(f2)}$$

By applying the two unused end conditions $y''(L) = 0$ and $y'''(L) = 0$ in Equation (f2), we may determine the constants c_1 and c_2 to be

$$c_1 = -\frac{PL}{EI} \quad \text{and} \quad c = \frac{P}{EI}$$

We thus have the solution for the deflection of the beam $y(x)$ in the following form:

$$y(x) = -\frac{PL}{2EI} x^2 + \frac{P}{6EI} x^3 = -\frac{P}{6EI}(-x^3 + 3Lx^2)$$

Example 6.17

This example will demonstrate the use of Laplace transform technique in dealing with statically indeterminate beam analyses. Indeterminate beams often are called "redundant beams" in the engineering community. Indeterminate beam analysis involves the solution of problems with the number of reactions exceeding the number of equilibrium equations, such as in the bending of a beam with both ends rigidly supported, as illustrated in Figure 6.7.

The beam in Figure 6.7 is indeterminate because there are two unknowns—the vertical reaction and the bending moment—at each end. This total number of four unknowns cannot be determined by the two available conditions obtained from the static equilibrium: $\sum F_y = 0$

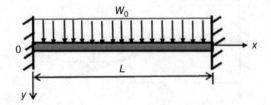

Figure 6.7 Uniformly loaded beam with built-in support at both ends.

and $\sum M_0 = 0$, where F_y includes the applied forces and reactions of the beam and the M_0 are the moments of the forces and reactions at a designated point of interest in the beam such as at the two end supports.

Conventional methods for the solution of determinant beam bending are tedious to execute. However, the Euler–Bernoulli equation for beam bending in Equation 6.22 can be used for both statically determinant and indeterminate beams analyses.

This example demonstrates how Laplace transform can be used in solving problems of this type in conjunction with the special functions described in Section 2.4.2.

Solution:
As in the two preceding examples, we need to convert the range that the function $W(x)$ and the variable x cover from the present $(0,L)$ to $(0,\infty)$ in order to validate the use of the Laplace transform. We may express the loading function to the beam for the Euler–Bernoulli equation as

$$W(x) = W_0[u(x) - u(x - L)] \tag{a}$$

where $u(x)$ and $u(x-L)$ are the unit step functions as defined in Section 2.4.2, and W_0 is the intensity of the uniformly distributed load applied to the beam in Figure 6.7. We will thus have the following differential equation for the induced deflection function of the beam under the applied uniformly distributed load:

$$\frac{d^4y(x)}{dx^4} = \frac{W(x)}{EI} = \frac{W_0}{EI}[u(x) - u(x - L)] \qquad 0 \le x < \infty \tag{b}$$

in which E and I are respectively the Young's modulus of the beam material and the section moment of inertia of the beam.

The situation in Figure 6.7 requires the following end (boundary) conditions:

$$y(x)|_{x=0} = y(0) = 0$$

$$\left.\frac{dy(x)}{dx}\right|_{x=0} = y'(0) = 0 \tag{c1}$$

$$y(x)|_{x=L} = y(L) = 0$$

$$\left.\frac{dy(x)}{dx}\right|_{x=L} = y'(L) = 0 \tag{c2}$$

Let $Y(s)$ be the Laplace transform of function $y(x)$ as defined in Equation (d) in Example 6.15. We will have the following expression for $Y(s)$ after applying the Laplace transform to Equation (b):

$$s^4 Y(s) - s^3 y(0) - s^2 y'(0) - s y''(0) - y'''(0) = \frac{W_0}{EI}\left(\frac{1 - e^{sL}}{s}\right)$$

As we did in Examples 6.15 and 6.16, after substituting the end conditions in Equation (c1) and letting $y''(0) = c_1$ and $y'''(0) = c_2$, where c_1 and c_2 are constants, we will have the expression of $Y(s)$ as

$$Y(s) = \frac{W_0}{EI}\frac{1 - e^{sL}}{s^5} + \frac{c_1}{s^3} + \frac{c_2}{s^4} \tag{d}$$

The solution for the deflection function $y(x)$ in Equation (a) is obtained by inverse Laplace transformation of $Y(s)$ in Equation (d), resulting in

$$y(x) = L^{-1}\left(\frac{W_0}{EI}\frac{1}{s^5}\right) - L^{-1}\left(\frac{W_0}{EI}\frac{e^{sL}}{s^5}\right) + c_1 L^{-1}\left(\frac{1}{s^3}\right) c_2 L^{-1}\left(\frac{1}{s^4}\right)$$

$$= \frac{W_0}{EI}\frac{x^4}{4!} - \frac{W_0}{EI}\frac{1}{4!}u(x - L) + \frac{c_1 x^2}{2!} + \frac{c_2 x^3}{3!} \tag{e}$$

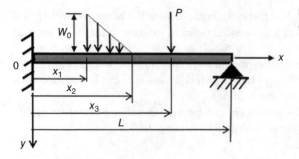

Figure 6.8 Bending of beams subjected to complicated loading and end conditions.

The solution in Equation (e) is applicable to the range $0 \leq x < \infty$, but our interest is in the range $0 \leq x \leq L$. The unit step function $u(x-L)$ appearing in the second term in Equation (e) will automatically drop out for $x > L$, according to the definition of the step function, which will give the solution in Equation (e) truncated by the term associated with this unit step function. Consequently, we will have the solution for the deflection function $y(x)$ as

$$y(x) = \frac{W_0}{EI} \frac{x^4}{24} + \frac{c_1 x^2}{2} \frac{c_2 x^3}{6} \tag{f}$$

The two arbitrary constants c_1 and c_2 in Equation (f) can be determined from the end conditions in Equation (c2), which yield

$$c_1 = \frac{W_0 L^2}{12EI} \quad \text{and} \quad c_2 = -\frac{W_0 L}{2EI}$$

We thus have the solution for the deflection function of the beam bending in Figure 6.7:

$$y(x) = \frac{W_0}{24EI} x^2 (x - L)^2 \tag{g}$$

As we have learned from Section 2.2.5, the bending moment $M(x)$ and shear force $V(x)$ distributions along the beam induced by the uniform distributed load W_0 may be obtained by The derivatives of the deflection function obtained in Equation (g) as

$$M(x) = EI \frac{d^2 y(x)}{dx^2} \quad \text{and} \quad V(x) = \beta \frac{d^3 y(x)}{dx^3}$$

in which $y(x)$ is as in Equation (g) and β is a constant. The normal bending stress distribution along the beam length can be computed from the bending moment distribution $M(x)$ obtained from the above analysis.

Similar approaches may be followed for other beam bending analyses involving complicated loading and end conditions, such as that illustrated in Figure 6.8.

6.7 Solution of Partial Differential Equations Using Laplace Transforms

We learned how to transform partial derivatives in Section 6.5.2 and derived expressions for the Laplace transform parameter; that is, $F^*(x,s)$ for the function $f(x,t)$ with variable t being transformed to the s-parameter domain, as defined in Equations 6.15 and 6.16) Mathematical expressions for partial derivatives were derived in the forms

$$L_t \left[\frac{\partial f(x,t)}{\partial x} \right] = \frac{\partial F^*(x,s)}{\partial x} = \frac{dF^*(x,s)}{dx} \tag{6.17}$$

and

$$L_t \left[\frac{\partial f(x,t)}{\partial t} \right] = sF^*(x,s) - f(x,0) \tag{6.20}$$

We notice that the Laplace transform of partial derivatives such as that expressed in Equations 6.19–6.21b follows a similar form to the recurrent relations in Equation 6.14 derived for the Laplace transform of ordinary derivatives.

We may thus use the Laplace transform method to solve partial differential equations following similar procedures as to those described in Section 6.6.

Example 6.18

Solve the following partial differential equation using the Laplace transform method:

$$\frac{\partial U(x,t)}{\partial x} = 2 \frac{\partial U(x,t)}{\partial t} + U(x,t) \qquad \text{with} \quad 0 \le t < \infty \tag{a}$$

with the initial and boundary conditions

$$U(x,t)|_{t=0} = U(x,0) = 6e^{-3x} \tag{b}$$

and the function $U(x,t)$ exists at $x > 0$ and $t > 0$

Solution:

The Laplace transform of the function $U(x,t)$ is

$$L_t[U(x,t)] = \int_0^\infty e^{-st} U(x,t)\, dt = U^*(x,s) \tag{c}$$

From this and Equations 6.17 and 6.20, we have

$$L_t \left[\frac{\partial U(x,t)}{\partial x} \right] = \frac{dU^*(x,s)}{dx}$$

$$L_t \left[\frac{\partial U(x,t)}{\partial t} \right] = -U(x,0) + sU^*(x,s)$$

The application of Laplace transformation to all terms in Equation (a) will result in the following first-order differential equation for $U^*(x,s)$:

$$\frac{dU^*(x,s)}{dx} = 2[-U(x,0) + sU^*(x,s)] + U^*(x,s)$$

With the initial condition $U(x,0) = 6e^{-3x}$ in Equation (b), we establish the differential equation of $U^*(x,s)$ as

$$\frac{dU^*(x,s)}{dx} = 2[-6e^{-3x} + sU^*(x,s)] + U^*(x,s) \tag{d}$$

The solution of the first order differential equation in Equation (d) is

$$U^*(x,s) = \frac{6}{s+2} e^{-3x} \tag{e}$$

We will obtain the solution $U(x,t)$ in Equation (a) by inverting the Laplace transform expression $U(x,s)$ in Equation (e) to obtain

$$U(x,t) = L_t^{-1}[U^*(x,s)] = L_t^{-1} \left[\frac{6}{s+2} e^{-3x} \right] = 6e^{-2t-3x}$$

Example 6.19

Solve the following partial differential equation using the Laplace transform method:

$$\frac{\partial^2 U(x,t)}{\partial x^2} = \frac{\partial U(x,t)}{\partial t} \quad \text{with} \quad 0 < x < \infty \quad \text{and} \quad 0 < t < \infty \tag{a}$$

with the initial condition (IC)

$$U(x,0) = 0 \tag{b1}$$

and the boundary condition (BC)

$$U(0,t) = 1 \tag{b2}$$

Solution:

We will learn in Chapter 9 that Equation (a) is often referred to as a "diffusion equation" in which the function $U(x,t)$ represents the concentration of a medium in a diffusion process. In other cases, Equation (a) represents transient heat conduction in a solid, with $U(x,t)$ being the temperature in the solid.

Because the variables x and t in the function $U(x,t)$ are not bounded in space and time, as shown in Equation (a), we may perform Laplace transformation of this function with either of these variables. However, we will perform the following computations based on Laplace transformation on the variable t. Consequently, the definition of the Laplace transform in Equation (c) in Example 6.18 is adopted in this example.

Applying the Laplace transform on both sides of Equation (a) results in

$$\frac{\partial^2 U^*(x,s)}{\partial x} = sU^*(x,s) - U(x,0)$$

Substituting the IC in Equation (b1) into the above equation, we get

$$\frac{\partial^2 U^*(x,s)}{\partial x} = sU^*(x,s) \tag{c}$$

We recognize that Equation (c) is a second-order ordinary differential equation in the remaining variable x. Consequently, we may express it in the following form:

$$\frac{d^2 U^*(x,s)}{dx^2} - sU^*(x,s) = 0 \tag{d}$$

The BC of Equation (a) is specified to be $U(0,t) = 1$ in Equation (b2). This condition needs to be related to the condition for Equation (d) by transforming it using the same Laplace transform, as follows:

$$L_t[U(0,t)] = L_t(1) = \int_0^\infty e^{-st}(1)\,dt = \frac{1}{s} \quad \text{at} \quad x = 0$$

We thus have a condition for Equation (d):

$$U^*(0,s) = \frac{1}{s} \tag{e}$$

The solution of the second-order differential equation in Equation (d) is

$$U^*(x,s) = c_1 e^{\sqrt{s}\,x} + c_2 e^{-\sqrt{s}\,x} \tag{f}$$

where c_1 and c_2 in Equation (f) are arbitrary constants to be determined by the explicit condition in Equation (e) and another implicit condition that will be described as follows.

We realize from Equation (f) that $x \to \infty$ will lead to $U^*(x,s) \to \infty$, and hence $U(x,t)|_{x\to\infty} \to \infty$, which is not a realistic solution for problems of this nature in engineering analysis. The only way that this situation may be avoided is to set the constant $c_1 = 0$ in Equation (f). By doing that, the solution of $U^*(x,s)$ in Equation (f) is reduced to

$$U^*(x, s) = c_2 e^{-\sqrt{s}x} \tag{g}$$

The remaining constant c_2 may be determined by the condition given in Equation (e) with $c_2 = 1/s$. We thus have the solution $U^*(x,s)$ as

$$U^*(x, s) = \frac{1}{s} e^{-\sqrt{s}x} \tag{h}$$

The solution of Equation (a) is obtained by the inverse of the Laplace transform of $U^*(x,s)$ in Equation (h), resulting in

$$U(x, t) = L_t^{-1}[U^*(x, s)] = L_t^{-1}\left(\frac{e^{-\sqrt{s}x}}{s}\right) = \text{erfc}\left(\frac{x}{2\sqrt{t}}\right) \tag{j}$$

From Laplace transform tables (Zwillinger, 2003):

$$L^{-1}\left[\frac{1}{s}e^{-\sqrt{s}k}\right] \rightarrow f(t) = \text{erfc}\left(\frac{k}{2\sqrt{t}}\right)$$

The complementary error function erfc(u) in Equation (j) has the form

$$\text{erfc}\left(\frac{x}{2\sqrt{t}}\right) = \frac{2}{\sqrt{\pi}} \int_{x/(2\sqrt{t})}^{\infty} e^{-\varphi^2}\, d\varphi$$

Both the complementary error function and the error function are described in Section 2.4.1.

6.8 Problems

6.1 Find the Laplace transform of each function by direct integration:
a) $f(t) = e^{3t}$
b) $f(t) = 4t^2 - 3$
c) $f(t) = \sinh 2t$

6.2 Use the shift property and the Laplace transform table in Appendix 1 to find the Laplace transform of each function:
a) $f(t) = 3te^{3t}$
b) $f(t) = e^{-2t}\cos 4t$
c) $f(t) = 4e^{-2t}\cosh t$

6.3 Use the Laplace transforms table in Appendix 1 to find the function $f(t)$ corresponding to each expressions of Laplace transform parameter:

a) $F(s) = \frac{1}{s}\left(\frac{2}{s^2} + \frac{1}{s} - 2\right)$

b) $F(s) = \frac{3s + 1}{s^2 - 4s - 5}$

c) $F(s) = \frac{4se^{-2\pi s}}{s^2 + 2s + 5}$

d) $F(s) = \dfrac{2s + 3}{(s^2 + 4s + 13)^2}$

6.4 Use the convolution theorem to find the inverse Laplace transform of each of the following expressions:

a) $F(s) = \dfrac{1}{(s^2 + a^2)^2}$ in which a is a constant

b) $F(s) = \dfrac{1}{s^2(s^2 - 1)}$

c) $F(s) = \dfrac{1}{s^2(s^2 + \omega^2)}$ in which ω is a constant

6.5 Use the shifting property and unit step functions to find the Laplace transforms of the following functions, with graphical illustrations of each function:

a) $f(t) = u_2(t)$
b) $f(t) = u_4(t) \sin \pi t$
c) $f(t) = t/2 - (t/2)u_4(t)$
d) $f(t) = \sin t$ with $0 < t < \pi$ and $f(t) = \sin 2t$ with $\pi < t$

6.6 Solve the following ordinary differential equations using the Laplace transform method:

a) $\dfrac{d^2y(t)}{dt^2} + 4y(t) = 0$ with $y(0) = 0$ and $\left.\dfrac{dy(t)}{dt}\right|_{t=0} = 10$

b) $\dfrac{d^2y(t)}{dt^2} + 4y(t) = 2\cos t$ with $y(0) = 0$ and $\left.\dfrac{dy(t)}{dt}\right|_{t=0} = 0$

c) $\dfrac{d^2y(t)}{dt^2} - 2\dfrac{dy(t)}{dt} - 8y(t) = 0$ with $y(0) = 1$ and $\left.\dfrac{dy(t)}{dt}\right|_{t=0} = 0$

with $0 < t < \infty$ for all the equations.

6.7 Use the Laplace transform to find the deflection function $y(x)$ of the beam subjected to partial uniformly distributed load as shown in Figure 6.9.

6.8 Use the Laplace transform method to find the deflection function $y(x)$ of the beam subject to variable distributed load as illustrated in Figure 6.10.

6.9 Solve the amplitude of vibration $y(t)$ of the systems illustrated in Figure 6.11 using the Laplace transform method. In the system, the damping coefficient, $c = 40$ kg/s and the applied force with function $F(t) = 10 \sin 6t$ newtons. The mass vibrates from rest

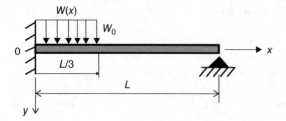

Figure 6.9 Simple beam subjected to partial uniformly distributed load.

Figure 6.10 Simple beam subjected to variable distributed load.

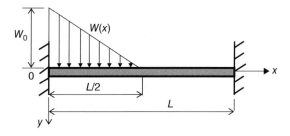

Figure 6.11 A damped mechanical vibration.

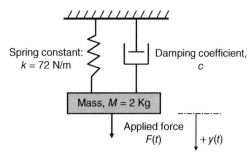

position; that is, $y(0) = 0$ and $y'(0) = 0$. The differential equation for the amplitude $y(t)$ has the form:

$$M\frac{d^2y(t)}{dt^2} + c\frac{dy(t)}{dt} + ky(t) = F(t)$$

7

Application of First-order Differential Equations in Engineering Analysis

Chapter Learning Objectives

- Learn to solve typical first-order ordinary differential equations of both homogeneous and nonhomogeneous types with or without specified conditions.
- Learn the definitions of essential physical quantities in fluid mechanics analyses.
- Learn the Bernoulli equation relating the driving pressure and the velocities of fluids in motion.
- Learn to use the Bernoulli's equation to derive differential equations describing the flow of noncompressible fluids in large tanks and funnels of different geometries.
- Learn how to find time required to drain liquids from containers of given geometry and dimensions.
- Learn the Fourier law of heat conduction in solids and Newton's cooling law for convective heat transfer in fluids.
- Learn how to derive differential equations to predict times required to heat or cool small solids by surrounding fluids.
- Learn to derive differential equations describing the motion of rigid bodies under the influence of gravitation.

7.1   Introduction

As we have learned in Section 2.5, differential equations are equations that involve "derivatives." They are used extensively in mathematical modeling of engineering and physical problems.

There are generally two types of differential equations used in engineering analysis:

1) *Ordinary differential equations (ODEs):* Equations with functions that involve only one variable and "ordinary" derivatives as described in Section 2.2.5.
2) *Partial differential equations (PDEs):* Equations with functions that involve more than one variable and "partial" derivatives as described in Section 2.2.5.

It also explained in Section 2.5 that differential equations are derived from relevant laws of physics. Laws of physics used frequently in mechanical engineering analyses include

- Newton's laws for statics, dynamics, and kinematics of solids.
- Fourier's law for heat conduction in solids.
- Newton's cooling law for convective heat transfer in fluids.

Applied Engineering Analysis, First Edition. Tai-Ran Hsu.
© 2018 John Wiley & Sons Ltd. Published 2018 by John Wiley & Sons Ltd.
Companion Website: www.wiley.com/go/hsu/applied

- Fick's law for diffusion of dissimilar substances in mixing.
- Bernoulli's principle for fluids in motion.

The three fundamental laws of physics that most engineering analyses involve are

- The law of conservation of mass.
- The law of conservation of energy.
- The law of conservation of momentum.

These laws serve as the bases for deriving all other laws of physics for specific engineering disciplines.

In this chapter, we will review methods commonly used to solve first-order differential equations. We will find that although they appear simple in form these differential equations can be used to solve a number of engineering problems in fluid mechanics, heat transfer, and kinematics of rigid bodies.

7.2 Solution Methods for First-order Ordinary Differential Equations

Solution methods for first-order ordinary differential equations are taught in freshman and sophomore year math courses in most engineering institutions. In this section we will refresh the solution methods that are most commonly used.

7.2.1 Solution Methods for Separable Differential Equations

Equation 7.1 shows a typical first-order differential equation, for which the sought solution is $u(x)$:

$$h(u)\frac{du(x)}{dx} = g(x) \tag{7.1}$$

in which $h(u)$ and $g(x)$ are given functions.

We may solve Equation 7.1 by rearranging the terms in the following form:

$$h(u)\,du = g(x)\,dx$$

We observe the following about the above expression: the left-hand side (LHS) involves a function of u (or constants) only, and the right-hand side (RHS) consists of a function of the variable x (or constants) only. Such a separation of functions with u and the variable x on different sides of the equation allows us to obtain the following solution of $u(x)$ by integrating both sides of the equality:

$$\int h(u)\,du = \int g(x)\,dx + c \tag{7.2}$$

where c is the integration constant, to be determined according to the specified conditions for the problem.

Example 7.1

Solve the following first-order ordinary differential equation:

$$x\frac{du(x)}{dx} + u^2 = 4 \tag{a}$$

Solution:

Rearrange the terms in Equation (a) into the separated form with the LHS involving the function $u(x)$ and the RHS involving the variable x:

$$\frac{du}{4 - u(x)^2} = \frac{dx}{x}$$

The solution of this equation, after integration on both sides, is

$$\int \frac{du}{4 - u^2} = \int \frac{dx}{x} + c \tag{b}$$

where c is an arbitrary constant of integration.

We may either use integration tables available in mathematical handbooks, such as in Zwillinger (2003), or use an electronic calculator to perform the integrations in Equation (b) to give

$$\frac{1}{4} \ln \frac{2 + u}{2 - u} = \ln x + c$$

and the solution u in the above expression may be rewritten as

$$u(x) = \frac{2(c'x^4 - 1)}{c'x^4 + 1}$$

with c' being another arbitrary constant to be determined by the specified conditions.

7.2.2   Solution of Linear, Homogeneous Equations

A typical form of this type of equations is

$$\frac{du(x)}{dx} + p(x)\,u(x) = 0 \tag{7.3}$$

where $u(x)$ is the sought solution and $p(x)$ is a function of variable x. Upon rearranging the terms, we may express Equation 7.3 in the form

$$\frac{du}{u} = -p(x)\,dx$$

Integrating both sides of the above equation leads to

$$\int \frac{du}{u} = -\int p(x)\,dx + c$$

in which c is an integration constant. After a further integration, we have the solution $u(x)$ in logarithmic form:

$$\ln u(x) = -\int p(x)\,dx + c$$

where c is the integration constant to be determined by the prescribed conditions of the particular problem.

The exact form of the sought solution can be derived by applying exponential operator to both sides of the above expression, which leads to the following for the solution $u(x)$:

$$|u(x)| = e^{-\int p(x)\,dx + c} = e^c e^{-\int p(x)\,dx}$$

$$= \frac{e^c}{e^{\int p(x)\,dx}} = \frac{e^c}{F(x)} \tag{7.4}$$

$$= \frac{K}{F(x)}$$

in which the function $F(x)$ has the form

$$F(x) = e^{\int p(x)\,dx} \tag{7.5}$$

and K is a constant.

The function $F(x)$ in Equation 7.5 is referred to as the integration factor.

Example 7.2

Solve the following first-order ordinary differential equation:

$$\frac{du(x)}{dx} = u(x)\sin x \tag{a}$$

Solution:

We may rearrange the terms in Equation (a) in the following way:

$$\frac{du(x)}{dx} - (\sin x)u(x) = 0$$

Comparing the above equation with Equation 7.3, we have $p(x) = -\sin x$, which leads to the integral

$$\int p(x)\,dx = \cos x$$

and hence we have

$$F(x) = e^{\int p(x)\,dx} = e^{\cos x}$$

from Equation 7.5.

The solution of the differential equation in Equation (a) is available in Equation 7.4, or in the form

$$|u(x)| = \frac{e^c}{F(x)} = Ke^{-\cos x}$$

where K is an arbitrary constant to be determined according to the appropriate prescribed conditions.

7.2.3 Solution of Linear, Nonhomogeneous Equations

The differential equation

$$\frac{du(x)}{dx} + p(x)u(x) = g(x) \tag{7.6}$$

is classified as a nonhomogeneous equation by virtue of having a function $g(x)$, or sometimes a real number, on the right-hand side of the equation.

The solution of the nonhomogeneous ordinary differential equation in Equation 7.6 is

$$u(x) = \frac{1}{F(x)}\int F(x)\,g(x)\,dx + \frac{K}{F(x)} \tag{7.7}$$

in which $F(x) = e^{\int p(x)\,dx}$ as shown in Equation 7.5 is called the "integration factor," and K is the integration constant.

Example 7.3

Solve the following first-order nonhomogeneous differential equation:

$$x^2\frac{du(x)}{dx} + 2u(x) = 5x$$

Solution:

By rearranging the terms we obtain

$$\frac{du(x)}{dx} + \frac{2}{x^2}u(x) = \frac{5}{x} \tag{a}$$

We may compare the terms in Equation (a) with those in the typical form in Equation 7.6 and find

$$p(x) = \frac{2}{x^2} \quad \text{and} \quad g(x) = \frac{5}{x}$$

The integration factor in Equation 7.5 is

$$F(x) = e^{\int p(x)\,dx} = e^{\int (2/x^2)\,dx} = e^{-2/x}$$

The solution of the Equation (a), $u(x)$, is obtained by substituting the above integration factor into Equation 7.7:

$$u(x) = \frac{1}{F(x)}\int F(x)g(x)\,dx + \frac{K}{F(x)}$$

$$= \frac{1}{e^{-2/x}}\int e^{-2/x}\left(\frac{5}{x}\right)dx + \frac{K}{e^{-2/x}}$$

$$= \frac{5}{e^{-2/x}}\int x^{-1}e^{-2/x}\,dx + Ke^{2/x}$$

The integration constant K in the solution of this ODE is determined according to the specified conditions, as will be demonstrated in Example 7.4.

Example 7.4

Solve the following differential equation:

$$\frac{du(x)}{dx} + 2u(x) = 2 \tag{a}$$

with the condition

$$u(0) = 2 \tag{b}$$

Comparing Equation (a) with the typical form in Equation 7.6, we have $p(x) = 2$ and $g(x) = 2$. Thus, from Equation 7.5, we have

$$F(x) = e^{\int p(x)\,dx} = e^{\int 2\,dx} = e^{2x}$$

Following Equation 7.7, we have the solution of Equation (a) as

$$u(x) = \frac{1}{F(x)}\int F(x)g(x)\,dx + \frac{K}{F(x)}$$

$$= \frac{1}{e^{2x}}\int (e^{2x})(2)\,dx + \frac{K}{e^{2x}}$$

$$= 1 + \frac{K}{e^{2x}} \tag{c}$$

The integration constant K may be determined using the specified condition in Equation (b) with $u(x) = 2$ at $x = 0$. Thus, substituting this condition into Equation (c), we get

$$1 + \frac{K}{e^{2x}}\bigg|_{x=0} = 2$$

from which we may obtain $K = 1$.

Consequently, we have the complete solution of $u(x)$ in Equation (a):

$$u(x) = 1 + \frac{1}{e^{2x}} = 1 + e^{-2x}$$

7.3 Application of First-order Differential Equations in Fluid Mechanics Analysis

Differential equations have been derived to model a variety of fluid dynamics problems that relate the motion of compressible or incompressible fluids subjected to various driving forces. In this section, we will demonstrate such applications in relatively simple cases that involve the flow of incompressible fluids, in particular, the flow of fixed amount of liquids: for example, water in containers of specific geometry, such as pipes and closed or open channels, and the time required to drain liquid contained in tanks or reservoirs or funnels. These problems are often related to the food and chemical processing industries, and in many cases to semiconductor manufacturing, in which large quantity of water or other solvents are used in fabrication processes. First, we will review some fundamental concepts from which appropriate differential equations are derived.

7.3.1 Fundamental Concepts

Fluids differ from solids in the fact that fluids have mass and volume but no shape. The shape of a fluid is determined by how the fluid is contained. The motion of fluids is governed by the containers in which the fluids flow, and the flow of fluids in containers is maintained by driving pressure or by gravitational force.

The following terms are frequently used in modeling the physical behavior of fluid flowing in a container, such as a conduit as illustrated in Figure 7.1.

The total mass flow: $\qquad\qquad Q = \rho A v \, \Delta t \qquad$ (kg) $\qquad\qquad\qquad$ (7.8a)

Total mass flow rate: $\qquad\qquad \dot{Q} = \frac{Q}{\Delta t} = \rho A v \qquad$ (kg/s) $\qquad\qquad$ (7.8b)

Total volumetric flow rate: $\qquad \dot{V} = \frac{\dot{Q}}{\rho} = A v \qquad$ (m^3/s) $\qquad\qquad$ (7.8c)

Total volumetric flow: $\qquad\quad V = \dot{V} \, \Delta t = A v \, \Delta t \qquad$ (m^3) $\qquad\qquad$ (7.8d)

In the above expressions, ρ is the mass density of fluid (kg/m^3), A is the cross-sectional area (m^2), v is the velocity (m/s), and Δt is the duration of fluid flow (s). The base units associated with the above quantities are kilograms (kg), meters (m), and seconds (s).

If the velocity in a moving fluid varies with time—that is, $v = v(t)$—then the change of volumetric flow becomes

$$\Delta V = A v(t) \, \Delta t \qquad\qquad\qquad\qquad (7.9)$$

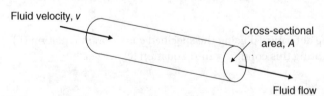

Fluid velocity, v

Cross-sectional area, A

Fluid flow

Figure 7.1 Fluid flow in a conduit.

7.3.2 The Bernoulli Equation

The kinematics of a moving fluid can best be described by the Bernoulli equation. Let us look at the situation illustrated in Figure 7.2, in which a fluid flows along a path of a streamline from state 1 to state 2. The subscripts 1 and 2 associated with the terms representing the pressure (P), velocity (v) and elevation (y) in Figure 7.2 denote the fluid in states 1 and 2, respectively.

On the basis of the law of conservation of energy (the same physical principle on which the first law of thermodynamics is derived), the following Bernoulli equation was derived:

$$\frac{v_1^2}{2g} + \frac{p_1}{\rho g} + y_1 = \frac{v_2^2}{2g} + \frac{p_2}{\rho g} + y_2$$

The Bernoulli equation can be used to model the flow of liquids in a large reservoir with the approximations described below; from it we may derive the exit velocity v_2 of the liquid from a large reservoir or tank as illustrated in Figure 7.3.

By rearranging the terms in the above Bernoulli's equation, we can obtain

$$\frac{v_1^2 - v_2^2}{2g} + \frac{p_1 - p_2}{\rho g} + (y_1 - y_2) = 0 \tag{7.10}$$

But since the pressures at liquid levels in the reservoir (p_1) and the fluid at the exit (p_2) is not much different (approximately equal to atmospheric pressure), and the velocity of the flowing liquid at the top of the reservoir at stage 1 is much lower than that at the exit located at the bottom of the reservoir in state 2, we may thus approximate the velocity of the fluid at this stage

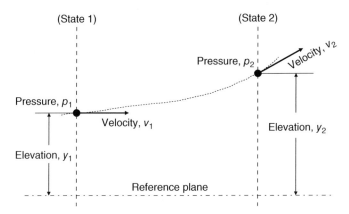

Figure 7.2 Fluid in motion from state 1 to state 2.

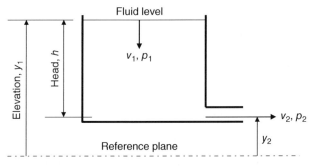

Figure 7.3 Fluid flow in a large reservoir.

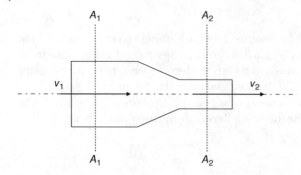

Figure 7.4 Fluid flow through various cross-sections.

at $v_1 \approx 0$ with $v_1 \ll v_2$. We may thus derive the exit velocity from Equation 7.10 by using these approximations in that equation to yield the following useful expression for the exit velocity of the liquid from a large reservoir:

$$-\frac{v_2^2}{2g} + 0 + h = 0$$

in which the head of the liquid level in the reservoir is expressed as $h = y_1 - y_2$.

We may thus express the exit velocity of the liquid from the reservoir as

$$v_2 = \sqrt{2gh} \tag{7.11}$$

This derivation includes an assumption that the friction between the moving liquid and the container wall is negligible.

7.3.3 The Continuity Equation

The continuity equation is frequently used in fluid dynamics analysis for pipelines of varying cross-section. The equation is derived from the principle of conservation of mass.

Consider the states of fluid flow in the situation depicted in Figure 7.4, in which fluid flows from a section of a pipeline with larger cross-sectional area A_1 to a section with smaller cross-sectional area A_2. The law of conservation of mass requires that $\dot{Q}_1 = \dot{Q}_2$, or $\rho A_1 v_1 = \rho A_2 v_2$. Consequently, for an incompressible fluid with constant mass density ρ, we have

$$q = A_1 v_1 = A_2 v_2 \tag{7.12}$$

The rate of volumetric flow q in Equation 7.12 has units of m^3/s.

7.4 Liquid Flow in Reservoirs, Tanks, and Funnels

The following examples illustrate the application of first-order differential equations in modeling the drainage of liquids in cylindrical containers such as large tanks typically used to contain water, chemical solvents, and so on. The container in the first example has the geometry and dimensions shown in Figure 7.5.

As illustrated on the left of Figure 7.5, the tank has a diameter D, with initial water level of h_0. A small drainage tube of diameter d is attached at the bottom of the tank.

We need to find a mathematical model (a differential equation in this case) to describe the drainage of the tank with water leaving the tank through the small drainage tube at the bottom of the tank. The solution for the instantaneous water level in the tank during the drainage process, $h(t)$, can enable prediction of the time required to empty the tank.

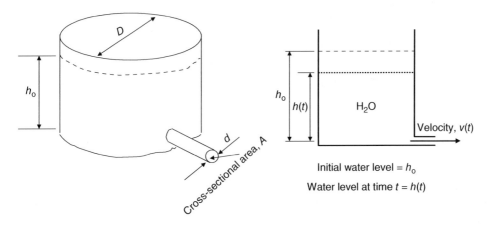

Figure 7.5 Drainage of water from a cylindrical tank.

7.4.1 Derivation of Differential Equations

We recognize that water leaves the tank in "volume," and the departing water volume is related to the "velocity" as expressed in Equation (7.8c). The amount of water that leaves the tank during the time increment or interval Δt can thus be described in words by the physical expression

> (linear distance traveled by the volume of water) × (the cross-sectional area of the passage; i.e., the cross-sectional area A of the drainage pipe)

With the assumption of negligible friction between the moving water and the container wall, the total volume leaving the tank during time Δt can thus be modeled mathematically as $\Delta V = q\,\Delta t = Av(t)\,\Delta t$ in m^3, as given in Equation 7.9, with the duration of water flow Δt in seconds.

From the Bernoulli law as presented in Equation 7.11, we have the exit velocity $v(t)$ from the tank:

$$v(t) = \sqrt{2g\,h(t)} \qquad (\text{m/s})$$

where g is the gravitational acceleration = 9.81 m/s^2.

The volume of water leaving the tank through the exit tube during Δt is thus

$$\Delta V = Av(t)\,\Delta t = \left(\frac{\pi d^2}{4} \right) \sqrt{2g\,h(t)}\,\Delta t$$

We realize at this point that whatever amount of water is leaving the tank through the drainage tube must be equal to the *same* amount of water supplied by the tank in relation to the drop of water level in the tank.

Now, let us examine the reduction of water level in the tank with $\Delta h(t)$ during the same time interval Δt, as illustrated in Figure 7.6.

The equivalent volume reduction in the tank during the time period Δt is equal to the volume of the "disk" of water shown in gray in Figure 7.6, which has a diameter D and is infinitesimally thin with thickness of $\Delta h(t)$. Consequently, the volume of water supplied by the tank can be expressed as

$$\Delta V = -\frac{\pi D^2}{4}\,\Delta h(t)$$

It must be remembered to apply a negative (−) sign in the above expression for ΔV because the water level $\Delta h(t)$ in the tank *reduces* (thus the reduction of volume of water in the tank) with the *increase* of the time variable t.

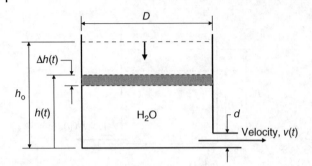

Figure 7.6 Drop of water level in the tank during drainage.

The law of conservation of mass requires the total volume of water leaving the tank during the time increment Δt to be equal to the total volume of water supplied by the tank. We thus arrive at the equality

$$-\frac{\pi D^2}{4}\,\Delta h(t) = \frac{\pi d^2}{4}\,\sqrt{2g\,h(t)}\,\Delta t$$

or in a different form:

$$\frac{\Delta h(t)}{\Delta t} = -[h(t)]^{1/2}\left(\frac{d^2}{D^2}\right)\sqrt{2g}$$

Because the drainage of the tank is a *continuous process*, which implies that $\Delta t \to 0$, we may replace the finite increment denoted by Δ in the above equation by the infinitesimally small increment designated by d as $\Delta t \to 0$. We will then have the following differential equation describing the drainage process:

$$\frac{dh(t)}{dt} = -\sqrt{2g}\left(\frac{d^2}{D^2}\right)\sqrt{h(t)} \tag{7.13}$$

Equation 7.13 can be used to determine the instantaneous water level $h(t)$ in the water tank at any given time t during the drainage process.

7.4.2 Solution of Differential Equations

The first-order ODE in Equation 7.13 for the instantaneous water level $h(t)$ in the tank at any given time t during the drainage can be solved by the technique for separable equations described in Section 7.2.1. Thus, rearranging the terms in Equation 7.13, we have

$$\frac{dh(t)}{\sqrt{h(t)}} = -\sqrt{2g}\left(\frac{d^2}{D^2}\right)dt$$

with the given initial condition $h(t)|_{t=0} = h(0) = h_0$, where h_0 is the given initial water level in the tank.

The solution of this separable equation can be obtained by integrating the terms on both sides:

$$\int h^{-1/2}dh = -\sqrt{2g}\left(\frac{d^2}{D^2}\right)\int dt + c$$

where c is a constant of integration to be determined by the initial conditions.

After carrying out the integration, we will have the instantaneous water level in the tank $h(t)$ as

$$2[h(t)]^{1/2} = -\sqrt{2g}\left(\frac{d^2}{D^2}\right)t + c$$

The integration constant, c, can be determined using the initial condition $h(0) = h_0$, resulting in $c = 2\sqrt{h_0}$.

We thus have the complete solution for the instantaneous water level $h(t)$:

$$h(t) = \left[-\sqrt{\frac{g}{2}} \left(\frac{d^2}{D^2} \right) t + \sqrt{h_0} \right]^2 \tag{7.14}$$

The time required to empty the tank, t_e, may be obtained using the physical situation that $h(t_e) = 0$ in Equation 7.14 to give

$$0 = \left[-\sqrt{\frac{g}{2}} \left(\frac{d^2}{D^2} \right) t_e + \sqrt{h_0} \right]^2$$

We may solve for t_e from the above equation to get

$$t_e = \frac{D^2}{d^2} \sqrt{\frac{2h_0}{g}} \quad \text{seconds} \tag{7.15}$$

Example 7.5

A numerical example of the time required to drain the water from a tank with the following conditions can be illustrated using the expression in Equation 7.15 with the data:

Tank diameter, $D = 12'' = 1$ ft.
Drain pipe diameter, $d = 1'' = 1/12$ ft.
Initial water level in the tank, $h_0 = 12'' = 1$ ft.
Gravitational acceleration, $g = 32.2$ ft/s^2.

The time required to empty the tank is

$$t_e = \left(\frac{1}{\frac{1}{12}} \right)^2 \sqrt{\frac{2 \times 1}{32.2}} = 35.89 \quad \text{seconds}$$

7.4.3 Drainage of Tapered Funnels

Tapered funnels are tools commonly used in the food and beverage industries in filling wine and soda bottles and the like and in other processes in industrial production. In many automatic filling operations, empty bottles have to be filled with beverages or other liquids both accurately and in the minimal amount of time. A similar principle to that used earlier in deriving the first-order differential equations for the drainage of straight cylindrical tanks can be used to represent similar drainage problems involving tapered containers or funnels.

We will illustrate such derivation with funnels of the geometry and dimensions illustrated in Figure 7.7. The mathematical model offers an effective tool for engineers to design a funnel that can fill bottles with predetermined volume of contents in a required filling time.

In Figure 7.7, the tapered cylindrical funnel has a taper angle θ and an initial liquid level H. These two parameters determine the volumetric content of the funnel. The funnel has a small opening at the bottom with a diameter d. For convenience in the subsequent analysis, we assume that the origin of the coordinate y that is used to indicate the instantaneous liquid level of the funnel is located at "0" rather than at the exit "A." This approximation is justifiable if the distance between "0" and "A" as shown in Figure 7.7 is small.

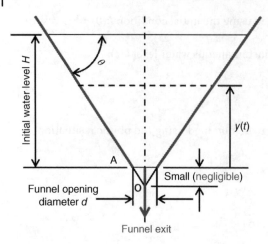

Figure 7.7 Drainage of a Tapered Funnel.

Following a mathematical analysis as in the example in Section 7.4.1 for straight cylindrical tanks, we can make the following statement for the current state of a tapered funnel:

> The total volume of liquid leaving the funnel during interval Δt (ΔV_{exit}) = the total volume of liquid supplied by the funnel during Δt (ΔV_{funnel})

According to Equation 7.9, we may express the total volume of liquid leaving the funnel during time interval Δt as

$$\Delta V_{exit} = A_{exit} v_e \Delta t$$

From Equation 7.11, we have the exit velocity of the liquid (or water in this case) $v_e = \sqrt{2g\,y(t)}$, which leads to

$$\Delta V_{exit} = \frac{\pi d^2}{4} \sqrt{2g\,y(t)}\, \Delta t \tag{7.16}$$

The total volume of water supplied by the funnel during Δt is ΔV_{funnel}, which is the volume of the grey cross-sectional area in Figure 7.8 (or the volume of a "thin water" disk with radius $r(y)$ and thickness Δy) as given by

$$\Delta V_{funnel} = -\{\pi[r(y)]^2\}(\Delta y) \tag{7.17}$$

The equivalence of the two volumetric flows in Equations 7.16 and 7.17 requires a mathematical relation between the instantaneous radius $r(y)$ of the thin "disk" of water in the tapered funnel in Equation 7.17 and linear distance of the liquid level reduction in the funnel $y(t)$ in Equation 7.16. The relationship between $y(t)$ and $r(y)$ may be obtained from the geometry of the funnel, as illustrated in Figure 7.8, in which

$$r(y) = \frac{y(t)}{\tan\theta} \tag{7.18}$$

By substituting the relation in Equation 7.18 into Equation 7.17, and equating the resulting form to ΔV_{exit} in Equation 7.16, we obtain the following equation for the instantaneous liquid level in the funnel, $y(t)$:

$$\frac{\pi d^2}{4} \sqrt{2g\,y(t)}\, \Delta t = -\pi \frac{[y(t)]^2}{\tan^2\theta} \Delta y \tag{7.19}$$

Figure 7.8 Instantaneous radius of the funnel vs. the instantaneous water level.

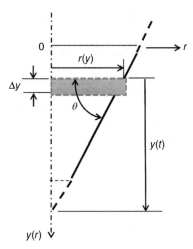

from which we may derive the differential equation for the instantaneous liquid level in a tapered funnel using the condition that

$$\lim_{\Delta t \to 0} \frac{\Delta y}{\Delta t} = \frac{dy(t)}{dt}$$

because of the continuous nature of the drainage process. Consequently, we have the following differential equation for the instantaneous liquid level $y(t)$:

$$\frac{[y(t)]^{3/2}}{\tan^2\theta} \frac{dy(t)}{dt} = -\frac{d^2}{4} \sqrt{2g} \tag{7.20}$$

Example 7.6

Determine the time required to drain a tapered funnel with a taper angle of 45° as shown in Figure 7.9. The funnel contains water with an initial level $H = 150$ mm.

Solution:

According to Equation 7.12, the rate of volumetric flow of water leaving the funnel, q_e (or $\dot{V}$, in ft^3/s) is $q_e = A_e v_{exit}$, in which the cross-sectional area of the funnel at the exit is $A_e = \pi d^2/4$.

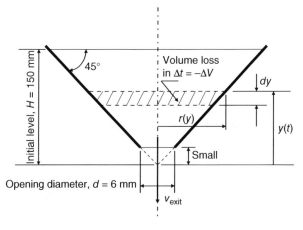

Figure 7.9 Draining of a tapered funnel.

And from the Bernoulli equation and Equation 7.11, we have the exit velocity at the bottom of the funnel as $v_{exit} = \sqrt{2g\,y(t)}$. We thus have the total volumetric flow of water continuously leaving the funnel during time interval, Δt (or dt in a continuous flow process) as

$$\Delta V_e = q_e\Delta t = \frac{\pi d^2}{4}\sqrt{2g\,y(t)}\,\Delta t \tag{a}$$

The water leaving the funnel must be supplied by the water "upstream" in the funnel. Consequently, this water supply can be related to the drop of water level upstream in the funnel as given by the following equation for the loss of volume of water:

$$\Delta V_{lost} = -(\pi r^2)dy = -\pi y^2\,dy \tag{b}$$

Referring to Figure 7.8 with $\theta = 45°$, we can relate the instantaneous radius of the funnel, $r(y)$ to the instantaneous water level $y(t)$ as $r = y/(\tan 45°) = y$.

By the law of conservation of mass, we should have ΔV_e in Equation (a) $= \Delta V_{lost}$ in Equation (b), which leads to the following equality:

$$\frac{\pi d^2}{4}\sqrt{2g\,y(t)}\,dt = -\pi y^2\,dy$$

from which we have the differential equation for determining the instantaneous water level in the funnel $y(t)$ as follows:

$$\frac{y^2}{\sqrt{y}}\frac{dy}{dt} + \frac{d^2}{4}\sqrt{2g} = 0 \tag{c}$$

This differential equation in Equation (c) is separable and yields

$$\frac{y^2}{\sqrt{y}}\,dy = -\frac{d^2}{4}\sqrt{2g}\,dt$$

We may get the solution of Equation (c) by integrating both sides of the separated equation as shown in the following expression:

$$\frac{2}{5}y^{5/2} = -\frac{d^2}{4}\sqrt{2g}\,t + c$$

in which c is the integration constant to be determined by the given initial condition of $y(0) = H$, where H is the specified initial water level in the funnel. The integration constant c can thus be determined as $c = (2/5)H^{5/2}$.

The exact solution to the problem is

$$\frac{2}{5}(y^{5/2} - H^{5/2}) = -\frac{d^2}{4}\sqrt{2g}\,t$$

or

$$[y(t)]^{5/2} = -\frac{5d^2}{8}\sqrt{2g}\,t + H^{5/2} \tag{d}$$

To evaluate the time t_e required to drain (that is, empty) the funnel, one may set $y(t_e) = 0$ in Equation (d), from which we may find the value of t_e:

$$t_e = \frac{8H^{5/2}}{5d^2\sqrt{2g}} \tag{e}$$

We compute $t_e = 87.44$ seconds to drain a funnel with the dimensions shown in Figure 7.9 with initial water level $H = 150$ mm and gravitational acceleration $g = 9810$ mm/s^2.

Example 7.7
Consider a circular funnel made of a straight section at its top and a lower tapered section, as illustrated in Figure 7.10a and with the dimensions indicated in Figure 7.10b. Use the integration

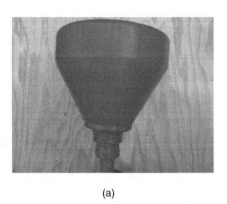
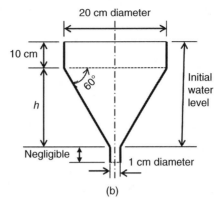

(a) (b)

Figure 7.10 A two-section funnel system. (a) A multisection funnel. (b) The dimensions of a two-section funnel.

method to determine (a) the volume of water it contains when full, and (b) the time required to drain this funnel from its initial level as indicated in Figure 7.10b.

Solution:
We will first establish the initial water level in the funnel system by computing the length of the tapered section of the funnel in Figure 7.10b. This is obtained from the taper angle $\theta = 60°$ and the radius of the circular straight portion of the funnel of 10 cm via the relation $h = 10 \tan 60° = 17.32$ cm. We thus have the initial water level in the funnel as

$$y(t)|_{t=0} = y(0) = H_0 = 10 + 17.32 = 27.35 \text{ cm}$$

in which $y(t)$ is the water level in the funnel, with the origin of the coordinate $y = 0$ located at the lower end of the tapered section of the funnel.

The funnel system consists of two portions, with the straight cylindrical portion (portion I) at the top and the tapered portion (Portion II) as the lower portion of the funnel system. Both sections share the same exit at the bottom of the system.

a) **The volume of the compound funnel** The volume V_1 of the straight section can be obtained using Equation (2.17) as follows:

$$V_1 = \int_0^{10} \pi \left(\frac{20}{2}\right)^2 dy$$
$$= 100\pi y|_0^{10} = 1000\pi$$
$$= 3140 \text{ cm}^3$$

The volume of the tapered portion of the funnel system (V_2) can be computed using Equation 2.16 with the profile of the funnel defined by the x–y coordinates shown in Figure 7.11. We thus compute the volume V_2 as

$$V_2 = \pi \int_0^{17.32} [y(x)]^2 dx$$
$$= \pi \int_0^{17.32} (0.549x + 0.5)^2 dx = 606.3\pi$$
$$= 1903.78 \text{ cm}^3$$

The total volume of the compound funnel is $V = V_1 + V_2 = 5043.78 \text{ cm}^3$.

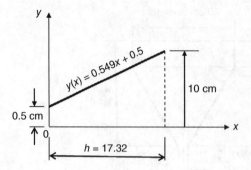

Figure 7.11 Profile of a tapered funnel.

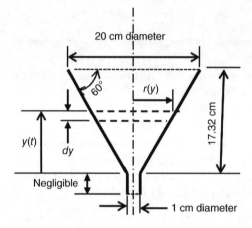

Figure 7.12 Drainage of a tapered section of a compound funnel.

b) **Time required to drain the funnel** Because the funnel is made of two sections of distinct geometry, we will compute the times required to drain each section, from which we may obtain the total time required to drain the entire funnel.

We may use the formula derived for draining a straight cylindrical tank presented in Equation 7.15 for t_{e1} by setting the diameter of the exit hose to be 1 cm as shown in Figure 7.10b. We thus calculate t_{e1} with $D = 20$ cm, $d = 1$ cm, $h_0 = 10$ cm and $g = 981$ cm/s², with the result

$$t_{e1} = \frac{D^2}{d^2}\sqrt{\frac{2h_0}{g}} = \left(\frac{20}{1}\right)^2\sqrt{\frac{2 \times 10}{981}} = 57 \text{ seconds} \tag{a}$$

Finding the time required to drain the lower tapered section of the funnel system requires first the following derivation of the differential equation for the instantaneous water level, as presented in Figure 7.12.

We may use Equation 7.20, derived for a tapered funnel with taper angle θ:

$$\frac{[y(t)]^{3/2}}{\tan^2\theta}\frac{dy(t)}{dt} = -\frac{d^2}{4}\sqrt{2g} \tag{7.20}$$

With $\theta = 60°$, $d = 1$ cm, and $g = 981$ cm/s², as shown in Figure 7.12 for the present problem, we will have the differential equation for the instantaneous water level $y(t)$ in the following form:

$$\frac{[y(t)]^{3/2}}{\tan^2(60°)}\frac{dy(t)}{dt} = -\frac{(1)^2}{4}\sqrt{2 \times 981}$$

or

$$[y(t)]^{3/2}\frac{dy(t)}{dt} = -33.22 \tag{b}$$

The solution of the differential equation in Equation (b) is obtained by integrating both sides of the equation to yield

$$[y(t)]^{5/2} = -83.047t + c$$

where c is the integration constant, to be determined from the given initial condition of $y(0) = 17.32$ cm, leading to $c = (17.32)^{2.5} = 1248.45$, we thus obtain the instantaneous water level in the tapered section of the funnel system as

$$[y(t)]^{5/2} = -83.047t + 1248.45 \tag{c}$$

The physical situation for an empty tapered funnel corresponds to a mathematical model having $y(t_{e2}) = 0$ in Equation (c), from which we may solve for t_{e2} from the following equation:

$$[y(t_{e2})]^{5/2} = 0 = -83.047t_{e2} + 1248.45$$

from which we obtain

$$t_{e2} = \frac{1248.45}{83.047} = 15.03 \text{ seconds}$$

We may thus solve for the total time required to drain the compound funnel:

$$t_e = t_{e1} + t_{e2} = 57 + 15.03 = 72.03 \text{ seconds}$$

Example 7.8

Design a circular funnel with a taper angle of 60° to fill the wine bottle described in Example 2.10 in Chapter 2. The designer will configure the dimensions, as illustrated in Figure 7.13a, and set the radius and the length H of the funnel, and specify the diameter of the exit to be 1.5 cm. Determine the time required to fill the bottle if it has the dimensions given in Figure 2.33.

Solution:

a) **Determine the dimensions of the funnel** We will first establish the volume of wine that can be held in the bottle to be 841.52 cm³ as determined in Example 2.10. The design objective is for the funnel to have the same volume. However, we need to determine the maximum radius R of the funnel in relation to the length H through the given taper angle of the funnel. This is given by tan 60° = H/R, from which we may write

$$R = H/\tan 60° = 0.577H \tag{a}$$

Equation (a) will enable us to determine both R and H for the funnel with a containment volume of 841.52 cm³.

Referring to Figure 7.13b, we may determine the volume of the funnel using Equation 2.17 with $x(y) = R(y) = 0.577y$ as

$$V_f = \pi \int_0^H [R(y)]^2 \, dy = \pi \int_0^H (0.577y)^2 \, dy$$

$$= (0.577)^2 \pi \left.\frac{y^3}{3}\right|_0^H$$

$$= 0.3485H^3$$

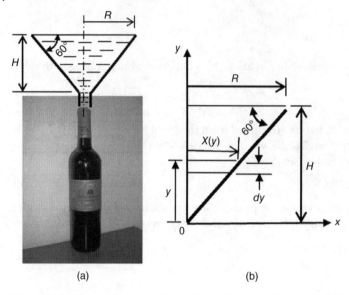

(a)   (b)

Figure 7.13 Filling a wine bottle from a circular funnel. (a) Funnel used for filling a wine bottle. (b) Volume of the tapered cone of the funnel.

But the volume of the funnel is equal to the volume of the bottle; we thus have $V_f = 0.3464H^3 = 841.52$, from which we have that the length of the funnel $H = 13.416$ cm. We may also obtain the diameter D of the funnel to be

$$D = 2R = 2 \times (0.577 \times 13.416) = 15.4821 \text{ cm}$$

b) **Determine the time required to fill the bottle** Because the volume of the bottle is equal to the volume of the funnel in this case, we may observe that the time required to fill the bottle is equal the time required to drain the funnel. Consequently, we may use Equation 7.20, which was derived to determine the required time to drain a tapered funnel. Thus, substituting $\theta = 60°$ and $d = 1.5$ cm into Equation 7.20, we will obtain the differential equation for the instantaneous wine level in the funnel $y(t)$ during the drainage process:

$$\frac{[y(t)]^{3/2}}{(\tan 60°)^2} \frac{dy(t)}{dt} = -\frac{(1.5)^2}{4} \sqrt{2 \times 981}$$

yielding

$$[y(t)]^{3/2} \frac{dy(t)}{dt} = -74.7513 \tag{b}$$

The solution $y(t)$ of Equation (b) may be obtained by integrating both sides of Equation (b):

$$[y(t)]^{2.5} = -186.8783t + C \tag{c}$$

The integration constant c may be determined from the initial condition $y(0) = H = 13.416$ cm in part (a) of the solution, giving $c = 634.9654$. We thus have the solution $y(t)$ in Equation (c) as

$$[y(t)]^{2.5} = -186.8783t + 634.9654 \tag{d}$$

The time t_e required to drain the funnel is equal to the time required to fill the wine bottle in Figure 7.13a. We may thus solve for t_e using Equation (d) to give

$$y(t_e) = -186.8783(t_e) + 634.9654 = 0 \tag{e}$$

By solving for t_e from Equation (e) we obtain the time required to fill the bottle as $t_e =$ 3.42 seconds.

7.5 Application of First-order Differential Equations in Heat Transfer Analysis

In this section, we will use another two laws of physics to derive the differential equations for heat transfer analyses. These are (1) *Fourier's law* for heat conduction in solids and (2) *Newton's cooling law* for heat convection in fluids. We will show that the first-order differential equations derived from these two laws can be used to solve a variety of practical problems, such as in the design of heat spreaders and reliability testing equipment for thermal cycling involving the performance design of refrigerators and cooling chambers and other heat treatment equipment.

7.5.1 Fourier's Law of Heat Conduction in Solids

Let us first look at a simple case in which heat flows from the left face of a solid slab to its right face, as illustrated in Figure 7.14.

The natural phenomenon of that heat flow in a substance is possible only in the presence of temperature gradients, with heat flowing from locations at higher temperature to locations at lower temperature. Consequently, heat will flow from the left side to the right side of the slab if we maintain a state of $T_a > T_b$ with T_a and T_b being the temperature at the left and right faces of the slab, respectively, as in Figure 7.14.

One may well envisage that the total amount of heat flow through the slab, designated as Q, is proportional to following factors of the cross-sectional area A for the heat flow, the temperature difference between the two faces T_a and T_b, and the time t allowing for the heat to flow. However, the total amount of heat flow is inversely proportional to the distance that the heat travels; that is, the thickness of the slab d. Mathematically, one may express the above qualitative correlations in the following form

$$Q \propto \frac{A(T_a - T_b)t}{d}$$

or in an equivalent equation with a proportionality constant k:

$$Q = k\frac{A(T_a - T_b)t}{d} \tag{7.21}$$

where the constant k is the *thermal conductivity* of the solid, with unit Btu/in-s-°F in the traditional system, or W/m-°C in the SI or metric system.

Figure 7.14 Conduction of heat in a solid slab.

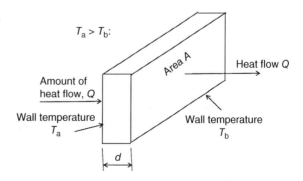

The thermal conductivity k in Equation 7.21 is a material property. The value of k for solids generally increases with the temperature. However, for most engineering materials at realistic operating temperature ranges, the value of k is regarded as constant. It relates to how conductive of heat the material is. Materials handbooks (Avallone *et al.*, 2006; Kreith, 1998) show that gold, silver, copper, and aluminum have superior heat conducting capabilities, with k-values in decreasing order. These materials have higher k-values than many other materials such as steel. Metallic materials are much better heat conducting materials than semiconductve materials such as silicon (Si), silicon carbide (SiC), and silicon dioxide (SiO_2). Ceramics such as aluminum oxide (Al_2O_3) are used as thermal insulating materials because of their very low k-values.

Equation 7.21 provides us with a way to relate the total heat flow in a planar slab with the temperature gradient $\Delta T = T_a - T_b$. A more realistic characterization of heat flow in solids is in terms of the *heat flux q*, defined as the *heat flow per unit area and time*. Thus, from Equation 7.21, we may express the heat flux in the slab as

$$q = \frac{Q}{At} = k\frac{(T_a - T_b)}{d} \tag{7.22}$$

The unit for heat flux q in Equation 7.22 may be expressed as W/m^2 or $J/m^2\text{-s}$, or $N/m\text{-s}$ in the SI unit system.

We will derive Fourier's law of heat conduction, which relates the heat flux and temperature variation in solids. In some ways, this law is analogous to the situation in which the amount of water flowing between two points depends upon the difference of their respective elevations.

7.5.2 Mathematical Expression of Fourier's Law

Let us revisit the situation depicted in Figure 7.14 in which heat flows from the left surface of the slab at a higher temperature T_a to the right surface at a lower temperature T_b. We will use this situation to derive a one-dimensional mathematical form of Fourier's law. The total amount heat flow in the slab may be computed by using Equation 7.21, and the heat flux q transmitted across the two faces of the slab is obtained from Equation 7.22. We further realize that heat flows from the left surface to the right surface because of the positive temperature gradient $\Delta T = T_a - T_b > 0$. A similar positive temperature gradient ΔT must also exist inside the solid slab to allow the passage of heat from the left surface to the right surface. The equation for heat flux transmission across the interior of the solid slab may be derived on the same principle; that equation will be the mathematical form of Fourier's law for heat conduction in solids.

We will derive the equation for Fourier's law of heat conduction in solids by letting the x-coordinate represent the thickness of the slab, as shown in Figure 7.15. The associated

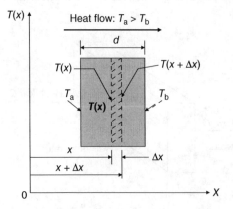

Figure 7.15 Heat flux in a solid slab.

temperature variation across the thickness of the slab is expressed by a function $T(x)$ with x being the direction of heat flow. Let us now "zoom in" to two extremely closely adjacent cross-sections inside the slab located at coordinates x and $(x + \Delta x)$, where Δx is the increment of the coordinate of the cross-section inside the solid slab.

According to Equation 7.22, the heat flux across the two faces separated by the "thickness" Δx can be expressed as

$$q = k\frac{T(x) - T(x + \Delta x)}{\Delta x} = -k\frac{T(x + \Delta x) - T(x)}{\Delta x} \tag{7.23}$$

Since the temperature variation in the slab varies as a *continuous* function $T(x)$ with $\Delta x \to 0$, we have the following relation according to Equation 7.23 for the continuous variation of function $T(x)$ with the variable x:

$$q = q(x) = -k\frac{dT(x)}{dx} \tag{7.24}$$

where the temperature gradient is $\Delta T = T(x + \Delta x) - T(x)$ with $\Delta x \to 0$.

Equation 7.24 is the mathematical form of Fourier's law of heat conduction in one-dimensional heat flow along the coordinate x. One recognizes this law as a first-order differential equation for the rate of change of temperature in a solid in the direction along the positive x-coordinate.

Example 7.9

A 1 meter long metal rod is thermally insulated around its circumference. The terminal temperatures of the rod are measured to be 100°C and 20°C as illustrated in Figure 7.16. Determine the heat flux in the rod for the cases where it is made of copper and of aluminum.

Solution:

We can find from a materials handbook that the thermal conductivities (k) for copper and aluminum are $k_{Cu} = 3.95$ W/cm-°C and $k_{Al} = 2.36$ W/cm-°C, respectively.

The differential equation for the problem can be expressed in a slightly different form from Equation 7.24 as

$$\frac{dT(x)}{dx} = -\frac{q}{k_{Cu}} = -\frac{q}{3.95} \tag{a}$$

for the copper rod. The constant q is the heat flux in the rod.

The appropriate boundary condition is

$$T(0) = 100°C \tag{b}$$

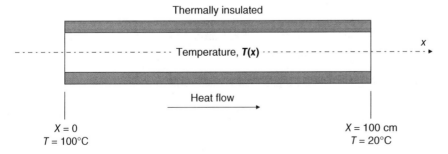

Figure 7.16 Heat flow in a thermally insulated rod.

Solution of the first-order differential equation in Equation (a) is straightforward by integration:

$$T(x) = -\frac{q}{3.95}x + c$$

where c is the integration constant. It can be determined from the specified boundary condition in Equation (b) that $c = 100$. We thus have the temperature distribution in the rod as

$$T(x) = -\frac{q}{3.95}x + 100 \tag{c}$$

Since we have the other condition that $T(100) = 20°C$, we may determine the heat flux, q from Equation (c): $q = 3.16 \text{ W/cm}^2$.

Following the same procedure we find the corresponding heat flux in an aluminum rod to be 1.89 W/cm^2.

Example 7.10

The power supplied in kilowatts (kW) for the heat that is conducted in a solid is proportional to the cross-sectional area for heat flow and the temperature gradient for a given thermal conductivity k (kW/m-°C). For a long, circumferentially insulated rod such as shown in Figure 7.16, we may use the expression in Equation 7.24 for equivalent heat flux (q).

For this example, we have that heat is transferred at the rate of 10 kW at the left end of the rod. Determine the temperature distribution in the rod if the right end of the rod at $x = 2$ m is held constant at 50°C. The cross-sectional area is 1200 mm^2 and the thermal conductivity $k = 100 \text{ kW/m-°C}$.

Solution:

A conductor of circular cross-section receives a heat supply of power 10 kW at its left end causing the heat to flow in the rod from the left toward the right, with no heat dissipating from its circumference. The heat flow in the rod in the longitudinal direction induces a temperature gradient along the same direction expressed by $T(x)$. The heat flow per unit time is equal to qA, where q is the heat flux as given in Equation 7.24 and A is the cross-sectional area for heat flow. The situation is graphically depicted in Figure 7.17.

We have the total heat flow in the rod per unit time being

$$Q = qA = -kA\frac{dT(x)}{dx} \tag{a}$$

with the end condition

$$T(x) = 50°C \quad \text{at} \quad x = 2 \text{ m} \tag{b}$$

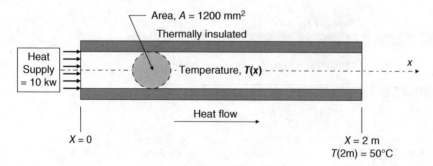

Figure 7.17 Temperature variation in a rod with to heat flow.

From Equation (a) we get the differential equation for the temperature distribution $T(x)$ in the following form:

$$\frac{dT(x)}{dx} = -\frac{Q}{kA} = -\frac{10}{100(1200 \times 10^{-6})} = -83.33\ °C/m \qquad (c)$$

Solving Equation (c) leads to

$$T(x) = -83.33x + c \qquad (d)$$

in which c is the arbitrary integration constant to be determined by the end condition specified in Equation (b), resulting in a value of $c = 216.67$.

We thus have the temperature variation in the rod as

$$T(x) = 216.67 - 83.33x$$

7.5.3 Heat Flux in a Three-dimensional Space

Heat flux as expressed in Equation 7.24 is a vector quantity as defined in Chapter 3. This vector may be graphically illustrated in Figure 7.18. It can be related to the associated temperature gradient by *Fourier's law of heat conduction* as will be shown below.

Referring to Figure 7.18, the heat flux vector in a solid situated in a space defined by a rectangular coordinate system can be expressed as

$$q(\mathbf{r}, t) = -k\nabla T(\mathbf{r}, t) \qquad (7.25)$$

where $\mathbf{r}$ represent the x-, y-, and z-coordinates in a rectangular coordinate system. Note that the same equation occurred as Equation 3.31a in Chapter 3. From Equations 7.22 and 7.24, the unit for heat flux can be W/m², J/m²-s, or N/m-s.

Equation 7.25 is the mathematical expression of the general Fourier's law of heat conduction. Expanding that equation into rectangular coordinates leads to the following expression (7.26) for the magnitude of the resultant heat flux in a solid:

$$q(x, y, z, t) = \sqrt{q_x^2 + q_y^2 + q_z^2} \qquad (7.26)$$

in which

$$q_x = -k_x\frac{\partial T(x, y, z, t)}{\partial x} \qquad (7.27a)$$

$$q_y = -k_y\frac{\partial T(x, y, z, t)}{\partial y} \qquad (7.27b)$$

$$q_z = -k_z\frac{\partial T(x, y, z, t)}{\partial z} \qquad (7.27c)$$

Figure 7.18 Heat flux in a solid.

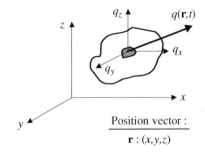

Position vector :

$\mathbf{r} : (x, y, z)$

where q_x, q_y, and q_z are the respective heat flux components in the x-, y-, and z-directions as shown in Figure 7.18. The terms k_x, k_y, and k_z are the respective thermal conductivities in the solid in the x-, y-, and z-directions. For isotropic materials with $k_x = k_y = k_z = k$, one will readily find that Fourier's law in Equation 7.24 is a special case of the general expression in Equation 7.25 for one-dimensional heat conduction in solids.

The negative sign that appears in Equation 7.25 often causes confusion to engineers. This sign appears when the heat flow in the solid is in the same direction as the *outward* normal of the surface. An outward normal direction is the normal line to the surface at which heat crosses but is pointing away from that surface. Figure 7.19 illustrates the appropriate sign to be used in Fourier's law of heat conduction. Example 7.11 will illustrate a rule with which the proper sign for the corresponding heat flux may be assigned. A generalized rule developed by the author for assigning proper signs to the heat flux is offered in the following description.

Two factors determine whether a positive (+ve) or negative (−ve) sign should be used in Fourier's law in Equations 7.25 and 7.27:

1) The sign of the "outward normal **n**" perpendicular to the face of the solid through which the heat flows.
2) The "direction of heat flow," either in the same direction as the outward normal line (**n**), or in the opposite direction.

The situation is described graphically in Figure 7.19.

With the coordinates (x,y) set as shown in Figure 7.19, we perceive that the outward normal **n** of the surface that the heat crosses is "positive"—that is "+n" because both its components along the x- and y-coordinate are in the positive directions. Engineers can use Table 7.1 to assign a +ve or −ve sign to the Fourier law in Equations 7.27.

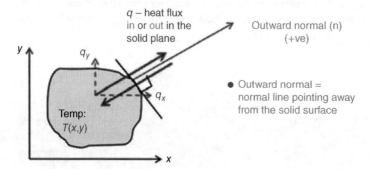

Figure 7.19 Two-dimensional heat flow in a solid.

Table 7.1 Rules for assigning signs in Fourier's law of heat conduction.

Sign of outward normal (n)	q along n?	Sign of q in Fourier's law
+	Yes	−
+	No	+
−	Yes	+
−	No	−

The use of Table 7.1 begins with engineers determining the sign of the outward normal (**n**) to the surface from which the heat flows. The sign applicable to Fourier's law is then determined by the corresponding situations listed in the middle column of the table, with the appropriate sign listed in the right-hand column in the table.

Example 7.11

Show the form of Fourier's law describing the heat flow across the four edges of a rectangular block as illustrated in Figure 7.20. The temperature distribution in the block $T(x,y)$ is either specified or is determined from another analysis. The directions of heat flow across the four edges of the block are determined by the temperature differences across these edges as shown in Figure 7.20.

Solution:

We will use the guideline presented in Table 7.1 to assign appropriate signs to the heat fluxes crossing the four edges of the block with their respective directions of heat flow illustrated in Figure 7.20.

For the heat flux q_1 crossing the left-hand face of the block, the outward normal carries a negative sign $(-\mathbf{n})$ but the heat flow is opposite to the direction of the outward normal. This situation fits Case 4 in Table 7.1 and a negative sign is applied in Fourier's law associated with q_1. We may assign positive or negative signs to the heat fluxes crossing the other three edges following the same procedure. The results are shown in Figure 7.21.

Example 7.12

Heat spreaders are frequently used to direct heat generated in heat sources to the surrounding air or other media. There are many applications of using heat spreaders to mitigate the

Figure 7.20 Heat flow in a rectangular block.

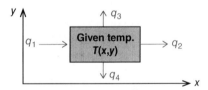

Sign of outward normal, n		q along n?	Sign of q in Fourier law
Case 1:	+	Yes	−
Case 2:	+	No	+
Case 3:	−	Yes	+
Case 4:	−	No	−

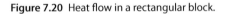

$$q_1 = -k\,\frac{\partial_T(x,y)}{\partial_x}$$

$$q_3 = -k\,\frac{\partial_T(x,y)}{\partial_y}$$

$$q_2 = -k\,\frac{\partial_T(x,y)}{\partial_x}$$

$$q_4 = +k\,\frac{\partial_T(x,y)}{\partial_y}$$

Temperature in solid: **T(x,y)**

Figure 7.21 Application of signs in Fourier's law of heat conduction in solids.

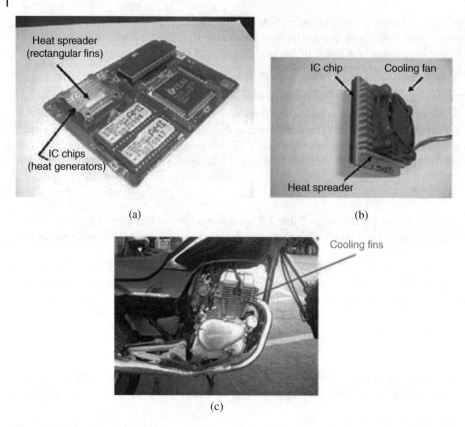

Figure 7.22 Heat spreaders of integrated circuits and an internal combustion engine. (a) A typical printed circuit board. (b) Dissipation of heat from an integrated circuit (IC) chip. (c) A motorcycle engine with cooling fins. *See color section.*

temperature rise in "heat generators" and prevent them from being over heated. Figures 7.22a and b show heat spreaders in the form of "fins" made of heat-conducting materials such as aluminum or copper. In microelectronics devices, these fins are attached to the heat-generating integrated circuit or microchip. Another common application of heat spreaders is the cooling fins attached to the cylinder heads of internal combustion engines on motorcycles as shown in Figure 7.22c.

In the case of heat spreaders on microchips shown in Figures 7.22a and b, the heat generated from the microchip flows to the exterior surfaces and the metal fins (the spreaders) dissipate the heat to the surrounding space, where cooling air is provided by small electric fans to facilitate removal of the dissipated heat.

In this example, we assume that a heat spreader with fins of triangular cross-section is used to dissipate heat from a heat-generating integrated circuit chip. The cross-section of the spreader fin is illustrated in Figure 7.23a. The temperature in the cross-section of the fins is determined from a separate analysis indicating a function $T(x, y) = 100 + 5xy^2 - 3x^2y$ in degrees Celsius, where x and y have the unit cm. The coordinates x and y are defined in Figure 3.23a.

Find the heat fluxes dissipating into the surrounding air at 20°C from the *mid-points* of the three edges if the thermal conductivity of the heat spreader material has a value of $k = 0.021$ W/cm-°C.

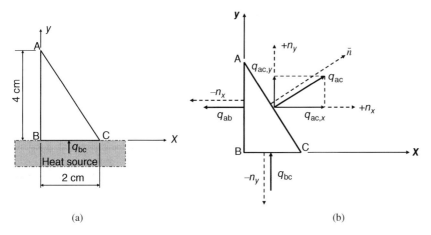

Figure 7.23 Heat flow in a cooling fin of a heat spreader. (a) Cross-section of half of a fin. (b) The coordinate system for analysis.

Solution:
Heat generated in the integrated circuit (IC) enters the heat spreader across the edge BC: the flux q_{bc} in Figure 7.23a. The heat that enters the fin spreader is dissipated to the surrounding air through the surfaces denoted by edges AB and AC. The heat fluxes across these three surfaces are illustrated in Figure 7.23b.

We are now ready to compute the heat fluxes across the three surfaces of the heat spreader fin in Figure 7.23.

a) **Heat flux across surface BC** As illustrated in Figure 7.21, the outward normal line to this surface carries a negative sign but the heat flux q_{bc} is in the opposite direction to the outward normal. This corresponds to Case 4 in Table 7.1. We will thus have the heat flux entering the surface BC as

$$q_{bc} = -k\frac{\partial T(x,y)}{\partial y}\bigg|_{y=0} = -0.021\frac{\partial(100 + 5xy^2 - 3x^2y)}{\partial y}\bigg|_{y=0}$$
$$= -0.021(10xy - 3x^2)|_{y=0}$$
$$= 0.063x^2 \ \text{W/cm}^2$$

The heat flux passing at mid-point on edge BC at $x = 1\,\text{cm}$ is thus $0.063 \times (1^2) = 0.063\,\text{W/cm}^2$.

b) **Heat flux across surface AB** We will first need to establish the direction of heat flow across this surface. This can be done by checking the temperature at the two terminal points of the edge AB.

At point B ($x = 0$ and $y = 0$) the corresponding temperature is

$$(100 + 5xy^2 - 3x^2y)|_{\substack{x=0\\y=0}} = 100°\text{C} > 20°\text{C}, \ \text{the ambient temperature}$$

Likewise, we may find the temperature at point A to be 100°C, which is also greater than the ambient temperature of 20°C. We thus determine that heat leaves the surface AB to the surrounding as illustrated in Figure 7.23b, and this corresponds to Case 3 in Table 7.1.

We thus have the heat flux expressed by the following expression:

$$q_{ab} = k \frac{\partial T(x, y)}{\partial x}\bigg|_{x=0}$$

$$= 0.021 \frac{\partial(100 + 5xy^2 - 3x^2 y)}{\partial x}\bigg|_{x=0}$$

$$= 0.105y^2 \quad \text{W/cm}^2$$

The mid-point of edge AB is located at $y = 2$ cm, which leads to the heat flux leaving this mid-point as $0.105 \times (2^2) = 0.42$ W/cm^2.

c) **Heat flux across surface AC** Again, we need first to check the direction of heat flux across this surface in order to determine the proper sign in the expression of Fourier's law for the heat flux.

As usual, we have the temperature at terminal point C ($x = 2, y = 0$) as 100°C, which is greater than the ambient temperature. The same temperature is obtained at the other terminal point A ($x = 0, y = 4$). We thus conclude that heat leaves this surface as illustrated in Figure 7.23b.

The situation for determining the direction of heat flux across surface AC is little more complicated than that of the other two surfaces, because this surface is inclined on the x–y plane as illustrated in Figure 7.23. The heat flux across the surface AC, q_{ac}, is a vector quantity that is neither along the x-coordinate nor along the y-coordinate. We thus need first to decompose this vector into two components, $q_{ac,x}$ and $q_{ac,y}$, as shown in Figure 7.23b. The corresponding outward normal lines (i.e., n_x and n_y) are positive for both these components as illustrated in the same figure. We thus conclude that these two heat flux components correspond to Case 1 in Table 7.1. Consequently, they can be expressed as follows:

$$q_{ac,x} = -k \frac{\partial T(x, y)}{\partial x}\bigg|_{\substack{x=1 \\ y=2}}$$

$$= -0.021(5y^2 - 6xy)|_{\substack{x=1 \\ y=2}}$$

$$= -0.021(20 - 12)$$

$$= -0.168 \text{ W/cm}^2$$

$$q_{ac,y} = -k \frac{\partial T(x, y)}{\partial y}\bigg|_{\substack{x=1 \\ y=2}}$$

$$= -0.021(10xy - 3x^2)|_{\substack{x=1 \\ y=2}}$$

$$= -0.021(20 - 3)$$

$$= -0.357 \quad \text{W/cm}^2$$

The resultant magnitude of these two vector components will yield the total heat flux leaving at the mid-point of surface AC using Equation 7.26:

$$q_{ac} = q_{ac,x} + q_{ac,y}$$

$$= \sqrt{(-0.168)^2 + (-0.357)^2}$$

$$= 0.3945 \quad \text{W/cm}^2$$

7.5.4 Newton's Cooling Law for Heat Convection

The mode of heat transmission in a fluid is quite different from that of conduction in solids in that convective heat transfer is associated with motion of the fluid. This motion of the heat transfer medium (i.e., the fluid) may be created by external forces, when it is termed *forced convection*, or by the buoyancy forces developed due to differences in specific gravity of the medium caused by the temperature difference in the medium, which is termed *natural convection*. The law that governs heat convection in fluid is *Newton's cooling law*, which stipulates that the amount of heat flow between two points A and B in a fluid with respective temperatures T_a and T_b, such as illustrated in Figure 7.24, is proportional to the difference of these two temperatures.

Mathematically, Newton's cooling law can be expressed as

$$q \propto (T_a - T_b) = h(T_a - T_b) \tag{7.28}$$

in which q is the heat flux transmitted in the fluid by the temperature difference $(T_a - T_b)$, and h is the heat transfer coefficient, with unit $W/m^2\text{-}°C$.

The heat transfer coefficient h in Equation 7.28 is normally determined empirically. Its value is imbedded in the dimensionless Nusselt number $Nu = (hL)/k$, where L is the "significant," or "characteristic" length of fluid flow and k is the thermal conductivity of the fluid. The Nusselt number is related to other dimensionless numbers such as the Reynolds number (Re) and the Prandtl number (Pr) for forced convective heat transfer in empirical formula as shown in Equation (2.4) in Chapter 2, and is related to another dimensionless number, the Grashof number (Gr), in natural convective heat transfer. The Nusselt number is related to the Reynolds number in both forced and natural convective heat transfer, and the latter is related to the velocity of the moving fluid in the form $Re = (\rho L v)/\mu$, where ρ is the mass density of the fluid, v is the velocity of the moving fluid, and μ is the dynamic viscosity of the fluid. We may thus assert that the heat transfer coefficient h in Equation 7.28 is proportional to the velocity of the fluid in convective heat transfer. Physically, this means that the faster the fluid motion is, the higher is the value of the heat transfer coefficient h, and accordingly the more effective is the convective heat transfer as given by Equation 7.28.

7.5.5 Heat Transfer between Solids and Fluids

The physical phenomenon of interfacial heat transfer between solids and fluids can be best explained by a person's common perception of weather. The solid in such case is the person's face and also his or her skin surface exposed to the surrounding air; the warm human body is the "solid" that give its heat to the cold surrounding air in winter, while the body receives heat from the hot surrounding air in summer. The person's feeling of being cold in winter and hot in summer is a measure of heat transference between the human body (the solid) and the surrounding air (the fluid) in heat transfer processes. An interesting feature associated with the transmission of heat between solid and surrounding fluid is the velocity of the surrounding fluid (the ambient air) in such processes; the well-known term "wind chill effect" is often used by meteorologist to reflect a person's actual response to the weather. He or she will feel

Figure 7.24 Convective heat flow in a bulk fluid.

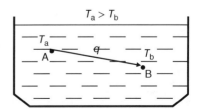

much colder if the cold surrounding air moves at a higher speed. Likewise, a person will feel more comfortable if a fan is used to blow cool surrounding air across the person's face in hot summer weather. The "wind-chill effect" may be mathematically modeled by Equation 7.28, in which the heat transfer coefficient *h* is a significant factor in the effectiveness of convective heat transfer. This coefficient is proportional to the velocity of the fluid: a fast-moving fluid, such as a high-speed breeze created by a fan, will increase the value of *h* and create a more effective convective heat transfer to the surrounding fluid.

There are a number of applications of solid–fluid interface heat transfer in engineering practice. In the semiconductor industry, it is standard procedure to heat small IC chips in hot ovens in "burn-in" tests that are part of the reliability testing before the chips are shipped to market. The heating of the chips to a predetermined temperature is often for a prolonged duration at elevated temperature. In addition to the "burn-in" tests, these chips are also required to pass specific "thermal cycling" and "thermal shock" tests. Engineers are expected to design proper cooling and heating chambers that will cool small-sized solids, such as IC chips, in the cooling portion of a thermal cycling tests for the chip's reliability testing. Additionally, cooling chambers such as refrigerators have many practical applications in modern society. Conditions for heating and cooling of small solids are illustrated in Figure 7.25.

Design analysis of the cooling and heating chambers requires mathematical formulation combining Fourier's law of heat conduction for solids given in Equations 7.24 for one-dimensional heat flow together with Equations 7.25 to 7.27 for the general case, and Newton's cooling law for heat convection in fluids given in Equation 7.28. The formulation of the mathematical analysis of the situations illustrated in Figure 7.25a for heating and Figure 7.25b for cooling will be based on the general situation illustrated in Figure 7.26.

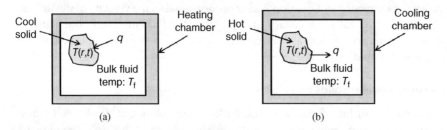

Figure 7.25 (a) Heating and (b) cooling of small solids.

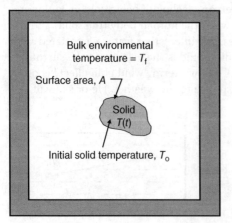

Figure 7.26 Heat flow between a solid and a fluid in a controlled-temperature enclosure.

Let us begin by considering a simple one-dimensional analytical case that involves a small solid at an initial temperature T_0 that is to be cooled or heated in an enclosure such as a cooling chamber or an oven with a surrounding fluid of controlled temperature T_f, as shown in Figure 7.26. It is desired to estimate how long it will take for the solid to reach a specified temperature provided by the enclosure with a given bulk fluid temperature T_f.

Our mathematical modeling of the cooling-heating of solids in environmental chambers begins with the following stipulations:

1) The solid is initially at a uniform temperature T_0.
2) Being small in size, the solid is of uniform temperature at all times, and this temperature changes with time: thus, the solid is at temperature $T(t)$ where t represents time.
3) The surrounding bulk fluid temperature is T_f at all times.
4) The change in the temperature of the solid is attributable only to the heat supplied or removed by the surrounding fluid.

We will use Newton's cooling law as expressed in Equation 7.28 for the heat transmission in the bulk fluid.

Let us take a look at a generic situation involving heat exchange between the solid surface and the surrounding fluid, with the heat flux q transferring heat from the solid surface at temperature $T(t)$ to the surrounding fluid at temperature T_f, as shown in Figure 7.27, in which $\bar{n}$ is the outward normal line that designates the direction of heat flow.

We may express the heat flux from the solid surface to the surrounding fluid by Newton's cooling law as follows:

$$q = h[T_s(t) - T_f] = h[T(t) - T_f] \tag{7.29}$$

where h is the heat transfer coefficient of the bulk surrounding fluid at the solid–fluid interface.

The reader will notice that we have made the temperature of the solid at the contacting surface, $T_s(t)$, be equal to the temperature of the solid, $T(t)$, in accordance with our stipulation of the uniformity of temperature in the solid. This is justified for small-size solids in which the interior temperature is close to the surface temperature.

The temperature of the solid $T(t)$ changes with time after it is submerged in the bulk fluid at a different temperature. The heat exchange between the solid and the surrounding fluid during a time interval Δt can be mathematically expressed as follows, using the first law of thermodynamics:

Change in internal energy of the small solid during Δt	$=$	Net heat flow from the small solid to the surrounding fluid during Δt

$$-\rho c V \, \Delta T(t) \quad = \quad Q = qA_s \, \Delta t = hA_s[T(t) - T_f] \Delta t$$

Figure 7.27 Heat transfer from a solid to a surrounding fluid.

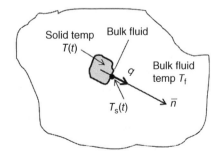

In this relationship, $\Delta T(t)$ is the change of temperature of the solid during the time interval Δt; ρ, c, V are respectively the mass density, specific heat, and volume of the solid; and A_s is the interface area between the solid and the surrounding fluid.

Rearranging the terms on both sides of the equation, we may derive the expression for the rate of change of the temperature in the solid:

$$\frac{\Delta T(t)}{\Delta t} = -\frac{h}{\rho c V} A_s[T(t) - T_f]$$

or

$$\frac{\Delta T(t)}{\Delta t} = -\alpha A_s[T(t) - T_f]$$

with the coefficient α being

$$\alpha = \frac{h}{\rho c V} \tag{7.30}$$

Since the heat transmission between the solid and the surrounding fluid is a continuous process, the finite increments of $\Delta T(t)$ and Δt can be replaced by infinitesimally small increments of $dT(t)$ and dt, respectively. We may thus derive the following first-order differential equation representing heat transmission between a small solid and surrounding fluid:

$$\frac{dT(t)}{dt} = -\alpha A[T(t) - T_f] \tag{7.31}$$

with the initial condition

$$T(t)|_{t=0} = T(0) = T_0$$

The SI units of the physical quantities in Equation 7.30 are W/m^2-°C for the heat transfer coefficient h; g/m^3 for the mass density of the fluid ρ; J/g-°C for the specific heat c of the solid; and m^3 for the volume V of the solid. The coefficient α in Equation 7.31 thus has the unit of /m^2-s.

Example 7.13

A small solid at temperature of 80°C is to be cooled in a cooling chamber maintained at 5°C. Determine the time required for the temperature of the solid to reach 8°C if the proportionality constant in Equation 7.31 is $\alpha = 0.002$/m^2-s with a contacting surface area $A = 0.2$ m^2.

Solution:

The situation described in the problem corresponds to the situation depicted in Figure 7.26 with $T_f = 5$°C and $T_0 = 80$°C. Also given are $\alpha = 0.002$/m^2-s and $A = 0.2$ m^2. Thus, by substituting these given conditions into Equation 7.31, we obtain the following differential equation and the initial condition for the solution of the instantaneous temperature of the solid, $T(t)$:

$$\frac{dT(t)}{dt} = -(0.002)(0.2)[T(t) - 5] \tag{a}$$

with the initial condition:

$$T(0) = T_0 = 80°C \tag{b}$$

The solution to the differential equation in Equation (a) can be obtained by integrating both sides of the equation to yield

$$\int \frac{dT(t)}{T(t) - 5} = -0.0004 \int dt + c_1 \tag{c}$$

where c_1 is an arbitrary integration constant.

The following expression is obtained after integration on both sides of Equation (c):

$$T(t) - 5 = e^{-0.0004t + c_1} = c e^{-0.0004t} \qquad \text{(d)}$$

The constant c in Equation (d) is determined by the condition given in Equation (b), resulting in $c = 75$.

The general solution to Equation (a) is thus:

$$T(t) = 5 + 75 e^{-0.0004t} \qquad \text{(e)}$$

Let t_e be the time required for the temperature of the solid to reach 8°C. The required time can be determined by setting $T(t)$ in Equation (e) to be

$$T(t_e) = 8 = 5 + 5 e^{-0.0004t_e} \qquad \text{(f)}$$

The required time, t_e can be obtained by solving the algebraic equation (f) to give $t_e = 8047.19$ seconds, or 134.12 minutes, or 2.24 hours.

The time required to heat or cool a solid at initial temperature T_0 by means of hot or cold bulk fluid is a process design issue. Engineers would normally adjust either the bulk fluid temperature T_f or the proportionality constant α in the chamber to shorten or extend the required time. There is usually a cost factor associated with lowering T_f and increasing the α-value to shorten the required time for the cooling in the case of Example 7.13. Increasing the α-value means an increasing of the h-value according to the relationship shown in Equation 7.30. The cost associated with the increasing h-value is that of providing more vigorous motion of the surrounding fluids with higher Reynolds numbers for higher h-values, such as is described toward the end of Section 7.5.4. This is a classic design issue involving proper balance of costs and benefits in the analysis.

Example 7.14

A Silicon Valley company produces microcomputer chips and is required to perform thermal cycling tests of the chips it has produced as a part of reliability testing. In this particular series of tests, each cycle involves heating and cooling of the chips between terminal temperatures and for the durations illustrated in Figure 7.28.

In the present example, each chip has an overall surface area of 8×10^{-4} m^2 with the heating and cooling chambers providing the conditions specified in Table 7.2.

The procedure for this thermal cycling testing is to first heat the chip from an initial temperature of 20°C to 100°C in the heating chamber for a period of t_h, followed by cooling the chip down to 20°C in the cooling chamber for a period of t_c. The subsequent cycles will have the chip's temperature cycling between $T_2 = 20$°C and $T_1 = 100$°C as illustrated in Figure 7.28.

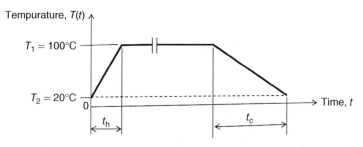

Figure 7.28 Temperature variations in one cycle of heating and cooling of chip under test.

Table 7.2 Conditions of the heating and cooling chambers.

Chamber	Proportionality coefficient (α), /m²-s	Chamber ambient temperature (T_f), °C
Heating	0.8	150
Cooling	0.2	−20

The process engineer's task is to determine the required times t_h and t_c for the heating and cooling of the chips in the first cycle of the planned thermal cycling testing as illustrated in Figure 7.28. He or she is also expected to propose plans on what physical arrangements could be made to shorten the computation times t_h and t_c in order to expedite the thermal cycling process.

Solution:
We will use the first-order differential equation in Equation 7.31 to solve for the instantaneous temperature in the chip $T(t)$ as follows:

$$\frac{dT(t)}{dt} = -\alpha A[T(t) - T_f] \tag{a}$$

a) **For the heating portion of the thermal cycling test** We have from Table 7.2 the values of $\alpha = 0.8/\text{m}^2$-s and $T_f = 150°C$ together with $A = 8 \times 10^{-4}$ m² and the initial temperature of the chip $T_0 = 20°C$. Substituting these conditions into Equation (a) results in the following differential equation for $T(t)$:

$$\frac{dT(t)}{dt} = -6.4 \times 10^{-4}[T(t) - 150] \tag{b}$$

with the condition

$$T(t)|_{t=0} = T(0) = T_0 = 20 \tag{c}$$

The solution of Equation (b) may be obtained by integrating both sides of the equation:

$$\int \frac{dT(t)}{T(t) - 150} = -6.4 \times 10^{-4} \int dt + c_1 \tag{d}$$

in which c_1 is the integration constant, to be determined from the initial condition in Equation (c). We thus obtain the following expression for function $T(t)$:

$$T(t) = 130e^{-6.4\times10^{-4}t} + 150 \tag{e}$$

The time required to heat the chips to 100°C may be determined using the relation $T(t_h) = T_1 = 100°C$, which leads to

$$100 = -130e^{-6.4\times10^{-4}t_h} + 150$$

We may solve for t_h from the above expression to get

$$t_h = \frac{\ln\left(\frac{150-100}{130}\right)}{-6.4 \times 10^{-4}} = 1493 \text{ seconds} \quad \text{or} \quad 24.88 \text{ minutes}$$

b) **For the cooling portion of the thermal cycling test** We have the proportionality coefficient $\alpha = 0.2/\text{m}^2$-s and $T_f = -20°C$ provided by the cooling chamber as indicated in Table 7.2. The chips have an initial temperature $T_0 = 100°C$ for the cooling process and they are to be cooled to $T(t_c) = 20°C$.

We will use the same Equation (a) for the solution $T(t)$ with the following expression:

$$\frac{dT(t)}{dt} = -0.2 \times (8 \times 10^{-4})[T(t) - (-20)]$$

$$= -1.6 \times 10^{-4}[T(t) + 20] \tag{f}$$

The solution for $T(t)$ in Equation (f) has the following form:

$$\ln[T(t) + 20] = -1.6 \times 10^{-4}t + c_2$$

in which c_2 is the integration constant, to be determined from the initial condition of $T(t)|_{t=0} = T(0) = T_1 = 100°C$, from which we have $c_2 = 4.7875$. We thus have the solution

$$T(t) = 120e^{-1.6 \times 10^{-4}t} - 20 \tag{g}$$

from which we may express the temperature of the chip at time t_c as

$$T(t_c) = 20 = 120e^{-1.6 \times 10^{-4}t_c} - 20$$

We may solve for t_c from the above equation and obtain $t_c = 6813$ seconds or 113.55 minutes, or 1.89 hour.

c) **Shortening the heating and cooling times** If it is required to shorten the $t_h = 24.88$ minutes for heating the chips and $t_c=1.89$ hours for cooling them, the options are as follows. (a) To shorten the time for heating, one may increase both the α-value and the bulk fluid temperature T_f. (b) To shorten the time for cooling, a plausible action would be to increase the α-value or lower the bulk fluid temperature T_f. Both options will increase the cost of testing because increasing α-values in the heating or cooling chambers requires increasing the speed of fluid moving over the solid surface, and increasing T_f for shorter t_h and reducing T_f for shorter t_c in the chambers can also be costly.

7.6 Rigid Body Dynamics under the Influence of Gravitation

In this section, we will demonstrate how the first-order ordinary differential equations for rigid body dynamic analysis under the influence of gravitation can be derived using Newton's second law.

Consider a rigid body of mass m that is thrown vertically up into the air with an initial velocity v_0, or that is falling from some height above the ground. The moving body encounters air resistance, which is proportional to the velocity $v(t)$, with t being the time into the motion, we may derive the following equations for the dynamic analysis:

a) The equation of motion in the coordinate systems shown in Figure 7.29.
b) An expression for the velocity of the mass at time t.
c) The time at which the rigid body reaches its maximum height with an initial velocity in the case of the body thrown up, or the velocity at the touchdown for the body falling from an initial height.

The forces that are associated with the rising and falling rigid bodies in Figure 7.29 include the following components:

1) The weight of the body, $w = mg$, where m is the mass of the body and g is the gravitational acceleration ($g = 9.81$ m/s^2). Both the weight of the body w and the gravitational acceleration g point toward the earth.

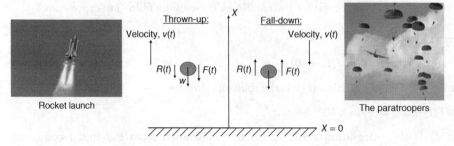

Figure 7.29 Rise and fall of rigid bodies.

2) The resistance encountered by the moving body from the surrounding medium, such as aerodynamic resistance, may be expressed as $R(t) = cv(t)$, where c is the proportionality constant determined by experiment and $v(t)$ is the instantaneous velocity of the moving body.

3) The dynamic (or inertial) force is $F(t) = ma$, where a is the acceleration (or deceleration, with a negative sign). The acceleration a is defined as the rate of change of velocity of a moving solid and it has the mathematical form $a = dv(t)/dt$, with $v(t)$ being the instantaneous velocity.

We should bear in mind that both $R(t)$ and $F(t)$ act on the moving body in the direction opposite to that of the motion.

We will use Newton's law on dynamic force equilibrium, $\sum F_x = 0$, to derive the required differential equations for the following cases.

a) **A body thrown up under the influence of gravitation** The following relation is derived from the force diagram shown at the left side of Figure 7.29:

$$\sum F_x = -F(t) - w - R(t) = 0$$

Substituting $R(t) = cv(t)$ and $F(t) = ma = m\,dv(t)/dt$ into the above expression, we will obtain the differential equation for the instantaneous velocity of the moving body:

$$\frac{dv(t)}{dt} + \frac{c}{m}v(t) = -g \tag{7.32}$$

The solution of Equation 7.32 can be obtained by comparing the terms in Equation 7.32 with the typical form of nonhomogeneous first-order differential equation as presented in Equation 7.6. Thus, $v(t)$ in Equation 7.32 can be expressed as

$$v(t) = \frac{1}{F(t)} \int F(t) g(t)\, dt + \frac{K}{F(t)}$$

in which $F(t) = e^{\int p(t)\,dt}$, $p(t) = c/m$, and $g(t) = -g$.

Consequently, the solution $v(t)$ in Equation 7.32 takes the form

$$v(t) = \frac{1}{e^{ct/m}} \int e^{ct/m}(-g)\, dt + \frac{K}{e^{ct/m}}$$

$$= -\frac{mg}{c} + Ke^{-ct/m} \tag{7.33}$$

in which K is constant to be determined from an appropriate initial condition.

For a given initial condition that $v(0) = v_0$, one may compute $K = v_0 + (mg/c)$. This will lead to the complete solution of $v(t)$ as

$$v(t) = -\frac{mg}{c} + \left(v_0 + \frac{mg}{c}\right) e^{-ct/m} \tag{7.34}$$

Equation 7.34 gives the instantaneous velocity of a rigid body thrown upward with an initial velocity of v_0.

The time for the body to reach the maximum height, t_m can be obtained by letting the velocity in Equation 7.34 be zero at that instant: mathematically, $v(t_m) = 0$. The following relation derived from Equation 7.34 is used to determine t_m:

$$0 = -\frac{mg}{c} + \left(v_0 + \frac{mg}{c} \right) e^{-(c/m)t_m}$$

Solve for t_m from the above expression to get

$$t = \frac{m}{c} \ln \left(1 + \frac{v_0 c}{mg} \right) \tag{7.35}$$

b) **A body in free fall** We have the situation depicted in the right-hand half of Figure 7.29. The differential equation for the instantaneous velocity of the falling body can be derived using the dynamic equilibrium of forces acting on the body at time t:

$$\sum F_x = F(t) - w + R(t) = 0$$

This expression leads to the differential equation for the instantaneous velocity, $v(t)$, as

$$\frac{dv(t)}{dt} + \frac{c}{m} v(t) = g \tag{7.36}$$

Note that the differential equation for a free-falling body in Equation 7.36 is similar to that for the case of a body thrown upwards in Equation 7.32, the only difference being in the sign of g in the right-hand side of the equation. A similar method to that used for solving Equation 7.32 can be used for the solution of Equation 7.36.

Example 7.15

An armed paratrooper with ammunition weighing 322 pounds jumps from a plane with zero initial velocity, as shown in Figure 7.30.

(https://www.google.com/search?q=free+copyright+pictures+of+paratroopers&tbm=isch&tbo=u&source=univ&sa=X&ved=0ahUKEwjj5e_2y4TVAhVFw4MKHaCRCaMQsAQIUA&biw=1697&bih=788#imgrc=zSkACnNLNrqJ0M:&spf=1499891706617)

The troopers encounter negligible side wind in their descent. However, they encounter air resistance whose magnitude is 15 times the square of the descent velocity $v(t)$; that is, $15[v(t)]^2$.

Figure 7.30 Paratroopers. (http://www.westandwithukraine.org/wp-content/uploads/2015/04/Paratroopers-3-596573.jpg)

Assume that the mass of the parachute is negligible. Do the following:

a) Derive the appropriate equation for the instantaneous descending velocity $v(t)$.
b) Solve the equation for the descent velocity.
c) Estimate the time required for the paratrooper to reach the ground from a height of 10 000 feet.
d) Estimate the impact velocity of the paratrooper upon touching the ground, and the associated momentum.

Solution:

a) **Derivation of the differential equation** The total mass of the falling body $m = 322/32.2 = 10$ slugs, and the air resistance $R(t) = cv(t) = 15[v(t)]^2$ as given. The instantaneous descending velocity $v(t)$ can be obtained by using Equation 7.36 as

$$\frac{dv(t)}{dt} + \frac{15[v(t)]^2}{10} = 32.2 \qquad (a)$$

Equation (a) may be written in the form

$$10\frac{dv(t)}{dt} = 322 - 15[v(t)]^2 \qquad (b)$$

with the initial condition

$$v(t)|_{t=0} = v(0) = 0 \qquad (c)$$

b) **Solution for the instantaneous velocity $v(t)$** Equation (b) is a separable differential equation and can be rearranged into the form

$$\frac{10}{322 - 15v^2}dv = dt \qquad (d)$$

where $v = v(t)$.

The solution of Equation (d) is obtained by integration on both sides of Equation (d), resulting in

$$10\int \frac{dv}{322 - 15v^2} = \int dt + c$$

where c is an integration constant.

The integrand on the left-hand-side of the above expression can be found from mathematical handbooks (Zwillinger, 2003) to be

$$\int \frac{dv}{322 - 15v^2} = \frac{1}{139}\left[\log\frac{4.634 + v}{4.634 - v}\right]$$

which gives the solution

$$\frac{10}{139}\log\frac{4.634 + v}{4.634 - v} = t + c \qquad (e)$$

The arbitrary constant, c in Equation (e) may be determined by using the initial condition of $v(0) = 0$ in Equation (c), which leads to $c = 0$. We thus have

$$\frac{10}{139}\log\frac{4.634 + v}{4.634 - v} = t$$

Consequently, one may express the solution of Equation (b) with $v = v(t)$ as

$$v(t) = \frac{4.634(e^{13.9t} - 1)}{e^{13.9t} + 1} \qquad (f)$$

(*Hint:* $\ln(z) = 2.3 \log(z)$.)

c) **Estimation of the time required for the paratrooper to reach the ground from a height of 10 000 feet** Let the descent distance of the paratrooper to be $X(t)$, where t is the time starting from the moment of his jumping out the carrier airplane. The relation $v(t) = dX(t)/dt$ leads to the following expression for the distance of descent:

$$X(t) = \int_0^t v(t)\, dt \tag{g}$$

in which $v(t)$ is available from Equation (f). We thus have

$$X(t) = \int_0^t \frac{4.634(e^{13.9t} - 1)}{e^{13.9t} + 1}\, dt$$
$$= [0.667 \ln(1 + e^{13.9t}) - 4.634t]|_0^t$$

The distance that the paratrooper has descended is determined to be

$$X(t) = 0.6667 \ln(1 + e^{13.9t}) - 4.634t - 0.4621 \tag{h}$$

If we let t_g be the required time to travel a distance of 10 000 feet, we will have

$$10\,000 = 0.6667 \ln(1 + e^{13.9t_g}) - 4.634t_g - 0.4621 \tag{j}$$

In the expression in Equation (j), we realize that t_g is a real number that has a value greater than zero (i.e., $t_g > 0$), and $e^{13.9t_g} \gg 1$ in Equation (j), which leads to the following approximation:

$$0.6667 \ln(1 + e^{13.9t_g}) \approx 0.6667 \ln(e^{13.9t_g})$$
$$= 0.6667 \times 13.9t_g$$
$$= 9.2671t_g$$

Substituting this into Equation (b), we obtain the following relation for t_g:

$$10\,000 = 4.6331t_g - 0.4621$$

which gives the solution $t_g = 2158.46$ seconds or 35.95 minutes.

d) **Estimation of the impact velocity of the paratrooper upon touching the ground, and the corresponding momentum** The impact velocity of the paratrooper is the "terminal" velocity of his descent from a height of 10 000 feet. We may take advantage of the expression in Equation (f) for the solution. Thus, substituting $t_g = 2158.46$ seconds (the time the paratrooper spent in traveling this distance, as obtained in part (c)) into Equation (f), we have for the impact velocity, V_f:

$$V_f = \frac{4.634(e^{13.9 \times 2158.46} - 1)}{e^{13.9 \times 2158.46} + 1} \approx 4.634 \text{ ft/s}$$

From this impact velocity we calculate the impact momentum, $M = \text{mass} \times \text{velocity}$, or $M = mV_f = (322/32.2)V_f = 46.34$ ft-lb, which is a significant amount that qualifies as a hard landing.

7.7 Problems

Solve the differential equations in the Problems 7.1 to 7.6.

7.1 $\dfrac{du(x)}{dx} = 10u(x)$

7.2 $\quad \dfrac{du(x)}{dx} = 2u(x) + 3$

7.3 $\quad x(x+2)\dfrac{du(x)}{dx} = [u(x)]^2$

7.4 $\quad x\dfrac{du(x)}{dx} + 2x = u(x)$

7.5 $\quad \dfrac{du(x)}{dx} = 2u(x) - 1 \quad$ with $\quad u(0) = 2$

7.6 $\quad x\dfrac{du(x)}{dx} + u(x) = 2x \quad$ with $\quad u(1) = 10$

7.7 Determine the time required to empty the water in a straight cylindrical tank with a diameter of 12 inches and initial water level of 20 inches. The diameter of the hose for draining the water from the tank is 1 inch.

7.8 A swimming pool 20 m long (L) × 10 m wide (W) × 2 m deep (D) is filled with water. A drainage pipe of diameter (d) 6 cm located at the bottom of the pool is used to drain the water. Determine the following:
a) The differential equation describing the draining of the pool.
b) The water level in the pool after one hour of the draining.
c) The time required for draining all the water from the pool.

7.9 Derive an expression to estimate the time required to drain a circular funnel with a slant angle of 30° as illustrated in Figure 7.31.

7.10 A food processing plant uses a flat tapered chute (i.e., with a rectangular cross-section) to fill bottles with juice. The geometry and dimensions of the funnel are illustrated in Figure 7.32. The chute was full when it was set to open at time $t = 0$.
a) What is volume of the juice in a full chute?
b) Establish the equation for the movement of the juice in the chute.
c) Compute the time required to empty the chute.
Assume that the friction between the flowing juice and the chute wall is negligible.

7.11 The square chute shown in Figure 7.33a is used to hold solvents for a chemical process plant. The dimensions of the chute are W = 20 cm, H = 10.8 cm, h = 1.2 cm, and

Figure 7.31 Water in a shallow tapered funnel.

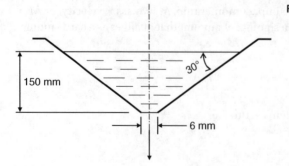

Figure 7.32 A flat chute for filling juice into bottles.

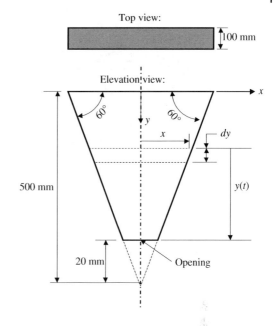

Figure 7.33 A square chute for liquid solvents. (a) A square chute. (b) Dimensions of the chute.

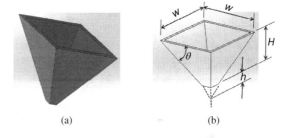

(a)   (b)

$\theta = 60°$, as shown in Figure 7.33b. Determine (a) the total volume of the contents of the chute, and (b) the time required to drain the chute.

7.12 An engineer is assigned to design a circular funnel for filling water bottle similar to the one described in Example 2.10. The funnel is to be attached to an automatic filling machine. The funnel has the dimensions shown in Figure 7.34. The water bottle will have similar geometry and dimensions to the wine bottle illustrated in Figure 2.33b. The straight portion of the upper section of the water bottle, however, is 4 cm in length rather than the 10 cm for the wine bottle in Figure 2.33b. Determine the following:
a) The volume of the water bottle.
b) The length H that will enable the funnel to hold enough water to fill the water bottle.
c) The time required to fill one water bottle on the filling machine.

7.13 Intravenous delivery (IV) bottles such as the one shown in Figure 7.35a are commonly used in hospitals. Use the idealized geometry of the IV bottle shown in Figure 7.35b to determine:
a) The volume of the liquid in the bottle.
b) The time required to empty the bottle.

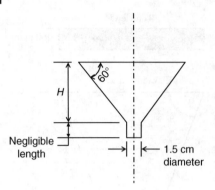

Figure 7.34 A tapered circular funnel for water bottle filling.

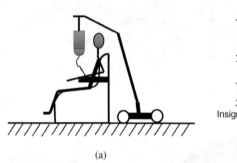

(a)

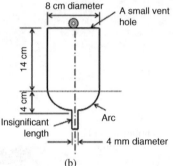

(b)

Figure 7.35 Drainage of an IV bottle. (a) Patient with IV bottle on stand. (b) Dimensions of idealized IV bottle.

    c) The length of the straight portion of the IV bottle in Figure 7.35b if the content is designed to be 1200 cc (cm³), and the time required to empty the bottle with the designed content?

7.14 A 1 m-long circular rod 50 mm in diameter has its circumference thermally insulated. One end of the rod is maintained at 20°C while heat is supplied at the other end. Determine (a) the required rate of heat supply q in kW to the rod for the temperature at its half-length to be 80°C, and (b) the temperature variation in the entire rod under this condition. The rod material has a thermal conductivity $k = 100$ kW/m-°C.

7.15 A chip containing an integrated circuit is undergoing reliability testing with heating–cooling cycles. It is first heated to 100°C in a heating chamber, followed by a cooling in a cooling chamber with its interior maintained at 8°C. Determine the value of the proportionality constant α in Equation 3.23 with a unit of /m²-s if the chip is to be cooled to 10°C in 2 hours. The surface area of the chip is 2×10^{-6} m². What could the process engineer do to shorten this time for the cooling to 1 hour with a new value of α, and how would he or she physically accomplish this and why?

7.16 It is customary to put integrated circuit packages through thermal cycling as a part of product reliability tests. A chip at room temperature (20°C) with surface area (A) of 8×10^{-4} m² is placed in a cooling chamber with its interior maintained at −60°C. As engineering students learn in class, the rate of temperature change of a small object such as the chip is proportional to the surface area and to the difference between the temperature of the surrounding medium and the object's temperature. Let the constant of proportionality be $\alpha = 0.4/m^2$-s.

a) Derive the expression for the temperature in the chip at any time t after it has been placed in the cooling chamber.

b) Determine the temperature in the chip 5 hours after it has been placed in the chamber.

c) Estimate the required time for the chip temperature to reach $-40°C$.

d) Which way would you adjust the proportionality constant, h (i.e., up or down) if we want to shorten the time that is required for the chip to reach the temperature of $-40°C$?

7.17 A microelectronic product is put to reliability tests involving thermal cycling. The product is heated to $100°C$ and at that time it is moved to the cooling chamber for the cooling portion of the cycle. The interior of the cooling chamber is maintained at $8°C$. The law of physics governing cooling of a solid is that the rate of temperature change of the solid is proportional to the surface area A and the difference between the temperature of the surrounding medium and the temperature of the solid, T, at any given time.

a) Determine the time for the temperature of the product to reach $20°C$ if the constant of proportionality $\alpha = 0.02/m^2$-s and the surface area of the product $A = 2 \times 10^{-6}$ m^2.

b) If the required time is reduced to 2 hours, what will be the corresponding value of the proportionality constant α?

7.18 A printed circuit board (PCB) from a local electronics manufacturing company is expected to go through thermal shock testing for its product reliability. The particular PCB has a surface area of $400\,cm^2$ and is initially at the ambient temperature of $20°C$. It is quickly placed into a cooling chamber that has its interior maintained at $-80°C$. The principle of convective heat transfer indicates that the rate of temperature change of the PCB is proportional to the surface area and the difference between the temperature of the surrounding medium and that of the PCB. If the constant of proportionality is $\alpha = 0.5/m^2$-s, determine the following:

a) The expression for the temperature in the PCB at any time t after it is placed in the cooling chamber.

b) The temperature in the PCB 20 minutes after it was placed in the chamber.

c) The time required for the PCB temperature to reach $-40°C$.

d) The α-value that is necessary if we want the PCB to reach a temperature of $-40°C$ in 5 minutes instead of the calculated value? What would be the engineering implications of the change of the α-value in the process?

7.19 A bullet with mass m is shot vertically upward in the air with an initial velocity V_0. Due to the gravitational effect g, and the air resistance, the bullet will reach a maximum height H_{max} and then fall vertically back to earth. Assume the resistance of the air to the moving bullet is proportional to the velocity.

a) Formulate the equation of motion of the bullet upward with velocity function $V(t)$, where t is the time into the motion in the upward direction.

b) Formulate the equation of motion of the bullet during the free fall stage.

c) Solve for $V(t)$ in both part (a) and part (b).

d) Derive the expression for the maximum height H_{max} that the bullet reaches in the upward motion.

The following information may be used in the numerical computation: the mass of the bullet (m) = 200 g; the initial velocity (V_0) = 50 m/s; gravitational acceleration (g) = 9.81 m/s²; the coefficient of air resistance for both upward and downward motion of the bullet is $c = 0.1$ N-s/m.

7.20 A crate containing bottles of fresh drinking water is airlifted to a remote area that was devastated by a Richter scale magnitude 7.8 earthquake. The crate weighs 1000 kg. It is dropped with a parachute from an aircraft 1500 m above the ground on a calm (i.e., "windless") day. The air resistance to the falling crate is approximately 20 times the square of the velocity of descent, $v(t)$. Assume that the mass of the parachute is negligible. Determine the following:

a) The equation and the appropriate conditions for the instantaneous descent velocity $v(t)$.

b) The mathematical expression for $v(t)$.

c) The time for the crate to reach the ground.

d) The impact velocity.

e) Whether the crate would survive the impact upon hitting the ground, and why, if it is designed to resist an impact force of 10 000 newtons with a "touchdown" time of 0.5 second.

8

Application of Second-order Ordinary Differential Equations in Mechanical Vibration Analysis

Chapter Learning Objectives

- Refresh your knowledge of the solution methods for typical second-order homogeneous and nonhomogeneous differential equations learned in previous mathematics courses.
- Learn to derive homogeneous second-order differential equations for free vibration analysis of simple mass–spring systems with and without damping effects.
- Learn to derive nonhomogeneous second-order differential equations for forced vibration analysis of simple mass–spring systems.
- Learn to use the solution of second-order nonhomogeneous differential equations to illustrate the resonant vibration of simple mass–spring systems and estimate the time for the rupture of the system in resonant vibration.
- Learn to use second-order nonhomogeneous differential equations to predict the amplitudes of the vibrating mass in near-resonant vibration and the physical consequences to the mass–spring systems.
- Learn the concept of modal analysis of machines and structures and the consequence of structural failure under the resonant and near-resonant vibration modes.

8.1 Introduction

Like the first-order differential equations presented in Chapter 7, second-order differential equations are also derived from the laws of physics. These equations are used to solve a variety of engineering problems relating to heat conduction in solids, stress analysis of structures such as bending of beams and buckling of slender columns, and mechanical vibration of rigid bodies, beam structures and long cables. However, we will focus our attention on their applications in structural vibrations due to limitations on text length.

Following the same practice as in Chapter 7, we will first review the techniques available for solving typical second-order differential equations. The solution methods presented in the subsequent sections are generic and effective techniques for engineering analysis.

8.2 Solution Method for Typical Homogeneous, Second-order Linear Differential Equations with Constant Coefficients

Let us consider a typical form of ordinary differential equations of second order:

$$\frac{d^2u(x)}{dx^2} + a\frac{du(x)}{dx} + bu(x) = 0 \tag{8.1}$$

Applied Engineering Analysis, First Edition. Tai-Ran Hsu.
© 2018 John Wiley & Sons Ltd. Published 2018 by John Wiley & Sons Ltd.
Companion Website: www.wiley.com/go/hsu/applied

in which $u(x)$ is the sought solution from Equation 8.1, and a and b are given constants of real numbers.

There are several ways one may solve Equation 8.1. The particular method that we will adopt here is to assume a solution of Equation 8.1 in the following form:

$$u(x) = e^{mx} \tag{8.2}$$

where the m is a constant to be determined in the following way.

Since the function $u(x)$ in Equation 8.2 is assumed to be the solution of Equation 8.1, it must satisfy that equation. Thus, by substituting the assumed solution in Equation 8.2 into Equation 8.1, we get

$$m^2 e^{mx} + a(m e^{mx}) + b(e^{mx}) = 0$$

We may factor out the common term e^{mx} in the above expression to give

$$e^{mx}(m^2 + am + b) = 0$$

Since the term e^{mx} cannot be zero (ask yourself why), we will arrive at the following quadratic equation for the unknown constant m:

$$m^2 + am + b = 0 \tag{8.3}$$

Equation 8.3 has the following two roots giving the solution m :

$$\left.\begin{array}{l} m_1 = -\dfrac{a}{2} + \dfrac{1}{2}\sqrt{a^2 - 4b} \\[2mm] m_2 = -\dfrac{a}{2} - \dfrac{1}{2}\sqrt{a^2 - 4b} \end{array}\right\} \tag{8.4}$$

We thus have the general solution for Equation 8.1 by including both m_1 and m_2 from Equation 8.4 in the solution in Equation 8.1:

$$u(x) = c_1 e^{m_1 x} + c_2 e^{m_2 x} \tag{8.5}$$

where c_1 and c_2 are the two arbitrary constants to be determined by the specified conditions often given with Equation 8.1.

Since both a and b in Equation 8.1 are given real-number constants, the value of $(a^2 - 4b)$ in $\sqrt{a^2 - 4b}$ in Equation 8.4 may be positive, negative or zero. These possibilities can produce radically different solutions of the differential equation as shown in Equation 8.5, as will be illustrated in following three different cases.

Case 1. $a^2 - 4b > 0$

In this case we recognize that both m_1 and m_2 in Equation 8.4 are real numbers. The solution of Equation 8.1 is

$$u(x) = e^{-ax/2}\left(c_1 e^{\sqrt{a^2 - 4b}(x/2)} + c_2 e^{-\sqrt{a^2 - 4b}(x/2)}\right) \tag{8.6}$$

Case 2. $a^2 - 4b < 0$

In this case both roots m_1 and m_2 in Equation 8.4 are complex numbers with

$$m_1 = -\frac{a}{2} + \frac{1}{2}\sqrt{(4b - a^2)}\, i$$

and

$$m_2 = -\frac{a}{2} - \frac{1}{2}\sqrt{(4b - a^2)}\, i$$

in which $i = \sqrt{-1}$. The substitution of m_1 and m_2 into Equation 8.5 will lead to the following expression:

$$u(x) = e^{-ax/2} \left(c_1 e^{(ix/2)\sqrt{4b-a^2}} + c_2 e^{-(ix/2)\sqrt{4b-a^2}} \right)$$

(8.7)

The solution $u(x)$ in Equation 8.7 has no practical value in engineering analysis, because it involves imaginary numbers, a more commonly used form involving trigonometric functions is used instead:

$$u(x) = e^{-ax/2} \left[A \sin \left(\frac{1}{2} \sqrt{4b - a^2} \right) x + B \cos \left(\frac{1}{2} \sqrt{4b - a^2} \right) x \right]$$

(8.8)

where A and B are arbitrary constants to be determined by the conditions given for Equation 8.1.

The expression in Equation 8.8 was derived by using the Biot relation, which has the form $e^{\pm i\theta} = \cos \theta \pm i \sin \theta$.

For the special case with coefficient $a = 0$ and b a negative number, the solution of Equation 8.1 becomes

$$u(x) = c_1 \cosh(2\sqrt{b})x + c_2 \sinh(2\sqrt{b})x$$

(8.9)

where c_1 and c_2 are arbitrary constants to be determined by the given conditions.

Case 3. $a^2 - 4b = 0$

We will quickly realize from Equation 8.4 that this case leads to $m_1 = m_2 = m = -a/2$, which will produce the solution of Equation 8.1 in the form $u(x) = (c_1 + c_2)e^{-ax/2}$ or $u(x) = ce^{-ax/2}$, with the constant coefficient $c = (c_1 + c_2)$. We notice that this form of solution contains only one arbitrary constant, whereas a second-order differential equation should have two arbitrary constants. We will thus need to find the missing term with another arbitrary constant in the general solution of Equation 8.1. Such an additional term with the "other" arbitrary constant may be found by letting the *additional root* of m in Equation 8.3 to be obtained by the following modified expression for the solution of Equation 8.1:

$$u_2(x) = V(x)e^{mx}$$

(8.10)

where $V(x)$ is an assumed function of x.

By substituting the above assumed solution $u_2(x)$ in Equation 8.10 into the differential equation in Equation 8.1, we will obtain the following equation for $V(x)$:

$$\frac{d^2 V(x)}{dx^2} + (2m + a)\frac{dV(x)}{dx} + (m^2 + am + b)V(x) = 0$$

(8.11)

One may recall from Equation 8.3 that $m^2 + am + b = 0$ and $m_1 = m_2 = m = -a/2$ for the case of $a^2 - 4b = 0$. The same equality also results in $2m + a = 0$. Thus, the only nonzero term left in Equation 8.11 is the second-order derivative of the function $V(x)$. Consequently, we have a simple differential equation for the function $V(x)$:

$$\frac{d^2 V(x)}{dx^2} = 0$$

We will obtain the solution of the above differential equation to be $V(x) = x$ after two sequential integrations.

The second solution for $u(x)$, by Equation 8.10, is thus found to be

$$V(x)e^{mx} = xe^{mx} = xe^{-ax/2}$$

which leads to the complete solution of the differential equation in Equation 8.1 with $a^2 - 4b = 0$:

$$u(x) = c_1 e^{-ax/2} + c_2 xe^{-ax/2} = (c_1 + c_2 x)e^{-ax/2}$$

(8.12)

Example 8.1

Solve the following differential equation:

$$\frac{d^2u(x)}{dx^2} + 5\frac{du(x)}{dx} + 6u(x) = 0 \tag{a}$$

Solution:

Comparison of Equation (a) and Equation 8.1 gives the equivalences $a = 5$ and $b = 6$, resulting in $a^2 - 4b = 5^2 - 4 \times 6 = 1 > 0$. So the Case 1 solution of Equation 8.6 applies in this problem, and the solution of Equation (a) becomes

$$u(x) = e^{-5x/2}(c_1 e^{x/2} + c_2 e^{-x/2})$$
$$= c_1 e^{-2x} + c_2 e^{-3x}$$

where c_1 and c_2 are arbitrary constants.

Example 8.2

Solve the following equation:

$$\frac{d^2u(x)}{dx^2} + 6\frac{du(x)}{dx} + 9u(x) = 0 \tag{a}$$

with the following specified conditions:

$$u(0) = 2 \quad \text{and} \quad \left.\frac{du(x)}{dx}\right|_{x=0} = 0 \tag{b}$$

Solution:

In Equation (a), we recognize that $a = 6$ and $b = 9$ after comparing Equation (a) with Equation 8.1. We thus obtain the roots m_1 and m_2 from Equation 8.4 to be

$$m_1 = -\frac{a}{2} + \frac{1}{2}\sqrt{a^2 - 4b} = -3 \quad \text{and} \quad m_2 = -\frac{a}{2} - \frac{1}{2}\sqrt{a^2 - 4b} = -3 \tag{c}$$

indicating that $m = m_1 = m_2 = -3$ and that Case 3 applies for the solution. Thus, according to Equation 8.12, we have the solution of the differential equation in Equation (a) as

$$u(x) = (c_1 + c_2 x)e^{-6x/2} = (c_1 + c_2 x)e^{-3x} \tag{d}$$

where the arbitrary constants c_1 and c_2 may be determined by the specified conditions in Equation (b): $u(0) = 2$ leads to $c_1 = 2$, and we get $c_2 = 6$ from the second condition. The complete solution of Equation (a) is thus

$$u(x) + 2(1 + 3x)e^{-3x}$$

8.3 Applications in Mechanical Vibration Analyses

8.3.1 What Is Mechanical Vibration?

Mechanical vibration is a form of *oscillatory* motion of a solid, a structure, a machine, or a vehicle induced by mechanical means. The extent of movement in these solids and structures is called the "*amplitude.*" The amplitudes of vibrating solids vary with time. Such variations may be either regular or in random fashion. Oscillatory motions of solids with their amplitudes vary with a fixed time interval called the "*period,*" and the reciprocal of the period is the "*frequency*" of the vibratory motion.

The effect of mechanical vibration on machines and structures can be trivial and cause mere discomfort to riders in a vibrating vehicle, or can be devastating and result in major structural failures.

Consequences of mechanical vibration can be immediate, such as in the case of resonant vibration with rapid increase of magnitudes of vibration, or damage can be induced cumulatively by long-term vibrations with low amplitudes. These prolonged low-amplitude vibrations may result in the failure of a machine or structure due to fatigue of the materials of which they are constructed.

8.3.2 Common Sources for Vibration

There are many causes of mechanical vibration in solids and in machines. Engineers need to be fully aware of such sources in order to deal with the consequences. Common sources for mechanical vibrations are

1) Application of time-varying mechanical force or pressure
2) Fluid-induced vibration due to intermittent forces of wind, tidal waves, etc.
3) Application of pressures associated with acoustics and ultrasonic waves
4) Random movements of supports; for example, seismic forces
5) Application of thermal, magnetic forces, etc.

8.3.3 Common Types of Vibration

There are different types of mechanical vibrations that design engineers should be aware of. Following are three common types of vibration that engineers may be involved in their analyses.

1) *Vibrating solids with constant amplitudes and frequencies:* In such cases, the solid vibrates in both directions from its initial equilibrium position with continuously varying amplitudes and phase following regular patterns such as illustrated in Figure 8.1a.
2) *Vibration with variable amplitude but constant frequency:* The solid vibrates in only one direction from its initial equilibrium position but with varying amplitude (Figure 8.1b).
3) *Random amplitudes and frequencies:* There is no clear pattern for the motion of the solid in the vibration as illustrated in Figure 8.1c. This pattern of vibration is commonly observed in the case of seismically induced vibration.

8.3.4 Classification of Mechanical Vibration Analyses

There are generally three distinct types of vibration analysis that engineers encounter in practice: (1) free vibration analysis, (2) damped vibration analysis, and (3) forced vibration analysis.

8.3.4.1 Free Vibration

Solids, whether they are rigid bodies such as vehicles, deformable structures, or machines and devices, all involve "mass" and an "elastic constraint" that supports the mass and are susceptible to mechanical vibrations. We may generalize the minimum requirement for a solid to vibrate from its initial static condition involves a mass and an elastic support. Accordingly, the simplest mathematical model for a mechanical vibrating system is what is termed a "mass–spring system" such as illustrated in Figure 8.2. Vibration of the mass in both configurations illustrated in Figure 8.2 is induced by an instantaneous disturbance—a short-lived disturbance to the mass. The inertia of the moving mass and the recoil of the supporting spring that together keep the mass moving continuously produce what is called *free vibration*.

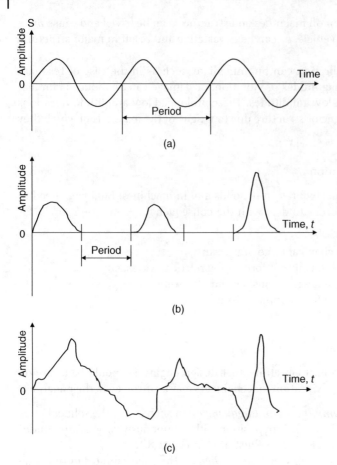

Figure 8.1 Forms of vibration. (a) Vibration with regular variable amplitudes and frequencies. (b) Vibration with irregular variable amplitudes but constant frequency. (c) Random vibrations.

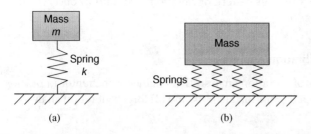

Figure 8.2 A mass–spring vibrational system. (a) A simple mass–spring system. (b) Mass–spring systems in general.

Free vibration of solids is thus the case in which the solid vibrates with no external forces or influences on the vibrating mass after the inception of the vibration. The vibration is caused solely by an initial disturbance. No excitation force is applied to the mass *during* the vibration.

8.3.4.2 Damped Vibration
Mechanical vibration with a damping effect is called *damped vibration.*

A retarding force, produced either by the internal friction of the molecules of the spring supports as the supports deform during the vibration of the mass, or by the friction force acting on the mass from the surrounding medium (such as the resistance from the ambient air), causes

Figure 8.3 A mass–spring–dashpot vibrational system.

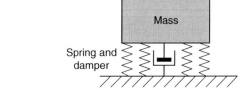

Figure 8.4 Forced vibration.

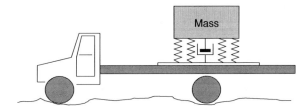

the reduction of the amplitude of the motion of the vibrating mass. These "frictional" forces are referred to as the *damping* force. A dashpot consisting of a piston in an air cylinder is used to simulate the damping effect in the vibration The model used for damped vibration analysis is illustrated in Figure 8.3.

8.3.4.3 Forced Vibration

In the case of *forced vibration*, an external force or external influence is applied to the vibrating mass at all times. Figure 8.4 illustrates one such situation, with external forces acing on the mass–spring–dashpot system, induced by the shaking of the support (the carrying truck) due to a bumpy road. The external force acting on the vibrating mass is called the *excitation force*. Magnitudes of the excitation forces may be random as in the case illustrated in Figure 8.4 or may exhibit a cyclic variation.

8.4 Mathematical Modeling of Free Mechanical Vibration: Simple Mass–Spring Systems

As mentioned in Section 8.3.4, the simplest mechanical vibration system is the "mass–spring system." The physical arrangement of this system includes a mass m that either rests on a spring support or is suspended from a spring with specific spring constant k, as illustrated in Figure 8.5. The spring constant is defined as the amount of force required to elongate or compress the spring by a unit length and is a property of the spring.

We can see that in either of the cases illustrated in Figure 8.5 the mass may be set to vibrate from its initial equilibrium state by a small external disturbance (e.g., a gentle push or pull on the mass in Figure 8.5). The elasticity of the spring, coupled with the inertia of the mass, produces the forces that cause the mass to vibrate vertically above and below its initial position following the initial disturbance.

The velocity of the moving mass is by no means constant with displacement from its initial equilibrium position. Imagine that the mass would move faster initially and then slow down due to the increasing recoil in spring force induced by its continuous elongation (or compression), and then to a standstill when it reaches an extreme position, at which point it reverses its direction of motion and accelerates and then decelerates until it reaches the other extreme position. The linear distance that the mass travels from its initial equilibrium position may be

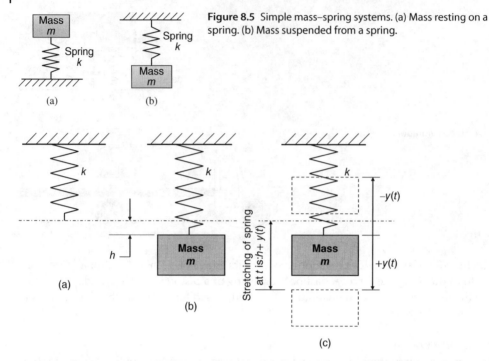

Figure 8.5 Simple mass–spring systems. (a) Mass resting on a spring. (b) Mass suspended from a spring.

Figure 8.6 Modeling a vibrating mass. (a) Free-hung spring. (b) Statically stretched spring. (c) A vibrating mass at time t.

denoted by $y(t)$, in which t is the time into the vibratory motion of the mass. The instantaneous position $y(t)$ varies along the vertical axis designated as the y-coordinate in the analysis.

We desire to derive a mathematical model such as a differential equation to determine the instantaneous position of the mass $y(t)$ during the vibration. From that, we may compute the corresponding velocity and acceleration (or deceleration). The acceleration (or deceleration) may be used to assess the associated dynamic forces on the vibrating mass, which is an important parameter in the design of a machine or of vehicle in motion.

Figure 8.6 illustrates the three stages of constructing a model for the mechanical vibration of a mass–spring system. Figure 8.6a represents a free-hung spring with a specified spring constant k. A mass m is then attached to the spring, which causes the spring to be stretched by an amount h as illustrated in Figure 8.6b. The force acting on the mass is a static force at this point, with the spring force $F_s = kh$ that is equal to the equivalent weight of the mass, or $F_s = W = mg$, as illustrated in Figure 8.7b.

We note from the free-body force diagram in Figure 8.7 that $F - W = 0$, which leads to the equality

$$kh = mg \tag{8.13}$$

We will now set the mass in Figure 8.6b in motion by a slight gentle downward push followed by a sudden release. We may expect the mass to bounce (vibrate) up and down from its initial equilibrium position as illustrated in Figure 8.6c, with the positive direction of the downward movement of the mass being $+y(t)$.

It is of interest to compute the instantaneous displacement of the vibrating mass $y(t)$ at any given time t during the vibratory motion of the mass. The equation that will enable us to solve for $y(t)$ may be derived using Newton's law with dynamic equilibrium of all applied forces on

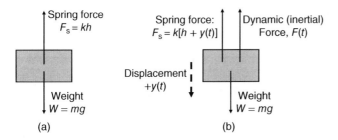

Figure 8.7 Forces acting on a vibrating mass. (a) Static equilibrium. (b) Dynamic equilibrium of a vibrating mass.

the mass at any given time t. Figure 8.7b illustrates all the forces acting on the mass at time t. We notice that the elastic force F_s exerted by the attached spring includes the force induced by its "extra" stretching by an amount $y(t)$, and a dynamic (or inertial) force that is associated with the movement of the mass at varying velocities. This dynamic force $F(t)$ can be expressed in terms of the distance travelled by the vibrating mass $y(t)$ in the form

$$F(t) = ma = m\frac{d^2y(t)}{dt^2}$$

in which a is the acceleration of the moving mass according to Newton's second law.

Thus, referring to Figure 8.7b, the equilibrium of the forces acting on the mass at time t should satisfy the following condition of having the summation of all forces acting on the mass in the y-direction equal to zero at given time t:

$$\sum F_y = -F(t) - F_s + W = 0$$

After substituting the appropriate forms of the force components into the above expression, we obtain the following:

$$-m\frac{d^2y(t)}{dt^2} - k[h + y(t)] + mg = 0$$

We know that $kh = mg$ as shown in Equation 8.13, and we thus obtain the differential equation for the instantaneous position of the vibrating mass $y(t)$ as

$$m\frac{d^2y(t)}{dt^2} + ky(t) = 0 \tag{8.14}$$

8.4.1 Solution of the Differential Equation

We recognize that Equation 8.14 is a special form of the typical second-order differential equation in Equation 8.1 with $a = 0$ and $b = k/m$ (a positive real number). The discriminator $a^2 - 4b = -4k/m < 0$ results in Case 2 as presented in Section 8.1. Consequently, the solution given in Equation 8.8 is used in the present case, leading to the following expression for the solution of Equation 8.14:

$$y(t) = A\cos\sqrt{\frac{k}{m}}t + B\sin\sqrt{\frac{k}{m}}t \tag{8.15}$$

in which A and B are arbitrary constants.

The solution in Equation 8.15 is often expressed in a more convenient form as

$$y(t) = A\cos\omega_0 t + B\sin\omega_0 t \tag{8.16}$$

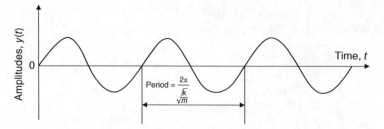

Figure 8.8 Graphical representation of a vibrating mass in simple harmonic motion.

in which $\omega_0 = \sqrt{k/m}$ is referred to as the *circular frequency* or *angular frequency* of the mass–spring system; it is also referred to as the *natural frequency* of this simple structure involving a mass constrained by a spring as illustrated in Figure 8.5. It has units of rad/s. The natural frequency ω_0 in Equation 8.16 also appeared in a similar mass–spring system presented in Section 4.8.3.

The corresponding frequency of the vibrating mass can thus be expressed as

$$f = \frac{\omega_0}{2\pi} = \frac{1}{2\pi}\sqrt{\frac{k}{m}} \tag{8.17}$$

The motion of the mass described by $y(t)$ in Equation 8.16 is referred to as "harmonic oscillation" involving sine and cosine functions. It can be graphically illustrated as in Figure 8.8.

A simple mass–spring structure such as illustrated in Figure 8.5 would seem to be rare in reality. However, engineers may change this perception once they realize that in reality the "spring" need not be in the form of "coils." In fact, any solid made to deform elastically can be viewed as a "spring." A steel rod that supports a solid mass as illustrated in Figure 8.9a is a "spring." Likewise, a cantilever beam that supports a solid of mass m shown in Figure 8.9b is also a "spring," as long as it deforms (or deflects) in response to the applied force (such as supplied by the weights in Figure 8.9) and reverts to its original state after the removal of the applied force.

The corresponding spring constants for these two nonconventional "springs" in Figure 8.9 can be defined in the same way as that for a coil spring. We can thus derive the spring constant for the "springs" in Figure 8.9 as

$$k_{\text{rod}} = \frac{EA}{L} \quad \text{for the rod spring} \tag{8.18a}$$

and

$$k_{\text{beam}} = \frac{3EI}{L^3} \quad \text{for the cantilever beam spring} \tag{8.18b}$$

where E is the Young's modulus of the rod or beam materials, A is the cross-sectional area of the rod in Figure 8.9a, and I is the area moment of inertia of the cantilever beam in Figure 8.9b.

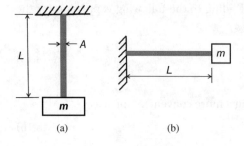

Figure 8.9 Non-coil springs. (a) Rod spring. (b) Cantilever spring.

(a)   (b)

These nonconventional springs are often used in special applications. For example, the cantilever beam spring is used in some microdevices, such as in early versions of inertia sensors for airbag deployment in automobiles [Hsu 2002, 2008]. Application of the concept of the rod spring will be described in the following example.

One may relate these situations to a number of structures and machines that can be modeled by a simple mass–spring system. The following example demonstrates such an application in an engineering analysis.

Example 8.3

An 800-pound machine is being lowered to the ground by a crane as illustrated in Figure 8.10. The steel cable that holds the machine has an equivalent spring constant $k = 6000 \text{ lb}_f/\text{in}$. The equivalent spring constant of a cable is defined as its elongation induced by an applied force in tension. The machine is being let down by the crane at a constant velocity of 20 ft/min when the motor that controls the descending cable suddenly jams. One may imagine that the machine would exhibit a "bouncing" motion up-and-down with visible amplitudes for a short period of time. Determine the following:

a) The frequency of vibration of the machine induced by the sudden seizure of the steel cable.
b) The maximum tension induced in the cable by this vibrating machine.
c) The maximum stress in the stranded cable is 0.25 inch in diameter.
d) Whether the cable would break with an allowable stress at 25 000 psi found from a materials handbook.

Solution:

We recognize that the present situation as depicted in Figure 8.10, involving a mass (the machine) suspended by an elastic steel cable is similar to that illustrated in Figure 8.5b. The steel cable is *elastic*, which means that physically it acts like a *spring* with a spring constant $k = 6000 \text{ lb}_f/\text{in}$.

a) The frequency of vibration of the machine is given in Equation 8.17. Numerically, The circular frequency is

$$\omega_0 = \sqrt{\frac{k}{m}} = \sqrt{\frac{6000 \times 12}{800/32.2}} = 53.83 \ \text{rad/s}$$

and the frequency of vibration is

$$f = \frac{\omega_0}{2\pi} = 8.57 \ \text{cycles/s}$$

b) The maximum tension in the cable is determined by the maximum elongation of the steel cable: that is, the maximum amplitude of the vibration of the machine after the cable

Figure 8.10 Hoisting a machine by a steel cable.

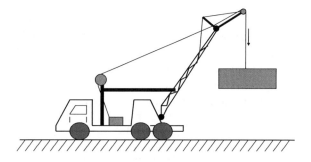

becomes seized plus the elongation of the cable by the dead weight of the machine. To obtain the amplitude of the vibrating mass, we need to solve a differential equation that has the form shown in Equation 8.14 with the appropriate conditions:

$$m\frac{d^2y(t)}{dt^2} + ky(t) = 0 \tag{8.18}$$

where in this case $m = W/g = 800/32.2$ slugs and $k = 6000 \times 12$ lb_f/ft. The initial conditions $y(0) = 0$, and $y'(0) = 20$ ft/min $= 0.3333$ ft/s are applicable.

The solution of Equation 8.18 is available in the form of Equation 8.16:

$$y(t) = A\cos 53.83t + B\sin 53.83t \tag{a}$$

where A and B are arbitrary constants to be determined by the given initial conditions. The first initial condition of $y(0) = 0$ results in $A = 0$, and the second in $B = 0.0062$.

We thus have the amplitude of vibration of the machine in the unit of feet as

$$y(t) = 0.0062\sin 53.83t \tag{b}$$

The maximum amplitude of vibration clearly is the coefficient of the sine function in the above expression, or $y_{max} = 0.0062$ ft.

The corresponding maximum tension in the cable is

$$T_m = ky_{max} + W$$
$$= (6000 \times 12) \times 0.0062 + 800 = 1246 \ lb_f$$

c) The corresponding maximum stress in the cable is

$$\sigma_{max} = \frac{T_m}{A} - \frac{1246}{\frac{\pi(0.25)^2}{4}} = 25\,396 \ \text{psi}$$

d) We note that the maximum tensile stress in the cable is 25 396 psi, which exceeds the allowable stress of the cable material found from a materials handbook by 396 psi. One may thus interpret the situation as critical.

8.5 Modeling of Damped Free Mechanical Vibration: Simple Mass–Spring Systems

8.5.1 The Physical Model

In Figure 8.8 the motion of a vibrating mass is shown as a perpetual oscillation of the mass within fixed bounds of amplitude variation with elapsed time. One may well appreciate that this is unrealistic and that the amplitude of vibration of the mass will continuously decrease with time, and eventually the vibrating mass will return to standstill at its original position. It is logical to ask what produces the latter situation in reality since, according to theory, the mass would oscillate indefinitely between fixed bounds of amplitudes as shown in Figure 8.8. The answer is that *friction* forces are the principal reason for damping of the vibration and the mass eventually returning to its original equilibrium state. These forces arise from friction between the moving mass and the surrounding air, and the internal friction between the molecules in the deforming spring.

The mathematical model for this more realistic damped vibration has to include a *damper* in the form of a dashpot, which consists of a piston–cylinder assembly as illustrated in Figure 8.11a and symbolized in the mass–spring system shown in Figure 11b.

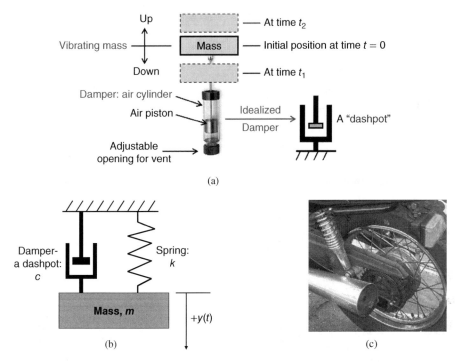

Figure 8.11 Physical model for damped vibration of a mass–spring system. (a) Vibrating mass supported by a dashpot. (b) Mass supported by spring and dashpot. (c) Mass–dashpot assembly for a motorcycle suspension.

Figure 8.11c shows the use of a "coilover" as frequently seen in the rear wheel suspension system of a motorcycle. The coilover consists of a dashpot that is enclosed by a coil spring as shown in the photograph. It is a common support used in the suspension systems of many vehicles.

8.5.2 The Differential Equation

The differential equation for the damped vibration may be derived by applying a damping force to the mass in the mass–spring system as illustrated in Figure 8.11b. The damping force associated with the damper is similar to the air resistance against a moving mass as described in Chapter 7, in which we formulated the air resistance $R(t)$ as

$$R(t) \propto \text{Velocity of moving mass in the form } \frac{dy(t)}{dt}$$

We may thus express the damping force as an additional force acting on the vibrating mass as follows:

$$R(t) = c\frac{dy(t)}{dt} \tag{8.19}$$

where c is the damping coefficient of the dashpot or damper. The value of c is determined by experiment, and is customary supplied by the manufacturers of the dashpots.

The forces acting on the vibrating mass at time t are shown in Figure 8.12.

For the system in a dynamic equilibrium at time t, we should have the following relationship according to Newton's second law:

$$\sum F_y = -F(t) - R(t) - F_s + W = 0$$

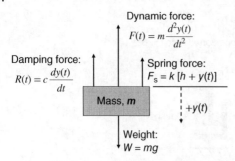

Dynamic force:
$$F(t) = m\frac{d^2y(t)}{dt^2}$$

Figure 8.12 Forces acting on a vibrating mass with damping.

Damping force:
$$R(t) = c\frac{dy(t)}{dt}$$

Spring force:
$$F_s = k\,[h + y(t)]$$

Mass, *m*

$+y(t)$

Weight:
$W = mg$

This expression of dynamic force equilibrium leads to the following mathematical expression:

$$m\frac{d^2y(t)}{dt^2} + c\frac{dy(t)}{dt} + k\,y(t) + kh - mg = 0$$

Since $kh = mg$ as we established in Figure 8.6b, we will thus have the differential equation for a damped vibration of a mass–spring as

$$m\frac{d^2y(t)}{dt^2} + c\frac{dy(t)}{dt} + k\,y(t) = 0 \tag{8.20}$$

8.5.3 Solution of the Differential Equation

A comparison of the terms in Equation 8.20 and those in Equation 8.1 will give $a = c/m$ and $b = k/m$, in which a and b are the constant coefficients in Equation 8.1.

We may use Equation 8.4 to express the coefficients m_1 and m_2 for the current case as follows:

$$m_1 = -\frac{a}{2} + \frac{\sqrt{a^2 - 4b}}{2} = -\frac{c}{2m} + \frac{\sqrt{c^2 - 4mk}}{2m} \tag{8.21a}$$

$$m_2 = -\frac{a}{2} - \frac{\sqrt{a^2 - 4b}}{2} = -\frac{c}{2m} - \frac{\sqrt{c^2 - 4mk}}{2m} \tag{8.21b}$$

The solution $y(t)$ in Equation 8.20 thus becomes

$$y(t) = c_1 e^{m_1 t} + c_2 e^{m_2 t} \tag{8.22}$$

in which c_1 and c_2 are arbitrary constants.

As indicated in Section 8.1, there are three distinct cases for the solution of $y(t)$ in Equation 8.20 depending on the sign of the arguments in the square roots in the expressions for m_1 and m_2 in Equations 8.21a and 8.21b.

Case 1. $c_2 - 4mk > 0$—the "over-damping situation"
In this case, both roots m_1 and m_2 in Equations 8.21a and 8.21b are real numbers. Consequently, the solution of the Equation 8.20 has the form

$$y(t) = e^{-(c/2m)t}(A\,e^{\Omega t} + B\,e^{-\Omega t}) \tag{8.23}$$

where A, B = arbitrary constants, and $\Omega = \sqrt{c^2 - 4mk}/(2m)$.

Graphical representation of the solution in Equation 8.23 is shown in Figure 8.13, on which one may make the following observations:

a) There is no oscillatory motion of the mass because of the absence of harmonic functions sine and cosine in the solution in Equation 8.23.

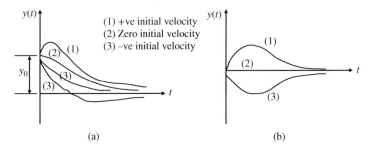

Figure 8.13 Amplitudes in over-damped vibration. (a) With +ve initial displacement, y_0. (b) With negligible initial displacement.

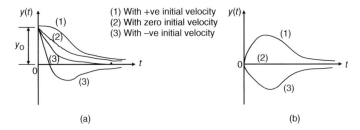

Figure 8.14 Amplitudes in critically damped vibration. (a) With +ve initial displacement. (b) With negligible initial displacement.

b) There can be an initial increase in the displacement, followed by continuous decays in the amplitudes of vibration.
c) There is rapid reduction in amplitudes of vibration initially, followed by more gradual reduction.

Case 2. $c_2 - 4mk = 0$—"Critical damping"

In this case the two roots as shown in Equations 8.21a and 8.21b are identical: $m_1 = m_2$. The solution of Equation 8.20 has the following form similar to that shown in Equation 8.12:

$$y(t) = e^{-(c/2m)t}(A + Bt) \tag{8.24}$$

where A and B are arbitrary constants.

Graphical representation of the solution in Equation 8.24 is illustrated in Figure 8.14. In this case we may make the following observations on the physical behavior:

a) There is no oscillatory motion of the mass for the same reason as in Case 1.
b) Amplitude reduces with time, but takes longer to diminish.

Case 3. $c_2 - 4mk < 0$—"under-damping"

We realize here that the two roots, m_1 and m_2 are both complex numbers. Consequently, the solution of Equation 8.20 takes a form similar to that shown in Equation 8.8. Thus by letting $\Omega = \sqrt{4mk - c^2}/(2m)$, we obtain the solution of $y(t)$ in Equation 8.20 as

$$y(t) = e^{-(c/2m)t}(c_1 \cos \Omega t + c_2 \sin \Omega t) \tag{8.25}$$

in which c_1 and c_2 are arbitrary constants.

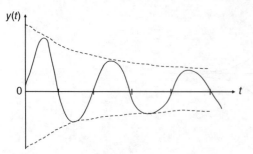

Figure 8.15 Amplitudes in under-damped vibration.

Graphical representation of Equation 8.25 is illustrated in Figure 8.15. We may characterize this case of "under-damping" as follows:

a) It is the only case of damped vibration having oscillatory motion of the mass.
b) The amplitude of oscillatory motion of the mass takes a long time to diminish.

The three cases of damped vibration of a mass–spring systems as illustrated in Figure 8.13, Figure 8.14, and Figure 8.15 illustrate interesting phenomena that lead to interesting questions for engineers, such as: "If you were to design a suspension system for a vehicle using combined support of spring and dashpot, such the coilover shown in Figure 8.11c, which of the above three cases would you choose?"

8.6 Solution of Nonhomogeneous, Second-order Linear Differential Equations with Constant Coefficients

8.6.1 Typical Equation and Solutions

There are times when engineers have to face situations in which the physical phenomena that they need to model are best fitted with nonhomogeneous second-order differential equations, rather the homogeneous equations exemplified by Equation 8.1. The solution method that we will adopt and is presented in this section for the nonhomogeneous differential equation is to extend the method that we have presented in Section 8.1 for typical second-order homogeneous differential equations.

A typical second-order nonhomogeneous differential equation can be written in a form similar to Equation 8.1:

$$\frac{d^2u(x)}{dx^2} + a\frac{du(x)}{dx} + bu(x) = g(x) \tag{8.26}$$

We see that Equation 8.26 differs from Equation 8.1 in a nonzero term on the right-hand side of the equation, which makes it nonhomogeneous.

The solution of Equation 8.26 consists of *two* parts:

$$u(x) = u_h(x) + u_p(x) \tag{8.27}$$

in which $u_h(x)$ is the complementary solution and $u_p(x)$ is the particular solution of Equation 8.26

8.6.2 The Complementary and Particular Solutions

The complementary solution, $u_h(x)$ in Equation 8.27 may be obtained by solving Equation 8.26 in its homogeneous form as shown in Equation 8.28.

$$\frac{d^2u_h(x)}{dx^2} + a\frac{du_h(x)}{dx} + bu_h(x) = 0 \tag{8.28}$$

We recognize that Equation 8.28 is similar to Equation 8.1. Hence $u_h(x)$ can be solved from Equation 8.28 using the method that is outlined in Section 8.1.

8.6.3   The Particular Solutions

The general solution of Equation 8.26 requires the particular solution $u_p(x)$ as indicated in Equation 8.27. There is no established rule to follow for the solution $u_p(x)$. However, a general guideline is that the form of $u_p(x)$ is similar to that of the nonhomogeneous portion $g(x)$ in Equation 8.26. The procedure for finding the particular solution of Equation 8.26 is to assume a function for $u_p(x)$ that is similar to the function $g(x)$ in Equation 8.26 but with unknown coefficients. In other words, we may assume the particular solution $u_p(x)$ to be a "close cousin" but not the "twin" of $g(x)$ in Equation 8.26. A guideline for choosing assumed particular solutions $u_p(x)$ is presented in Table 8.1. The constant coefficients associated with the assumed particular solution $u_p(x)$ in the table will be determined later by substituting the assumed $u_p(x)$ function into the original differential equation in Equation 8.26 as

$$\frac{d^2 u_p(x)}{dx^2} + a\frac{du_p(x)}{dx} + bu_p(x) = g(x) \tag{8.29}$$

Equation 8.29 with the assumed function for $u_p(x)$ will convert the nonhomogeneous differential equation into an algebraic equation with the unknown constants in the assumed solution of $u_p(x)$. The assumed constants may be determined by comparing the coefficients of the terms on both sides of these resultant algebraic equations.

The following guidelines may be helpful for engineers in assuming the solution of $u_p(x)$ in Equation 8.27.

a) *For the case $g(x)$ a polynomial function*: We may assume a solution $u_p(x)$ in the form of a polynomial of the same order as the function $g(x)$:

$$u_p(x) = A_0 + A_1 x + A_2 x^2 + \cdots + A_n x^n \tag{8.30a}$$

in which $A_0, A_1, A_2, \ldots, A_n$ are unknown constants. The highest order of the assumed polynomial function in Equation 8.30a is dictated by the highest order in the given function $g(x)$ in Equation 8.26.

Table 8.1 Guidelines for choosing assumed particular solution $u_p(x)$.

Nonhomogeneous function $g(x)$ with specified coefficients	Assumed particular solution $u_p(x)$ with assumed unknown coefficients
Polynomial function of order n $g(x) = ax^4 + bx^3 + cx^2 + dx + e$ (order 4)	*Polynomial function of order n* $u_p(x) = A_0 + A_1 x + A_2 x^2 + A_3 x^3 + A_4 x^4$ (order 4)
Trigonometric functions $g(x) = a\sin(ax)$, or $g(x) = b\cos(ax)$, or $g(x) = a\cos(ax) + b\sin(\beta x)$	*Trigonometric functions* $u_p(x) = A_1 \cos(ax) + A_2 \sin(\beta x)$
Exponential functions $g(x) = ae^{bx}$	*Exponential function* $u_p(x) = A_1 e^{bx}$
Combination of functions $g(x) = ax^3 + b\cos(\beta x) + ce^{-dx}$	*Combination of functions* $u_p(x) = (A_0 + A_1 x + A_2 x^2 + A_3 x^3)$ $+ [A_4 \cos(ax) + A_5 \sin(ax)] + A_6 e^{-dx}$

b) *For the case $g(x) = Ce^{kx}$, an exponential function:* We may assume that

$$u_p(x) = A_1 e^{kx} \tag{8.30b}$$

with A_1 being the unknown constant.

c) *For the case with $g(x)$ to be a trigonometric function:* The function may be either a sine or a cosine function or a sum of these functions in the form $a \cos \alpha x + b \sin \beta x$. We assume a particular solution to be

$$u_p(x) = A_1 \cos \alpha x + A_2 \sin \beta x \tag{8.30c}$$

where A_1 and A_2 are the unknown constants.

The following examples illustrate how the guideline in Table 8.1 for selecting $u_p(x)$ can be used to determine the form of the particular solution.

Example 8.4

Solve the following nonhomogeneous differential equation:

$$\frac{d^2y(x)}{dx^2} - \frac{dy(x)}{dx} - 2y(x) = 4x^2 \tag{a}$$

Solution:

Equation (a) is a nonhomogeneous equation, so we will express the general solution as

$$y(x) = y_h(x) + y_p(x) \tag{b}$$

in which $y_h(x)$ is the complementary solution of the equation

$$\frac{d^2y_h(x)}{dx^2} - \frac{dy_h(x)}{dx} - 2y_h(x) = 0$$

leading to a solution

$$y_h(x) = c_1 e^{-x} + c_2 e^{2x} \tag{c}$$

using the method given in Section 8.1.

The constants c_1 and c_2 in Equation (c) are arbitrary constants to be determined by the specified conditions. Determination of these arbitrary constants can be done *only after the complete general solution is obtained.*

Since the nonhomogeneous part of the equation $g(x)$ in Equation (a) is $4x^2$, which is a *second-order polynomial function*, we thus assume a particular solution, $u_p(x)$ to be a polynomial of the same order, as stipulated in Table 8.1, We thus assume the following polynomial function of the same order (2) for the assumed function of $u_p(x)$:

$$y_p(x) = A_0 + A_1 x + A_2 x^2 \tag{d}$$

in which A_0, A_1, and A_2 are unknown constant coefficients.

The particular solution $y_p(x)$ in Equation (d) should satisfy the differential equation in Equation (a) as shown below:

$$\frac{d^2y_p(x)}{dx^2} - \frac{dy_p(x)}{dx} - 2y_p(x) = 4x^2 \tag{e}$$

The above differential equation will be converted into an algebraic equation after substituting the assumed $y_p(x)$ in Equation (d), resulting in

$$2A_2 - (A_1 + 2A_2 x) - 2(A_0 + A_1 x + A_2 x^2) = 4x^2 \tag{f}$$

By comparing the coefficients of the terms on both sides of Equation (f), we may solve for $A_0 = -3$, $A_1 = 2$, and $A_2 = -2$.

We thus have the particular solution:

$$y_p(x) = -2x^2 + 2x - 3 \tag{g}$$

The complete solution of Equation (a) is the sum of the complementary solution in Equation (c) and the particular solution in Equation (g):

$$y(x) = y_h(x) + y_p(x) = c_1 e^{-x} + c_2 e^{2x} - 2x^2 + 2x - 3$$

The arbitrary constants c_1 and c_2 in the above general solution can now be determined from the specified conditions for the differential equation in Equation (a).

Example 8.5

Solve the following differential equation:

$$\frac{d^2 y(x)}{dx^2} - \frac{dy(x)}{dx} - 2y(x) = e^{3x} \tag{a}$$

Solution:

We notice that the nonhomogeneous part of Equation (a) is an exponential function, and the left-hand side—the homogeneous part—is identical to that in Equation (a) in Example 8.4. Consequently, we will have the same complementary solution as in Example 8.4 in the form:

$$y_h(x) = c_1 e^{-x} + c_2 e^{2x} \tag{b}$$

The particular solution for Equation (a) in this example should be an exponential function in accordance with Table 8.1:

$$y_p(x) = A_1 e^{3x} \tag{c}$$

In Equation (c), A_1 is a constant that can be determined by substituting the assumed particular solution in Equation (c) into Equation (a) to yield:

$$\frac{d^2 y_p(x)}{dx^2} - \frac{dy_p(x)}{dx} - 2y_p(x) = \frac{d^2(A_1 e^{3x})}{dx^2} - \frac{d(A_1 e^{3x})}{dx} - 2(A_1 e^{3x})$$
$$= e^{3x}$$

We will get the following algebraic equation after differentiating each term in the above equation:

$$9A_1 e^{3x} - 3A_1 e^{3x} - 2A_1 e^{3x} = e^{3x}$$

from which we get $A_1 = 1/4$. We thus have

$$y_p(x) = \frac{1}{4} e^{3x} \tag{d}$$

The general solution of the differential equation in Equation (a) is thus the sum of the complementary solution in Equation (b) and the particular solution in Equation (d):

$$y(x) = y_h(x) + y_p(x) = c_1 e^{-x} + c_2 e^{2x} + \frac{1}{4} e^{3x}$$

The two arbitrary constants c_1 and c_2 can be determined from the specified conditions for the differential equation (a).

Example 8.6

Solve the following nonhomogeneous differential equation:

$$\frac{d^2y(x)}{dx^2} - \frac{dy(x)}{dx} - 2y(x) = \sin 2x \tag{a}$$

Solution:

Again, we will have the same form of complementary solution as those in Examples 8.4 and 8.5 because the left-hand side of Equation (a) in this example is identical to that in the differential equations in the two previous examples. Consequently, the complementary solution will be

$$y_h(x) = c_1 e^{-x} + c_2 e^{2x} \tag{b}$$

Since the nonhomogeneous part of the differential equation in Equation (a) is a trigonometric function, we will assume the same form for the particular solution, $u_p(x)$:

$$y_p(x) = A_1 \sin 2x + A_2 \cos 2x \tag{c}$$

where A_1 and A_2 are assumed constant coefficients to be determined by substituting $y_p(x)$ from Equation (c) into Equation (a), resulting in

$$(-4A_1 \sin 2x - 4A_2 \cos 2x) - (2A_1 \cos 2x - 2A_2 \sin 2x) - 2(A_1 \sin 2x + A_2 \cos 2x) = \sin 2x$$

After rearranging the terms in the above expression, we get

$$(-6A_1 + 2A_2) \sin 2x + (-2A_1 - 6A_2) \cos 2x = \sin 2x$$

By comparing the coefficients of the terms on both sides of the expression, we get $-6A_1 + 2A_2 = 1$ and $-2A_1 - 6A_2 = 0$, from which we solve for $A_1 = -3/20$ and $A_2 = 1/20$, with which we may express the particular solution in the form

$$u_p(x) = -\frac{3}{20} \sin 2x + \frac{1}{20} \cos 2x$$

Consequently, the general solution of the differential equation in Equation (a) is

$$y(x) = y_h(x) + y_p(x) = c_1 e^{-x} + c_2 e^{2x} + \left(-\frac{3}{20} \sin 2x + \frac{1}{20} \cos 2x \right)$$

Example 8.7

Solve the following nonhomogeneous differential equation:

$$\frac{d^2u(x)}{dx^2} + \frac{du(x)}{dx} + 2u(x) = 4e^x + 2x^2 \tag{a}$$

Solution:

Following the same procedure as outlined in Examples 8.4, 8.5, and 8.6, we solve for the complementary solution, $u_h(x)$ from the homogeneous part Equation (a):

$$\frac{d^2u_h(x)}{dx^2} + \frac{du_h(x)}{dx} + 2u_h(x) = 0 \tag{b}$$

Using the solution method outlined in Section 8.1, the two roots, m_1 and m_2 take the forms of the expressions in Equation 8.4:

$$m_1 = -\frac{1}{2} + \frac{\sqrt{7}}{2}i \quad \text{and} \quad m_2 = -\frac{1}{2} - \frac{\sqrt{7}}{2}i$$

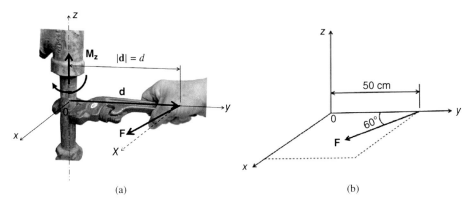

(a)   (b)

Figure 3.17 Cross product of two vectors giving the torque applied to a pipe. (a) Physical interpretation. (b) Numerical situation.

(a)   (b)

Figure 3.20 Lombard Street in San Francisco. (a) A narrow and winding street. (b) Breath-taking driving. (*Source*: (a) © Gaurav1146 Wikimedia Commons.)

Applied Engineering Analysis, First Edition. Tai-Ran Hsu.
© 2018 John Wiley & Sons Ltd. Published 2018 by John Wiley & Sons Ltd.
Companion Website: www.wiley.com/go/hsu/applied

Figure 5.1 A mechanism with periodic motions. (a) Riding horse on a merry-go-round. (b) Up-and-down motion of a mechanical pony.

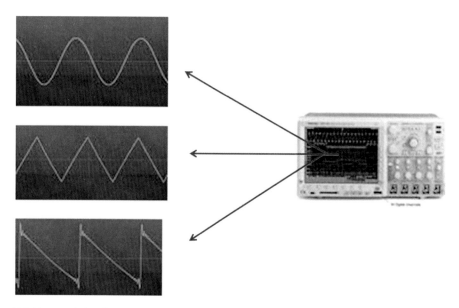

Figure 5.5 Measured signals representing periodic physical phenomena.

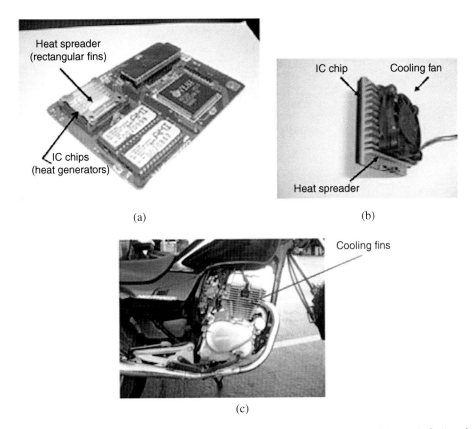

(a)

(b)

(c)

Figure 7.22 Heat spreaders of integrated circuits and an internal combustion engine. (a) A typical printed circuit board. (b) Dissipation of heat from an integrated circuit (IC) chip. (c) A motorcycle engine with cooling fins.

(a)

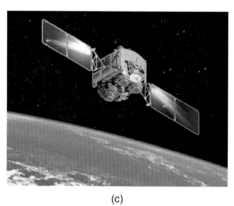

(c)

Figure 8.32 Structures of complex geometry that may be subjected to intermittent loading. (a) A billboard. (c) A satellite in orbit.

Figure 9.12 Long cables in electric power transmission structures.

Figure 9.13 Guy wire support for a tall radio transmission tower.

Figure 9.14 Golden Gate suspension bridge.

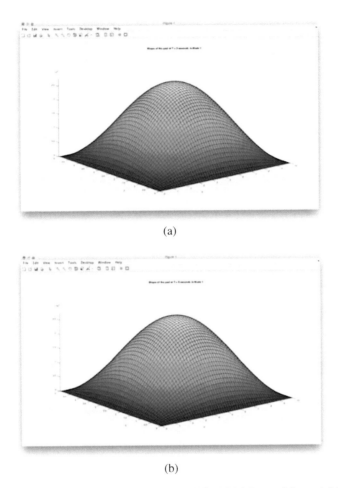

(a)

(b)

Figure 9.26 Shape of the mouse pad in mode 1 vibration. (a) The initial shape of the pad. (b) Mode shape at time 1/8 second.

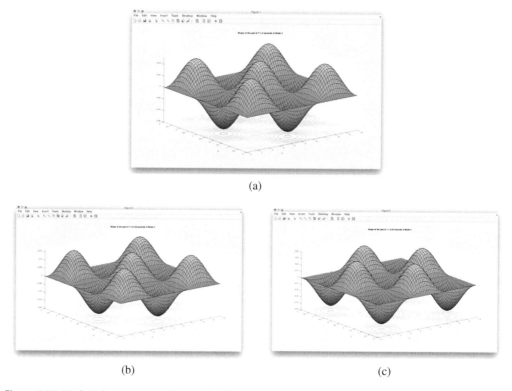

(a)

(b)   (c)

Figure 9.27 Mode 3 shapes in vibration of a flexible plate. (a) At time $t = 0$. At time $t = 1/8$ second. (c) At time $t = 1/4$ second.

which leads to the complementary solution

$$u_h(x) = e^{-x/2} \left(c_1 \sin \frac{\sqrt{7}}{2} x + c_2 \cos \frac{\sqrt{7}}{2} x \right) \tag{c}$$

The nonhomogeneous part of Equation (a) is a combination of a polynomial function of order 2 and an exponential function. We will thus assume a particular solution that contains both these functions:

$$u_p(x) = Ae^x + (Bx^2 + Cx + D) \tag{d}$$

in which A, B, C, and D are constants.

One may determine these constants after substituting the assumed $u_p(x)$ from Equation (d) into Equation (a) to give $A = 1$, $B = 1$, $C = -1$, and $D = -1/2$, which leads to the particular solution

$$u_p(x) = e^x + x^2 - x - \frac{1}{2} \tag{d}$$

The general solution of Equation (a) thus has the form

$$u(x) = u_h(x) + u_p(x)$$

$$= e^{-x/2} \left(c_1 \sin \frac{\sqrt{7}}{2} x + c_2 \cos \frac{\sqrt{7}}{2} x \right) + \left(e^x + x^2 - x - \frac{1}{2} \right)$$

8.6.4 Special Case for Solution of Nonhomogeneous Second-order Differential Equations

The method used to solve nonhomogeneous ordinary differential equations as outlined so far works well for the problems posed up to now. However, there are occasions in which *at least one function in the complementary solution of the differential equation is identical to that of the function in the nonhomogeneous part $g(x)$*. In such a case, we have to modify the assumed particular solution, as will be demonstrated in Example 8.8.

Example 8.8
Solve the following differential equation:

$$\frac{d^2 u(x)}{dx^2} + 4u(x) = 2 \sin 2x \tag{a}$$

Solution:
We will solve for the complementary solution from the homogeneous part of Equation (a) by letting

$$\frac{d^2 u_h(x)}{dx^2} + 4u_h(x) = 0 \tag{b}$$

The two constants m_1 and m_2 in the assumed solution of $u(x) = e^{mx}$ in Equation 8.2 are $m_1 = +2i$ and $m_2 = -2i$ for the form of homogeneous equation of Equation (b) as presented in Section 8.1, resulting in the following complementary solution:

$$u_h(x) = c_1 \cos 2x + c_2 \sin 2x \tag{c}$$

We note that the second term in the above expression for $u_h(x)$ in Equation (c) has the identical sine function as in the given nonhomogeneous part $g(x)$ of the given differential equation in Equation (a) in this example.

The usual approach of deriving the particular solution by assuming $u_p(x) = A_1 \cos 2x + A_2 \sin 2x$ following the guideline in Table 8.1 will lead to the following ambiguous equality after substituting the assumed expression into Equation (a):

$$(0) \cos 2x + (0) \sin 2x = 2 \sin 2x$$

Apparently the assumed form of $u_p(x)$ obtained by following the guideline in Table 8.1 can no longer be applied in this particular case, and a modified approach to assuming the particular solution needs to be found.

Alternatively, we may assume a modified particular solution to take the form

$$u_p(x) = x(A_1 \cos 2x + A_2 \sin 2x) \qquad (d)$$

in which A_1 and A_2 are unknown constant coefficients.

With this modified $u_p(x)$ in Equation (d), we may arrive at the following expression after substituting it into Equation (a):

$$(-4A_1 x \cos 2x - 2A_1 \sin 2x - 2A_1 \sin 2x$$
$$- 4A_2 x \sin 2x + 2A_2 \cos 2x + 2A_2 \cos 2x)$$
$$+ (4A_1 x \cos 2x + 4A_2 x \sin 2x)$$
$$= 2 \sin 2x$$

from which we may solve for $A_1 = -1/2$ and $A_2 = 0$.

The particular solution with the assumed form in Equation (d) thus results in the solution

$$u_p(x) = -\frac{x}{2} \cos 2x$$

The general solution of Equation (a) thus has the form

$$u(x) = u_h(x) + u_p(x)$$
$$= c_1 \cos 2x + c_2 \sin 2x - \frac{x}{2} \cos 2x$$

8.7   Application in Forced Vibration Analysis

As mentioned in Case 3 in Section 8.3.4, there are occasions when a vibrating machine or device is subject to periodic forces with a frequency ω at all times. This situation is called *forced vibration*.

8.7.1   Derivation of the Differential Equation

We will derive differential equations describing forced vibration in a mass–spring systems as illustrated in Figure 8.16.

In Figure 8.16, the applied periodic force is represented by a function $F(t)$, which in most cases is in the form of harmonic functions involving sine or cosine functions or combined sine and cosine functions. The equation of motion of the mass at any given time t can be derived from dynamic equilibrium of forces acting on the mass according to Figure 8.16 as

$$\sum F_y = 0 \rightarrow -F_d - k[h + y(t)] + W + F(t) = 0$$

In the above equations, F_d is the dynamic (or inertial) force, h is the static deflection of the spring due to the weight of the solid, and $F(t)$ is the applied excitation force.

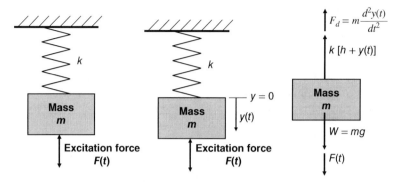

Figure 8.16 Forced vibration of a mass–spring system.

Figure 8.17 Excitation force acting on the mass in a mass–spring system.

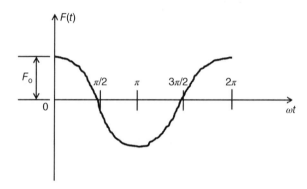

By following the same procedure as in Section 8.4, we may arrive at the following nonhomogeneous second-order differential equation describing the motion of the solid:

$$m\frac{d^2y(t)}{dt^2} + ky(t) = F(t) \tag{8.31}$$

The dynamic motion of the mass can be characterized by the instantaneous amplitude of the vibrating mass $y(t)$, depending on the form of the excitation force $F(t)$ in Equation 8.31.

In order to demonstrate the dynamic behavior of a vibrating mass subjected to an excitation force $F(t)$ described by a harmonic function, we assume that the excitation force $F(t)$ follows the form:

$$F(t) = F_0 \cos \omega t \tag{8.32}$$

where F_0 is the maximum magnitude of the applied cyclic force, and ω is the circular *excitation frequency* of the applied force. Graphical representation of the excitation force is illustrated in Figure 8.17.

Substituting the force function $F(t)$ that appears in Equation 8.32 into the equation of motion in Equation 8.31, we have the following nonhomogeneous second-order differential equation for the amplitude of vibration of mass m in Figure 8.16:

$$m\frac{d^2y(t)}{dt^2} + ky(t) = F_0 \cos \omega t \tag{8.33}$$

The solution of $y(t)$ in Equation 8.33 may be obtained by following the procedure for solving nonhomogeneous differential equations as presented in Section 8.6.

We realize that the solution of Equation 8.33 has the form of $y(t) = y_h(t) + y_p(t)$, and the homogeneous solution $y_h(t)$ has the following form derived in Section 8.4.1:

$$y_h(t) = c_1 \cos \omega_0 t + c_2 \sin \omega_0 t$$

in which c_1 and c_2 are arbitrary constants to be determined by specified conditions, and the natural frequency of the mass–spring system is related to $\omega_0 = \sqrt{k/m}$ as indicated in Section 8.4.1.

The particular solution, $y_p(t)$ takes the following form according to Table 8.1:

$$y_p(t) = A_1 \cos \omega t + A_2 \sin \omega t \tag{8.34}$$

where ω is the circular frequency of the excitation force $F(t)$ in Equation 8.32.

Upon substituting the $y_p(t)$ in Equation 8.34 into Equation 8.33, we will get the following expression:

$$m(-A_1\omega^2 \cos \omega t - A_2\omega^2 \sin \omega t) + k(A_1 \cos \omega t + A_2 \sin \omega t) = F_0 \cos \omega t$$

from which we solve for

$$A_1 = \frac{F_0}{k - m\omega^2} \quad \text{and} \quad A_2 = 0$$

We thus have the complete solution of $y(t)$ in Equation 8.33:

$$y(t) = c_1 \cos \omega_0 t + c_2 \sin \omega_0 t + \frac{F_0}{k - m\omega^2} \cos \omega t$$

The above expression for the solution of $y(t)$ may be expressed in a different form as in Equation 8.35 using the relationship $\omega_0 \sqrt{k/m}$, or $k = m\omega_0^2$:

$$y(t) = c_1 \cos \omega_0 t + c_2 \sin \omega_0 t + \frac{F_0}{m(\omega_0^2 - \omega^2)} \cos \omega t \tag{8.35}$$

Close inspection of the solution in Equation 8.35 will reveal that the first part of the solution (i.e., the complementary solution) is oscillatory with the amplitudes kept within fixed bounds by nature. The second part (i.e., the particular solution), however, depends on the value of $(\omega_0^2 - \omega^2)$. One may further observe that when the two circular frequencies become indiscriminating—i.e., when $\omega = \omega_0$ in a situation where the applied excitation force frequency equals the natural frequency of the mass–spring system—the solution in Equation 8.35 becomes unbounded with $y(t) \to \infty$. This unbounded solution implies that the amplitude of the vibrating mass reaches "infinity" regardless of time t, as indicated in Equation 8.35. Such behavior obviously cannot happen in reality. The solution in Equation 8.35, although mathematically correct, is physically unrealistic. Hence, a form of solution other than that shown in Equation 8.35 needs to be derived for the case $\omega = \omega_0$.

8.7.2 Resonant Vibration

Here, we will derive the solution for the amplitudes of vibration of a mass m suspended by a spring with a spring constant k. The vibrating mass is subjected to a cyclic force that can be described as $F(t) = F_0 \cos \omega t$, where F_0 is the maximum value of $F(t)$ and t designates the time at which the mass vibrates and ω is the frequency of the applied force.

The governing equation for the solution of the amplitude of the vibrating mass subjected to a cyclic force may be solved by Equation 3.33 with the circular natural frequency of the mass–spring system being ω_0. This circular natural frequency may be related to the actual natural frequency f as shown in Equation 8.17.

The solution to Equation 8.33, $y(t)$, consists of two parts with the complementary solution being $y_h(t) = c_1 \cos \omega_0 t + c_2 \sin \omega_0 t$, where c_1 and c_2 are arbitrary constants, and the particular solution as shown in Equation 8.27. We may substitute the circular frequency ω_0 in the above solution by ω under the condition $\omega_0 = \omega$ in the present case. We may, therefore, express the complementary solution of Equation 3.33 in the form of $y_h(t) = c_1 \cos \omega t + c_2 \sin \omega t$, where the first term in this solution coincides with the applied force function at the right-hand side of Equation 3.33. Such a coincidence warrants a special case in assuming the particular solution $y_p(t)$ in the solution $y(t)$ of Equation 3.33 as presented in Section 8.6.4. We will thus assume that the particular solution of this differential equation takes the form:

$$y_p(t) = t(A_1 \cos \omega t + A_2 \sin \omega t)$$

we will obtain the following expression after substituting the above special form of particular solution into Equation 8.33:

$$(-A_1 \omega^2 t \cos \omega t - 2A_1 \omega \sin \omega t - A_2 \omega^2 t \sin \omega t + 2A_2 \omega \cos \omega t)$$
$$+ \omega_0^2 (A_1 t \cos \omega t + A_2 t \sin \omega t)$$
$$= F_0 \cos \omega t$$

from which, we obtain $A_1 = 0$ and $A_2 = F_0/2m\omega$. These coefficients will lead to the following solution to the governing Equation (3.33)

$$y_p(t) = \frac{F_0}{2m\omega} t \sin \omega t$$

The general solution for Equation 8.33 with the special condition of $\omega_0 = \omega$ thus takes the form

$$y(t) = c_1 \cos \omega_0 t + c_2 \sin \omega_0 t + \frac{F_0}{2m\omega} t \sin \omega t \qquad (8.36)$$

This situation with $\omega = \omega_0$ in Equation 8.33 is called *resonant vibration* of a mass–spring system. We recognize that the first part of the solution in Equation 8.36 remains oscillatory between specific bounds by nature. The second part of the solution, however, indicates an indefinite continuous increase of the amplitude of the vibrating mass with time t. Graphical solution of Equation 8.36 is qualitatively illustrated in Figure 8.18.

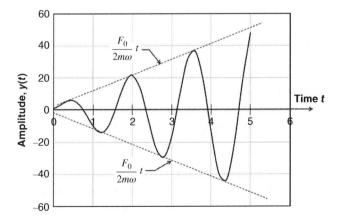

Figure 8.18 Amplitude of a resonant vibration.

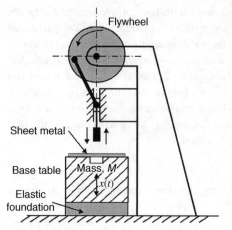

Figure 8.19 Working principle of a sheet metal stamping machine.

The amplitude of the vibrating mass (or a structure) in the case of $\omega = \omega_0$ increases rapidly with time in a resonant vibration. This phenomenon often causes the disintegration of a structure because of the rapid increase in elongation of the elastic support (the spring, in the case of a mass–spring system) in a short time after its inception. Design engineers should always avoid the occurrence of resonant vibration by avoiding having the situation in which the frequency of the applied cyclic force coincides with the natural frequency of the structure.

Example 8.9

A metal stamping machine shown in Figure 8.19 is used to punch holes in sheet metals or to produce dishes or shallow cups from flat sheet metal. The machine operates on the vertical movements of a plunger with an attached die that is designed for the shape of the stamped product. The plunger is driven by a massive flywheel as illustrated in the same figure. The sheet metal lies flat on a heavy base table with a mass $m = 2000$ kg attached to an elastic foundation that has an equivalent spring constant $k = 2 \times 10^5$ N/m. One may observe that the movement of the plunger, and thus the force exerted on the sheet metal during stamping, is periodic if the driving flywheel rotates at a constant angular velocity. In the present example, the force acting on the sheet metal, and thus on the base table, is represented by a function $F(t) = 2000 \sin(10t)$, in which t designates time in seconds. The unit for the force is newton (N). If the base table is initially depressed downward an amount of 0.005 m by the stamping force, do the following:

a) Derive the differential equation and the appropriate conditions that will describe the instantaneous position $x(t)$ of the machine base.
b) Assess whether this is a resonant vibration situation. Explain why this force function would induce a resonant vibration of the table.
c) Solve the differential equation by letting $x(t) = x_h(t) + x_p(t)$, where $x_h(t)$ is the complementary solution and $x_p(t)$ is the particular solution. Give the expressions for $x_h(t)$ and $x_p(t)$.
d) If this is in fact a resonant vibration, estimate how long will it take the elastic support to break as a result of having exceeded the allowable elongation of the elastic foundation of 0.03 m.

Solution:

The situation may be simulated by a simple mass–spring systems as illustrated in Figure 8.20.

We have the mass of the base $m = 2000$ kg, the equivalent spring constant of the elastic support $k = 2 \times 10^5$ N/m, and the applied excitation force $F(t) = 2000 \sin 10t$ with a circular excitation frequency $\omega = 10$ rad/s.

Figure 8.20 A simulated punch machine.

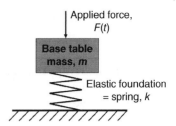

a) **The governing differential equation** We may use Equation 8.31 to derive the differential equation for the amplitude of the vibrating support of the machine $x(t)$ with the prescribed force function:

$$2000\frac{d^2x(t)}{dt^2} + 2 \times 10^5 x(t) = 2000 \sin 10t \tag{a}$$

The appropriate initial conditions of the stamping operations of the machine expressed in mathematical terms are

$$x(0) = 0.005 \text{ m} \quad \text{and} \quad \left.\frac{dx(t)}{dt}\right|_{t=0} = 0 \tag{b}$$

b) **Is this resonant vibration?** With the given values for the mass of the supporting base of the machine and the equivalent spring constant of the elastic support, we may calculate the natural frequency of the equivalent mass–spring system in Figure 8.20:

$$\omega_0 = \sqrt{\frac{k}{m}} = \sqrt{\frac{2 \times 10^5}{2 \times 10^3}} = 10 \text{ rad/s} \tag{c}$$

Since the computed ω_0 in Equation (c) has the same value as the given excitation frequency ω of the excitation force $F(t)$, we may assert that there will be a resonant vibration of the base table.

c) **The solution of Equation (a) for the instantaneous position, or amplitude of vibration $x(t)$ of the supporting mass m** There are two parts in the solution $x(t)$ in Equation (a): the complementary solution $x_h(t)$, and the particular solution $x_p(t)$. The complementary solution is obtained by solving Equation (a) as a homogeneous differential equation:

$$x_h(t) = c_1 \cos 10t + c_2 \sin 10t \tag{d}$$

This being a case with $\omega = \omega_0$, the particular solution for the present case is obtained by a special assumed form of

$$x_p(t) = t \, (A_1 \cos 10t + A_2 \sin 10t) \tag{e}$$

By substituting this assumed particular solution into Equation (a), we get the following expression:

$$-20A_1 \sin 10t + 20A_2 \cos 10t = \sin 10t$$

from which we solve for $A_1 = -1/20$ and $A_2 = 0$.
Consequently, we have the complete solution of Equation (a) as

$$x(t) = x_h(t) + x_p(t)$$
$$= c_1 \cos 10t + c_2 \sin 10t - \frac{t}{20} \cos 10t$$

The two arbitrary constants c_1 and c_2 in the above solution may be determined from the two conditions specified in Equation (b), resulting in $c_1 = 0.005$ and $c_2 = 1/200$.

We thus have the solution of the amplitudes of vibration $x(t)$ in the form

$$x(t) = 0.005 \cos 10t + \frac{1}{200} \sin 10t - \frac{t}{20} \cos 10t$$

d) **Time to breaking of the elastic foundation (t_f)** The elastic foundation is expected to break at 0.03 m elongation; that is, $x(t_f) = 0.03$. We may use the following equation to solve for the time for breaking t_f:

$$0.03 = 0.005 \cos 10t_f + \frac{1}{200} \sin 10t_f - \frac{t_f}{20} \cos 10t_f$$

$$= \left(0.005 - \frac{t_f}{20}\right) \cos 10t_f + \frac{1}{200} \sin 10t_f$$

The solution of the above equation is $t_f = 0.7$ s, obtained by a "short-cut" solution method of letting $\cos 10t_f = -1$ (and thus $\sin 10t_f = 0$). A more accurate solution of $t_f = 0.862$ s was obtained by using the combined solution methods of (1) Microsoft Excel software described later in Section 10.3.1 and (2) the Newton–Raphson method described in Section 10.3.2, as will be presented in Example 10.4 in Chapter 10.

Example 8.10
A light-duty motorcycle is supported by a suspension system that involves two coilovers of the sort illustrated in Figure 8.11c. The vehicle is rolling over a wavy road surface that can be described by a sine function as illustrated in Figure 8.21. The combined mass of the motor cycle and the rider is $m = 120$ kg. The coilover has a spring constant $k = 12\ 000$ N/m and a dashpot with a damping coefficient $c = 1920$ N-s/m. The rough road surface exerts an equivalent vertical cyclic force on the mass that can be approximated by the function $F(t) = 1200\ \sin(10t)$ N, in which t is the time into the vehicle vibration induced by the rough road surface. Determine the following:

a) The appropriate differential equation for the amplitude of vibration of the mass $y(t)$.
b) The differential equation with graphical representation of the solution using the following initial conditions:

$$y(0) = 0 \quad \text{and} \quad \left.\frac{dy(t)}{dt}\right|_{t=0} = 5 \text{ m/s}$$

c) The position of the mass 2 seconds after the initiation of the vibration.
d) The mode of vibration of the vehicle if the damper ceases to function. Give your reason for your predicted consequences.
e) A graphical representation of the amplitudes of the vibrating vehicle in the case (d) using the same initial conditions as stipulated in (b).

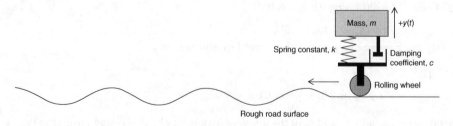

Figure 8.21 Vehicle cruising over rough road surface.

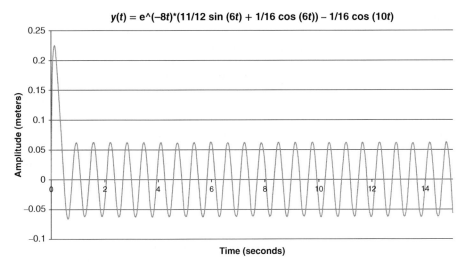

Figure 8.22 Graphical representation of the amplitude of vibration of a motor cycle.

Solution:

a) The appropriate differential equation for the amplitude of the vibrating vehicle is expressed following Equation 8.31 including the damping force to be:

$$120\frac{d^2y(t)}{dt^2} + 1920\frac{dy(t)}{dt} + 12000y(t) = 1200\sin 10t$$

or

$$\frac{d^2y(t)}{dt^2} + 16\frac{dy(t)}{dt} + 100y(t) = 10\sin 10t \tag{a}$$

with the conditions

$$y(0) = 0 \quad \text{and} \quad \left.\frac{dy(t)}{dt}\right|_{t=0} = 5 \text{ m/s} \tag{b}$$

b) The solution of Equation (a) is expressed in Equation 8.27:

$$y(t) = y_h(t) + y_p(t)$$

where $y_h(t)$ is the complementary solution and $y_p(t)$ is the particular solution.
We obtained the complementary solution by following the procedure described in Section 8.1:

$$y_h(t) = e^{-8t}(c_1 \cos 6t + c_2 \sin 6t) \tag{c}$$

The particular solution $y_p(t)$ can be obtained by assuming a form

$$y_p(t) = A_1 \cos 10t + A_2 \sin 10t$$

in which A_1 and A_2 are constants to be determined by substituting the expression assumed for $y_p(t)$ into Equation (a), resulting in $A_1 = -1/16$ and $A_2 = 0$.
Thus, the particular solution $y_p(t)$ is

$$y_p(t) = -\frac{1}{16}\cos 10t \tag{d}$$

The complete solution of Equation (a) is the sum of expressions in (c) and (d):

$$y(t) = e^{-8t}(c_1 \cos 6t + c_2 \sin 6t) - \frac{1}{16}\cos 10t \tag{e}$$

The arbitrary constant coefficients c_1 and c_2 in the complete solution in Equation (e) are determined using the specified conditions in Equation (b), yielding $c_1 = 1/16$ and $c_2 = 11/12$, which lead to the solution of Equation (a) as

$$y(t) = e^{-8t}\left(\frac{1}{16}\cos 6t + \frac{11}{12}\sin 6t\right) - \frac{1}{16}\cos 10t \tag{f}$$

Graphical representation of the amplitudes of the vibrating vehicle in Equation (f) is shown in Figure 8.22.

We may observe from Figure 8.22 that the amplitude of the vibration of the vehicle rises rapidly within the first 0.5 seconds into the vibration and decays quickly thereafter due to the damper, and is then followed by oscillatory motion of the spring in the suspension system.

c) The amplitude of vibration of the vehicle at 2 seconds into the vibration: The position of the vibrating mass at $t = 2$ seconds from Equation (f) with $y(2) = -\cos 20$ (in radians)/16 = 0.0255 m or 2.55 cm.

d) If the damper in the coilover ceases to function: With a dysfunctional damper, the vehicle would be supported only by the spring in the suspension system. Mathematically, this would mean the damping coefficient $c = 0$, and hence the disappearance of the term $dy(t)/dt$ in Equation (a). Consequently, Equation (a) will have a form

$$\frac{d^2y(t)}{dt^2} + \omega_0^2 y(t) = 10\sin\omega t \tag{g}$$

In Equation (g), ω_0 is the natural frequency of the mass–spring structure with $\omega_0 = 10$ rad/s; the excitation frequency ω, the frequency of the applied force, also equals 10 rad/s. The vehicle will thus be in resonant vibration since $\omega = \omega_0 = 10$ rad/s.

The solution of the differential equation (g) with the same assumed initial conditions as in Equation (b) has the form of $y(t) = y_h(t) + y_p(t)$ in which both $y_h(t)$ and $y_p(t)$ have the form

$$y_h(t) = c_1\cos\omega_0 t + c_2\sin\omega_0 t \tag{h}$$

The particular solution for this case of resonant vibration will have the form $y_p(t) = t(A_1\cos\omega t + A_2\sin\omega t)$ for Equation (g), yielding the constant coefficients $A_1 = -1/2$ and $A_2 = 0$, or

$$y_p(t) = -\frac{t}{2}\cos\omega t \tag{j}$$

The complete solution of Equation (g) is thus

$$y(t) = c_1\cos\omega_0 t + c_2\sin\omega_0 t - \frac{t}{2}\cos\omega t \tag{k}$$

with $\omega = \omega_0 = 10$ rad/s.

The arbitrary constant c_1 and c_2 in Equation (k) are determined using the same initial conditions as in Equation (b) with $c_1 = 0$ and $c_2 = 11/20$. We thus have the solution of Equation (g) as

$$y(t) = \frac{11}{20}\sin\omega_0 t - \frac{t}{2}\cos\omega t$$

for the case of a coilover with dysfunctional damper, the instantaneous position of the mass, or the amplitudes of the oscillatory motion of the mass $y(t)$, will be

$$y(t) = \frac{11}{20}\sin 10t - \frac{t}{2}\cos 10t \tag{m}$$

e) Graphical representation of the solution in Equation (m) is shown in Figure 8.23. We may observe that the amplitude of the vibrating motorcycle in resonance increases rapidly with time, similarly to what is illustrated in Figure 8.18.

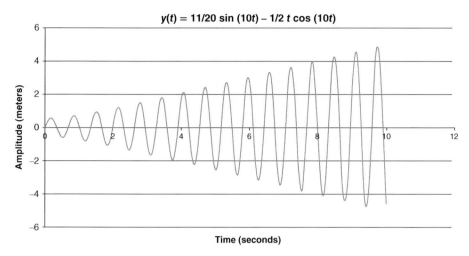

Figure 8.23 Amplitude of vibration of a vehicle in resonance.

8.8 Near Resonant Vibration

Resonant vibration, in which the excitation frequency ω of the externally applied cyclic force coincides with a natural frequency of the engineering system with ω_n ($n = 0, 1, 2, 3, \ldots$), can be devastating because of the rapid increase of amplitudes as illustrated in Figure 8.18. There are situations in which $\omega \neq \omega_n$ but the two frequencies are approximately equal: $\omega \approx \omega_n$ ($n = 0, 1, 2, 3, \ldots$). In such cases, devastation of the system or structure may not occur, but violent vibration will persist.

Let us re-visit the situation of a simple mass–spring system as illustrated in Figure 8.16 with the governing differential equation given in Equation 8.33. The general solution of this equation is given in Equation 8.35:

$$y(t) = c_1 \cos \omega_0 t + c_2 \sin \omega_0 t + \frac{F_0}{M(\omega_0^2 - \omega^2)} \cos \omega t \tag{8.35}$$

Upon imposing the initial conditions $y(0) = 0$ and $dy(t)/dt|_{t=0} = 0$, the arbitrary constants c_1 and c_2 are determined, and the complete solution in Equation 8.35 takes the form

$$y(t) = \frac{F_0}{M(\omega_0^2 - \omega^2)} [\cos(\omega t) - \cos(\omega_0 t)] \tag{8.37}$$

Equation 8.37 may be re-written using a trigonometric identity (Zwillinger, 2003):

$$\cos \alpha - \cos \beta = -\sin \frac{1}{2}(\alpha + \beta) \sin \frac{1}{2}(\alpha - \beta)$$

Equation 8.37 can thus be expressed as

$$y(t) = \frac{2F_0}{M(\omega_0^2 - \omega^2)} \sin \left[(\omega_0 + \omega)\frac{t}{2} \right] \ \sin \left[(\omega_0 - \omega)\frac{t}{2} \right] \tag{8.38}$$

Since the two frequencies ω and ω_0 are very similar in this case, meaning $\omega \approx \omega_0$, $(\omega_0 - \omega)$ is very small in Equation 8.38. We may thus use the following approximations:

$$\frac{\omega_0 + \omega}{2} \approx \omega \tag{8.39a}$$

and

$$\frac{\omega_0 - \omega}{2} = \varepsilon \tag{8.39b}$$

The circular frequency ε in Equation 8.39b is much smaller than the excitation frequency ω in Equation 8.39a. Consequently, we have the solution of the amplitude $y(t)$ for $\omega \approx \omega_0$ in the following form:

$$y(t) = \left[\frac{2F_0}{M(\omega_0^2 - \omega^2)}\right] \sin(\varepsilon t) \sin(\omega t) \tag{8.40}$$

We observe that the quantity in the square bracket in Equation 8.40 represents the maximum amplitude for the vibrating mass with amplitude varying as the product of two sine functions.

We may rearrange the terms in Equation 8.40 to the following form for interpretation of the physical phenomenon of near-resonant vibration of a mass–spring system:

$$y(t) = \left[\frac{2F_0}{M(\omega_0^2 - \omega^2)} \sin(\varepsilon t)\right] \sin(\omega t)$$

and we observe from that equation that the function $\sin(\varepsilon t)$ regulates the periodic variation of the maximum amplitude of the vibrating mass while the function $\sin(\omega t)$ dictates the period of the variation of the maximum amplitudes. Thus, the graphic representation of the instantaneous position of the mass $y(t)$, or the amplitude of the vibration with $\omega \approx \omega_0$, is shown in Figure 8.24, and the oscillation has a period of $t = 2\pi/\varepsilon$. The period of the cycles of maximum amplitude from the above expression is $T = 2\pi/\omega$, as also shown in Figure 8.24.

A "rough" sketch in Figure 8.24 shows the amplitudes of vibration of the mass in a mass–spring system in a near-resonant vibration situation, in which the mass vibrating with amplitudes in the "beats" with the variation of the magnitudes of these beats by envelopes represented by the sine function ($\sin(\omega t)$) in Equation (8.40).

The periods of the "beats" and of the maximum amplitudes beats in Figure 8.24 are

$$t = \frac{2\pi}{\varepsilon} \quad \text{for the "beats"} \tag{8.41a}$$

$$T = \frac{2\pi}{\omega} \quad \text{for maximum amplitudes} \tag{8.41b}$$

The maximum amplitude of the vibration of the mass is

$$y(t) = \left[\frac{2F_0}{M(\omega_0^2 - \omega^2)}\right] \tag{8.42}$$

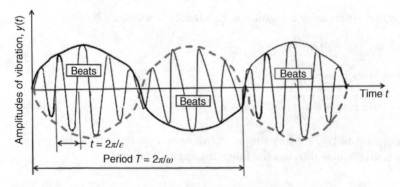

Figure 8.24 Amplitudes in near-resonance vibrations.

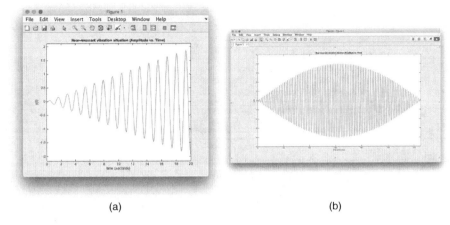

(a)   (b)

Figure 8.25 Amplitudes of a vibrating mass in the initial period and over one cycle. (a) Instantaneous amplitudes over 25 seconds. (b) Amplitudes in one complete cycle.

Near-resonant vibration of solids with $\omega \approx \omega_0$ will not cause unbounded amplitudes of vibration with time such as seen in the case of resonant vibration. However, many near-resonant situations can cause violent oscillation with "fast" beats that make machines and structures vulnerable to failure by material fatigue of the elastic supporting mechanisms, such as the spring that supports the vibrating mass in a mass–spring system.

Graphic display of the near-resonant situation such as that expressed in Equation 8.40 is available with the MatLAB software package with input/output as presented in Case 1, Appendix 4. The following numerical data were used in generating the numerical results: the mass of the vibrating solid $M = 1000$ kg with a natural frequency $\omega_0 = 4.95$ rad/s. The applied periodic force was $F(t) = F_0 \cos(\omega t)$ with $F_0 = 1000$ N and $\omega = 5$ rad/s.

Figure 8.25a illustrates the oscillating amplitude of the mass within the first 25 seconds into the vibration, with a complete cycle shown in Figure 8.25b. The amplitudes of the vibrating mass over two cycles are shown in Figure 8.26, which shows a similar behavior to that sketched in Figure 8.24.

Example 8.11

A stamping machine such as that illustrated in Figure 8.19 is used to produce shallow metal cups from flat sheet metal. The machine base has a mass of 1000 kg and it is bolted to an elastic foundation that has an equivalent spring constant $k = 25\ 000$ N/m. A measurement of the stamping force indicated a force function of $F(t) = 1000 \sin(4.95t)$ (note the different force function in this example) applied to the machine during the stamping process. The variable t in the force function is time in seconds. The unit for the force function is newton (N). If the elastic foundation is initially in a static equilibrium condition before the cyclic stamping force $F(t)$ is applied to the machine, determine the following:

a) The differential equation and the appropriate conditions that describe the instantaneous position $x(t)$ of the base.
b) The differential equation for the amplitudes of the vibrating machine $x(t)$.
c) A graphical illustration of the amplitudes of the vibrating machine base versus time.
d) The maximum deflection of the elastic foundation.
e) Whether the elastic foundation would break if its maximum allowed elongation is 5 cm (or 0.05 m).
f) The time to breakage of the elastic foundation should it occur.

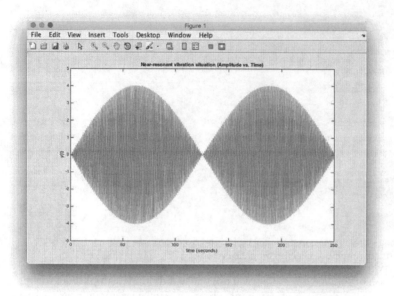

Figure 8.26 Amplitudes of a vibrating mass in near-resonant vibration in two cycles.

Solution:

a) According to Equation 8.31, we have the following differential equation with appropriate conditions for the present problem:

$$1000\frac{d^2x(t)}{dt^2} + 25\,000\,x(t) = 1000\sin(4.95t) \tag{a}$$

with the conditions

$$x(0) = 0 \quad \text{and} \quad \frac{dx(t)}{dt}\bigg|_{t=0} = 0 \tag{b}$$

b) We will follow the established procedure described in Section 8.6.1 for the solution of the differential equation in Equation (a) with the specified conditions in Equation (b). The solution $x(t)$ of the Equation (a) is

$$x(t) = -4.02\cos(4.95t)\sin(0.025t) \tag{c}$$

c) Graphical representation of the solution $x(t)$ in Equation (c) for the first 6 seconds is shown in Figure 8.27, whereas the amplitude of vibration for one complete cycle is as shown in Figure 8.28, in which the "beats" for one cycle are illustrated.
The graphical representation of the vibrating mass over three complete cycles is illustrated in Figure 8.29, and is seen to be similar to the qualitative illustration of the amplitudes of near-resonant vibration of a mass sketched in Figure 8.24.

d) The maximum deflection of the vibrating machine and thus the maximum elongation of the elastic foundation is 4.02 m as shown in Figures 8.28 and 8.29.

e) The maximum elongation of the elastic foundation, at 4.02 m, far exceeds the maximum allowed elongation of 0.05 m, so the elastic foundation will break.

f) The time at which the elastic foundation breaks appears to be 0.5 seconds as indicated in Figure 8.27.

Figure 8.27 Amplitude of the vibrating machine in the first 6 seconds.

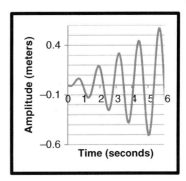

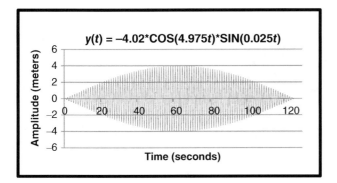

Figure 8.28 Amplitude of the vibrating machine during the first full cycle.

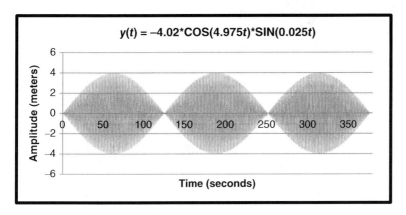

Figure 8.29 Amplitude of the vibrating machine in the first three cycles.

8.9 Natural Frequencies of Structures and Modal Analysis

In Section 8.4 we expressed the frequency of the vibrating mass in a simple mass–spring system in Equation 8.17. This frequency is often referred to as the "natural frequency" of a mass–spring system. For machine structures of more complicated geometry, we may consider that these structures are made of an aggregation of great many "mass particles" (such as "molecules with molecular masses") interconnected by many "elastic bonds" (e.g., molecular bonds). Deformation of solids result in displacement of molecules inside the solids, and any displacement of molecules will induce molecular forces as illustrated in Figure 8.30. One may also consider that

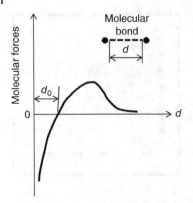

Figure 8.30 Variation of molecular forces with distance.

Figure 8.31 The Golden Gate Bridge.

there are millions of molecular masses interconnected by many millions more molecular bonds in solid structures in reality. Consequently, there are many natural frequencies inherent in these structures.

The "natural frequency," which may be related to the mass and elastic bonds of structures in the concept of "stiffness," is governed by the geometry and the material properties and these frequencies are "inherited" along with the structures.. These frequencies being "inherited" with the structures are called "natural frequencies" of the machine structure. The number of natural frequencies associated with a structure depends on the number of *degrees-of-freedom* of the mass in the structure. For structures involving a single mass and spring as illustrated in Figure 8.5, there is only one natural frequency with which the mass has only one way to move. For structures of complicated geometry such as the Golden Gate Bridge across the San Francisco Bay (shown in Figure 8.31), there are millions of molecules interconnected by millions of "elastic" bonds as in all solid structures, and accordingly millions of inherent "natural frequencies." Natural frequencies of real machines and structures are usually expressed as ω_n, with $n = 1, 2, 3, \ldots$. The subscript associated with these natural frequencies is referred to as the *mode number*.

The natural frequency of a machine or structure is important to design engineers when dealing with machines or structures that have to be designed to withstand possible intermittent or periodic loads. Matching of the frequency of the applied load with any mode of the natural frequency of the machine or structure will result in resonant vibration, which is often detrimental to the integrity of structures.

Figure 8.32 Structures of complex geometry that may be subjected to intermittent loading. (a) A billboard. (b) A jet fighter. (c) A satellite in orbit. *See color section.*

Modal analysis involves the determination of the natural frequencies of a structure by free vibration analysis. As already mentioned, for machines and structures of complex geometry such as those illustrated in Figure 8.32, there are various modes of natural frequencies, denoted ω_n ($n = 1, 2, 3, \ldots$) where n is the mode number. *The number of natural frequencies of the structure is the same as the number of degrees-of-freedom (DOF) of the structure.* Structures with complex geometry such as jet aircraft fighter shown in Figure 8.33b or the Golden Gate Bridge in Figure 8.31 have, in theory, an infinite number of DOF, and therefore they have infinite number of natural frequencies. Design engineers need to find the first few modes of the natural frequencies, which have more grave consequences for structure integrity than those of higher mode numbers. It is their responsibility to alert the potential users of these machines to the need to avoid applying intermittent or cyclic loads at excitation frequencies that match any mode of the natural frequencies of the machine that they have computed from their modal analysis.

Modal analysis is a powerful design tool for engineers; one valuable outcome of such analysis is the *mode shapes*, which indicate the shape the structure adopts in each of its possible modes of vibration. Figure 8.33 shows two typical mode shapes of a thin flat plate, with Figure 8.33a and b illustrating modes 1 and 3, respectively. These mode shapes illustrate possible shapes of

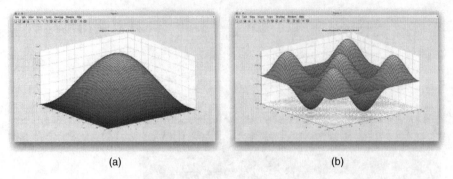

(a)   (b)

Figure 8.33 Mode shapes of a flat plate. (a) Mode 1. (b) Mode 3.

a flat plate in lateral vibration as described in case 2 of Appendix 4. The grid lines that divide the plate structure were used in a numerical analysis using MatLAB software.

Engineers may use mode shapes to avoid placing delicate elements at locations at which large amplitudes occur in the event of vibration of the plate.

8.10 Problems

8.1 Solve the following homogeneous second-order differential equations using the method offered in Section 8.1:

a) $\dfrac{d^2u(x)}{dx^2} + 9u(x) = 0$

b) $4\dfrac{d^2u(x)}{dx^2} + u(x) = 0$

c) $\dfrac{d^2u(x)}{dx^2} - 4u(x) = 0$ with $u(0) = 2$ and $\left.\dfrac{du(x)}{dx}\right|_{x=0} = 1$

d) $\dfrac{d^2u(x)}{dx^2} + 5\dfrac{du(x)}{dx} + 6u(x) = 0$ with $u(0) = 2$ and $\left.\dfrac{du(x)}{dx}\right|_{x=0} = 0$

8.2 Determine the motion of a mass moving toward the origin with a force of attraction that is proportional to the distance from the origin. Assume that a 10 kg mass starts at rest at a distance of 10 m and that the constant of proportionality is 10 N/m. What is the speed of the mass at 5 m from the origin?

8.3 A 4 kg mass is hanging from a spring with a spring constant $k = 100$ N/m. Sketch on the same plot the two specific solutions found from: (a) $y(0) = 0.5$ m, $y'(0) = 0$, and (b) $y(0) = 0$, $y'(0) = 10$ m/s. The coordinate y is measured from the equilibrium position of the mass.

8.4 Sketch on the same plot the motion of a 2 kg mass and that of a 10 kg mass suspended by a 50 N/m spring if motion starts from the equilibrium position with $y'(0) = 10$ m/s.

8.5 A body weighs 50 N and is hung by a spring with a spring constant of 50 N/m. A dash-pot is attached to the body. If the body is raised 0.02 m from its equilibrium position and released from rest, determine the amplitude of vibration if (a) $c = 17.7$ kg/s and (b) $c = 40$ kg/s, where c is the damping coefficient of the dashpot. What is the maximum amplitude of vibration in each case?

8.6 Find the amplitude of vibration of a mass in an over-damped motion of a spring–mass system with mass $M = 2$ kg, damping coefficient $c = 32$ kg/s, and $k = 100$ N/m from the initial states with $y(0) = 0$ and $y'(0) = 10$ m/s. Express your answer in the form of Equation 8.23.

8.7 Find the displacement $y(t)$ for a mass of 5 kg hanging from a spring with $k = 100$ N/m if there is a dashpot attached having damping coefficient $c = 30$ kg/s. The initial conditions are $y(0) = 0.0$ 1 m and $y'(0) = 0$.

8.8 You are an engineer who is assigned to design a wheel suspension system for a lightweight passenger vehicle. The suspension system involves a pair of coil springs with a shock absorber (Figure 8.34a). The shock absorber, which functions as a damper, is situated inside the coil spring as shown in Figure 8.34b. Each spring-damper is subjected to a mass of 270 kg. The spring constant $k = 70\,000$ N/m according to the manufacturer. The damping effect by the tire is assumed to be negligible. Determine the following:
a) The minimum damping coefficient of the shock absorber you would choose so as to avoid oscillatory motion of the mass when the wheel runs over a small obstacle that is 2 cm high, as illustrated in Figure 8.34b.
b) The expression for the amplitude of vibration of the mass after the vehicle runs over this small road block.
c) The amplitude of vibration of the mass 1 millisecond and 1 second after the vehicle runs over the same road block.

8.9 An automobile with four wheels is about to pass over a speed bump as shown in the insert in Figure 8.35a. The velocity of the vehicle at that time is 10 km/h. The vehicle is supported by a suspension system that consists of one coilover per wheel (a coilover is a combination of a shock absorber, or a damper, enclosed by a coil spring). The analytical model of the suspension system is illustrated in Figure 8.35b. The mass of the automobile carried by each wheel is 250 kg. The spring constant of the coilover is $k = 50\,000$ N/m

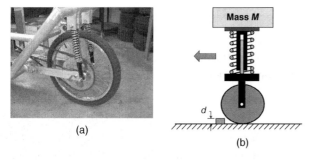

(a)

(b)

Figure 8.34 Suspension system of a light-duty vehicle. (a) Wheel suspension. (b) A small road block, $d = 2$ cm.

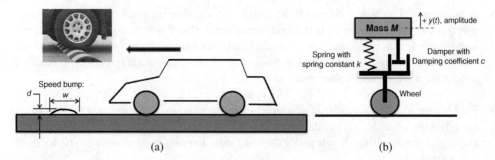

Figure 8.35 Vibration of a vehicle induced by passing over a speed bump. (a) A speed bump. (b) Analytical model of the suspension system.

with a damping coefficient c. The damping effect by the flexible tires of the wheels is assumed to be negligible. Determine the following:

a) An estimate of the time required (in seconds) for either the front wheels or rear wheels to pass the speed bump. This time is considered to be small so as to justify the assumption that the disturbance to the suspension system of the vehicle by the speed bump is instantaneous. The cross-section of the speed bump is $d = 6\,cm$ and $w = 18\,cm$ as indicated in Figure 8.35a.

b) The value of minimum damping coefficient c of the shock absorber for nonoscillatory motion of the vehicle after passing the speed bump.

c) The expression for the amplitude of vibration of the mass after the vehicle runs over the speed bump, and also the maximum amplitude of vibration of the mass.

d) The time from the start of the vibration at which the amplitude of vibration is 1 mm.

8.10 Find solutions of the following nonhomogeneous second-order differential equations:

a) $\dfrac{d^2u(x)}{dx^2} + 4\dfrac{du(x)}{dx} - 5u(x) = x^2 + 5$ with $u(0) = 0$ and $\dfrac{du(x)}{dx}\bigg|_{x=0} = 0$

b) $\dfrac{d^2u(x)}{dx^2} + 7\dfrac{du(x)}{dx} + 10u(x) = \cos 2x$ with $u(0) = 0$ and $\dfrac{du(x)}{dx}\bigg|_{x=0} = 0$

8.11 Solve the following differential equation:

$$m\frac{d^2y(t)}{dt^2} + c\frac{dy(t)}{dt} + mg = 0$$

and determine how high the solid with a mass $m = 2\,kg$ will rise with an initial velocity of 100 m/s in an upward direction. The damping coefficient $c = 0.4$ kg/s.

8.12 A solid weighing 100 N is dropped from rest. The drag is assumed to be proportional to the first power of the velocity with the constant of proportionality being 0.5. Approximate the time necessary for the solid to attain terminal velocity. Define the terminal velocity to be equal to $0.99V_\infty$, where V_∞ is the velocity attained at $t \to \infty$.

8.13 A 2 kg mass is suspended by a spring with a spring constant $k = 32$ N/m. A force of 0.1 sin(4t) is applied to the mass. Calculate the time required for failure to occur if the spring breaks when the amplitude of the oscillation exceeds 0.05 m. The motion starts from rest and the damping effect is negligible.

8.14 A 20 N weight is suspended by a frictionless spring with $k = 98$ N/m. A force of $2 \cos(7t)$ acts on the weight. Calculate the frequency of the "beats" and find the maximum amplitude of motion, which starts from rest.

8.15 A device with 2 kg mass is suspended by an elastic cable with an equivalent spring constant $k = 32$ N/m. A force $F(t) = 0.32(\cos 4t)$ N is applied at the mass at $t > 0$, where t is time in seconds. The mass is initially pulled down by an amount of 0.01 m from its static equilibrium position, and is released at rest. (a) Derive the equation of motion of the device mass at time $t > 0$. (b) Derive the expression for the position of the mass, $y(t)$, in the direction of the coordinate y with its positive direction downward. (c) Estimate the time required for the cable to break if its maximum allowable elongation of the spring is 0.1 m.

8.16 A machine with a 4 kg mass is attached to an elastic support with an equivalent spring constant of 64 N/m. A periodic force, $F(t)$ with t time in seconds, is applied to the machine during an operation. The applied force $F(t)$ can be described by a harmonic function: $F(t) = F_0 \cos(\omega t)$ where $F_0 = 4$ N and the excitation frequency $\omega = 4$ rad/s.
   a) Derive an expression for the amplitudes of the vibrating machine under the influence of the applied force.
   b) Assess whether the elastic support would break with a maximum allowable elongation of 0.05 m.
   c) Estimate the time required for the support to break should it ever happen.
   Assume that the machine is initially depressed by 0.005 m from its equilibrium state, and that operation begins from that position of rest.

9

Applications of Partial Differential Equations in Mechanical Engineering Analysis

Chapter Learning Objectives

- Learn the physical meaning of partial derivatives of functions.
- Learn that there are different order of partial derivatives describing the rate of changes of functions representing real physical quantities.
- Learn the two commonly used techniques for solving partial differential equations by (1) Integral transform methods that include the Laplace transform for physical problems covering half-space, and the Fourier transform method for problems that cover the entire space; (2) the separation of variables technique.
- Learn the use of the separation of variables technique to solve partial differential equations relating to heat conduction in solids and vibration of solids in multidimensional systems.

9.1 Introduction

Partial differential equations such as that shown in Equation 2.5 are the equations that involve partial derivatives described in Section 2.2.5. A partial derivative represents the rate of change of a function (a physical quantity in engineering analysis) with respect to one of several variables that the function is associated with.

The independent variables in partial derivatives can be (1) *spatial* variables represented by (x,y,z) in a rectangular coordinate system or (r,θ,z) in a cylindrical polar coordinate system and (2) *temporal* variables represented by time t.

Partial differential equations can be categorized as "boundary-value problems" or "initial-value problems", or "initial-boundary value problems". *Boundary-value problems* are the ones for which the complete solution of the partial differential equation is possible with specific boundary conditions. *Initial-value problems* are those partial differential equations for which the complete solution of the equation is possible with specific information at one particular instant. In reality, however, solutions to most problems require both boundary conditions and initial conditions to be specified. Mathematical modeling of real physical conditions to boundary and initial conditions for solution of partial differential equations is thus important part of the effort in obtaining the solutions.

9.2 Partial Derivatives

Mathematical formulation of partial derivatives is more complicated than for those derivatives for ordinary functions involving only one variable as defined in Section 2.2.5. The complication in expressing partial derivatives is due to the fact that the value of the function in partial

Applied Engineering Analysis, First Edition. Tai-Ran Hsu.
© 2018 John Wiley & Sons Ltd. Published 2018 by John Wiley & Sons Ltd.
Companion Website: www.wiley.com/go/hsu/applied

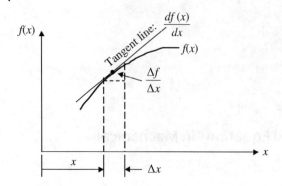

Figure 9.1 Graphic representation of an derivative of ordinary function $f(x)$.

derivatives is determined by variation of more than one independent variable associated with the function.

Figure 9.1 represents the derivative of a continuous ordinary function $f(x)$ with respect to its only variable x. The derivative of the function $f(x)$ described in a rectangular coordinate system is expressed in Equation 2.9 as

$$\frac{df(x)}{dx} = \lim_{\Delta x \to 0} \frac{f(x + \Delta x) - f(x)}{\Delta x} \tag{2.9}$$

For the function which varies with more than one independent variable—e.g., x and t expressed as $f(x,t)$—we need to express the derivative of this function with *both* of the independent variables x and t separately. The derivatives of the function $f(x,t)$ with respect to each of these two independent variables become "partial derivatives" as defined in the following.

The partial derivative of function $f(x,t)$ with respect to x only is defined as

$$\frac{\partial f(x,t)}{\partial x} = \lim_{\Delta x \to 0} \frac{f(x + \Delta x, t) - f(x,t)}{\Delta x} \tag{9.1}$$

In this case, the independent variable t is treated as a "constant" in the derivation.

Likewise, the derivative of function $f(x,t)$ with respect to the other variable, t, is defined as

$$\frac{\partial f(x,t)}{\partial t} = \lim_{\Delta t \to 0} \frac{f(x, t + \Delta t) - f(x,t)}{\Delta t} \tag{9.2}$$

with now the variable x being treated as a "constant" in the derivation.

The partial derivatives of higher order can be expressed in a similar way as for ordinary functions. For instance, the second-order partial derivatives of the function have the forms

$$\frac{\partial^2 f(x,t)}{\partial x^2} = \lim_{\Delta x \to 0} \frac{\dfrac{\partial f(x + \Delta x, t)}{\partial x} - \dfrac{\partial f(x,t)}{\partial x}}{\Delta x} \tag{9.3}$$

and

$$\frac{\partial^2 f(x,t)}{\partial t^2} = \lim_{\Delta t \to 0} \frac{\dfrac{\partial f(x, t + \Delta t)}{\partial t} - \dfrac{\partial f(x,t)}{\partial t}}{\Delta t} \tag{9.4}$$

There exists another second-order partial derivative with cross differentiations in the form

$$\frac{\partial^2 f(x,t)}{\partial x \partial t} = \frac{\partial^2 f(x,t)}{\partial t \partial x} \tag{9.5}$$

9.3   Solution Methods for Partial Differential Equations

Partial differential equations are those equations that involve partial derivatives. There are four common methods available for the solution of partial differential equations:

1) *Use appropriate "trial functions" for the solution of the equation.* Trial functions are usually in the form of polynomial function involving independent variables with unknown coefficients. These coefficients are determined in a similar way as we did for the particular solutions of second-order ordinary differential equations as described in Section 8.6. This method is rarely used in practice due to lack of a set of rules and guidelines for assuming the appropriate forms of trial functions for the solutions of partial differential equations.

2) *Use integral transforms to transform independent variables in the partial differential equations.* Such a procedure effectively converts partial differential equations into ordinary differential equations. The solution obtained from the ordinary differential equation, such as those presented in Chapters 7 and 8, needs to be inversely transformed back to the original variable domains for the complete solution. The following are three integral transform methods available for solving partial differential equations:

   a) Laplace transform for variables varying in the range $(0,\infty)$.
   b) Fourier transform for variables varying in the range $(-\infty, +\infty)$.
   c) The Hankel transform for transforming the variable in the radial coordinate (the r-coordinate) in a cylindrical polar coordinate system (r,θ,z)

   The integral transform method was used to solve a variety of heat conduction problems (Ozisik 1968).

3) *Separation of variables technique*—a popular solution method for partial differential equations. The principle of this method will be described in the subsequent Section 9.3.1, with applications in the solution of partial differential equations for heat conduction in solids and vibration analysis of long cables and membranes. Not all partial differential equations are separable with respect to the involved variables, as will be demonstrated with some chapter-end problems.

4) *Numerical solution methods*—the most commonly used methods are the finite-difference (FDM) and finite-element (FEM) methods. The principle of both of these solution methods will be presented in the subsequent Chapters 10 and 11.

9.3.1   The Separation of Variables Method

The essence of this method is to "separate" the independent variables, such as x, y, z, and t involved in the functions appearing in partial differential equations.

We will illustrate the principle of this solution technique with a function $F(x,y,t)$ in a partial differential equation. The process begins with an assumption of the original function $F(x,y,t)$, to be a product of three functions, each involving only one of the three independent variables, as expressed in Equation (9.6).:

$$F(x, y, t) = f_1(x)f_2(y)f_3(t) \tag{9.6}$$

in which $f_1(x)$, $f_2(y)$, and $f_3(t)$ are functions of the variables x, y, and t, respectively.

The expression in Equation 9.6 has effectively *separated* the three independent variables in the original function $F(x,y,t)$ into the product of three separate functions; each consists of only one of the three independent variables.

The three separate function f_1, f_2, and f_3 in Equation 9.6 will be obtained by solving three individual ordinary differential equations involving "separation constants." The solution of f_1, f_2, and f_3 from these ordinary differential equations will be related to the original function

F according to what is shown in Equation 9.6, which is the solution to the original partial differential equation.

9.3.2 Laplace Transform Method for Solution of Partial Differential Equations

The use of Laplace transform for solving partial differential equations was presented in Section 6.7 in Chapter 6, with an example on solving a partial differential equation using this technique in Example 6.18. This method is restricted to functions that are valid for the variables covering the range $(0,\infty)$. Many physical quantities that can be represented by functions involving time variable t with $t > 0$ justify the use of this solution method.

9.3.3 Fourier Transform Method for Solution of Partial Differential Equations

Like Laplace transform method, the Fourier transform is another integral transform method for solving partial differential equations in engineering analysis. The condition for using this technique in solving a partial differential equation, however, is that the variable that is transformed should cover the entire domain $(-\infty, \infty)$.

The Fourier transform for a function $f(x)$ is expressed in mathematical form as

$$\Im[f(x)] = \int_{-\infty}^{\infty} f(x)e^{-i\omega x}\,dx = F(\omega) \tag{9.7}$$

where i is the imaginary number $\sqrt{-1}$ and ω is the transformation parameter (similar to the parameter s in the Laplace transform).

The inverse Fourier transform is obtained as Equation 9.8:

$$\Im^{-1}[F(\omega)] = \frac{1}{2\pi}\int_{-\infty}^{\infty} F(\omega)e^{i\omega x}\,d\omega \tag{9.8}$$

We recognize that the form of the Fourier transform in Equation 9.7 is similar to that of the Laplace transform in Equation 6.1, with differences in the integration limits and the exponents in the integral.

Like the Laplace transform of functions, the Fourier transform can also apply to derivatives of the functions, with variables covering the range $(-\infty,\infty)$. The Fourier transform of the derivative of function $f(x)$ may be obtained from a rather simple formulae as shown in Equation 9.9:

$$\Im[f^{n}(x)] = \int_{-\infty}^{\infty} \left(\frac{d^{n}f(x)}{dx^{n}}\right)e^{-i\omega x}\,dx = (i\omega)^{n}F(\omega) \tag{9.9}$$

where n is the order of the derivatives being transformed.

We may thus derive the following expressions for the first- and second-order derivatives of function $f(x)$:

$$\Im[f'(x)] = i\omega F(\omega) \quad \text{and} \quad \Im[f''(x)] = (i\omega)^{2}F(\omega) = -\omega^{2}F(\omega), \quad \text{etc.} \tag{9.10}$$

Example 9.1

Find the Fourier transform of a function that is defined as

$$f(x) = \begin{cases} 0 & x < -a \\ h & -a < x < a \\ 0 & x > a \end{cases} \qquad \text{with} -\infty < x < \infty \tag{a}$$

This function is graphically represented in Figure 9.2.

Figure 9.2 Function for Fourier Transformation.

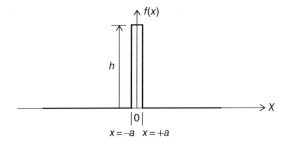

Table 9.1 Fourier transform of selected functions

Function for Fourier transform $f(x)$	After Fourier transform $F(\omega)$
(1) $f(x-a)$	$F(\omega)e^{-i\omega a}$
(2) $\delta(x)^*$	1
(3) $u(x)^*$	$(i\omega)^{-1}$
(4) $e^{-\alpha|x|}$ $\alpha > 0$	$\dfrac{2\alpha}{\alpha^2 + \omega^2}$
(5) $u(x)\sin ax$	$\dfrac{a}{a^2 - \omega^2}$
(6) $u(x)\cos ax$	$\dfrac{i\omega}{a^2 - \omega^2}$

$^*\delta(x)$ is the delta function, or impulse function and $u(x)$ is the unit step function. Both of these functions are defined in Section 2.4.2.

Solution:
The Fourier transformation of the function $f(x)$ in Equation (a) may be performed by substituting the function into the definition of the Fourier transform in Equation 9.7 as follows:

$$\mathfrak{I}[f(x)] = \int_{-\infty}^{\infty} f(x)\, e^{-i\omega x}\, dx$$

$$= \int_{-\infty}^{-a}(0)e^{-i\omega x}\, dx + \int_{-a}^{a}(h)e^{-i\omega x}\, dx + \int_{a}^{\infty}(0)e^{-i\omega x}\, dx = \frac{2h\sin(\omega a)}{\omega} = F(\omega)$$

Table 9.1 presents a useful formulae for Fourier transforms of a some selected functions.

Example 9.2
Solve the following partial differential equation using Fourier transform method:

$$\frac{\partial^2 T(x,t)}{\partial x^2} = \alpha^2 \frac{\partial T(x,t)}{\partial t} \qquad -\infty < x < \infty \tag{9.11}$$

where the coefficient α is a constant. The equation satisfies the following condition:

$$T(x,t)|_{t=0} = T(x,0) = f(x) \tag{9.12}$$

Solution:
Since the variable x varies in the range $(-\infty,\infty)$, we may transform it in the function $T(x,t)$ to the parametric domain of ω via a Fourier transform as defined in Equation 9.7. Thus letting

$T^*(\omega, t)$ be the Fourier-transformed function of $T(x,t)$ in Equation 9.7:

$$T^*(\omega, t) = \Im[T(x, t)] = \int_{-\infty}^{\infty} T(x, t) e^{-i\omega x} \, dx \qquad (a)$$

Applying the above integral to the left-hand side of Equation 9.11 yields

$$\Im\left[\frac{\partial^2 T(x, t)}{\partial x^2}\right] = \int_{-\infty}^{\infty} \left(\frac{\partial^2 T(x, t)}{\partial x^2}\right) e^{-i\omega x} \, dx$$
$$= -\omega^2 T^*(\omega, t)$$

from the ordinary derivative in Equation 9.10.

Likewise, the Fourier transform of the right-hand side of Equation 9.11 will lead to

$$\Im\left[\alpha^2 \frac{\partial T(x, t)}{\partial t}\right] = \alpha^2 \int_{-\infty}^{\infty} \left(\frac{\partial Tx, t}{dt}\right) e^{-i\omega x} \, dx$$
$$= \alpha^2 \frac{\partial}{\partial t} \int_{-\infty}^{\infty} T(x, t) e^{-i\omega x} \, dx$$
$$= \alpha^2 \frac{\partial T^*(\omega, t)}{\partial t}$$

We have thus transformed Equation 9.11 by equating the Fourier transforms of the right-hand-side and left-hand-side terms to yield

$$-\omega^2 T^*(\omega, t) = \alpha^2 \frac{dT^*(\omega, t)}{dt} \qquad (b)$$

One will note that in Equation (b) after performing Fourier transform of the variable x on both sides of Equation 9.11, the partial derivatives in that equation have been replaced by the function $T^*(\omega, t)$ and an ordinary derivative of the function $T^*(\omega, t)$ on the right side of the equation.

Equation (b) is a first-order ordinary differential equation, and the method of obtaining the general solution of this equation is described in Chapter 7.

At this point, we need to transform the specified condition in Equation 9.12 by the Fourier transform defined in Equation (a), or by the expression

$$T^*(\omega, 0) = \Im[T(x, 0)]$$
$$= \int_{-\infty}^{\infty} T(x, 0) e^{-i\omega x} \, dx$$
$$= \int_{-\infty}^{\infty} f(x) e^{-i\omega x} \, dx = g(\omega) \qquad (c)$$

in which $f(x)$ is the given condition. Thus the expression $g(\omega)$ can be obtained as the integral in Equation (c).

We will thus have the solution of the function $T^*(\omega, t)$ as

$$T^*(\omega, t) = g(\omega) e^{-(\omega^2/\alpha^2)t} \qquad (d)$$

The solution of the partial differential equation in Equation 9.11 with the specified condition in Equation 9.12 may be obtained by inverting the transform $T^*(\omega, t)$ to $T(x,t)$ using Equation 9.8 through the following expression:

$$T(x, t) = \frac{1}{2\pi} \int_{-\infty}^{\infty} T^*(\omega, t) e^{i\omega x} \, d\omega = \frac{1}{2\pi} \int_{-\infty}^{\infty} [g(\omega)] e^{-(\omega^2/\alpha^2)t} e^{i\omega x} \, d\omega \qquad (e)$$

where $g(\omega)$ is available in Equation (c) to be the Fourier-transformed specified condition of $T(x,0)$ in Equation 9.12.

9.4 Partial Differential Equations for Heat Conduction in Solids

9.4.1 Heat Conduction in Engineering Analysis

We have learned from Chapter 7 that temperature variations in a medium are induced by heat transmission. This variation is called the *temperature field*. Excessive heat flow can induce high-temperature fields in the medium, which may result in the following three major negative consequences to the materials:

1) High temperature weakens materials. Many properties of engineering materials change their values with temperatures. For instance, there can be significant reduction of Young's modulus of common engineering materials. Young's modulus relates to the stiffness of the material. Common sense indicates, for instance, that metals become "softer" due to such reduction of Young's modulus when heated. The ultimate strength of the material also reduces with increasing temperature. These changes have become major concerns in design analyses by engineers.
2) The basic laws of thermodynamics indicate that higher operating efficiency of thermal engines—such as internal combustion engines, gas turbines, and steam and nuclear power reactors—is achievable by operating these machines at higher temperatures. Unfortunately, high operating temperatures not only weaken the material strength as described in (1), but also introduce thermal stresses if the temperature is not uniformly distributed in machinery systems with improper mechanical constraints. These thermal stresses that are induced by heat need to be accounted for in all design analyses. Credible and reliable heat transfer analysis is thus a critical part of engineering analysis.
3) Another potentially serious concern in engineering analysis that relates to thermal effects is the possible creep deformation of structural materials. Creep is the phenomenon of continuous deformation of materials with time without them being subjected to additional loads. Creep deformation often occurs in materials at elevated temperature above half of the homologous melting point. We may thus envision that materials such as solder alloys with low melting points used in microelectronic devices are vulnerable to creep failures. The unexpected creep deformation induced by high temperature is also the source of functional problems for high-performance gas turbines and high-precision machines and devices, and is also known to be problematic for some IT (information technology) devices that require high precision in assemblies.

We thus appreciate that heat conduction analysis involving temperature fields in solid structures is an important part of engineering analysis.

9.4.2 Derivation of Partial Differential Equations for Heat Conduction Analysis

We have learned in Chapter 2 that all differential equations used in engineering analysis are derived from laws of physics, and the equations for heat conduction in solids are no exception. The law on which the derivation presented here is based is the law of conservation of energy and its derivative—the first law of thermodynamics.

Referring to Figure 9.3, a solid with a control volume is subjected to heat flow with incoming heat in the form of heat flux $\mathbf{q}(\mathbf{r},t)$ into a small element shown as the small open circle in the figure. The heat leaving the element is $\mathbf{q}(\mathbf{r}+\Delta\mathbf{r},t)$, where $\mathbf{r}$ designates the spatial variables of (x,y,z) in a rectangular coordinate system or (r,θ,z) in a cylindrical polar coordinate system.

We may develop an expression for the energy balance based on the law of conservation energy as illustrated in Figure 9.4. We may use the Fourier law of heat conduction defined in Equation 7.25 to represent the heat entering and leaving the element in Figure 9.4, and the energy storage

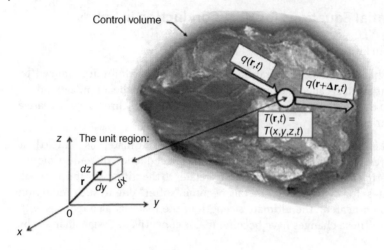

Figure 9.3 Flow of heat in a solid.

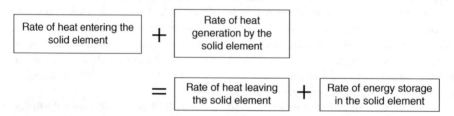

Figure 9.4 Energy balance of heat flow in a solid.

in the element may be related to temperature rise in the form of change of internal energy Δu in the solid. Mathematically, we can express Δu in the form of $\Delta u = \rho c\, \Delta T$, in which ρ is the mass density and c is the specific heat of the solid. ΔT in the expression denotes the temperature rise from its reference state. We may thus establish the following partial differential equation for the situation depicted in Figures 9.3 and the diagram in Figure 9.4:

$$\rho c \frac{\partial T(\mathbf{r}, t)}{\partial t} = \nabla \cdot [k \nabla T(\mathbf{r}, t)] + Q(\mathbf{r}, t) \tag{9.13}$$

where the symbol ∇ in Equation 9.13 is the divergence defined in Section 3.5.3. $Q(\mathbf{r},t)$ is the heat generation by the solid per unit volume and time.

Examples of heat generation $Q(\mathbf{r},t)$ by materials might include the heat generated by the nuclear fission of uranium fuel in nuclear reactor cores, or $i^2 R$ ohmic heating of a material in electronic circuits or devices with the passage of electric current i through a resistance R.

9.4.3 Heat Conduction Equation in Rectangular Coordinate Systems

The general heat conduction equation in Equation 9.13 will take the following form with $T(\mathbf{r}, t) = T(x, y, z, t)$:

$$\rho c \frac{\partial T}{\partial t} = \frac{\partial}{\partial x}\left[k_x \frac{\partial T}{\partial x}\right] + \frac{\partial}{\partial y}\left[k_y \frac{\partial T}{\partial y}\right] + \frac{\partial}{\partial z}\left[k_z \frac{\partial T}{\partial z}\right] + Q(x, y, z, t) \tag{9.14a}$$

in which k_x, k_y, and k_z are the thermal conductivities of the solid along the x-, y-, and z-coordinate, respectively.

9.4.4 Heat Conduction Equation in a Cylindrical Polar Coordinate System

Heat conduction equation in this coordinate system is obtained by expanding Equation 9.13 as follows with $T(\mathbf{r}, t) = \text{T}(r, \theta, z, t)$:

$$\rho c \frac{\partial T}{\partial t} = \frac{\partial}{\partial r}\left[k_r \frac{\partial T}{\partial r}\right] + \frac{1}{r}\left[k_r \frac{\partial T}{\partial r}\right] + \frac{1}{r^2}\left[\frac{\partial}{\partial \theta}k_\theta \frac{\partial T}{\partial \theta}\right] + \frac{\partial}{\partial z}\left[k_z \frac{\partial T}{\partial z}\right] + Q(r, \theta, z, t) \qquad (9.14b)$$

where k_r, k_θ, and k_z are thermal conductivities of the material along the r-, θ- and z-coordinates respectively.

9.4.5 General Heat Conduction Equation

Thermal conductivities k_x, k_y, and k_z in Equation 9.14a and k_r, k_θ, and k_z in Equation 9.14b are used for heat conduction analysis of solids with their thermophysical properties varying in different directions, such as for fiber filament composites. For most engineering analyses, such variations of thermophysical properties do not exist. Consequently, a generalize heat conduction equation may be expressed as follows:

$$\nabla^2 T(\mathbf{r}, t) + \frac{Q(\mathbf{r}, t)}{k} + \frac{1}{\alpha}\frac{\partial T(\mathbf{r}, t)}{\partial t} \qquad (9.15)$$

where k = thermal conductivity of the material and $Q(\mathbf{r}, t)$ is the heat generated by the material per unit volume and time.

The symbol α in Equation 9.15 is the "thermal diffusivity" of the material, its value being $\alpha = k/\rho c$; it is often used as a measure of how "fast" heat can flow by conduction in a solid.

9.4.6 Initial and Boundary Conditions

Solution of the heat conduction equation in Equation 9.15 involves determining a number of arbitrary constants according to specific initial and boundary conditions. These conditions are necessary to translate the real physical conditions into mathematical expressions. Proper specification and translation of these conditions are important steps in formulating numerical solutions, either by the finite-difference method or by the finite-element method, as will be described in Chapters 10 and 11, respectively.

Initial conditions are required only when dealing with transient heat transfer problems in which the temperature field in a solid changes with the elapsing time. These conditions specify the temperature distribution in the solid before and at the onset of the changing thermal conditions that create temperature distributions. The common initial condition in a solid can be expressed mathematically as

$$T(\mathbf{r}, t)|_{t=0} = T(\mathbf{r}, 0) = T_0(\mathbf{r}) \qquad (9.16)$$

where the temperature field $T_0(\mathbf{r})$ is a function of the spatial coordinates $\mathbf{r}$ only.

In many practical applications, the initial temperature distribution $T_0(\mathbf{r})$ in Equation 9.16 can be assigned a constant value such as room temperature at 20°C for uniform temperature (isothermal) conditions.

Specific boundary conditions are required, however, in the analysis of all transient and steady-state problems involving solids of finite shape. Several types of boundary conditions are commonly used, as will be described below.

a) *Prescribed surface temperature, $T_s(t)$.* Quite often, in practice, the temperature at the surface of the solid structure is measured by either attaching thermocouples to the structure surface or by some noncontact methods such as infrared thermal imaging scanning camera. The mathematical expression for this case takes the form

$$T(\mathbf{r}, t)|_{\mathbf{r}=\mathbf{r}_s} = T_s(t) \tag{9.17a}$$

where $\mathbf{r}_s$ is the coordinates of the boundary surface where temperature is specified as $T_s(t)$. This type of boundary condition with prescribed surface temperatures is called a Dirichlet condition by mathematicians.

b) *Prescribed heat flux boundary condition, $q_s(t)$.* Many structures have their surfaces exposed to a heat source or a heat sink. One such example is the heat treatment of a large forged piece, for example, a turbine shaft in an autoclave in which heat is being supplied to the piece through its outside surface. The mathematical translation of the heat flux to or from a solid surface can be readily carried out by using the Fourier law of heat conduction as defined in Equation 7.25. The mathematical formulation of the heat flux across a solid boundary surface can be expressed as

$$\left. \frac{\partial T(\mathbf{r}, t)}{\partial \mathbf{n}_i} \right|_{\mathbf{r}=\mathbf{r}_s} = -\frac{q_s(\mathbf{r}_s, t)}{k} \tag{9.17b}$$

where k is the thermal conductivity of the solid material. The symbol $\partial/\partial \mathbf{n}_i$ is the differentiation along the outward-drawn normal to the boundary surface S_i. The term $q_s(\mathbf{r}_s, t)$ in Equation 9.17b is the specified heat flux across this surface defined by the coordinate $\mathbf{r}_s$ in the same direction as $\mathbf{n}_i$. This type of boundary condition is referred to as the Neuman condition by mathematicians. We may express Equation 9.17b for the boundaries that are impermeable to heat flow, or a boundary that is thermally insulated as

$$\left. \frac{\partial T(\bar{\mathbf{r}}, t)}{\partial \mathbf{n}} \right|_{\mathbf{r}=r_s} = 0 \tag{9.17c}$$

c) *Convective boundary conditions.* Many engineering applications involve boundary surfaces of a solid in contact with fluids in gaseous or liquid states such as described in Section 7.5.5. A special expression has to be derived for the boundary conditions of this type.

One common phenomenon occurs when a solid is submerged in a stationary or moving fluid at a different temperature; a boundary layer is developed at the interface of the solid and fluid

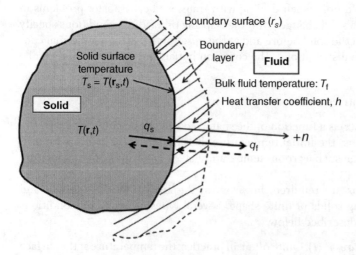

Figure 9.5 Heat transfer at the interface of a solid submerged in fluid.

as illustrated in Figure 9.5. This boundary layer acts as a "barrier" for heat transfer between the solid and the surrounding fluid, resulting in $T_s \neq T_f$, where T_s is the temperature at the surface of the solid and T_f is the contacting bulk fluid temperature. The thickness of the boundary layer is related to the temperature difference between the two media and the velocity of the moving fluid in contact in the case of forced convective heat transfer in the surrounding fluid. The resistance induced by the boundary layer is related to the heat transfer coefficient h as described in Sections 7.5.4 and 7.5.5.

The mathematical expression for the boundary condition for the solid submerged in fluid can be derived by equating the heat flux q_s entering the boundary surface $\mathbf{r}_s$ of the solid with the heat conduction in the solid and the heat flux q_f leaving the same boundary surface by convective heat transfer. Mathematical expressions for heat conduction follow Fourier's law of heat conduction and the formula for convective heat transfer is expressed by Newton' cooling law in Section 7.5.4. We thus have the equality

$$-k\frac{\partial T(\mathbf{r},t)}{\partial n}\bigg|_{\mathbf{r}=\mathbf{r}_s} = h[T(\mathbf{r}_s,t) - T_f]$$

which leads to the following boundary condition for heat conduction analysis of the solid:

$$\frac{\partial T(\mathbf{r},t)}{\partial n}\bigg|_{\mathbf{r}=\mathbf{r}_s} + \frac{h}{k}T(\mathbf{r},t)|_{\mathbf{r}=\mathbf{r}_s} = \frac{h}{k}T_f \qquad (9.17d)$$

where h is the heat transfer coefficient of the surrounding fluid and k is the thermal conductivity of the solid.

The "mixed boundary condition" expressed in Equation 9.17d actually could be used for problems involving prescribed surface temperatures (or Dirichlet conditions) in Equation 9.17a with $h \rightarrow \infty$, which makes $T(\mathbf{r},t)|_{\mathbf{r}=\mathbf{r}_s} = T_s = T_f$: that is, the surface temperature of the solid (T_s) is equal to the bulk fluid temperature of the surrounding fluid (T_f). We may also show that letting $h = 0$ in Equation 9.17d will lead to a thermally insulated boundary condition with $q_s = 0$ in Equation 9.17b.

Example 9.3

Find the boundary conditions of a long, thick-walled pipe for hot steam flow at a bulk temperature T_s and heat transfer coefficient h_s. The outside wall of the pipe is in contact with cold air at a temperature of T_a and a heat transfer coefficient h_a, as illustrated in Figure 9.6.

Figure 9.6 Steam flow in a thick wall pipe.

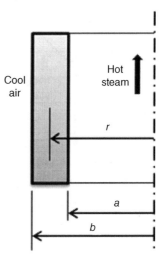

Cool air

Hot steam

r

a

b

Solution:

We may hypothesize that heat will flow in the positive radial direction (r) in a long pipe such as in this example because the temperature gradient crosses the pipe wall. We thus recognize two boundary surfaces in this example at the inner surface with $r = a$ and the outside surface at $r = b$.

We may use Equation 9.17d to establish the conditions at both boundary surfaces as follows.

At inner boundary with $r = a$:

$$k\frac{dT(r)}{dr}\bigg|_{r=a} - \frac{h_s}{k}T(r)\bigg|_{r=a} = \frac{h_s}{k}T_s$$

At the outside boundary with $r = b$:

$$k\frac{dT(r)}{dr}\bigg|_{r=b} + \frac{h_a}{k}T(r)|_{r=b} = \frac{h_a}{k}T_a$$

In the above expressions, k is the thermal conductivity of the pipe material.

Example 9.4

Find the temperature distribution in a long, thick-walled pipe with inner and outside radii a and b, respectively, by using the three types of boundary conditions in Equations 9.17a, 9.17b, and 9.17d. Conditions for establishing the mathematical expressions for these boundary conditions with hot steam inside the pipe and cool surrounding air outside the pipe are indicated in Figure 9.7.

Solution:

We recognize that the physical situation of the current example is the same as that in Example 9.3: that the principal direction of heat flow is from the hot steam inside the pipe to the outside cool air in the radial direction. We may thus hypothesize that the temperature distribution in the pipe wall is represented by $T(r)$ as illustrated in Figure 9.7.

We may select the relevant terms from the heat conduction equation derived for the cylindrical polar coordinate system expressed in Equation 9.14b as follows:

$$\frac{d^2T(r)}{dr^2} + \frac{1}{r}\frac{dT(r)}{dr} = 0 \tag{a}$$

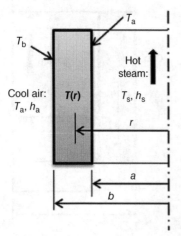

Figure 9.7 Temperature variations in the pipe wall.

Equation (a) is a nonlinear second-order homogeneous differential equation. One may obtain the solution of this equation by expressing Equation (a) in the following form:

$$\frac{d}{dr}\left[r\frac{dT(r)}{dr}\right] = 0 \tag{b}$$

The solution of temperature distribution $T(r)$ in Equation (b) may be obtained by integrating Equation (b) twice and result in the following solution:

$$T(r) = c_1 \ln(r) + c_2 \tag{c}$$

The two arbitrary constants c_1 and c_2 in Equation (c) can be determined by three different types of boundary conditions given in Equations 9.17a, 9.17b, and 9.17d.

With prescribed boundary conditions in Equation 9.17a We have T_a as the temperature at the inner surface with $T(a) = T_a$, and $T(b) = T_b$ at the outside surface of the pipe.

Substituting the above boundary conditions into Equation (c), we have

$$c = \frac{T_a - T_b}{\ln(a/b)} \quad \text{and} \quad c_2 = T_a - \frac{T_a - T_b}{\ln(a/b)}\ln(a)$$

The temperature distribution in the pipe wall is thus obtained as

$$T(r) = T - \frac{T_a - T_b}{\ln(a/b)}\ln\left(\frac{r}{a}\right) \tag{d}$$

With prescribed heat flux qa across the inner surface and Tb at the outside surface We need to solve the following differential equation with the aforementioned boundary conditions:

$$\frac{d^2 T(r)}{dr^2} + \frac{1}{r}\frac{dT(r)}{dr} = 0 \tag{a}$$

with the following boundary conditions according to Equation 9.17b:

$$\left.\frac{dT(r)}{dr}\right|_{r=a} = -\frac{q_a}{k} \tag{e}$$

where q_a is the specific heat flux input to the inner surface of the pipe wall.

The boundary condition at the outside surface of the pipe is established by using Equation 9.17a as

$$T(r)|_{r=b} = T(b) = T_b \tag{f}$$

We may determine the constants c_1 and c_2 by substituting the boundary conditions in Equations (e) and (f) into Equation (c):

$$c = -\frac{aq_a}{k} \quad \text{and} \quad c = T + \frac{aq_a}{k}\ln(b)$$

The temperature distribution in the pipe wall $T(r)$ can thus be obtained by substituting the constants c_1 and c_2 in the above expressions into Equation (c), resulting in the following solution:

$$T(r) = T - \frac{aq_a}{k}\ln\left(\frac{r}{b}\right) \tag{g}$$

With mixed boundary conditions Referring to Figure 9.7 and Equation 9.17d, the two boundary conditions for this case are

$$\left.\frac{dT(r)}{dr}\right|_{r=a} - \frac{h_s}{k}T(r)|_{r=a} = \frac{h_s}{k}T_s \tag{h}$$

$$\left.\frac{dT(r)}{dr}\right|_{r=b} + \frac{h_a}{k}T(r)|_{r=b} = \frac{h_a}{k}T_a \tag{j}$$

The constants c_1 and c_2 in the solution of Equation (a) may be determined by applying the boundary conditions in Equations (h) and (j), resulting in

$$c_1 = \frac{h_s h_a (T_a - T_s)}{\dfrac{kh_a}{a} + \dfrac{kh_s}{b} + h_s h_a \ln\left(\dfrac{b}{a}\right)}$$

and

$$c_2 = T_a - \frac{h_s h_a (T_a - T_s)}{\dfrac{kh_a}{a} + \dfrac{kh_s}{b} + h_s h_a \ln\left(\dfrac{b}{a}\right)} \left(\frac{k}{h_a b} + \ln(b)\right)$$

The temperature distribution in the pipe wall may be obtained by substituting the constants c_1 and c_2 in the above expressions into the solution in Equation (c).

9.5 Solution of Partial Differential Equations for Transient Heat Conduction Analysis

The partial differential equation in Equation 9.15 is valid for the general case of heat conduction in solids including transient cases in which the induced temperature field $T(r,t)$ varies with time t. In this section, we will demonstrate how the separation of variables technique described in Section 9.3 can be used to solve problems of this type in both rectangular and cylindrical polar coordinate systems.

9.5.1 Transient Heat Conduction Analysis in Rectangular Coordinate System

The case that we will present here is a large, flat slab with thermal conductivity k. The slab has thickness L as illustrated in Figure 9.8. It has an initial temperature given by a specified function $f(x)$, and the temperatures of both its faces are maintained at temperature T_f at time $t > 0$. We need to determine the temperature variation in the slab with respect to time t.

We recognize that the physical situation of this problem is relevant to the cooling of a slab from its initial temperature distribution according to an initial temperature function $f(x)$ to a new state in which both its faces are exposed to a uniform temperature T_f for $t > 0$. It is thus clear that the heat flow will be along the x-coordinate—that is, the transient temperature in the slab may be expressed as $T(x,t)$.

One may imagine that the temperature in the slab will vary continuously with time t, until the temperature in the slab reaches a uniform temperature T_f. The purpose of our subsequent

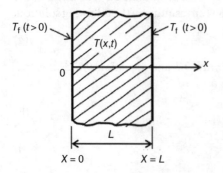

Figure 9.8 Conduction of heat across thickness of a slab.

analysis, however, is to find the transient temperature $T(x,t)$ in the slab before it reaches the ultimate uniform temperature of T_f.

The governing differential equation for this physical situation may be deduced from Equations 9.14a and 9.15 with the thermal conductivity of the slab material $k_x = k_y = k_z = k$ for an isotropic material. The terms $Q(x,y,z,t)$ in Equation 9.14a and $Q(\mathbf{r},t)$ in Equation 9.15 are omitted because the slab does not generate heat by itself. Consequently, the equation that describes the specified physical situation becomes

$$\frac{\partial^2 T(x,t)}{\partial x^2} = \frac{1}{\alpha} \frac{\partial T(x,t)}{\partial t} \tag{9.18}$$

where α is the thermal diffusivity of the slab material.

The solution of Equation 9.18 will satisfy the following specified conditions:

Initial condition (IC):

$$T(x,t)|_{t=0} = T(x,0) = f(x) \quad \text{a given function} \tag{9.19a}$$

Boundary conditions (BCs):

$$T(x,t)|_{x=0} = T(0,t) = T_f \qquad t > 0 \tag{9.19b}$$

$$T(x,t)|_{x=L} = T(L,t) = T_f \qquad t > 0 \tag{9.19c}$$

Equations 9.18 and Equations 9.19a, 9.19b, and 9.19c are proper mathematical expressions for the problem. However, the solution of Equation 9.18 may be simplified by converting the nonhomogeneous BCs in Equations 9.19b and 9.19c to homogeneous ones by the substitution of $u(x,t)$ for $T(x,t)$:

$$u(x,t) = T(x,t) - T_f \tag{9.20}$$

Equations 9.18 and 9.19 will thus be converted to the following equivalent system of differential equations with modified IC and BCs by the substitution of Equation 9.20 into Equations 9.18 and Equations 9.19a, 9.19b, and 9.19c:

$$\frac{\partial^2 u(x,t)}{\partial x^2} = \frac{1}{\alpha} \frac{\partial u(x,t)}{\partial t} \tag{9.21}$$

with the following converted initial condition:

$$u(x,t)|_{t=0} = u(x,0) = f(x) - T_f \tag{a}$$

and for the BCs:

$$u(x,t)|_{x=0} = u(0,t) = T(x,t)|_{x=0} - T_f = T_f - T_f = 0 \tag{b}$$

$$u(x,t)|_{x=L} = u(L,t) = T(x,t)|_{x=L} - T_f = T_f - T_f = 0 \tag{c}$$

The solution of Equation 9.21 satisfying the IC and BCs in Equations (a), (b), and (c) can be obtained using either the Laplace transform with t as the variable in the transformation, as presented in Section 6.7, or using the separation of variables technique as described in Section 9.3.1. We will use the latter technique for obtaining the solution of Equation 9.21.

The solution process begins with the separation of the variables in the function for the solution. In the present case, the variables to be separated are variables x and t in the function $u(x,t)$. We will proceed by letting

$$u(x,t) = X(x)\tau(t) \tag{9.22}$$

in which function $X(x)$ is a function with variable x only and the other function $\tau(t)$ has a variable t only.

Substituting the relationship in Equation 9.22 into Equation 9.21 gives the expression

$$\frac{\partial^2[X(x)\tau(t)]}{\partial x^2} = \frac{1}{\alpha}\frac{\partial[X(x)\tau(t)]}{\partial t}$$

The above expression leads to the following equalities:

$$\tau(t)\frac{\partial^2 X(x)}{\partial x^2} = \frac{1}{\alpha}X(x)\frac{\partial\tau(t)}{\partial t}$$

or (d)

$$\tau(t)\frac{d^2 X(x)}{dx^2} = \frac{1}{\alpha}X(x)\frac{d\tau(t)}{dt}$$

The second expression in Equation (d) will yield the following special form of equality:

$$\frac{1}{X(x)}\frac{d^2 X(x)}{dx^2} = \frac{1}{\alpha}\frac{1}{\tau(t)}\frac{d\tau(t)}{dt}$$

We observe that the left-hand side (LHS) of the expression contains only variable x, whereas the right-hand side (RHS) of the expression consists of another independent variable t. The only way this type of equality can exist is to have LHS = RHS = constant, and this constant needs to be a negative number to be meaningful in the solution. Consequently, we will introduce a "separation constant" $-\beta^2$ for the above expression as follows:

$$\frac{1}{X(x)}\frac{d^2 X(x)}{dx^2} = \frac{1}{\alpha}\frac{1}{\tau(t)}\frac{d\tau(t)}{dt} = -\beta^2 \tag{9.23}$$

One may derive two separate ordinary differential equations by equating the LHS $= -\beta^2$ in Equation 9.23 to give

$$\frac{d^2 X(x)}{dx^2} + \beta^2 X(x) = 0 \tag{9.24}$$

and another differential equation by letting the terms in the RHS $= -\beta^2$ in Equation 9.23 to give to attend these first two classes

$$\frac{d\tau(t)}{dt} + \alpha\beta^2\tau(t) = 0 \tag{9.25}$$

We have thus successfully separated the variables x and t from the function $u(x,t)$ in Equation 9.21 into two ordinary differential equations in Equations 9.24 and 9.25 for the solutions of functions $X(x)$ and $\tau(t)$ in Equation 9.22.

The solution $X(x)$ and $\tau(t)$ in Equations 9.24 and 9.25 requires the specific condition for both these equations. We will derive these conditions for Equation 9.24 by the following procedures.

Let us first substitute the expression $u(0,t) = 0$ in equation (b) into Equation 9.22, resulting in: $u(0, t) = X(0)\tau(t) = 0$. We see that function $\tau(t) \neq 0$; we will thus have $X(0) = 0$. The condition $X(L) = 0$ can be derived in a similar way. We will thus have the specified conditions for Equation 9.24:

$$X(0) = 0 \tag{e1}$$
$$X(L) = 0 \tag{e2}$$

Equation 9.24 is a linear homogeneous second-order differential equation, and the solution $X(x)$ is available from Section 8.2 in the form

$$X(x) = A\cos\beta x + B\sin\beta x \tag{f}$$

in which A and B are arbitrary constants, and β is the unknown separation constant. We will use the specified condition in Equation (e1) to determine $A = 0$, and the condition in Equation (e2) leads to $X(L) = B \sin \beta L = 0$. The latter expression provides two possibilities, either the constant $B = 0$ or $\sin \beta L = 0$. Since the constant A in the solution of $X(x)$ is already equal to zero, having $B = 0$ will lead to $X(x) = 0$, and therefore $u(x,t) = 0$ according to Equation 9.22. The solution $u(x,t) = 0$ is a trivial solution of Equation 9.22 and it is considered to be unrealistic. We are thus left with the only possibility of letting

$$\sin \beta L = 0 \tag{9.26}$$

We will quickly realize that there are multiple values of the separation constant β that satisfy Equation 9.26. These are $\beta = n\pi$, with $n = 1, 2, 3, \ldots$. Alternatively, we may express the separation constant in the form

$$\beta_n = \frac{n\pi}{L} \qquad (n = 1, 2, 3, \ldots) \tag{9.27}$$

We may thus express the solution of Equation 9.24, that is, the function $X(x)$, in the following form:

$$X(x) = B_1 \sin \frac{\pi x}{L} + B_2 \sin \frac{2\pi x}{L} + B_3 \sin \frac{3\pi x}{L} + \cdots$$

$$= \sum_{n=1}^{\infty} B_n \sin \frac{n\pi x}{L} \qquad (n = 1, 2, 3, \ldots) \tag{9.28}$$

In Equation 9.28, the constant coefficients $B_1, B_2, B_3, \ldots$ correspond to the constant coefficient B in the solution of $X(x)$ in Equation (f). We need to include all valid solutions of $X(x)$ in the solutions of linear differential equation expressed in Equation 9.24, with those expressed in Equation 9.28.

We will note that the separation constant β in Equation 9.27 has multiple values excluding the case of $n = 0$, and all other values are multiples of integer numbers. These facts make $\sin \beta L = 0$ in Equation 9.26 the "characteristic equation" of the differential equation in Equation 9.24, and the solutions $\beta L = n\pi$ ($n = 1, 2, 3, \ldots$) the eigenvalues as defined in Section 4.8.

We are now ready to solve for the other function $\tau(t)$ in Equation 9.25. That equation is a first-order differential equation and the solution $\tau(t)$ may be found by the solution methods presented in Section 7.2. Thus, we will have the solution for Equation 9.25 in the following form:

$$\tau(t) = C_n e^{-\alpha \beta_n^2 t} \tag{9.29}$$

where C_n with $n = 1, 2, 3, \ldots$ are multi-valued integration constants because of the multi-valued β_n in the solution.

The general solution of Equation 9.21 can thus be obtained by substituting the solutions $X(x)$ in Equation 9.28 and $\tau(t)$ in Equation 9.29 into Equation 9.22 to give

$$u(x, t) = \sum_{n=1}^{\infty} C_n B_n e^{-\alpha \beta_n t} \sin \frac{n\pi x}{L}$$

$$= \sum_{n=1}^{\infty} b_n e^{-\alpha \beta_n t} \sin \frac{n\pi x}{L} \tag{9.30}$$

We have lumped the product of constants $C_n B_n$ and make this product of multi-valued constants to a single multi-valued constant b_n in Equation 9.30. The constants b_n ($n = 1, 2, 3, \ldots$) may be determined by the remaining specified condition in Equation (a) with $u(x,0) = f(x) - T_f$:

$$u(x, 0) = f(x) - T_f = \sum_{n=1}^{\infty} b_n \sin \frac{n\pi x}{L} \tag{9.31}$$

Constants b_n may be determined by the following steps involving the use of the orthogonality property of integrals of trigonometric functions as follows.

Step 1: Multiply both sides of Equation 9.31 with the function $\sin(n\pi x/L)$:

$$\left(\sin \frac{n\pi x}{L}\right)[f(x) - T_f] = \left(\sin \frac{n\pi x}{L}\right) \sum_{n=1}^{\infty} b_n \sin \frac{n\pi x}{L}$$

$$= \sum_{n=1}^{\infty} b_n \left(\sin \frac{n\pi x}{L}\right) \sin \frac{n\pi x}{L} \tag{g}$$

Step 2: Integrate both sides of Equation (g) with integration limits of $(0, L)$:

$$\int_0^L \left(\sin \frac{n\pi x}{L}\right)[f(x) - T_f]\,dx = \int_0^L \sum_{n=1}^{\infty} b_n \left(\sin \frac{n\pi x}{L}\right) \sin \frac{n\pi x}{L}\,dx$$

$$= \sum_{n=1}^{\infty} \int_0^L b_n \left(\sin \frac{n\pi x}{L}\right)^2 dx \tag{h}$$

Step 3: Make use of the orthogonality of the harmonic functions such as sine and cosine functions through the following relationships:

$$\int_0^p \sin \frac{n\pi x}{p} \sin \frac{m\pi x}{p}\,dx = \begin{cases} 0 & \text{if } m \neq n \\ p/2 & \text{if } m = n \end{cases} \tag{9.32}$$

which imply that

$$\int_0^p \left(\sin \frac{n\pi x}{p}\right)^2 dx = \frac{p}{2}$$

with $m = n$ in Equation 9.32. We thus have from Equation (h)

$$\int_0^L \left(\sin \frac{n\pi x}{L}\right)[f(x) - T_f]\,dx = b_n \left(\frac{L}{2}\right),$$

which leads to the unknown coefficients b_n being

$$b_n = \frac{2}{L} \int_0^L [f(x) - T_f] \sin \frac{n\pi x}{L}\,dx \tag{9.33}$$

We thus have the solution of Equation 9.21 as

$$u(x, t) = \frac{2}{L} \sum_{n=1}^{\infty} \left\{ \int_0^L [f(x) - T_f] \sin \frac{n\pi x}{L}\,dx \right\} e^{-\alpha \beta_n^2 t} \sin \frac{n\pi x}{L}$$

The solution of $T(x,t)$ in Equation 9.18 for the temperature distribution in the slab can thus be obtained by the relationship in Equation 9.20 to take the form

$$T(x, t) = T_f + \frac{2}{L} \sum_{n=1}^{\infty} \left\{ \int_0^L [f(x) - T_f] \sin \frac{n\pi x}{L}\,dx \right\} e^{-\alpha \beta_n^2 t} \sin \frac{n\pi x}{L} \tag{9.34}$$

where T_f is the temperature of both surfaces in Figure 9.8, L is the thickness of the slab, $f(x)$ represents the initial temperature distribution in the slab, and $\beta_n = n\pi/L$ with $n = 1, 2, 3, \ldots$ are the separation constants.

One would envisage that the slab, with the initial temperature distribution $T(x, t)|_{t=0} = f(x)$, would after having both its surfaces maintained at temperature T_f for time $t > 0$ ultimately reach a uniform temperature T_f. This solution reflects such a physical situation with $t \to \infty$ for which $T(x,\infty) = T_f$ will be obtained as the solution of Equation 9.34.

9.5.2 Transient Heat Conduction Analysis in the Cylindrical Polar Coordinate System

We will present the next case of solving heat conduction problems using the separation of variables technique by replacing the flat slab in Section 9.5.1 with a solid cylinder of radius a as shown in Figure 9.9. The cylinder initially has a temperature distribution of $f(r)$. It is submerged in an agitated fluid with bulk fluid temperature T_f. Because of the vigorous agitation of the fast-moving surrounding fluid, the heat transfer coefficient h approaches ∞, resulting in a circumference temperature of the cylinder being the same as the surrounding bulk fluid temperature T_f for time $t > 0$ as in Equation 9.17d. The initial temperature distribution in the cylinder $T(r)$ will thus vary with time t after it is submerged in the fluid at $t > 0$. We need to find the temperature distribution $T(r,t)$ in the cylinder for $t > 0$.

The reader may relate the current case to a common industrial application in the quenching of hot solids in a cold fluid bath. We assume that the cylinder is long enough to justify a simplified case of having the principal heat flow in the radial direction (r).

The applicable partial differential equation for the current application may be deduced from Equation 9.14b by dropping the third and other terms in the right-hand side of the equation, resulting in

$$\frac{1}{\alpha}\frac{\partial T(r,t)}{\partial t} = \frac{\partial^2 T(r,t)}{\partial r^2} + \frac{1}{r}\frac{\partial T(r,t)}{\partial r} \tag{9.35}$$

where $\alpha = k/\rho c$ is the thermal diffusivity of the cylinder material, with ρ and c being the mass density and specific heat of the cylinder material, respectively.

The appropriate conditions for the current problem are

Initial condition:

$$T(r,t)|_{t=0} = T(r,0) = f(r) \tag{a}$$

The function $f(r)$ in Equation (a) is the specified temperature distribution in the radial direction at time $t = 0$.

Boundary conditions:

$$T(r,t)|_{r=a} = T(a,t) = T_f \qquad t > 0 \tag{b1}$$

where T_f is the given bulk fluid temperature.

Figure 9.9 Heat Conduction in a Long Solid Cylinder.

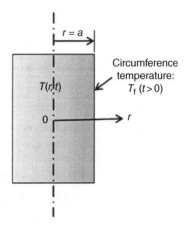

The other boundary condition at $r = 0$, the center of the solid cylinder, has to be a finite value. The mathematical expression of this condition is

$$T(r,t)|_{r=0} = T(0,t) \neq \infty \quad \text{or} \quad T(0,t) = \text{finite value} \tag{b2}$$

We will convert Equation 9.35 from $T(r,t)$ to $\underline{u}(r,t)$ as we did in Section 9.5.1. We thus have

$$u(r,t) = T(r,t) - T_f \tag{c}$$

Substituting the above expression into Equation 9.35 and the associated initial and boundary conditions in Equations (a) and (b), we obtain the following converted partial differential equation

$$\frac{1}{\alpha} \frac{\partial u(r,t)}{\partial t} = \frac{\partial^2 u(r,t)}{\partial r^2} + \frac{1}{r} \frac{\partial u(r,t)}{\partial r} \tag{9.36}$$

with the converted specified conditions:

$$u(r,t)|_{t=0} = u(r,0) = f(r) - T_f \quad \text{for } t = 0 \tag{d}$$
$$u(r,t)|_{r=a} = u(a,t) = 0 \quad \text{for } t > 0 \tag{e}$$

We may solve Equation 9.36 using either the Laplace transform method with variable t being the transform parameter according to the procedure presented in Section 6.7, or we may use the separation of variables technique as we did in Section 9.3.1. We will learn to use the latter technique for solving partial differential equations in the cylindrical polar coordinate system.

Letting the solution of function $u(x,t)$ in Equation 9.36 with two independent variables r and t be separated according to the following relation:

$$u(r,t) = R(r)\tau(t) \tag{9.37}$$

where the function $R(r)$ is a function of variable r only and the function $\tau(t)$ is a function of the other variable t only. Upon substituting this relation into Equation 9.36 we get

$$\frac{1}{\alpha} \frac{\partial[R(r)\tau(t)]}{\partial t} = \frac{\partial^2[R(r)\tau(t)]}{\partial r^2} + \frac{1}{r} \frac{\partial[R(r)\tau(t)]}{\partial r}$$

or $\tag{f}$

$$\frac{R(r)}{\alpha} \frac{\partial \tau(t)}{\partial t} = \tau(t) \frac{\partial^2 R(r)}{\partial r^2} + \frac{\tau(t)}{r} \frac{\partial R(r)}{\partial r}$$

Equation (f) legitimates the conversion of the partial derivatives of $R(r)$ and $\tau(t)$ to ordinary derivatives because both these functions involve only one variable. Consequently, we may establish the following equality as:

$$\frac{R(r)}{\alpha} \frac{d\tau(t)}{dt} = \tau(t) \frac{d^2 R(r)}{dr^2} + \frac{\tau(t)}{r} \frac{dR(r)}{dr}$$

After rearranging the terms in the above expression, we get the following equality:

$$\frac{1}{\alpha\tau(t)} \frac{d\tau(t)}{dt} = \frac{1}{R(r)} \left[\frac{d^2 R(r)}{dr^2} + \frac{1}{r} \frac{dR(r)}{dr} \right] \tag{g}$$

We note that the left-hand side of Equation (g) is a function of variable t only and the right-hand side involves another function of the other variable r only. The only way this equality can hold is to have both sides equal to a constant (a negative constant to be realistic). Thus we have the following expression:

$$\frac{1}{\alpha\tau(t)} \frac{d\tau(t)}{dt} = \frac{1}{R(r)} \left[\frac{d^2 R(r)}{dr^2} + \frac{1}{r} \frac{dR(r)}{dr} \right] = -\beta^2 \tag{9.38}$$

Equation 9.38 will result in two separate ordinary differential equations:

$$\frac{d\tau(t)}{dt} + \alpha\beta^2\tau(t) = 0 \tag{9.39}$$

and

$$r^2\frac{d^2R(r)}{dr^2} + r\frac{dR(r)}{dr} + \beta^2 r^2 R(r) = 0 \tag{9.40}$$

The solution of Equation 9.39 is identical to Equation 9.29 in the form

$$\tau(t) = c_n e^{-\alpha\beta_b^2 t} \tag{h}$$

where c_n with $n = 1, 2, 3, \ldots\ldots$ is a multi-valued integration constant.

Equation 9.40 is a homogeneous second-order ordinary differential equation but without constant coefficients as in the cases in Section 8.2. Therefore the solution method presented in that section cannot be used for the solution of Equation 9.40. However, a close look at this equation reveals that it is a special case of the Bessel equation in Equation (2.27) with order $n = 0$. Consequently, the solution of Equation 9.40 can be expressed by the Bessel functions given in Equation 2.28 with $n = 0$ in the following form:

$$R(r) = AJ_0(\beta r) + BY_0(\beta r) \tag{9.41}$$

where A and B are arbitrary constants to be determined by appropriate boundary conditions, and $J_0(\beta r)$ and $Y_0(\beta r)$ are the Bessel functions of first and second kind with order zero, respectively.

The two constants A and B in Equation 9.41 can be determined using the two boundary conditions specified in Equations (b1) for $r = a$, and Equation (b2) for $r = 0$. Application of the latter condition for $T(0,t)$ to be finite results in $B = 0$ for the reason that the Bessel function of second kind $Y_0(\beta r)|_{r=0} = Y(0) \rightarrow -\infty$, which violates the condition that the value of $T(x,t)$ cannot be $\pm\infty$. Consequently, the only way we can avoid $T(0,t) \rightarrow -\infty$ is to let the constant $B = 0$ in Equation 9.41.

We thus have the solution of Equation 9.40 as

$$R(r) = AJ_0(\beta r) \tag{j}$$

The remaining arbitrary constant A in the solution in Equation (j) can be evaluated by the boundary condition in Equation (e) that

$$R(a) = AJ_0(\beta a) = 0$$

The above expression offers two possible solutions: we may either let the constant $A = 0$ or have $J_0(\beta a) = 0$. We recall that constant B in the expression for $R(r)$ in Equation 9.41 is already set to zero. Setting the other constant $A = 0$ will make $R(r)$ and thus $u(r,t)$ and $T(r,t)$ zero in the cylinder at all times, which violates physical reality. Consequently, we need to let

$$J_0(\beta a) = 0 \tag{9.42}$$

Equation 9.42 offers the values of the separation constant β in Equation 9.38 because $J_0(x) = 0$ is an equation that has multiple roots like $\sin(\beta L) = 0$ in Equation 9.26. The roots of the equation $J_0(\beta a) = 0$ may be found either from Figure 2.45a or from mathematical handbooks. For instance, the first six approximate solutions for βa in Equation 9.42 can be found to be 2.4, 5.52, 8.65, 11.79, 14.93, and 18.07 (Zwillinger, 2003), or from commercial software such as "Mathematica" or "MatLAB" as will be described in Chapter 10.

Example 9.5

Find the first four roots of Equation 9.42 with $a = 20$ units.

Solution:

We are required to find the first four roots of the equation of $J_0(20\beta) = 0$.

We will find the first four approximate roots of the equation $J_0(x) = 0$ to be $x = 2.4, 5.52, 8.65$, and 11.79 (Zwillinger, 2003). This equation may be related to the equation $J_0(20\beta) = 0$ with $x = 20\beta$. We will thus have the first four roots of the equation $J_0(20\beta) = 0$ with $\beta = x/20$, with $\beta_1 = 2.4/20 = 0.12$, $\beta_2 = 5.52/20 = 0.276$, $\beta_3 = 8.65/20 = 0.4325$, and $\beta_4 = 11.79/20 = 0.5895$.

We may thus express the separation variable β in Equation 9.38 as β_n ($n = 1, 2, 3, \ldots$), and the solution of $R(r)$ in Equation 9.40 takes the form

$$R(r) = A_n J_0(\beta_n r) \quad \text{with } n = 1, 2, 3, \ldots$$

We recognize that Equation 9.42 is the "characteristic equation" of the differential equation in Equation 9.40 with its roots β_n with $n = 1, 2, 3, \ldots$ being the eigenvalues as described in Section 4.8.

The complete solution of $R(r)$ including all possible solutions becomes

$$R(r) = \sum_{n=1}^{\infty} A_n J_0(\beta_n r) \tag{9.43}$$

The solution $u(r,t)$ in Equation 9.36 can thus be obtained by the relationship in Equation 9.37 after substitution of $\tau(t)$ in Equation (h) and $R(r)$ in Equation 9.43, resulting in

$$u(r, t) = \sum_{n=1}^{\infty} b_n e^{-\alpha \beta_n^2 t} J_0(\beta_n r) \tag{9.44}$$

We have "lumped" coefficients c_n and A_n in the respective expressions in Equation (h) and Equation 9.43 into another multi-valued constant b_n in Equation 9.44.

The determination of the constant b_n in Equation 9.44 requires the use of the initial condition specified in Equation (d), which leads to

$$u(r, 0) = f(r) - T_f$$

$$= \sum_{n=1}^{\infty} b_n J_0(\beta_n r)$$

$$= b_1 J_0(\beta_1 r) + b_2 J_0(\beta_2 r) + b_3 J_0(\beta_3 r) + \cdots \tag{9.45}$$

We will multiply both sides of Equation 9.45 by the following summation of Bessel functions:

$$[r J_0(\beta_1 r) + r J_0(\beta_2 r) + r J_0(\beta_3 r) + \cdots]$$

leading to the following expression:

$$[r J_0(\beta_1 r) + r J_0(\beta_2 r) + r J_0(\beta_3 r) + \cdots][f(r) - T_f]$$

$$= [r J_0(\beta_1 r) + r J_0(\beta_2 r) + r J_0(\beta_3 r) + \cdots][b_1 J_0(\beta_1 r) + b_2 J_0(\beta_2 r) + b_3 J_0(\beta_3 r) + \cdots]$$

We may express the above relation in the following form:

$$r J_0(\beta_n r)[f(r) - T_f] = r J_0(\beta_n r)[b_1 J_0(\beta_1 r) + b_2 J_0(\beta_2 r) + b_3 J_0(\beta_3 r) + \cdots] \tag{k}$$

We will integrate both sides of the expression in Equation (k) with respect to variable r:

$$\int_0^a rJ_0(\beta_n r)[f(r) - T_f]dr$$

$$= \int_0^a rJ_0(\beta_n r)[b_1 J_0(\beta_1 r) + b_2 J_0(\beta_2 r) + b_3 J_0(\beta_3 r) + \cdots]dr \qquad \text{for } n = 1, 2, 3, \ldots$$

$$= \int_0^a rb_n[J_0(\beta_n r)]^2\, dr + \int_0^a rb_1[J_0(\beta_1 r)J_0(\beta_2 r) + J_0(\beta_1 r)J_0(\beta_3 r) + \cdots]dr + \cdots \qquad \text{(l)}$$

According to Fourier–Bessel expansions and integrals (Gray and Mathews, 1966):

$$\int_0^a rJ_0(\beta_m r)J_0(\beta_n r)\, dr = \begin{cases} 0 & \text{if } \beta_m \neq \beta_n \\ \int_0^a r[J_0(\beta_n r)]^2\, dr & \text{if } \beta_m = \beta_n \end{cases}$$

Thus, we have from the above expression that only the first integral on the right-hand side of the above expression has nonzero value. Consequently, we have the expression after integration of the above expression:

$$\int_0^a rJ_0(\beta_n r)[f(r) - T_f]\, dr = b_n \int_0^a r[J_0(\beta_n r)]^2 dr \qquad \text{(m)}$$

But we find that integration of the expression in Equation (m) will result in

$$\int_0^a rJ_0(\beta_n r)[f(r) - T_f]\, dr = b_n \int_0^a r[J_0(\beta_n r)]^2 dr$$

$$= b_n \frac{a^2}{2}[J_0^2(\beta_n a) + J_1^2(\beta_n a)] \qquad \text{(n)}$$

Since we have already established the relation that $J_0(\beta_n a) = 0$ in Equation 9.42, we will have

$$\int_0^a rJ_0(\beta_n r)[f(r) - T_f]\, dr = b_n \frac{a^2}{2}[J_0^2(\beta_n a) + J_1^2(\beta_n a)]$$

$$= b_n \frac{a^2}{2}J_1^2(\beta_n a)$$

which means that the coefficients b_n in Equation 9.44 are

$$b_n = \frac{2}{a^2 J_1^2(\beta_n a)} \int_0^a r[f(r) - T_f]J_0(\beta_n r)\, dr \qquad \text{(9.46)}$$

We may thus obtain the solution of Equation 9.35 via the relation in Equation 9.37 to have the form:

$$T(r, t) = T_f + \sum_{n=1}^{\infty} b_n e^{-\alpha \beta_n^2 t} J_0(\beta_n r) \qquad \text{(9.47)}$$

in which the coefficient b_n with $n = 1, 2, 3, \ldots$ may be computed from Equation 9.46. As in the case in Section 9.5.1, the physical situation in the present case is that the ultimate temperature of the cylinder becomes equal to the surrounding fluid temperature T_f at time $t \rightarrow \infty$.

We observe from this case that Bessel functions often appear in the solution of partial differential equations in the cylindrical polar coordinate system.

9.6 Solution of Partial Differential Equations for Steady-state Heat Conduction Analysis

Often, we are required to find the temperature distributions in solids with stable heat flow patterns, which make the temperature distributions in solids independent of time variation — that is, steady-state heat conduction. Mathematical representation of this situation is available by use of the partial differential equation without the term related to time variable t. Consequently, the partial differential equation in Equation 9.15 is reduced to the following form:

$$\nabla^2 T(\mathbf{r}) + \frac{Q(\mathbf{r})}{k} = 0 \tag{9.48}$$

where the position vector $\mathbf{r}$ represents (x,y,z) in the rectangular coordinate system, or (r,θ,z) in the cylindrical polar coordinate system. Equation 9.48 is further reduced to the "Laplace equation" in the following form if no heat is generated by the solid:

$$\nabla^2 T(\mathbf{r}) = 0 \tag{9.49}$$

9.6.1 Steady-state Heat Conduction Analysis in the Rectangular Coordinate System

We will demonstrate the use of the Laplace equation in Equation 9.49 for the temperature distribution in a square plate with the temperature at its three edges maintained constant at $0°C$ as illustrated in Figure 9.10.

We recognize the situation in which the temperature distribution in the plate is stable with both its flat faces thermally insulated and heat flows in the x-y plane. The induced temperature distribution in the plate is expressed by the function $T(x,y)$ as shown in Figure 9.10. Being a steady-state heat flow, the analysis fits the representation of the physical situation by Equation 9.49. We may thus expand this equation to the following form in the x-y plane:

$$\frac{\partial^2 T(x, y)}{\partial x^2} + \frac{\partial^2 T(x, y)}{\partial y^2} = 0 \tag{9.50}$$

with $0 \le x \le 100$ and $0 \le y \le 100$.

The following boundary conditions apply in the present case:

$$T(x, y)|_{x=0} = T(0, y) = 0 \tag{a1}$$

$$T(x, y)|_{x=100} = T(100, y) = 0 \tag{a2}$$

$$T(x, y)|_{y=0} = T(x, 0) = 0 \tag{a3}$$

$$T(x, y)|_{y=100} = T(x, 100) = 100 \tag{a4}$$

Figure 9.10 Temperature distribution in a plate.

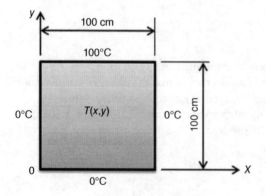

Because both of the variables x and y are within finite bounds, none of the two integral transform methods can be used for the solution of Equation 9.50; we will therefore use the separation of variable technique for the solution. Consequently, we express the solution $T(x,y)$ in Equation 9.50 in the form

$$T(x, y) = X(x)\,Y(y) \tag{b}$$

in which the function $X(x)$ involves only variable x, and the function $Y(y)$ involves variable y only.

Substituting the relationship in Equation (b) into Equation 9.50 and rearranging terms yields the following:

$$\frac{1}{X(x)}\frac{d^2 X(x)}{dx^2} = -\frac{1}{Y(y)}\frac{d^2 Y(y)}{dy^2}$$

We will use the same argument as we did in Sections 9.5.1 and 9.5.2 that the only way the above equality can exist is to have both sides of the equality to be equal to the same constant. We will thus have the following equality:

$$\frac{1}{X(x)}\frac{d^2 X(x)}{dx^2} = -\frac{1}{Y(y)}\frac{d^2 Y(y)}{dy^2} = -\beta^2 \tag{c}$$

in which β is the separation constant.

We may derive two ordinary differential equations from Equation (c), one by equating the left-hand side of Equation (c) with the separation constant, and another by equating the right-hand side of the expression with the same separation constant β:

$$\frac{d^2 X(x)}{dx^2} + \beta^2 X(x) = 0 \tag{d}$$

and

$$\frac{d^2 Y(y)}{dy^2} - \beta^2 Y(y) = 0 \tag{e}$$

The relationship in Equation (b) also leads to the following boundary conditions for Equations (d) and (e):

$$X(0) = 0 \tag{f1}$$
$$X(100) = 0 \tag{f2}$$
$$Y(0) = 0 \tag{f3}$$

Equation (d) corresponds to the typical second-order linear differential equation with constant coefficients as presented in Section 8.2. It has a solution in the form

$$X(x) = A \cos \beta x + B \sin \beta x \tag{g}$$

Applying the first condition in Equation (f1) to Equation (g) will result in $A = 0$ and the second condition in Equation (f2) will lead to the expression:

$$X(100) = B \sin(100\beta) = 0 \tag{h}$$

The equality in Equation (h) offers the possibilities of having either the coefficient $B = 0$ or $\sin(100\beta) = 0$. Because the coefficient A in Equation (g) is already zero, we cannot allow constant $B = 0$. We thus have $\sin(100\beta) = 0$, which leads to $100\beta = n\pi$ with $n = 1, 2, 3, \ldots$, or $\beta = n\pi/100$ with $n = 1, 2, 3, \ldots$. We may thus express the separation constant β in a multi-valued form as

$$\beta_n = \frac{n\pi}{100} \qquad \text{with } n = 1, 2, 3, \ldots$$

As in the similar situations in the previous two cases in Section 9.5, since $\sin(100\beta) = 0$ in Equation (h) is the "characteristic equation" for the differential equation in Equation (d), the solutions $\beta = \beta_n$ with $n = 1, 2, 3, \ldots$ are eigenvalues as defined in Section 4.8. We thus have the solution of $X(x)$ in the form

$$X(x) = B_n \sin \beta_n x \tag{j}$$

with the constant coefficients B_n and the eigenvalues β_n where $n = 1, 2, 3, \ldots$.

The solution $Y(y)$ in Equation (e) can also be obtained using the method presented in Section 8.2, but we will adopt using the hyperbolic sine and cosine functions instead of exponential functions in the solution. The expression for function $Y(y)$ is

$$Y(y) = C \cosh \beta y + D \sinh \beta y \tag{k}$$

where C and D are arbitrary constants.

The boundary condition in Equation (f3) leads to the coefficient $C = 0$; thus the form of the function $Y(y)$ will be

$$Y(y) = D \sinh \beta y$$

Because the separation constant β in the above expression is multi-valued, we may express the solution of $Y(y)$ in the following form:

$$Y(y) = D_n \sinh \beta_n y \tag{m}$$

We have obtain the solution $T(x,y)$ of Equation 9.50 after substituting the expressions for $X(x)$ in Equation (j) and $Y(y)$ in Equation (m) into Equation (b), which results in

$$
\begin{aligned}
T(x, y) &= \sum_{n=1}^{\infty} X(x)Y(y) \\
&= \sum_{n=1}^{\infty} B_n D_n \left(\sin \frac{n\pi}{100} x \right) \left(\sinh \frac{n\pi}{100} y \right) \\
&= \sum_{n=1}^{\infty} b_n \left(\sin \frac{n\pi}{100} x \right) \left(\sinh \frac{n\pi}{100} y \right)
\end{aligned}
\tag{n}
$$

The unknown coefficients b_n in Equation (n) may be determined using the remaining boundary condition in Equation (a4) that $T(x,100) = 100$, which leads to

$$T(x, 100) = 100 = \sum_{n=1}^{\infty} b_n \left(\sin \frac{n\pi}{100} x \right) (\sinh n\pi)$$

or

$$100 = \sum_{n=1}^{\infty} (b_n \sinh n\pi) \sin \frac{n\pi}{100} x \tag{p}$$

By following a similar procedure to determination of the coefficient b_n in Equation 9.31 and making use of the orthogonality properties of the harmonic sine and cosine functions in Equation 9.32, we find the expression for the coefficients b_n in the present case to be

$$b_n \sinh n\pi = \frac{2}{100} \int_0^{100} 100 \sin \frac{n\pi}{100} x \, dx = -\frac{200}{n\pi} (\cos n\pi - 1)$$

We thus have

$$b_n = -\frac{200(\cos n\pi - 1)}{n\pi \sinh n\pi} \quad \text{with } n = 1, 2, 3, \ldots \tag{q}$$

Figure 9.11 Steady-state heat conduction in a solid cylinder.

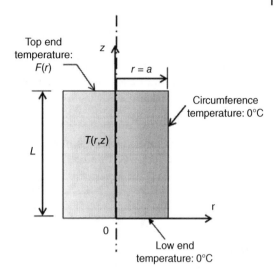

The solution of the temperature distribution in the flat plate in Equation 9.50 with the coefficients b_n expressed in Equation (q) thus has the form

$$T(x, y) = \frac{200}{\pi} \sum_{n=1}^{\infty} \frac{(1 - \cos n\pi)}{n \sinh n\pi} \sin \frac{n\pi}{100} x \sinh \frac{n\pi}{100} y \tag{9.51}$$

One may substitute $\cos n\pi = (-1)^n$ with $n = 1, 2, 3, \ldots$ in Equation 9.51 and compute the temperature at the geometric center of the plate as $T(50,50) = 25.2°C$.

9.6.2 Steady-state Heat Conduction Analysis in the Cylindrical Polar Coordinate System

We will demonstrate the use of separation of variables technique for the solution of steady-state heat conduction in a solid in a cylindrical polar coordinate system. This coordinate system is used for engineering analysis involving solids of circular geometry such as disks and cylinders. As demonstrated in Section 9.5.2, in analyses of solids of this geometry, Bessel functions often appear in the solution.

The case we have here involves a solid cylinder of radius a and length L, with temperature at the circumference and the bottom end maintained at $0°C$ and the temperature at the top surface subjected to a temperature distribution following a specified function $T(r)$ as shown in Figure 9.11.

We recognize the physical situation in which heat flows from the top end of the cylinder in both the radial and longitudinal directions. We may thus represent the temperature in the cylinder by $T(r,z)$ in a cylindrical polar coordinate system as illustrated in Figure 9.11.

The governing partial differential equation for $T(r,z)$ in steady-state heat conduction as described above may be obtained by selecting the appropriate terms in Equation 9.14b in a cylindrical polar coordinate system in the following form:

$$\frac{\partial^2 T(r, z)}{\partial r^2} + \frac{1}{r} \frac{\partial T(r, z)}{\partial r} + \frac{\partial^2 T(r, z)}{\partial z^2} = 0 \tag{9.52}$$

with specified boundary conditions:

$$T(a, z) = 0 \tag{a1}$$

$$T(0, z) \neq \infty \tag{a2}$$

$$T(r,0) = 0 \tag{a3}$$

$$T(r,L) = F(r) \tag{a4}$$

Because both the variables associated with the temperature distribution $T(r,z)$ are valid for finite ranges, we cannot use either of the integral transform methods presented in Section 9.3; we will use the separation of variables technique for the solution of Equation 9.52 by letting

$$T(r,z) = R(r)Z(z) \tag{b}$$

We obtain the following expression after substituting the assumed relation in Equation (b) into Equation 9.52, followed by a rearrangement of terms:

$$\frac{1}{R(r)} \frac{\partial^2 R(r)}{\partial r^2} + \frac{1}{rR(r)} \frac{\partial R(r)}{\partial r} = -\frac{1}{Z(z)} \frac{\partial^2 Z(z)}{\partial z^2} \tag{c}$$

We note that the left-hand side of Equation (c) contains function $R(r)$ with variable r only, whereas the right-hand side contains a function $Z(z)$ with variable z only. The only way this equality can be valid is making the functions on each side of the above equation equal to the same constant. Consequently, we have the following expression in Equation (d) after the separation of variables r and z:

$$\frac{1}{R(r)} \frac{d^2 R(r)}{dr^2} + \frac{1}{rR(r)} \frac{dR(r)}{dr} = -\frac{1}{Z(z)} \frac{d^2 Z(z)}{dz^2} = -\beta^2 \tag{d}$$

where β is the separation constant.

Note that we have converted the partial differentiations in Equation 9.52 with respect to both independent variables r and z into two ordinary differential equations in Equation (d).

The constant β appearing in Equation (d) is referred to as the separation constant for Equation 9.52. We will derive two ordinary differential equations by equating the left-hand side of the expression in Equation (d) with $-\beta^2$, and another ordinary differential equation by equating the right-hand side of the expression in Equation (d) with $-\beta^2$. These two ordinary differential equations are shown below:

$$r\frac{d^2 R(r)}{dr^2} + \frac{dR(r)}{dr} + r\beta^2 R(r) = 0 \tag{e}$$

and

$$\frac{d^2 Z(z)}{dz^2} - \beta^2 Z(z) = 0 \tag{f}$$

We may express the specified boundary conditions for Equation 9.52 in Equations (a1), (a2), and (a3) as the following equivalent conditions using the relationship shown in Equation (b):

$$R(a) = 0 \tag{g1}$$

$$R(0) \neq \infty \tag{g2}$$

$$Z(0) = 0 \tag{g3}$$

The solution of Equation (e) involves Bessel functions as in the case in Section 9.5.2 and in Equation 9.41:

$$R(r) = A J_0(\beta r) + B Y_0(\beta r) \tag{9.41}$$

The condition specified in Equation (g2) results in having the constant coefficient $B=0$, because the second term in the above solution in Equation 9.41 cannot be allowed in the expression because $Y_0(0) \to -\infty$, which leads to an unrealistic physical situation for $T(r,z)$. Consequently, the solution $R(r)$ of Equation (e) has the form

$$R(r) = A J_0(\beta r)$$

The subsequent application of the condition in Equation (g1) to the above relation results in $R(a) = 0 = AJ_0(\beta a)$, in which we may either have $A = 0$ or $J_0(\beta a) = 0$. It is clear that we cannot let the constant $A = 0$ because the other constant B in the solution of $R(r)$ in Equation 9.41 already equals zero. Letting $A = 0$ will make $R(r) = 0$ at all times, which will lead to a trivial solution of $T(r,z)$. We thus take the equation for the separation constant β to be the solutions of $J_0(\beta a) = 0$. This equation is the characteristic equation of Equation (e) and it will lead to multiple values of the separation constant β, as the equation $J_0(x) = 0$ has in theory an infinite number of roots x_1, x_2, x_3, ..., as demonstrated in Example 9.5. We may thus express the separation constant β in the form β_n with $n = 1, 2, 3, ...$, and the solution for the function $R(r)$ should include all possible values of β_n

$$R(r) = A_n J_0(\beta_n r) \tag{h}$$

in which A_n are the constant coefficients corresponding to the values of β_n obtained from the following equation:

$$J_0(\beta_n a) = 0 \tag{j}$$

with $n = 1, 2, 3, ...$ The solution of Equation (f) can be expressed as

$$Z(z) = C \cosh(\beta z) + D \sinh(\beta z) \tag{k}$$

Substituting the condition $Z(0) = 0$ for Equation (a3) into Equation (k) leads to the constant $C = 0$. We will thus have

$$Z(z) = D \sinh(\beta z)$$

Since the solution $Z(z)$ in the above expression also involves the separation constants β_n with $n = 1, 2, 3, ...$, we need to express $Z(z)$ accordingly with β_n in the form

$$Z(z) = D_n \sinh(\beta_n z) \tag{m}$$

We can thus express the solution $T(r,z)$ in Equation 9.52 in the following form with the solutions of $R(r)$ in Equation (h) and $Z(z)$ in Equation (m):

$$T(r, z) = [A_n J_0(\beta_n r)][D_n \sinh(\beta_n z)] \quad \text{with } n = 1, 2, 3, ...$$

We need to include all valid solutions corresponding to the values of β_n and express the complete solution of $T(r,z)$ in the following form:

$$T(r, z) = \sum_{n=1}^{\infty} b_n [J_0(\beta_n r)](\sinh \beta_n z) \tag{9.53}$$

We note that the multi-valued constant coefficient b_n appearing in Equation 9.53 replaces the product of two other multi-valued constants A_n and D_n as in the analyses in preceding cases.

The constant coefficients b_n in Equation 9.53 may be determined by using the Fourier–Bessel series as we did in Section 9.5.2 with the use of the remaining boundary condition in Equation (a4). We will have the following expression:

$$b_n \sinh \beta_n L = \frac{2F(r)}{L^2 J_1^2(\beta_n L)} \int_0^L r J_0(\beta_n r) \, dr$$

from which we obtain the expression for b_n:

$$b_n = \frac{2F(r)}{L^2 J_1^2(\beta_n L) \sinh \beta_n L} \int_0^L r J_0(\beta_n r) \, dr \quad \text{with } n = 1, 2, 3, ... \tag{n}$$

We thus have the solution of temperature distribution in the solid cylinder $T(r,z)$ in Equation 9.53 with the coefficient b_n expressed in Equation (n) with specified temperature variation $F(r)$ at the top edge, and the separation constants β_n being the roots of Equation (j).

9.7 Partial Differential Equations for Transverse Vibration of Cable Structures

Engineers often are required to carry out vibration analysis of long flexible cable structures subjected to external forces. Cable structures are common in places such as power transmission lines shown in Figure 9.12, guy wire supports in Figure 9.13, and suspension bridges in Figure 9.14.

These cable structures, flexible in nature, are vulnerable to resonant vibrations as described in Chapter 8. The galloping motions that often develop in resonant vibration of these structures can result in devastating structural failure. Rupture of long power transmission cables resulting from such vibration often occurs in places with cold climate in which freezing rain can build up heavy icicles on the cables overnight. The significant weight gain (and thus an increase of mass) of the cable causes the cable to vibrate violently with gale-force winds, and in many cases can result in breaking of the cables. Many transmission towers have also been destroyed due to violent vibration of the transmission cable lines, resulting in millions of dollars of property loss to the power companies in these regions. This is thus an important subject for mechanical and structural engineers in the safe design of structures of this type.

9.7.1 Derivation of Partial Differential Equations for Free Vibration of Cable Structures

Mathematical modeling of vibration analysis of cable structures begins with the illustration of the cable movements as in Figure 9.15, in which a cable of length L is at a static equilibrium condition with both its ends fixed at supports and with an initial sag due to its own weight and the associated tension in the cable. This initial shape of the free-hung cable is represented by function $f(x)$ as illustrated in the figure.

Formulation of the differential equation for vibration of long cables is based on the law of conservation of momentum, and in particular, the use of Newton's second law in dynamics.

Like all engineering analyses, the mathematical modeling of vibrating flexible cable structures begins with a number of assumptions and idealizations such as were described in Stage 2 in Section 1.4. Some of these assumptions are presented below.

Figure 9.12 Long cables in electric power transmission structures. *See color section.*

Figure 9.13 Guy wire support for a tall radio transmission tower. *See color section.*

Figure 9.14 Golden Gate suspension bridge. *See color section.*

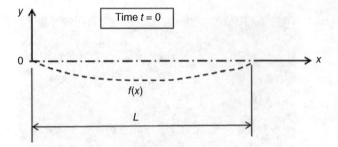

Figure 9.15 A long cable in a static equilibrium state.

1) The cable is as flexible as a string. This means that the cable has no strength to resist a bending moment and there is no shear force associated with its deflection. The mathematical model that we will develop subsequently will be more relevant to the case with long cables for their inherent flexibility.
2) There exists a tension in the string in its free-hung static state as shown in Figure 9.15. This tension is so large that the weight, but not the mass, of the cable is neglected in the analysis.
3) Each small segment of the cable along its length, that is, the segment with length Δx, moves only in the vertical direction during vibration.
4) The vertical movement of the cable along the length is small, so the slope of the deflection curve is small.
5) The mass of the cable along the length is constant, that is, the cable is made of the same material along its length.

Figure 9.16 illustrates the vibration of the cable produced by a small instantaneous disturbance in the vertical direction (or y-direction)—a situation similar to the free vibration that we dealt with for mass–spring systems in Chapter 8.

Let the mass per unit length of the string be designated by m. The total mass of string in an incremental length Δx will thus be $(m\,\Delta x)$. The condition for a dynamic equilibrium according to Newton's second law presented in the *equation of motion* obeys the relationship

$$\underset{\sum F}{\text{Total applied forces}} = \underset{m}{\text{Mass}} \times \underset{a}{\text{Acceleration}}$$

Let us consider the forces acting on the small segment of the string shown in Detail A of Figure 9.16b in the free-body diagram in Figure 9.17. From Figure 9.17, we may express the equation of motion of this string segment in the y-direction as

$$\sum F_y = -(P + \Delta P)\sin(\alpha + \Delta \alpha) - P \sin \alpha + (m\,\Delta x)\frac{\partial^2}{\partial t^2}\left(u + \frac{\Delta u}{2}\right) = 0$$

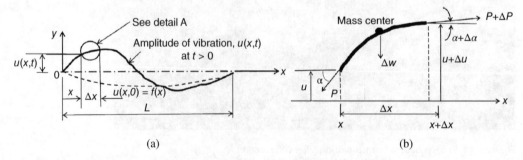

(a)

(b)

Figure 9.16 A vibrating long cable. (a) Shape of the vibrating cable at time t. (b) Forces on the cable in Detail A.

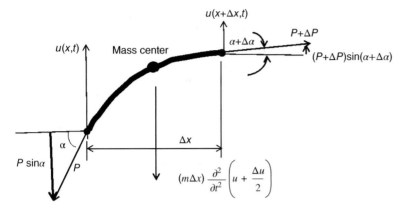

Figure 9.17 Dynamic forces acting on a cable segment.

We note that ΔP is small, so that $\Delta P \sin(\alpha+\Delta\alpha) \to 0$; we thus have

$$-P \sin(\alpha + \Delta\alpha) - P \sin\alpha + (m\,\Delta x)\frac{\partial^2}{\partial t^2}\left(u + \frac{\Delta u}{2}\right) = 0$$

For the small angle α and thus very small $\Delta\alpha$, the following relationship exists:

$$\sin(\alpha + \Delta\alpha) \approx \tan(\alpha + \Delta\alpha) = \frac{\partial u(x + \Delta x, t)}{\partial x}$$

holds and

$$\sin\alpha \approx \tan\alpha = \frac{\partial u(x, t)}{\partial x}$$

the above equation of motion can be expressed as

$$P\left[\frac{\partial u(x + \Delta x, t)}{\partial x} - \frac{\partial u(x, t)}{\partial x}\right] = (m\,\Delta x)\frac{\partial^2}{\partial t^2}\left(u + \frac{\Delta u}{2}\right)$$

or, after dividing both sides by Δx, as

$$P\frac{\dfrac{\partial u(x + \Delta x, t)}{\partial x} - \dfrac{\partial u(x, t)}{\partial x}}{\Delta x} = m\frac{\partial^2}{\partial t^2}\left(u + \frac{\Delta u}{2}\right)$$

We recall the definition of the derivative of a function $y(x)$ in Section 2.2.5 as

$$\frac{dy(x)}{dx} = \lim_{\Delta x \to \infty} \frac{y(x + \Delta x) - y(x)}{\Delta x}$$

We may use this definition to arrive at the following relation for partial derivatives:

$$\lim_{\Delta x \to \infty} \frac{\dfrac{\partial u(x + \Delta x, t)}{\partial x} - \dfrac{\partial u(x, t)}{\partial x}}{\Delta x} = \frac{\partial^2 u(x, t)}{\partial x^2}$$

The equation of motion of the string can be further simplified to take the form

$$P\frac{\partial^2 u(x, t)}{\partial x^2} = m\frac{\partial^2}{\partial t^2}\left(u + \frac{\Delta u}{2}\right)$$

But since we have assumed that Δu is small, and $\Delta u/2$ is also small enough to be neglected, we thus have the equation for the instantaneous deflection of the string at x as

$$\frac{\partial^2 u(x, t)}{\partial t^2} = a^2\frac{\partial^2 u(x, t)}{\partial x^2} \tag{9.54}$$

in which $a = \sqrt{P/m}$ with P = tension in the string, with the unit of newton (N), and m = mass of the cable per unit length, with the unit of kg/m. The unit for the constant a in Equation 9.54 is thus m/s. This unit is derived from the relations that the tension P has the unit of newton, or kg-m/s^2.

Equation 9.54 is often referred to as the "wave equation" that describes wave motion, with $u(x,t)$ being the amplitude of the wave and the constant coefficient a being the wave propagation speed.

The solution $u(x,t)$ in Equation 9.54 for vibration of strings (or long flexible cables) represents the instantaneous deflection (or amplitude of vibration) of the cable in a free vibration state. Some common specified conditions used in solving this equation are as follows.

The initial condition at time $t = 0$:

$$u(x, t)|_{t=0} = u(x, 0) = f(x) \tag{9.55a}$$

and the initial state at static equilibrium with

$$\left.\frac{\partial u(x, t)}{\partial t}\right|_{t=0} = \dot{u}(x, 0) = 0 \tag{9.55b}$$

Also the following end (boundary) conditions with fixed ends:

$$u(x, t)|_{x=0} = u(0, t) = 0 \tag{9.56a}$$

and

$$u(x, t)|_{x=L} = u(L, t) = 0 \tag{9.56b}$$

9.7.2 Solution of Partial Differential Equation for Free Vibration of Cable Structures

We will use the separation of variables technique to solve Equation 9.54 for free vibration of a long flexible cable with the initial and end conditions specified in Equations 9.55 and 9.56.

We recognize that the cable is deformed by its own weight into a shape that can be described by a function $f(x)$ at time $t = 0$, and the vibration is induced to the cable by a small disturbance that causes the cable to vibrate in the direction perpendicular to the x-axis as shown in Figure 9.18. We will determine the amplitude of vibration of the cable $u(x,t)$ at time $t = 0^+$, that is, having the cable released from its initially deformed state $f(x)$. The instantaneous position of the cable, or the amplitude of vibration of the cable, is represented by a function $u(x,t)$. Physically this function $u(x,t)$ is the vertical displacement of any given point of the cable at distance x from the left end at time $t = 0$, as illustrated in Figure 9.18.

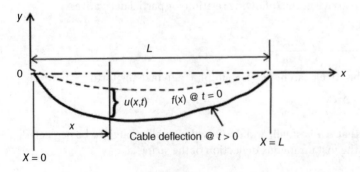

Figure 9.18 A long cable subjected to lateral vibration.

The situation in this case corresponds to the description of free vibration of a flexible cable, which makes Equation 9.54 relevant for the current problem. We thus have the partial differential equation for the problem in the form

$$\frac{\partial^2 u(x,t)}{\partial t^2} = a^2 \frac{\partial^2 u(x,t)}{\partial x^2} \tag{9.54}$$

We may use the Laplace transform method by transforming the variable t to the Laplace transform domain because the variable t covers the range $(0,\infty)$, which legitimizes the use of this method. We may also use the separation of variables technique for the solution of Equation 9.54. We will choose the latter technique for the solution by letting

$$u(x,t) = X(x)\,T(t) \tag{9.57}$$

in which the function $X(x)$ involves variable x only in Equation 9.57, and the other function $T(t)$ involves variable t only.

Substituting the expression in Equation 9.57 into Equation 9.54, one gets the following expression after rearranging the terms:

$$\frac{1}{a^2 T(t)} \frac{d^2 T(t)}{dt^2} = \frac{1}{X(x)} \frac{d^2 X(x)}{dx^2}$$

Inspection of the above relationship will reveal that the left-hand side (LHS) of the above expression is a function of variable t only, whereas the right-hand side (RHS) is a function of variable x only. The only possibility of having such a relationship to be valid is that both sides of the above equation are equal to the same constant (LHS = RHS = constant), or in mathematical form:

$$\frac{1}{a^2 T(t)} \frac{d^2 T(t)}{dt^2} = \frac{1}{X(x)} \frac{d^2 X(x)}{dx^2} = -\beta^2 \tag{9.58}$$

in which β is the "separation constant." A negative sign is attached to the square of β to ensure that both the LHS and RHS of Equation 9.58 are equal to a *negative* constant. The reason for this is that a positive constant will lead to an "unbounded" solution of Equation 9.54.

The relationship in Equation 9.58 is a critical step in the solution, as it allows us to obtain two separate ordinary differential equations, one for the function $X(x)$ and the other for the function $T(t)$. Thus, letting LHS of Equation 9.58 $= -\beta^2$, we will have the following differential equation for $T(t)$:

$$\frac{d^2 T(t)}{dt^2} + a^2 \beta^2 T(t) = 0 \tag{9.59}$$

The conditions associated with Equation 9.59 can be derived by substituting the expression in Equations 9.55a and 9.55b into Equation 9.57 to give the following forms for the initial conditions:

$$u(x,t)|_{t=0} = u(x,0) = X(x)T(0) = f(x) \tag{9.60a}$$

$$\left.\frac{dT(t)}{dt}\right|_{t=0} = 0 \tag{9.60b}$$

If we let the term on the right-hand side of Equation 9.58 also be equal to $-\beta^2$, we will have the other differential equation for $X(x)$:

$$\frac{d^2 X(x)}{dx^2} + \beta^2 X(x) = 0 \tag{9.61}$$

Again, the conditions associated with Equation 9.61 can be derived from Equations 9.56a and 9.56 using Equation 9.57, resulting in

$$X(0) = 0 \tag{9.62a}$$
$$X(L) = 0 \tag{9.62b}$$

One realizes that both differential equations in Equations 9.59 and 9.61 are homogeneous second-order ordinary differential equations with constant coefficients. The general solutions of this type of differential equation may be obtained using the method described in Section 8.2. We thus have the solution of Equation 9.59 as

$$T(t) = A\sin(\beta at) + B\cos(\beta at) \tag{9.63}$$

and the solution of Equation 9.61 as

$$X(x) = C\sin(\beta x) + D\cos(\beta x) \tag{9.64}$$

in which A, B, C, and D are arbitrary constants to be determined by the conditions stipulated in Equations 9.60a,b and 9.62a,b.

By substituting the expressions in Equations 9.63 and 9.64 into Equation 9.57, we will have the solution $u(x,t)$ of the problem:

$$u(x,t) = [A\sin(\beta at) + B\cos(\beta at)][C\sin(\beta x) + D\cos(\beta x)]$$

We are now ready to determine the arbitrary constants A, B, C, and D in the above solution. Let us first use the condition in Equation 9.62a: $X(0) = 0$ leads to

$$C\sin(\beta \times 0) + D\cos(\beta \times 0) = 0$$

from which we have $D = 0$. We thus have $X(x) = C\sin(\beta x)$. Now, if we use the other condition in Equation 9.62b, we will have

$$X(L) = 0 = C\sin(\beta L)$$

At this point, we have the choice of letting $C = 0$ or $\sin(\beta L) = 0$ from the above relationship. A close look at these choices will show that $C \neq 0$ if we want to avoid a trivial solution of $u(x,t)$. We thus have

$$\sin(\beta L) = 0 \tag{9.65}$$

Equation 9.65, like those in the previous cases in Sections 9.5 and 9.6, is a characteristic equation of the differential equation in Equation 9.61; the infinite number of valid solutions with $\beta = \pi, 2\pi, 3\pi, 4\pi, 5\pi, \ldots, n\pi$, in which n is an integer, are the eigenvalues as described in Section 4.8. We may thus obtain the values of the separation constant, β as

$$\beta_n = \frac{n\pi}{L} \quad (n = 0, 1, 2, 3, 4 \ldots) \tag{9.66}$$

The solution of the partial differential equation in Equation 9.54 thus becomes

$$u(x,t) = \left(A_n \sin\frac{n\pi}{L}at + B_n \cos\frac{n\pi}{L}at\right) C_n \sin\frac{n\pi}{L}x$$

By combining the constants, A_n, B_n, and C_n and for $n = 1, 2, 3, \ldots$ one may express the above expression in the following form with $a_n = A_n C_n$ and $b_n = B_n C_n$:

$$u(x,t) = \left(a_n \sin\frac{n\pi}{L}at + b_n \cos\frac{n\pi}{L}at\right) \sin\frac{n\pi}{L}x$$

with $n = 1, 2, 3, \ldots$, and a_n and b_n being the new arbitrary constants.

We may determine the constants, a_n and b_n, by using the condition given in Equation 9.55b in the above expression:

$$\left.\frac{\partial u(x,t)}{\partial t}\right|_{t=0} = 0 = \frac{n\pi a}{L}\left(a_n \cos\frac{n\pi at}{L} - b_n \sin\frac{n\pi at}{L}\right)\Bigg|_{t=0} \sin\frac{n\pi}{L}x$$

leading to $a_n\left(\frac{n\pi a}{L}\right)\sin\frac{n\pi}{L}x = 0$

Since $\sin(n\pi/L)x \neq 0$ in the above expression, we see that the coefficients $a_n = 0$. We will thus have

$$u(x,t) = \sum_{n=1}^{\infty} b_n \cos\frac{n\pi a}{L}t \sin\frac{n\pi a}{L}x$$

The only remaining undetermined multi-valued constant coefficients b_n in the above expression can be determined using the remaining unused condition in Equation 9.55a; that is, $u(x,0) = f(x)$, or

$$u(x,t)|_{t=0} = u(x,0) = f(x)$$

We thus have the following relationship:

$$f(x) = \sum_{n=1}^{\infty} b_n \sin\frac{n\pi x}{L} \tag{9.67}$$

with the function $f(x)$ on the LHS in Equation 9.67 being the given initial condition of the partial differential equation of Equation 9.54. The coefficients b_n ($n = 1,2,3, \ldots$) in the RHS of the equation will be determined by the procedures presented in Section 9.5.1, as follows:

1) Multiply both sides of Equation 9.67 by a function $\sin(n\pi x/L)$ as follows:

$$\sin\frac{n\pi x}{L}f(x) = \left(\sin\frac{n\pi x}{L}\right)\sum_{n=1}^{\infty} b_n \sin\frac{n\pi x}{L}$$

$$= \sum_{n=1}^{\infty} b_n \left(\sin\frac{n\pi x}{L}\right)\sin\frac{n\pi x}{L}$$

2) Integrate both sides of the above equation:

$$\int_0^L \sin\frac{n\pi x}{L}f(x)\,dx = \int_0^L \sum_{n=1}^{\infty} b_n\left(\sin\frac{n\pi x}{L}\right)\sin\frac{n\beta\pi x}{L}\,dx$$

$$= \sum_{n=1}^{\infty}\int_0^L b_n\left(\sin\frac{n\pi x}{L}\right)^2 dx$$

The orthogonality of integration of sine functions leads to the expression

$$\int_0^p \sin\frac{n\pi x}{p}\sin\frac{m\pi x}{p}\,dx = 0 \quad \text{if } m \neq n,$$

$$= \frac{p}{2} \quad \text{if } m = n$$

The above expression leads to the coefficients b_n in Equation 9.67 as

$$\int_0^L \left(\sin\frac{n\pi x}{L}\right)^2 dx = \frac{L}{2}$$

from which we will have

$$\int_0^L \sin\frac{n\pi x}{L}f(x)\,dx = b_n\left(\frac{L}{2}\right) \quad \text{with } n = 1,2,3,4,\ldots$$

We thus establish the expression for the multi-valued constant b_n as

$$b_n = \frac{2}{L}\int_0^L f(x)\sin\frac{n\pi x}{L}\,dx \quad \text{with } n = 1, 2, 3, 4, 5, \ldots \tag{9.68}$$

With the coefficients b_n determined by Equation 9.68, we may express the complete solution of Equation 9.54 as

$$u(x,t) = \sum_{n=1}^{\infty}\frac{2}{L}\left(\int_0^L f(x)\sin\frac{n\pi x}{L}\,dx\right)\cos\frac{n\pi at}{L}\sin\frac{n\pi x}{L} \tag{9.69}$$

9.7.3   Convergence of Series Solutions

Solutions of partial differential equations by the separation of variables technique such as presented in Sections 9.5 and 9.6 include summations of an infinite number of terms associated with the infinite number of roots of the transcendental equation (or characteristic equations as mentioned in Chapter 4). The solution in Equation 9.69 for the partial differential equation in (9.54) is also in the form of infinite series. Numerical solutions of these equations can be obtained by summing up the solutions with each assigned value of n, that is with $n = 1, 2, 3, \ldots$ up to a very large integer number. In normal circumstances, The series should converge fairly rapidly, so one needs only to sum up approximately a dozen terms using $n = 1, 2, \ldots, 12$. However, the effect of the rate of convergence of infinite series, such as that in Equation 9.69, on the accuracy of the analytical results remains a concern to engineers in their analyses.

We will demonstrate the convergence of the following fictitious series solution that is similar but not identical to that shown in Equation 9.69, with selected solution point at $x = 5$ and $t = 1$.

$$u(5,1) = \sum_{n=1}^{\infty}\frac{2}{20}\left[\int_0^{20}\left(0.25\sin\frac{n\pi}{20}x\right)dx\right]\cos\left(\frac{120n\pi}{20}t\right)\sin\frac{n\pi}{20}x$$

or in final form as

$$u(5,1) = \frac{1}{40}\sum_{n=1}^{\infty}\left[\cos(6n\pi)\sin\frac{n\pi}{4}\left(\int_0^{20}\sin\frac{n\pi}{20}x\,dx\right)\right]$$

We will let the numerical solution of $u(5,1)$ be the summation of all terms denoted by n:

$$u(5,1) = u_1 + u_2 + u_3 + u_4 + u_5 + \cdots + n_n$$

where $u_1, u_2, u_3, u_4, u_5, \ldots, u_n$ are valid solutions of $u(5,1)$ with $n = 1, 2, 3, 4, 5, \ldots, n$. Valid solutions with n up to 16 have been computed and their respective values are listed in the following table:

u_1	u_2	u_3	u_4	u_5	u_6	u_7	u_8	u_9	u_{10}	u_{11}	u_{12}	u_{13}	u_{14}	u_{15}	u_{16}
1.6	3.8	2.8	0	−9.38	3.22	9.38	0	−5.63	1.14	5.63	0	−4.02	5.77	4.02	0
E−2	E−2	E−2		E−3	E−3	E−3		E−3	E−3	E−3		E−3	E−4	E−3	

The contribution of the individual valid solutions with u_n ($n = 1, 2, 3, 4, 5, \ldots, 30$) is plotted in Figure 9.19.

We envisage from Figure 9.19 that the magnitude of individual terms in the infinite series diminishes rapidly with the increasing n-values. This trend indicates that contributions of individual terms in the infinite series solution such as in Equation 9.69 also diminish as the

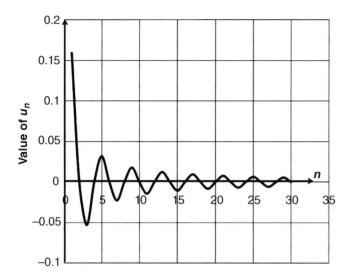

Figure 9.19 Convergence of an infinite series solution.

n-value increases. Reasonable accuracy of the solution for this particular case is considered to be obtained with n-value less than 20 in the numerical solution. It is plausible to assume that reasonable accuracy of the infinite series solution of long cable vibration may be achieved by including fewer than 20 terms in the series solutions.

9.7.4 Modes of Vibration of Cable Structures

We learned from the fundamentals of the mechanical vibration of solids in Section 8.9 that any solid structure made of *elastic material* with a given *geometry* possesses various *modes of vibration*. Each mode offers a unique "deformed shape" of the structure in the vibration. In the case of free vibration of cable structures, the "peaks" and "valleys" in the shape of the structure differ in different modes. The analysis that determines various modes of vibration of a structure is called *modal analysis*. This analysis is of great importance for any structure that is vulnerable to mechanical vibration, and from it engineers may determine how to design cable structures that are likely subjected to intermittent loads, such as wind loads, at certain locations in order to avoid resonant vibration. Additionally, modal analysis will also indicate *where* and *when* the structure may experience maximum amplitude of vibration. Such information could offer the user of the structures the option of not installing delicate attachments at these sensitive locations.

One should bear in mind that modal analysis is also a part of free-vibration analysis of structures with no externally applied load. Elastic material properties and the geometry of the structure are the only required conditions for such analyses.

In the following we will illustrate the first three modes of a vibrating string using the solution given in Equation 9.69 in the form

$$u(x,t) = \sum_{n=1}^{\infty} \frac{2}{L} \left(\int_0^L f(x) \sin \frac{n\pi x}{L} \, dx \right) \cos \frac{n\pi a t}{L} \sin \frac{n\pi x}{L} \tag{9.69}$$

or

$$u(x,t) = \sum_{n=1}^{\infty} b_n \cos \frac{n\pi a t}{L} \sin \frac{n\pi x}{L}$$

with

$$b_n = \frac{2}{L} \int_0^L f(x) \sin \frac{n\pi x}{L} dx \tag{9.68}$$

The *mode shapes of the vibrating cable in this case are determined by the assigned distinct n-value in Equation 9.69 with the coefficients b_n evaluated in Equation 9.68 with the same n-value.* Following are the shapes of the vibrating cable in the first three modes with $n = 1$ for mode 1, $n = 2$ for mode 2, and $n = 3$ for mode 3 vibration

The shape of the vibrating cable in **mode 1 with $n = 1$**

We will get the magnitudes of the vibrating cable $u_1(x,t)$ from Equation 9.69 and the coefficient b_1 from Equation 9.68 with $n = 1$ as follows:

$$u_1(x,t) = \left(b_1 \cos \frac{\pi a t}{L} \right) \sin \frac{n\pi}{L} x \text{ and } b_1 = \frac{2}{L} \int_0^L f(x) \sin \frac{\pi x}{L} dx = \text{constant } t$$

The above integral will result in a constant value of the coefficient b_1 with the given function $f(x)$. We may envisage that the shape of the vibrating cable will follow a sine function with its maximum amplitude at any given instant obtained by a $\sin[(\pi/L)x]$ function. Graphically, it can be illustrated in Figure 9.20. We note that the two ends of the string are fixed and have zero amplitude of vibration. The maximum amplitudes of vibration occur at the mid-span of the cable. The corresponding frequency of vibration for this mode of vibration is the coefficient of the argument in the cosine function divided by 2π in the general solution of $u(x,t)$ in Equation 9.69, or as shown in Equation 9.70 in the case of mode 1 vibration.

$$f_1 = \frac{\pi a / L}{2\pi} = \frac{a}{2L} = \frac{1}{2L} \sqrt{\frac{P}{m}} \tag{9.70}$$

in which P = the tension in the cable (N) and m = the mass of the cable per unit length (kg/m). The frequency f_1 in Equation 9.70 has a unit of Hertz (Hz).

The shape of the vibrating cable in mode 2 with $n = 2$ Likewise, the shape of the vibrating cable in mode 2 may be obtained in a similar way by using Equations 9.68 and 9.69 with $n = 2$, as shown in the following expressions:

$$u_2(x,t) = \left(b_2 \cos \frac{2\pi a}{L} t \right) \sin \frac{2\pi}{L} x \tag{9.71a}$$

where

$$b_2 = \frac{2}{L} \int_0^L f(x) \sin \frac{2\pi x}{L} dx = \text{constant } t \tag{9.71b}$$

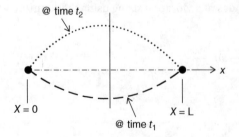

@ time t_2

X = 0   X = L

@ time t_1

Figure 9.20 Shapes of the cable in mode 1 vibration.

Figure 9.21 Shapes of the cable in mode 2 vibration.

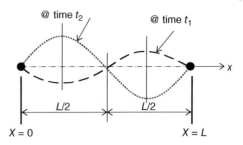

Figure 9.22 Shapes of the cable in mode 3 vibration.

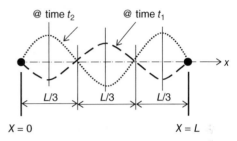

We recognize from Equation 9.68 that $u_2(t) = 0$ at $x = 0$, $L/2$, and L. Graphically, the amplitudes in Equation 9.71a can be illustrated in Figure 9.21.

It is interesting to note from Figure 9.21 that there are now three nodes at which the amplitude of vibration is zero. Physically, it means that there are two spots along the cable at which maximum amplitudes of vibration occur at $L/4$ and $3L/4$. The corresponding frequency is

$$f_2 = \frac{2\pi a/L}{2\pi} = \frac{a}{L} = \frac{1}{L}\sqrt{\frac{P}{m}} \tag{9.72}$$

We may find mode 3 of the vibrating cable from Equations 9.68 and 9.69 with $n = 3$ in the following expressions:

$$u_3(x, t) = \left(b_3 \cos\frac{3\pi a}{L}t\right)\sin\frac{3\pi}{L}x \tag{9.73a}$$

where

$$b_3 = \frac{2}{L}\int_0^L f(x)\sin\frac{3\pi x}{L}dx = \text{constant} \tag{9.73b}$$

We note from Equation 9.73a that there are four nodes in this mode of vibration at $x = 0$, $L/3$, $2L/3$, and L as illustrated in Figure 9.22. The corresponding frequency is

$$f_3 = \frac{3\pi a/L}{2\pi} = \frac{3a}{2L} = \frac{3}{2L}\sqrt{\frac{P}{m}} \tag{9.74}$$

Example 9.6

A flexible cable 10 m long is fixed at both ends with a tension of 500 N in the free-hung state. The cable has a diameter of 1 cm and a mass density of 2.7 g/cm^3. The cable begins to vibrate due to an instantaneous but small disturbance from its initial shape that can be described by the function $f(x) = 0.005x(1-x/10)$, in which x is the coordinate along the length of the cable with $x = 0$ at one of the two fixed ends. Determine the following:

a) The differential equation for the amplitudes of vibration of the cable represented by $u(x,t)$ in the unit of meters, in which t is the time into the vibration with the unit of seconds.

b) The mathematical expressions for the initial and end conditions.

c) The solution for $u(x,t)$ of the differential equation in the unit of meters.

d) The amplitude of the vibrating cable in mode 1, that is, $u_1(x,t)$ with the magnitude and location of the maximum deflection of the cable in this mode of vibration.

e) The numerical values of the frequencies of the first and second modes of vibration.

f) The physical significance of these mode shapes to the design engineer.

Solution:

The present problem involves the following specific conditions:

The length of the cable $L = 10$ m and its diameter $d = 1$ cm $= 0.01$ m.

The cable is made of aluminum with a mass density $\rho = 2.7$ g/cm^3.

The cable is subjected to a tension $P = 500$ N and with initial sag described by the function $f(x)$:

$$f(x) = 0.005x \left(1 - \frac{x}{10}\right)$$

a) The applicable differential equation for the amplitudes of vibration is Equation 9.54 with the constant coefficient a in that equation to be determined by $a = \sqrt{P/m}$, where $P =$ the tension in the cable $= 500$ N and $m =$ the mass per unit length, which needs to be computed from the given values. The mass per unit length of the cable is $m = M/L$, where $M =$ the total mass of the cable $M = \rho V$, with V being the volume of the cable. We will get the volume of the cable from the expression $V = [(\pi d^2/4)L] = 7.85 \times 10^{-4}$ m^3 with $d =$ the given diameter of the cable. We will thus have the total mass of the cable $M = \rho V = (2.7 \times 10^3)$ $(7.85 \times 10^{-4}) = 2.12$ kg, meaning the mass per unit length of the cable is 0.212 kg/m.

The constant coefficient a in Equation 9.54 according to the expression given above is

$$a = \sqrt{\frac{P}{m}} = \sqrt{\frac{500}{0.212}} = 48.56 \text{ m/s}$$

We will thus need to solve Equation 9.54 with $a^2 = 2358.5$ on the right-hand side of this equation:

$$\frac{\partial^2 u(x, t)}{\partial t^2} = 2358.5 \frac{\partial^2 u(x, t)}{\partial x^2} \tag{a}$$

b) The mathematical expressions of the initial and end conditions:

The problem requires satisfying the initial and end conditions expressed mathematically as

The initial conditions are

$$u(x, t)|_{t=0} = u(x, 0) = f(x) = 0.005x \left(1 - \frac{x}{10}\right) \tag{b1}$$

$$\frac{\partial u(x, t)}{\partial t}\bigg|_{t=0} = \dot{u}(x, 0) = 0 \tag{b2}$$

and the end conditions are

$$u(x, t)|_{x=0} = u(0, t) = 0 \tag{c1}$$

$$u(x, t)|_{x=L} = u(L, t) = u(10, t) = 0 \tag{c2}$$

c) The solution of $u(x,t)$ of Equation (a) satisfying the conditions in Equations (b1) and (b2) and Equations (c1) and (c2) will be obtained as follows. The solution of Equation (a) is similar to that of Equation 9.69 with $a = 48.56$ m/s in the following expression:

$$u(x, t) = \sum_{n=1}^{\infty} \frac{2}{10} \left\{ \left[\int_0^{10} 0.005x \left(1 - \frac{x}{10} \right) \sin \frac{n\pi x}{10} \, dx \right] \left[\cos \frac{n\pi(48.56)}{10} t \right] \left[\sin \frac{n\pi}{10} x \right] \right\}$$

or

$$u(x, t) = \sum_{n=1}^{\infty} \frac{1}{5} \left\{ \left[\int_0^{10} 0.005x \left(1 - \frac{x}{10} \right) \sin \frac{n\pi x}{10} \, dx \right] [\cos 15.25nt][\sin 0.314nx] \right\} \quad \text{(d)}$$

Using the integrals available from the handbook (Zwillinger, 2003),

$$\int x \sin \alpha x \, dx = \frac{1}{\alpha^2} \sin \alpha x - \frac{x}{\alpha} \cos \alpha x$$

and

$$\int x^2 \sin \alpha x \, dx = \frac{2x}{\alpha^2} \sin \alpha x + \frac{2 - \alpha^2 x^2}{\alpha^3} \cos \alpha x$$

we evaluate the integral in Equation (d) in the form

$$\int_0^{10} \left[0.005x \left(1 - \frac{x}{10} \right) \sin \frac{n\pi x}{10} \, dx \right] = -\frac{0.5(-1)^n}{n\pi} \left(1 + \frac{2 - n^2 \pi^2}{n^3 \pi^3} \right)$$

The solution in Equation (d) thus takes the form

$$u(x, t) = -0.1 \sum_{n=1}^{\infty} \frac{1}{n\pi} \left(1 + \frac{2 - n^2 \pi^2}{n^3 \pi^3} \right) (\cos 15.25nt)(\sin 0.314nx) \quad \text{(e)}$$

d) The amplitude of the vibrating cable in mode 1, that is, $u_1(x,t)$ with the magnitude and location of the maximum deflection of the cable in this mode of vibration. The required solution is obtained by letting $n = 1$ in Equation (e) as

$$u_1(x, t) = -0.1 \left[\frac{1}{\pi} \left(1 + \frac{2 - \pi}{\pi^3} \right) \right] \cos 15.25t \sin 0.314x$$
$$= -0.02376 \cos 15.25t \sin 0.314x \quad \text{(f)}$$

The maximum amplitude occurs at the mid-span of the cable at $x = 5$ m, and at the time given by $\cos 15.25t = 1.0$. We thus have the maximum amplitude $u_{1,\text{max}} = 0.023\,76$ m, or 2.376 cm, at $x = 5$ m and at the time given by $15.25t = \pi$, or time $t = \pi/15.25 = 0.2$second.

e) *The numerical values of the frequencies of the first and second mode of vibration:* We may use Equations 9.70 and 9.72 to compute the numerical values of the frequencies of the first and second modes of vibration as follows:

$$f_1 = \frac{1}{2L} = \frac{1}{2 \times 10} \sqrt{\frac{500}{0.212}} = 2.43 \text{ Hz} \quad \text{for mode 1}$$

$$f_2 = \frac{1}{L} \sqrt{\frac{P}{m}} = \frac{1}{10} \sqrt{\frac{500}{0.212}} = 4.86 \text{ Hz} \quad \text{for mode 2}$$

f) *The physical significance of these mode shapes to the design engineer:* Engineers will use the outcomes of the above modal analysis to advise the users of this cable structure on the possibility of potentially devastating resonant vibration of the cable structure should the frequency of an applied cyclic force, such as wind force, coincide with any of the natural frequencies, *such as* $f_1 = 2.43\,\text{Hz}, f_2 = 4.86\,\text{Hz}$, ... computed in part (e). The users will also be made aware of the locations where maximum amplitudes of vibration may occur as indicated by the mode shapes in the modal analysis. The users should avoid placing delicate attachments at these locations to avoid potential damage due to excessive vibration there.

9.8 Partial Differential Equations for Transverse Vibration of Membranes

Solids of plane geometry commonly appear in machines and structures. Thin plates can be as small as micrometers in thickness in printed electronic circuit boards or as large as floors in building structures. Like flexible cables, thin flexible plates are normally vulnerable to transverse vibration. In some cases, these plates may rupture due to resonant vibration, resulting in significant loss of property or even human life.

This section will present the derivation of partial differential equations that allow engineers to assess the amplitudes of free vibration of thin plates that are flexible enough to be simulated as thin membranes. Engineers may use this mathematical model for their modal analysis for safe design of these types of machine components and structures.

9.8.1 Derivation of the Partial Differential Equation

We will derive the mathematical model for the transverse vibration of thin plates with the following idealizations and assumptions:

1) The derivation of mathematical expressions is based on the lateral displacement of solids of plane geometry that are flexible and offer no resistance to bending. In reality, the structure corresponds to the description of "membranes."
2) The analysis considers thin plates with unsupported large spans that are sufficiently flexible in lateral deformation.
3) Because they are flexible, there is no shear stress in the thin plates in the subsequent analysis.
4) The thin plate is initially flat with its edges fixed. There is an initial sag represented by a function $f(x,y)$ sustained by in-plane tension P per unit length of the plate in all directions. The tension P is large enough to justify neglecting the weight of the plate.
5) Figure 9.23 shows the plate in the (x,y) plane with lateral displacement $z(x,y,t)$, the amplitude of vibration of the plate at the locations defined by the x–y coordinates and at time t.
6) Each element of the plate vibrates in the direction perpendicular to the plane surface of the plate in the z-coordinate as illustrated in Figure 9.24. The slopes of the deformed surface of the plate element are small.
7) The mass per unit area of the plate (m) is uniform throughout the plate.

Figure 9.24 is a free-body diagram showing all the forces acting on a small element of the plate during a lateral vibration. The situation satisfies a dynamic equilibrium condition with the summation of all forces present at time t being equal to zero. Mathematically, we may express this condition in the form

$$\sum F_z = 0$$

The induced dynamic force F by Newton's second law plays a major role in the formulation of the above equilibrium of forces. Mathematically, this force may be expressed as

$$F = m\frac{\partial^2 z(x, y, t)}{\partial t^2}$$

as illustrated in Figure 9.24.

Figure 9.23 Lateral deformations of thin plates in vibration.

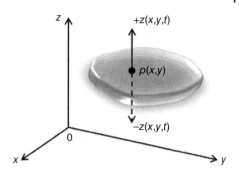

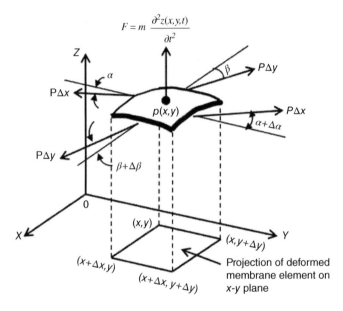

Figure 9.24 Forces on a vibrating element of a thin plate.

Referring to the small element shown in Figure 9.24, we have the following dynamic equilibrium conditions:

$$P\,\Delta x\sin(\alpha + \Delta\alpha) - P\,\Delta x\sin\alpha + P\,\Delta y\sin(\beta + \Delta\beta)$$
$$- P\,\Delta y\sin\beta - m(\Delta x\,\Delta y)\frac{\partial^2 z[x, y, t]}{\partial t^2} = 0 \tag{9.75}$$

where $m =$ mass per unit area of the plate material.

Since the slopes of the deformed plate element in both the x- and y-directions are very small as stipulated in "idealization number 6," which means that both angles α and β are small, leading to the following approximate relationships:

$$\sin\alpha \approx \tan\alpha = \frac{\partial z\left(x + \frac{\Delta x}{2}, y, t\right)}{\partial y}$$

$$\sin(\alpha + \Delta\alpha) \approx \tan(\alpha + \Delta\alpha) = \frac{\partial z\left(x + \frac{\Delta x}{2}, y + \Delta y, t\right)}{\partial y}$$

$$\sin\beta \approx \tan\beta = \frac{\partial z\left(x, y + \frac{\Delta y}{2}, t\right)}{\partial x}$$

$$\sin(\beta + \Delta\beta) \approx \tan(\beta + \Delta\beta) = \frac{\partial z\left(x + \Delta x, y + \frac{\Delta y}{2}, t\right)}{\partial x}$$

Substituting the above approximate relationships into Equation 9.75 results in the following expression:

$$P\Delta x \left[\frac{\partial z\left(x + \frac{\Delta x}{2}, y + \Delta y, t\right)}{\partial y} - \frac{\partial z\left(x + \frac{\Delta x}{2}, y, t\right)}{\partial y}\right]$$

$$+ P\Delta y \left[\frac{\partial z\left(x + \Delta x, y + \frac{\Delta y}{2}, t\right)}{\partial x} - \frac{\partial z\left(x, y + \frac{\Delta y}{2}, t\right)}{\partial x}\right]$$

$$- m\,\Delta x\,\Delta y\frac{\partial^2 z(x, y, t)}{\partial t^2}$$

$$= 0$$

The following expression is obtained by dividing the above expression by $\Delta x \Delta y$:

$$P\left[\frac{\dfrac{\partial z\left(x + \dfrac{\Delta x}{2}, y + \Delta y, t\right)}{\partial y} - \dfrac{\partial z\left(x + \dfrac{\Delta x}{2}, y, t\right)}{\partial y}}{\Delta y}\right.$$

$$\left. + \frac{\dfrac{\partial\left(x + \Delta x, y + \dfrac{\Delta y}{2}, t\right)}{\partial x} - \dfrac{\partial z\left(x, y + \dfrac{\Delta y}{2}, t\right)}{\partial x}}{\Delta x}\right]$$

$$- m\frac{\partial^2 z(x, y, t)}{\partial t^2}$$

$$= 0$$

Given that the lateral deformation of the plate varies continuously with locations on the plane defined by the x- and y-coordinates, we have the following relationships:

$$\lim_{\Delta y \to 0} \frac{\dfrac{\partial z\left(x + \dfrac{\Delta x}{2}, y + \Delta y, t\right)}{\partial y} - \dfrac{\partial z\left(x + \dfrac{\Delta x}{2}, y, t\right)}{\partial y}}{\Delta y} = \frac{\partial^2 z(x, y, t)}{\partial y^2}$$

and

$$\lim_{\Delta x \to 0} \frac{\dfrac{\partial z\left(x + \Delta x, y + \dfrac{\Delta y}{2}, t\right)}{\partial x} - \dfrac{\partial z\left(x, y + \dfrac{\Delta y}{2}, t\right)}{\partial x}}{\Delta x} = \frac{\partial^2 z(x, y, t)}{\partial x^2}$$

The equilibrium equation thus has the following form with $\Delta x \to 0$ and $\Delta y \to 0$ for continuous variation of the amplitude of vibration of the plate in both x- and y-coordinates with

$$P\left[\frac{\partial^2 z(x, y, t)}{\partial y^2} + \frac{\partial^2 z(x, y, t)}{\partial x^2}\right] - m\frac{\partial^2 z(x, y, t)}{\partial t^2} = 0$$

or in the form

$$\frac{\partial^2 z(x, y, t)}{\partial t^2} = a^2\left[\frac{\partial^2 z(x, y, t)}{\partial x^2} + \frac{\partial^2 z(x, y, t)}{\partial y^2}\right] \tag{9.76}$$

in which the constant a in Equation 9.76 has the same *form* as in Equation 9.54 but a different meaning:

$$a = \sqrt{\frac{P}{m}} \tag{9.77}$$

in which P is the tension per unit length (with unit N/m), and m is the mass per unit area (kg/m^2). The constant a thus has the unit of m/s.

9.8.2 Solution of the Partial Differential Equation for Plate Vibration

Solution of the transverse vibration of a flexible plate in Equation 9.76 is far more complicated than that in Equation 9.54 for transverse vibration analysis for long flexible cables, for the reason of having additional variables in the equation for plate vibrations. We will use the following case, involving transverse vibration of a flexible plate as illustrated in Figure 9.25, to show how the separation of variables technique may be used to solve Equation 9.76.

The applicable partial differential equation is

$$\frac{\partial^2 z(x, y, t)}{\partial t^2} = a^2\left[\frac{\partial^2 z(x, y, t)}{\partial x^2} + \frac{\partial^2 z(x, y, t)}{\partial y^2}\right] \tag{9.76}$$

where the constant $a = \sqrt{P/m}$, P is the constant tension per unit length in the plate and m is the mass per unit area of the plate material. The following boundary and initial conditions apply for the case illustrated in Figure 9.25.

A) The boundary conditions:

$$z(x, y, t)|_{x=0} = z(0, y, t) = 0 \tag{a1}$$

$$z(x, y, t)|_{x=b} = z(b, y, t) = 0 \tag{a2}$$

$$z(x, y, t|_{y=0} = z(x, 0, t) = 0 \tag{b1}$$

$$z(x, y, t)|_{y=c} = z(x, c, t) = 0 \tag{b2}$$

Figure 9.25 Plan view of a flexible plate undergoing transverse vibration.

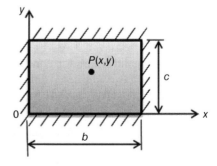

B) The initial conditions:

$$z(x, y, t)|_{t=0} = f(x, y) \tag{c1}$$

$$\frac{\partial z(x, y, t)}{\partial t}\bigg|_{t=0} = g(x, y) \tag{c2}$$

where $f(x,y)$ is a specified function describing the initial shape of the plate; for example, the sag of the plate before the vibration takes place. The function $g(x,y)$ is another specified function that describes the velocity of the plate at the inception of the vibration. In many cases, $g(x,y) = 0$ for the plate to be initially in static equilibrium.

We begin our solution of Equation 9.76 by letting the amplitude of vibration of the plate to be a product of three individual functions:

$$Z(x, y, t) = X(x) Y(y) T(t) \tag{9.78}$$

in which $X(x)$, $Y(y)$, and $T(t)$ are functions involving independent variables x, y, and t, respectively.

We will obtain the following equality upon substituting Equation 9.78 into Equation 9.76, as we did in a similar way in Section 9.5.1. Such substitution yields

$$\frac{1}{X(x)} \frac{d^2 X(x)}{dx^2} = -\frac{1}{Y(y)} \frac{d^2 Y(y)}{dy^2} + \frac{1}{a^2 T(t)} \frac{d^2 T(t)}{dt^2}$$

Inspection of the above expression reveals that the left-hand side of the equality relates to variable x only, whereas that variable x does not appear in the right-hand side of the equality. The only way such an equality can be valid is for both sides of the equality to equal the same constant. As is a general practice when using the separation of variables technique, a negative sign is assigned to the constant to avoid unbounded solutions of $z(x,y,t)$. We thus have the following relationships with the first separation constant λ:

$$\frac{1}{X(x)} \frac{d^2 X(x)}{dx^2} = -\frac{1}{Y(y)} \frac{d^2 Y(y)}{dy^2} + \frac{1}{a^2 T(t)} \frac{d^2 T(t)}{dt^2} = -\lambda^2 \tag{d}$$

These equalities lead to the following differential equation:

$$\frac{d^2 X(x)}{dx^2} + \lambda^2 X(x) = 0 \tag{e}$$

and another equality:

$$\frac{1}{Y(y)} \frac{d^2 Y(y)}{dy^2} = \frac{1}{a^2 T(t)} \frac{d^2 T(t)}{dt^2} + \lambda^2 \tag{f}$$

Using the same argument as we did with the equality in Equation (d), we derive the following equalities for functions involving the independent variables y and t:

$$\frac{1}{Y(y)} \frac{d^2 Y(y)}{dy^2} = \frac{1}{a^2 T(t)} \frac{d^2 T(t)}{dt^2} + \lambda^2 = -\mu^2 \tag{g}$$

where μ is second separation constant used in separating the variables y and t in Equation (f). We may derive two additional differential equations from the above equalities:

$$\frac{d^2 Y(y)}{dy^2} + \mu^2 Y(y) = 0 \tag{h}$$

and

$$\frac{d^2 T(t)}{dt^2} + a^2(\lambda^2 + \mu^2) T(t) = 0 \tag{j}$$

We recognize that all three differential equations for $X(x)$, $Y(y)$, and $T(t)$ in the respective Equations (e), (h), and (j) are linear second-order differential equations with constant coefficients. Solutions for these functions are available in Section 8.2. The general solutions of these functions can be expressed as follows:

$$X(x) = c_1 \cos \lambda x + c_2 \sin \lambda x \tag{k1}$$

$$Y(y) = c_3 \cos \mu y + c_4 \sin \mu y \tag{k2}$$

$$T(t) = c_5 \cos a \sqrt{\lambda^2 + \mu^2}\, t + c_6 \sin a \sqrt{\lambda^2 + \mu^2}\, t \tag{k3}$$

The constants c_1, c_2, c_3, ..., c_6 in Equations (k1), (k2), and (k3) are arbitrary constants to be determined by appropriate specified conditions in Equations (a1), (a2), (b1), (b2), (c1), and (c2).

The following conditions for the determination of constants in Equations (k1) and (k2) are derived by substituting the relationship in Equation 9.78 into Equations (a1), (a2), (b1), and (b2), yielding the following conditions for Equations (k1) and (k2). $X(0) = 0$ and $Y(0) = 0$ lead to $c_1 = c_3 = 0$. The other two conditions: $X(b) = 0$ and $Y(c) = 0$ lead to the expressions for the separation constants: $\lambda = m\pi/b$ with $m = 1, 2, 3, \ldots$, and $\mu = n\pi/c$ with $n = 1, 2, 3, \ldots$. We thus have the solutions of $X(x)$ and $Y(y)$ as follows:

$$X(x) = c_2 \sin \lambda x = c_{2,m} \sin \frac{m\pi x}{b} \qquad \text{with} \quad m = 1, 2, 3, \ldots \tag{m1}$$

$$Y(y) = c_4 \sin \mu y = c_{4,n} \sin \frac{n\pi y}{c} \qquad \text{with} \quad n = 1, 2, 3, \ldots \tag{m2}$$

The constant c_5 and c_6 in solution in Equation (k3) may be determined with the initial conditions specified in Equations (c1) and (c2) for $z(x,y,t)$:

$$z(x, y, t) = X(x)\, Y(y)\, T(t)$$

$$= \left(c_{2,m} \sin \frac{m\pi x}{b} \right) \left(c_{4,n} \sin \frac{n\pi y}{c} \right)$$

$$\left(c_5 \cos a \sqrt{\left(\frac{m\pi}{b}\right)^2 + \left(\frac{n\pi}{c}\right)^2}\, t + c_6 \sin a \sqrt{\left(\frac{m\pi}{b}\right)^2 + \left(\frac{n\pi}{c}\right)^2}\, t \right) \tag{n}$$

Let

$$\omega_{mn} = \sqrt{\lambda_m^2 + \mu_n^2} = \sqrt{\left(\frac{m\pi}{b}\right)^2 + \left(\frac{n\pi}{c}\right)^2} \tag{p}$$

The solution $z(x,y,t)$ may be expressed in the form

$$z(x, y, t) = A_{mn} \left(\sin \frac{m\pi x}{b} \right) \left(\sin \frac{n\pi y}{c} \right) (A_{mn} \cos a\omega_{mn} t + B_{mn} \sin a\omega_{mn} t) \tag{q}$$

with $m = 1, 2, 3, \ldots$, and $n = 1, 2, 3, \ldots$.

One would have observed that we made the products of constants $(c_{2,m})(c_{4,n})(c_5) = A_{mn}$ and $(c_{2,m})(c_{4,n})(c_6) = B_{mn}$ in Equation (p).

Since the solution in Equation (q) is valid for every value of m and n, the complete solution will be the sum of all valid solutions as expressed in the following:

$$z(x, y, t) = \sum_{m=1}^{\infty} \sum_{n=1}^{\infty} \left(\sin \frac{m\pi}{b} x \right) \left(\sin \frac{n\pi}{c} y \right) (A_{mn} \cos a\omega_{mn} t + B_{mn} \sin a\omega_{mn} t) \tag{9.79}$$

The multi-valued constant coefficients A_{mn} and B_{mn} in Equation 9.80a may be determined using the two initial conditions in Equations (c1) and (c2) following a similar procedure to that outlined in determining coefficient b_n in Section 9.7.2. However, in the current situation, we need to determine A_{mn} by multiplying the double summation in Equation 9.80a by the product $\sin(m\pi x/b)\sin(n\pi y/c)$ on both sides of the equation and then integrating over the ranges $0 \leq x \leq b$ and $0 \leq y \leq c$. This will give the following expressions for these two multi-valued constant coefficients:

$$A_{mn} = \frac{4}{bc}\int_0^c \int_0^b f(x,y)\left(\sin\frac{m\pi}{b}x\right)\left(\sin\frac{n\pi}{c}y\right)dx\,dy \tag{9.80a}$$

and

$$B_{mn} = \frac{4}{abc\omega_{mn}}\int_0^c \int_0^b g(x,y)\left(\sin\frac{m\pi}{b}x\right)\left(\sin\frac{n\pi}{c}y\right)dx\,dy \tag{9.80b}$$

The transverse vibration modes of flexible plates may be obtained by assigning $m = 1$ and $n = 1$ for mode 1, and $m = 2$ and $n = 2$ for mode 2 vibrations, and so on. The periods of the various modes of vibration may be computed from the expression: $T_n = 2\pi/\omega_{mn}$, with ω_{mn} as given by Equation (p).

9.8.3 Numerical Solution of the Partial Differential Equation for Plate Vibration

Numerical solution for the amplitude of transverse vibration of flexible plates given in Equation 9.79 with coefficients in Equations 9.80a and 9.80b is much more tedious and complicated than that for flexible cables described in Section 9.7.

We will present the finding of such a solution using a commercial software package called MatLAB for a case of free-vibration analysis of a flexible mouse pad such as shown in Figure 9.25 with dimensions $b = 10$ inches and $c = 5$ inches. The pad has thickness 0.185 inch and mass density $1.55\ \text{lb}_m/\text{in}^3$. The pad has an initial shape that can be described by the function $f(x,y) = xy(b-x)(c-y)$, or $f(x,y) = xy(10-x)(5-y)$, and is subjected to a constant tension of $P = 0.5\ \text{lb}_f/\text{in}$. Vibration of the pad begins from a state of static equilibrium, *with $g(x,y) = 0$ in* Equation (c2) in Section 9.8.2.

We first determine the constant coefficient a in Equation 9.77:

$$a = \sqrt{\frac{Pg}{\rho}} = \sqrt{\frac{(0.5\ \text{lb}_f/\text{in})(32.2\ \text{ft/s}^2)}{0.00155\ \text{lb}_m/\text{in}^2}}\left(\frac{12\ \text{in}}{\text{ft}}\right) = 353.05\ \text{in/s}$$

The frequency ω_{mn} required to compute the period T is given in Equation (p) in Section 9.8.2 with eigenvalues $\lambda_m = m\pi/10$ and $\mu_n = n\pi/5$ with $m, n = 1, 2, 3, \ldots$, as shown in the same section.

The mode shapes of this plate are computed from free lateral vibration analysis with the following expression:

$$z(x,y,t) = A_{mn}\left(\sin\frac{m\pi x}{10}\right)\left(\sin\frac{n\pi y}{5}\right)(A_{mn}\cos a\omega_{mn}t) \tag{9.81}$$

in which $m, n = 1, 2, 3, \ldots$.

The coefficient A_{mn} in the above expression has the form

$$A_{mn} = \frac{16(bc)^2[1+(-1)^{n+1}][1+(-1)^{m+1}]}{m^3 n^3 \pi^6} \tag{9.82}$$

with $b = 10$ inches and $c = 5$ inches.

Graphical representations of this numerical solution of the modal analysis of a flexible mouse pad are obtained with the MatLAB software package (version R2015a) made available by the

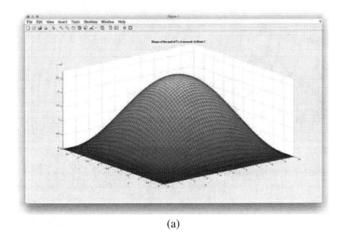

(a)

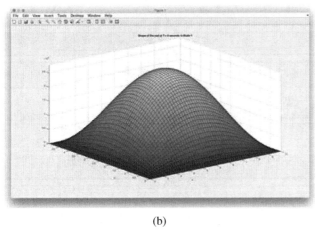

(b)

Figure 9.26 Shape of the mouse pad in mode 1 vibration. (a) The initial shape of the pad. (b) Mode shape at time 1/8 second. *See color section.*

College of Engineering at San José State University. An overview of this software will be given in Section 10.6.2 of Chapter 10, with input/output files for this analytical problem presented in Case 2 in Appendix 4.

Figure 9.26 shows the shape of the pad in mode 1 vibration with $m = n = 1$ in Figure 9.26a at time $t = 0$ and Figure 9.26b at $t = 1/8$ second. There is a significant drop in the maximum amplitude during this period of time.

Figure 9.27 shows graphically the shape of the pad in mode 3 vibration with $m = n = 3$ in three instants of $t = 0$, 1/8, and ¼ seconds. We observe that the pad will have three multiple peaks of vibrations along the x- and y-directions, respectively. Mode shapes and peak amplitudes of this mode at different instants are depicted in the same figure.

As described in Section 8.9, when engineers design a thin plate structure such as the printed circuit board, mode shapes obtained from modal analyses would allow them to avoid placing delicate electronic components at locations where maximum amplitude of vibration may occur such as that illustrated in Figures 9.26 and 9.27.

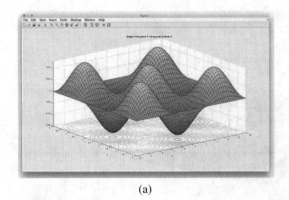

(a)

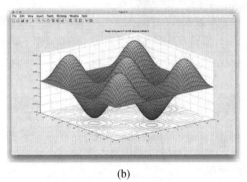

(b)

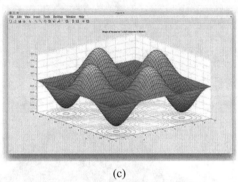

(c)

Figure 9.27 Mode 3 shapes in vibration of a flexible plate. (a) At time $t = 0$. At time $t = 1/8$ second. (c) At time $t = 1/4$ second. *See color section.*

9.9 Problems

9.1 Determine the first- and second-order partial derivatives of each of the following functions:

a) $f(x, y) = 5x^3 + 10x^2y + 8xy^2 + 7y^3$

b) $x(t) = \dfrac{1}{10} \cos 10t + \dfrac{1}{200} \sin 10t - \dfrac{t}{20} \cos 10t$

c) $u(x, t) = \displaystyle\sum_{n=1}^{\infty} b_n \cos \dfrac{n\pi at}{L} \sin \dfrac{n\pi x}{L}$, where b_n $(n = 1, 2, 3, \ldots, n)$ are constants.

9.2 Use the Fourier transform in Section 9.3.3 to solve Equation 9.54 for the amplitudes of vibration of a long flexible cable:

$$a^2 \frac{\partial^2 u(x, t)}{\partial x^2} = \frac{\partial^2 u(x, t)}{\partial t^2}$$

with $-\infty < x < \infty$ and $t > 0$ and where a is a constant. The specified initial conditions are

$$u(x, t)|_{t=0} = u(x, 0) = f(x)$$

$$\left. \frac{\partial u(x, t)}{\partial t} \right|_{t=0} = u'(x, 0) = g(x)$$

9.3 Determine whether the separation of variables technique may be used to solve the following partial differential equations:

a) $\dfrac{\partial u(x,t)}{\partial x} = \dfrac{\partial u(x,t)}{\partial t}$

b) $\dfrac{\partial u(x,t)}{\partial x} + \dfrac{\partial u(x,t)}{\partial t} - u(x,t) = 0$

c) $\dfrac{\partial^2 \varphi(x,y)}{\partial x^2} + \dfrac{\partial^2 \varphi(x,y)}{\partial x \partial y} + \dfrac{\partial^2 \varphi(x,y)}{\partial y^2} = 0$

d) $\dfrac{\partial^2 u(x,t)}{\partial x^2} = \dfrac{\partial^2 u(x,t)}{\partial t^2} + a^2 \qquad (a = \text{constant})$

9.4 Determine whether the temperature distribution in a long metal rod such as illustrated in Figure 7.16 can be obtained by solving the following differential equation:

$$\frac{\partial^2 T(x,t)}{\partial x^2} = \frac{1}{\alpha} \frac{\partial T(x,t)}{\partial t}$$

in which $\alpha = k/(\rho c)$ is the thermal diffusivity of the rod material, where $k =$ thermal conductivity in W/cm-°C, $\rho =$ mass density in g/cm^3, and $c =$ specific heat of the material in J/g-°C. Thermal diffusivity α for most engineering materials in normal temperature is a constant.

The prescribed conditions for this problem are

Initial condition (IC): $T(x,0) = f(x) = 100 - 40x$, a given function.

End conditions (BC): $T(0,\ t) = 0°C$ and $T(L,t) = 0°C$,where L is the length of the rod.

A numerical case involving $L = 1$ m is to be generated with the following material properties of copper:

Thermal conductivity $k = 3.93$ W/cm-°C

Mass density $\rho = 8.9$ g/cm^3

Specific heat $c = 0.386$ J/g-°C

Use the separation of variables technique to solve for the temperature $T(x,t)$ by letting $T(x,t) = X(x)\tau(t)$ where $X(x)$ and $\tau(t)$ are two distinct functions. Derive the equations for these two functions with appropriate conditions.

9.5 A rod of length L is made of a material with thermal conductivity k, initially with a temperature distribution along its length (the x-direction) that can be described by a function $f(x)$. Find the temperature distribution in the rod $T(x,t)$ at time $t > 0$ if all the surfaces and the two ends of the rod are thermally insulated.

9.6 Write the differential equation with appropriate conditions for the case in Problem 9.5 but with its circumferential surface at $r = b$ exposed to surrounding air with bulk temperature T_a and heat transfer coefficient h.

9.7 A rectangular plate used as a heat spreader with the temperatures at its four sides maintained as indicated in Figure 9.10. The temperature distribution in the plate $T(x,y)$ may be obtained by solving the partial differential equation

$$\frac{\partial^2 T(x, y)}{\partial x^2} + \frac{\partial^2 T(x, y)}{\partial y^2} = 0 \tag{9.50}$$

a) Show the mathematical forms of the boundary conditions required to solve Equation 9.50 according to the specified conditions indicated in Figure 9.28.
b) Solve Equation 9.50 with the specified boundary conditions in (a).
c) Determine the temperature at the center of the plate.
There is negligible heat flow from the flat faces of the plate in contact with stagnant air. All temperatures in Figure 9.28 have the unit of °C.

9.8 Solve Equation 9.54 for a long cable of length L in free vibration as illustrated in Figure 9.18 with the initial shape of the cable being $f(x) = x(L - x)$.

9.9 Do the same as in Problem 9.8 but with $f(x) = \sin(\pi x)/L$.

9.10 The amplitude $y(x,t)$ of free vibration of a beam of length L can be obtained by solving the following partial differential equation (called the Euler–Lagrange equation):

$$a^2 \frac{\partial^4 y(x, t)}{\partial x^4} = \frac{\partial^2 y(x, t)}{\partial t^2} \qquad \text{for} \quad 0 \le x \le L \quad \text{and} \quad t > 0$$

where $a^2 = EI/m$, in which E is the Young's modulus and m is the mass per unit length of the beam material, and I is the section moment of inertia of the beam cross-section. Solve the partial differential equation with the following conditions. The beam is simply supported with no bending moments (bending moments are proportional to the second order derivative of the deflection of the beam) and the initial conditions of the beam are specified by the following functions of $f(x)$ and $g(x)$:

$$y(x, t)|_{t=0} = y(x, 0) = f(x)$$

and

$$\left. \frac{\partial y(x, t)}{\partial t} \right|_{t=0} = g(x)$$

9.11 Solve Problem 9.10 with $f(x) = \sin(\pi x)/L$.

9.12 Solve Problem 9.11 but with both ends of the beam being rigidly held; that is, the slope of the beam at both ends is zero.

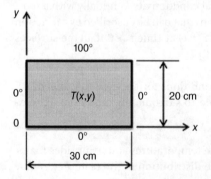

Figure 9.28 Temperature distribution in a flat plate.

10

Numerical Solution Methods for Engineering Analysis

Chapter Learning Objectives

- Learn the alternative ways of using numerical methods to solve nonlinear equations, perform integration, and solve differential equations.
- Learn the principles of various numerical techniques for solving nonlinear equations, performing integrations, and solving differential equations by the Runge–Kutta methods.
- Learn the fact that numerical methods offer approximate but credible accurate solutions to the problems that are not readily or possibly solved by closed-form solution methods.
- Learn the fact that numerical solutions are available to the users only at the preset solution points, and the accuracy of the solutions is largely dependent on the size of the increments of the variable selected for the solutions.
- Become familiar with the value of commercially available numerical solution software packages such as Mathematica and MatLAB.

10.1   Introduction

Numerical methods are techniques for solving the mathematical problems involved in engineering analysis that cannot be solved by closed-form solution methods such as those presented in the preceding chapters. In this chapter, we will learn the use of some of the available numerical methods that will not only enable engineers to solve many mathematical problems, but also allow engineers to minimize the need for the many hypotheses and idealization of the conditions, such as those stipulated in Section 1.4.

Numerical techniques have greatly expanded the types of problems that engineers can solve, as illustrated in a number of publications in the open literature (Chapra, 2012; Ferziger, 1998; Hoffman, 1992). The goal of this chapter is to present to readers the basic principles of some of these techniques that are frequently used in engineering analysis. The author's own experience indicates that engineers who understand the principles of the numerical methods are usually more effective and intelligent users of these methods than most technical personnel who are trained to carry out the same computational assignments using "turn-key" software packages such as the finite-difference method and the MatLAB software package in Appendix 4. This chapter will cover the principles of commonly used numerical techniques for (1) the solution of nonlinear polynomial and transcendental equations; (2) integration involving complex forms of functions; and (3) the solution of differential equations by the basic finite-difference schemes and the Runge–Kutta methods. The chapter will also cover the overviews of two popular commercial software packages called Mathematica and MatLAB.

Applied Engineering Analysis, First Edition. Tai-Ran Hsu.
© 2018 John Wiley & Sons Ltd. Published 2018 by John Wiley & Sons Ltd.
Companion Website: www.wiley.com/go/hsu/applied

The principal task in numerical methods for engineering analysis is to develop algorithms that involve arithmetic and logical operations so that such operations can be performed at incredible speed by digital computers with enormous data storage capacities. Because readers of this book are expected to be users of numerical methods, we will present only the principles that are relevant to the development of these algorithms, not the theories and the proof of these methods.

10.2 Engineering Analysis with Numerical Solutions

Most engineering problems require enormous computational effort when numerical methods are used. Digital computers are essential tools for obtaining numerical solutions. Digital computers have incredible power in computational speed and enormous memory capacity. Unfortunately, these machines have no intelligence of their own, and they are not capable of making independent judgment on their own. Additionally, engineers need to realize the fact that digital computers can only perform simple arithmetic operations with ($+$, $-$, $\times$, $\div$) and handle the logical flow of data. It cannot perform higher mathematical operations even in such simple cases as evaluating exponential and trigonometric functions without proper algorithms that convert the evaluation of these functions into simple arithmetic operations; thus, all complicated mathematical operations have to be converted into simple arithmetical operations. Numerical methods that enable engineers to develop algorithms for various mathematical functions and operations using digital computers have thus become essential knowledge and skills for solving many advanced engineering problems using mathematical tools.

Despite the fact that numerical techniques have greatly expanded the types of problems that engineers can handle as mentioned in Section 10.1, users need to be aware of many unique characteristics of these methods, as outlined in the following.

1) Numerical solutions are available only at selected (discrete) solution points of the domain that is being investigated, not at all points in the entire domain covered by the functions as is the case with analytical solution methods described in Chapters 7, 8, and 9.
2) Numerical methods are essentially "trial-and-error" processes. Typically, users need to know the initial and boundary conditions that the intended solution will cover. The selection of increments of the variable at which the solution points are positioned is critical in the solution of the problem. Unstable numerical solutions may result from improper selection of such increments, called the step sizes with solutions.
3) Most numerical solution methods result in some error in the solutions. Two types of error are inherent with numerical solutions:
   a) *Truncation errors* – because of the approximate nature of numerical solutions of many engineering problems, these solutions often consist of both lower-order terms and higher-order terms. The latter terms are often dropped in the computations in order to achieve computational efficiency, resulting in error in the solution.
   b) *Round-off errors* – Most digital computers handle numbers either with 7 places or 14 decimal places in numerical solutions. In the case of a 32-bit computer with double precision (i.e., numbers of 14 decimal places), any number after the 14th decimal point will be dropped. This may not sound like a big deal, but if a huge number of operations are involved in the computation, such error can accumulate and result in significant error in the end results.

Both of these types of error are of cumulative nature. Consequently, errors in numerical solution may grow to be significant in solutions obtained after many step increments.

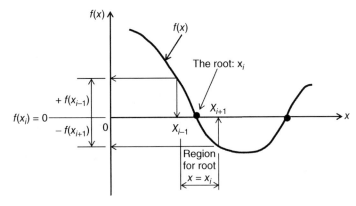

Figure 10.1 Root of a nonlinear equation $f(x) = 0$.

10.3 Solution of Nonlinear Equations

Often, engineers need to solve nonlinear equations in their analyses. These equations can be as simple as quadratic equations such as in Equation 8.3 with the two roots expressed in Equation 8.4. There is also the need to find roots of equations in higher-order polynomial functions such as shown in Equation 10.1, relating to the problem of locating the marking on a measuring cup, to be described in Example 10.3 and in Figure 10.4:

$$L^3 + 70.3L^2 + 1647.39L - 18656.72 = 0 \tag{10.1}$$

or solution of the time t_f required for the mass to rupture from its spring attachment in the case of resonant vibration analysis that was illustrated in Example 8.9:

$$\left(0.005 - \frac{t_f}{20}\right) \cos 10t_f + \frac{1}{200} \sin 10t_f - 0.03 = 0 \tag{10.2}$$

Solutions of nonlinear equations such as Equations 10.1 and 10.2 may be obtained by setting expressions $f(x) = 0$, and finding their roots (i.e., L in Equation 10.1 and t_f in Equation 10.2) located at the points of cross-over of the function $f(x)$ and the x-coordinate axis, as illustrated in Figure 10.1.

We will use the following two methods for the solution of the nonlinear equation given in Equation 10.2

10.3.1 Solution Using Microsoft Excel Software

Because the roots of an equation represented by $f(x) = 0$ are located at the intersections of the function $f(x)$ and the x-coordinate axis as illustrated in Figure 10.1, one may use the widely available tool of spreadsheets (such as Microsoft Office Excel) to locate the roots of the equation by evaluating the function $f(x)$ with for various values of the variable x. The range in which the roots of the equation $f(x) = 0$ lie can be identified as the values where $f(x)$ change sign from positive to negative or vice versa, as illustrated for the range $(x_{i-1}-x_{i+1})$ in Figure 10.1. More accurate values of the roots may be found by repeating the computation of the function values with smaller increments of the variable x within that range. This method, though sounds tedious, involves straightforward evaluation of the function $f(x)$ with an initially estimated value of x_i by using the Excel software that has enormous computational speed to identify the range in which the roots are located, as will be illustrated by Example 10.1.

Example 10.1

Solve the nonlinear polynomial Equation 10.3:

$$x^4 - 2x^3 + x^2 - 3x + 3 = 0 \qquad (10.3)$$

using a Microsoft Excel spreadsheet.

Solution:

As in many other numerical solution methods, we begin with an estimate of the root of the equation at $x = 0$ and assign an increment of 0.5 in variable x in our evaluation of the function $f(x) = x^4 - 2x^3 + x^2 - 3x + 3$. The spreadsheet indicates the values of $f(x)$ tabulated in the right-hand portion of Figure 10.2 with an assumed starting point at $x = 0$:

x	$f(x)$
0	3.00
0.5	1.56
1.0	0
1.5	−0.94
2.0	1.00
2.5	9.56
3.0	30.00
3.5	69.06
4.0	135.00

We note from the computed values of $f(x)$ with variable x in Figure 10.2 that there are two roots of the equation, one in the range of ($x = 1.0$ and 1.5) and the other in the range of ($x = 1.5$ and 2.0), because the sign of the function $f(x)$ changes across each these ranges of variable x. The first root of $x = 1$ is obvious because it results in $f(x) = 0$. The search for the second root by computation of the function $f(x)$ with smaller increment of x between $x = 1.5$ and $x = 2.0$ indicates a root at approximately $x = 1.8$, as illustrated in the graphic representation of the results in Figure 10.2.

10.3.2 The Newton–Raphson Method

Perhaps the most widely used method for finding the roots of nonlinear equations is the Newton–Raphson method. This method offers rapid convergence to the roots of many nonlinear equations from the initially estimated roots. The fast convergence to true roots from

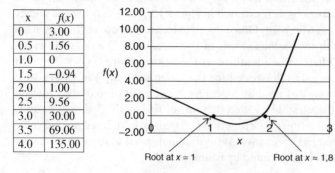

x	$f(x)$
0	3.00
0.5	1.56
1.0	0
1.5	−0.94
2.0	1.00
2.5	9.56
3.0	30.00
3.5	69.06
4.0	135.00

Figure 10.2 Roots of a nonlinear equation.

Figure 10.3 Newton–Raphson method for solving nonlinear equations.

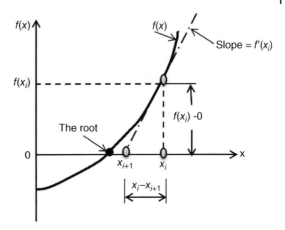

the estimated roots is achieved by means of computation of both the function $f(x_i)$ and the corresponding slope $f'(x_i)$ of the function at x_i, as illustrated in Figure 10.3.

Figure 10.3 illustrates the principle of the Newton–Raphson method of solving nonlinear equations. As in many other numerical solution methods, the user has to estimate a root at $x = x_i$ for the equation $f(x) = 0$, from which they may compute the function $f(x_i)$ and at the same time the slope of the curve generated by the function $f(x)$. This slope may be expressed $f'(x_i)$. The graphical representation of this situation indicates that the slope $f'(x_i)$ may be expressed by Equation 10.4:

$$f'(x_i) = \frac{f(x_i) - 0}{x_i - x_{i+1}} \tag{10.4}$$

which leads to the following expression for the next estimated root at $x = x_{i+1}$:

$$x_{i+1} = x_i - \frac{f(x_i)}{f'(x_i)} \tag{10.5}$$

One readily sees from Figure 10.3 that the computed approximate next root x_{i+1} is much closer to the real root (shown as a solid circle) than the previous estimated value at x_i.

Example 10.2

Solve the nonlinear polynomial Equation 10.3 in Example 10.1 using the Newton–Raphson method.

Solution:

In this example, we are required to find the roots of Equation 10.3, with the function

$$f(x) = x^4 - 2x^3 + x^2 - 3x + 3 \tag{a}$$

From this we may express the first-order derivative that represent the slope of the curve generated by the function $f(x)$ as

$$f'(x) = 4x^3 - 6x^2 + 2x - 3 \tag{b}$$

Substituting $f(x_i)$ and $f'(x_i)$ into Equation 10.5 for the Newton–Raphson method, we will have the following expression for finding the estimated roots beginning with $x = x_i$:

$$
\begin{aligned}
x_{i+1} &= x_i - \frac{f(x_i)}{f'(x_i)} \\
&= x_i - \frac{x_i^4 - 2x_i^3 + x_i^2 - 3x_i + 3}{4x_i^3 - 6x_i^2 + 2x_i - 3} \tag{c}
\end{aligned}
$$

Thus, estimating the first root at $x = x_1 = 0.5$ for $i = 1$, we will have the next estimated x value at x_2 with $i = i + 1$ from Equation (c) as

$$x_2 = 0.5 - \frac{(0.5)^4 - 2(0.5)^3 + (0.5)^2 - 3(0.5) + 3}{4(0.5)^3 - 6(0.5)^2 + 2(0.5) - 3}$$

$$= 1.02083$$

Iterating the same procedure, we will obtain convergence of the x-values to the root of the equation. The following tabulation shows the attempts made to find the first two roots of Equation (a).

Roots of a polynomial equation of order 4

Attempt number (i)	x_i	Computed x_{i+1}	Note
1	0.5	1.0208	Estimate of first root
2	1.0208	0.9998	
3	0.9998	1.0010	Converges to first root
4	4.0	3.1818	Estimate of second root
5	3.1818	2.5990	
6	2.4655	2.1247	
7	2.1247	1.9382	
8	1.9382	1.8723	Begins to converge to root
9	1.8723	1.8638	
10	1.8638	1.8637	Converges to second root

We note from the above tabulation that it took only three attempts to find the convergence to the first root at $x = 1.0$. However, it took six attempts to reach convergence to the second root at $x = 1.8637$ with an initial estimate of the root at $x = 4$. These two roots are indicated in Figure 10.2 with the plot of the polynomial function.

Example 10.3

Locate the line mark (L) for a contents volume of 500 milliliters (ml) in a measuring cup with the dimensions shown in Figure 10.4.

Solution:

We let the mark on the measuring cup for a contents volume of 500 ml be situated at L as shown in Figure 10.4. We let the volume of the measuring cup with the content at the height L to be 500 ml.

The volume of a solid of revolution of a given height may be determined from Equations 2.16 or 2.17 in Chapter 2. The profile of the measuring cup in Figure 10.4 is represented by a function $x(y) = 0.16y + 3.75$ in an x–y coordinate system in Figure 10.5.

The volume of the contents of the measuring cup at height L may be determined by the following integral following Equation 2.17:

$$V = \pi \int_0^L \left[x(y)\right]^2 dy = \pi \int_0^L (0.16y + 3.75)^2 dy$$

$$= 0.0268L^3 + 1.884L^2 + 44.15L$$

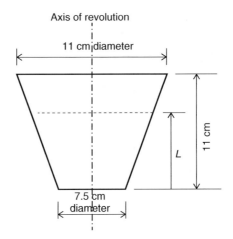

Figure 10.4 Dimensions of a measuring cup.

Figure 10.5 Profile of a measuring cup in the *x–y* coordinate system.

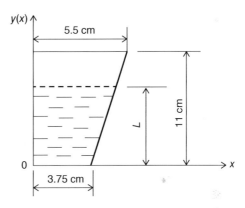

Since the volume of the measuring cup at height L is 500 ml or 500 cm^3, we have the following equation for the unknown quantity L:

$$500 = 0.0268L^3 + 1.884L^2 + 44.15L$$

or in alternative form:

$$L^3 + 70.3L^2 + 1647.39L - 18656.72 = 0 \tag{a}$$

We recognize that Equation (a) is a cubic equation, and one of the roots of this equation will be the length L.

We will use the Newton–Raphson method to solve the cubic equation in Equation (a) with the formula given in Equation 10.5 by letting $L = x$ in Equation (a). We thus have the following expressions for the Newton–Raphson method:

$$f(x) = x^3 + 70.3x^2 + 1647.39x - 18656.72 \tag{b}$$

and its derivative:

$$f'(x) = 3x^2 + 140.6x + 1647.39 \tag{c}$$

We estimate the root of Equation (a) to be $L = x_1 = 4.0$ with $i = 1$, and using Equation 10.5 make the next estimate of the root as

$$x_2 = x_1 - \frac{f(x_1)}{f'(x_1)} = 4 - \frac{f(4)}{f'(4)}$$

$$= 4 - \frac{(-10878.36)}{2257.79}$$

$$= 8.818$$

We find the subsequent estimations of the roots of Equation (c) to be rapidly converging:

$$x_3 = 8.818 - \frac{2022}{3120.47} = 8.17$$

$$x_4 = 8.17 - \frac{56.5825}{2996.3287} = 8.15$$

The last estimated root is $x_4 = 8.15$ which is close to the solution $x = 8.1566$ obtained from the online software Wolfram/Alpha Widgets (http://www.wolframalpha.com/widgets/view .jsp?id).

We may thus conclude that the mark line for 500 ml for the measuring cup in Figure 10.4 is located at the height $L = 8.15$ cm from the bottom of the cup.

Example 10.4

In Example 8.9 in Chapter 8, we derived an equation, Equation (a), to describe a mass that is attached to a spring that would break when its elongation reached 0.03 m during resonant vibration of the spring–mass system. We wish to determine the time t_f at which the spring breaks.

The equation derived in Example 8.9 for the breaking time t_f was

$$0.03 = 0.05 \cos 10t_f + \frac{1}{200} \sin 10t_f - \frac{t_f}{20} \cos 10t_f$$

or

$$\left(0.05 - \frac{t_f}{20}\right) \cos 10t_f + \frac{1}{200} \sin 10t_f - 0.03 = 0 \tag{a}$$

Solution:

We will use the Newton–Raphson method to solve for the unknown quantity t_f in Equation (a) by first assuming a solution of $t_f = 0.75$. We made this assumption of the solution based on a crudely approximated value of $t_f = 0.7$ in Example 8.9.

In order to illustrate how the Newton–Raphson method in Equation 10.5 is used to arrive at a convergent solution, we will replace the unknown quantity t_f in Equation (a) by x in the following form:

$$\left(0.05 - \frac{x}{20}\right) \cos 10x + \frac{1}{200} \sin 10x - 0.03 = 0 \tag{b}$$

with

$$f(x) = \left(0.05 - \frac{x}{20}\right) \cos 10x + \frac{1}{200} \sin 10x - 0.03 \tag{c}$$

and the derivative

$$f'(x) = -(0.05 - 0.5x) \sin 10x \tag{d}$$

Thus, the estimated root x_{i+1} obtained after the previously estimated root x_i may be computed by using the expression in Equation 10.5.

We will begin the solution process with our initial estimate of the root of Equation (b) as x_1 = 0.75, which leads to the following subsequent approximate root x_2:

$$x_2 = x_1 - \frac{f(x_1)}{f'(x_1)} = 0.75 - \frac{f(0.75)}{f'(0.75)}$$

$$= 0.75 - \frac{[(0.05 - 0.75/20)\cos(10 \times 0.75) + \sin(10 \times 0.75/200 - 0.03]}{[-(0.05 - 0.5 \times 0.75)\sin(10 \times 0.75)]}$$

$$= 0.869\,979$$

The result of this computation, $x_2 = 0.869\,979$ is presented as trial no. 1 in the following table. We make similar trials with assumed solution x_i and with a coarse increment of 0.5 as shown in the same table using the Microsoft Excel software.

We note that the value of the function $f(x)$ in the table change from $-0.003\,432$ with $x_i =$ 0.85 in trial no. 3 to $+0.008\,5058$ with $x_i = 0.9$ in trial no. 4. This change of the sign of the value of the function indicates that the first root of Equation (b) is between $x = 0.85$ and 0.90. Indeed, our subsequent trials with x_i values assigned within this range of solution using the Newton–Raphson method does indicate convergence to a more precise solution of $x = 0.861$ 933 as indicated in the last trial, no. 8.

Trial no.	Assigned x_i	$f(x)$	$f'(x)$	x_{i+1}	% Difference
1	0.75	−0.036 576	0.304 86	0.869 979	16
2	0.80	−0.019 961	0.346 275	0.857 644	7.21
3	0.85	−0.003 432	0.299 433	0.861 46	1.35
4	0.90	0.008 5058	0.164 847	0.848 402	−5.733 12
5	0.855	−0.001 959	0.289 694	0.861 762	0.790 885
6	0.857	−0.001 384	0.285 55	0.861 846	0.565 415
7	0.86	0.005 367	0.279 071	0.861 923	0.223 604
8	0.863	0.000 2905	0.272 28	0.861 933	−0.123 61

10.4 Numerical Integration Methods

Integration of functions over specific intervals of the variables that define the functions is a frequent requirement in engineering analysis. Some of the practical applications of integration are presented in Section 2.3 in Chapter 2. Exact evaluation of many definite integrals can be found in handbooks (Zwillinger, 2003) but many others with functions to be integrated are so complicated that analytical solutions for these integrals are not possible. Numerical methods are the only viable ways for such evaluations.

In this section, we will present three numerical integration methods: (1) the trapezoidal rule; (2) Simpson's one-third rule; and (3) Gaussian quadrature. The first two methods are commonly used for integration of nonlinear functions, and the third method is extensively used in numerical analysis of complex engineering analyses, such as the finite-element analysis.

We will focus our effort on refreshing the principles that are relevant to the development of algorithms of these particular numerical integration methods, but will not rehearse the underlying theories and their proofs. The reader will find derivation of the formulae for these numerical integration methods in reference books such as those by Chapra (2012), Ferziger (1998), and Hoffman (1992).

10.4.1 The Trapezoidal Rule for Numerical Integration

We learned in Section 2.2.6 that the value of a definite integral of a function $y(x)$ is equal to the area under the curve produced by this function between the upper and lower limits of the integration as illustrated in Figure 10.6. Mathematically, the integral of function $y(x)$ can be expressed as

$$I = \int_{x_a}^{x_b} y(x)\,dx = \text{Area } A \tag{10.6}$$

The value of the integral of a function may thus be determined by computing the area covered by the function between the two specified limits. For example, the value of the function $y(x)$ in Figure 10.7 may be approximated by the sum of the three trapezoidal plane areas A_1, A_2, and A_3.

The area of a trapezoidal plane may be evaluated by the formula of half of the sum of the length of two parallel sides multiplied by the distance between these two sides. Mathematically,

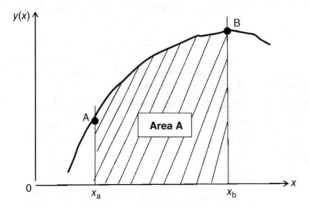

Figure 10.6 Graphical representation of integration of a continuous function.

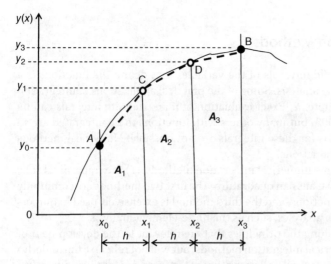

Figure 10.7 Approximation of the integral of a continuous function $y(x)$.

the plane areas A_1, A_2, and A_3 with equal distance h between two parallel sides in Figure 10.7 may be computed by the following formulae:

$$A_1 = \left(\frac{y_0 + y_1}{2}\right) h \qquad A_2 = \left(\frac{y_1 + y_2}{2}\right) h \qquad A_3 = \left(\frac{y_2 + y_3}{2}\right) h$$

The sum of A_1, A_2, and A_3 is given by

$$A_1 + A_2 + A_3 \approx \int_{x_0}^{x_3} y(x)\, dx \approx \left(\frac{y_0 + y_1}{2}\right) h + \left(\frac{y_1 + y_2}{2}\right) h + \left(\frac{y_2 + y_3}{2}\right) h$$

$$= \frac{h}{2}(y_0 + 2y_1 + 2y_2 + y_3) \tag{10.7}$$

in which h is the assigned increment of variable x, and y_0, y_1, y_2, and y_3 are the values of the function evaluated at x_0, x_1, x_2, and x_3, respectively.

Example 10.5

Use the trapezoidal rule to evaluate the integral $\int_{x_a}^{x_b} y(x)\, dx$ in which the function $y(x) = x\sqrt{(16 - x^2)^3}$ with $x_a = 0.5$ and $x_b = 3.5$ and with assigned increment $h = 1.0$

Solution:

We will demonstrate the use of trapezoidal rule for numerical integration by plotting the function $y(x)$ versus x as shown in Figure 10.8. The value of the integral that we need to determine is

$$I = \int_{0.5}^{3.5} x\sqrt{(16 - x^2)^3}\, dx$$

By using the trapezoidal rule with the three trapezoids illustrated in Figure 10.8, we may evaluate the integral I by using the expression in Equation 10.7 as

$$I = \int_{0.5}^{3.5} x\sqrt{(16 - x^2)^3}\, dx \approx \frac{h}{2}(y_0 + 2y_1 + 2y_2 + y_3)$$

$$= \frac{1}{2}(31.25 + 2 \times 76.48 + 2 \times 76.11 + 25.4)$$

$$= 180.92$$

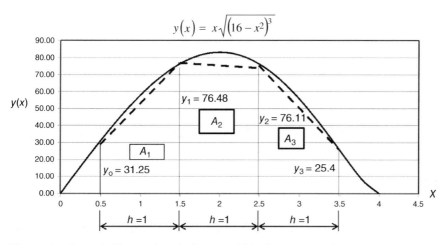

Figure 10.8 Numerical integration of a function $y(x)$ by three trapezoids.

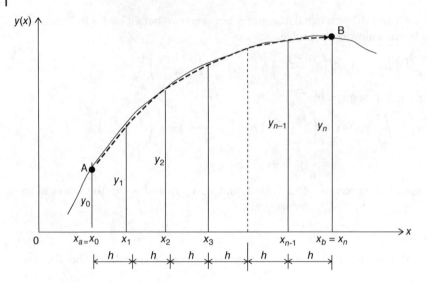

Figure 10.9 Integration of function $y(x)$ with multiple trapezoids.

This approximate value of the integral I obtained by the above formula is less than the analytical solution of 191.45 obtained from a mathematical handbook (Zwillinger, 2003). The difference between the value of the integral of 180.92 obtained using the trapezoidal rule and that of 191.45 from the analytical solution method is to be expected. This difference in results represents the errors inherent in any numerical method. In the trapezoidal rule method in this example, the discrepancy is introduced by the approximation of the curve representing the given function by straight line segments (shown dashed) that were the edges of the trapezoids in computing the approximate area under the curve in Figure 10.8. One may readily observe that this discrepancy between the curved edges and the straight edges shown as dashed lines can be reduced by the reduction of the size of the increment h, as can be observed from the graphic illustration in Figure 10.8. Consequently, closer approximation, or more accurate results from numerical integration, may be achieved by reducing the size of increment h in this particular numerical analysis.

Figure 10.9 shows the area under function $y(x)$, which can be approximated as the sum of the areas of $(n-1)$ trapezoids. Many more trapezoids under the function curve such as shown in Figure 10.9 entails many more increments of h along the x-coordinate between the upper and lower integration limits, with concomitant more accurate results.

The approximate value of the integral of function $y(x)$ in Figure 10.9 is equal to the sum of all the trapezoids in the figure according to the following equation derived from the same principle as in the earlier case with three trapezoids:

$$I = \int_{x_a}^{x_b} y(x)\,dx \approx \sum(A_1 + A_2 + A_3 + \cdots + A_{n-1})$$

$$= \frac{h}{2}(y_0 + 2y_1 + 2y_2 + 2y_3 + \cdots + 2y_{n-1} + y_n) \tag{10.8}$$

where h is the increment along the x-coordinate axis in the numerical integration.

Example 10.6
Evaluate the same integral as in Example 10.5 but with a reduced size of the increment $h = 0.5$.

Solution:

The value of the integral that we need to determine is

$$I = \int_{0.5}^{3.5} y(x)\,dx = \int_{0.5}^{3.5} x\sqrt{(16-x^2)^3}\,dx$$

with the function $y(x) = x\sqrt{(16-x^2)^3}$. The area covered by the function between $x = 0.5$ and $x = 3.5$ is approximated by what is shown in Figure 10.10.

We will first determine the function values $y(x)$ at $x = 0.5$, 1.0, 1.5, 2.0, 2.5, 3.0, and 3.5. Table 10.1 shows the function values, y_1, y_2, y_3, y_4, y_5, and y_6 required in the computation of the areas of the five trapezoids in Figure 10.10.

We may thus use Equation 10.8 to compute the sum of the areas of the five trapezoids in Figure 10.10 as

$$I \approx \frac{0.5}{2}(31.25 + 2 \times 58.09 + 2 \times 76.48 + 2 \times 83.18 + 2 \times 76.11 + 2 \times 55.56 + 25.42)$$

$$= 188.88$$

This value of 188.88 for the same integral now with $h = 0.5$ is much closer to the analytical value of 191.45 than that of 180 obtained with $h = 1.0$ in Example 10.5. It has thus been

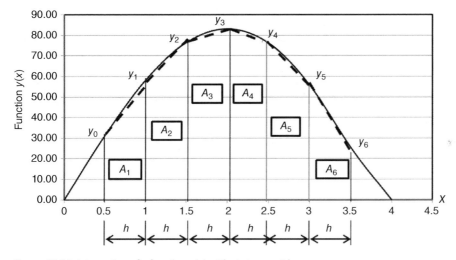

Figure 10.10 Integration of a function $y(x)$ with six trapezoids.

Table 10.1 Function values at designated points.

n	x_n	y_{n-1}
1	0.5	31.25
2	1.0	58.09
3	1.5	76.48
4	2.0	83.18
5	2.5	76.11
6	3.0	55.56
7	3.5	25.42

demonstrated the fact that the smaller the increment one uses in numerical integration, the more accurate a result will be obtained.

10.4.2 Numerical Integration by Simpson's One-third Rule

In Section 10.4.1, we refreshed the principle of evaluating an integral as the plane area under the curve representing the function (the integrand) in the integral between two limits of the variable in the integration. This area is graphically expressed in Figure 10.11a.

The value of an integral may be obtained by numerical methods such as the trapezoidal rule described in Section 10.4.1. The trapezoidal method is a simple but a relatively "crude" method that will result in an approximate value of the integral with minimal computational effort. Graphical representation of the trapezoidal method is illustrated in Figure 10.11b for the simplest possible numerical approximation of the area under the curve of $y(x)$ as the area of one trapezoid (shown as the cross-hatched area) under the curve. This method is popular and easy to use because the formula used to compute the plane area of trapezoids is well known to engineers.

Another popular numerical method for integration is by Simpson's rule, in particular, "Simpson's one-third rule." This rule differs from the trapezoidal method by assuming that the function $y(x)$ can be approximated in the range of interest by a parabolic function as shown in Figure 10.11c. The function $y(x) = ax^2 + bx + c$ that describes the sector of the function in Figure 10.11c contains the unknown constant coefficients a, b, and c which can be determined with the following simultaneous equations relating to the function values at the discrete variable

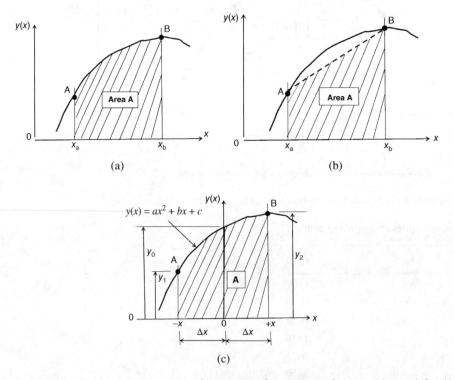

Figure 10.11 Graphical representation of integration of a continuous function. (a) Area defined by a function. (b) Approximation of the area by a trapezoid. (c) Approximation of the area by a parabolic function.

values $x = -x$, $x = 0$, and $x = +x$ as follows:

$$y_0 = a(-x)^2 + b(-x) + c$$

$$y_1 = c$$

$$y_2 = a(x)^2 + b(x) + c$$

from which we may solve for

$$a = \frac{1}{2x^2}(y_0 - 2y_1 + y_2) \tag{10.9a}$$

$$c = y_1 \tag{10.9b}$$

The value of the integral of the function $y(x)$ is equal to the plane area A in Figure 10.11c, or

$$I = \text{AREA}(-xABx) = \int_{-\Delta x}^{+\Delta x} y(x)\,dx = \int_{-\Delta x}^{+\Delta x} (ax^2 + bx + c)\,dx$$

$$= \left(\frac{ax^3}{3} + \frac{bx^2}{2} + cx \right)\Bigg|_{-\Delta x}^{+\Delta x}$$

$$= \frac{\Delta x}{3}[2a(\Delta x)^2 + 6c]$$

By substituting the constant coefficients a and c in Equations 10.9a and 10.9b, together with $\Delta x = x$ illustrated in Figure 10.11c, into the above expression, we get the following relation for Simpson's one-third rule for the integral I:

$$I = \int_{-\Delta x}^{+\Delta x} y(x)\,dx = \int_{-x}^{+x} (ax^2 + bx + c)\,dx = \frac{\Delta x}{3}(y_0 + 4y_1 + y_2) \tag{10.10}$$

Example 10.7
Use Simpson's one-third rule to find the numerical value of the integral in Example 10.5.

Solution:
We will use the three function values at $x = 0.5$, 2.0, and 3.5 to compute the value of the integral of the function in Example 10.5. In this case, the increment of the integration variable is $\Delta x = 1.5$. The integral is determined using Equation 10.10 for the graphic representation of the function $y(x)$ in Figure 10.12.

We may obtain the function values y_0, y_1, and y_2 at $x = 0.5$, 2, and 3.5 from Table 10.1 as $y_0 = 31.25$, $y_1 = 83.18$, and $y_2 = 25.42$. Integration of the function $y(x)$ in Example 10.5 can thus be carried out by substituting the values of y_0, y_1 and y_2 and the increment of x, $\Delta x = 1.5$, into Equation 10.10 to give

$$I = \int_{-\Delta x}^{+\Delta x} y(x)\,dx = \int_{-x}^{+x} (ax^2 + bx + c)\,dx$$

$$= \frac{\Delta x}{3}(y_0 + 4y_1 + y_2)$$

$$= \int_{0.5}^{3.5} y(x)\,dx$$

$$= \int_{0.5}^{3.5} x\sqrt{(16 - x^2)^3}\,dx$$

$$= \frac{1.5}{3}(31.25 + 4 \times 83.18 + 25.42) = 194.70$$

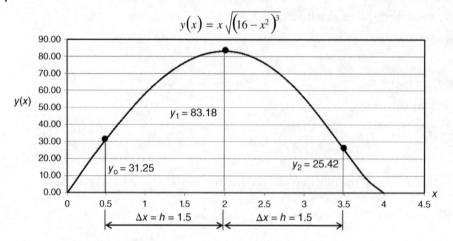

Figure 10.12 Numerical integration by Simpson's one-third rule.

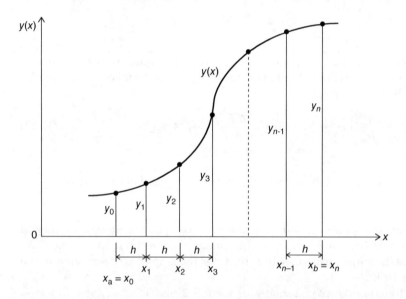

Figure 10.13 Integration of a nonlinear function $y(x)$ by Simpson's one-third rule.

The analytical or "exact" result of the above integral is 191.45 from a table of definite integrals (Zwillinger 2003), which we may compare with the results $I = 188.88$ using three trapezoids in Example 10.5 and $I = 194.70$ with three function values using Simpson's one-third rule.

We will use the same illustration in Figure 10.13 and Equation 10.10 to derive the general expression for Simpson's one-third rule for numerical integration.

We derived Equation 10.10 with the first three function values y_0, y_1 and y_2 at $x = x_0$, $x = x_1$, and $x = x_2$, respectively, with an assumed parabolic function connecting these three points. The next parabolic function for the next adjoining three-point segment requires that the function value y_2 be evaluated twice. The same happens to the second last point at x_{n-1} in the region for the integration at which the function value y_{n-1} in Figure 10.13 being evaluated twice. The coefficients associated with the y_i for n points in Figure 10.13 are given in the following tabulation:

x_i	0	1	2	3	4	5	6	...	$n-2$	$n-1$	n
Coefficient of y_i	1	4	2	4	2	4	2	...	2	4	1

We may thus formulate the general expression of Simpson's one-third rule as follows:

$$I = \int_{x_a}^{x_b} y(x) \, dx$$

$$= \frac{h}{3}(y_0 + 4y_1 + 2y_2 + 4y_3 + 2y_4 + \cdots + 2y_{n-2} + 4y_{n-1} + y_n) \tag{10.11}$$

We note from Equation 10.11 that the use of this relationship for Simpson's one-third rule requires even number of function values with odd number increments in the integration variable.

Example 10.8
Use Simpson's one-third rule in Equation 10.10 to evaluate the integral in Example 10.6:

$$I = \int_{0.5}^{3.5} y(x) \, dx = \int_{0.5}^{3.5} x\sqrt{(16 - x^2)^3} \, dx$$

Solution:
The trapezoidal method was used in numerical evaluation of the integral with seven function values as indicated in Table 10.1 in Example 10.6. Here we will require an even number of function values for Simpson's one-third rule as indicated in Equation 10.11. Consequently, we need to reduce the increment of x and increase the number of increments from seven in Example 10.6 to eight for the present example, with a slight reduction of increment from $\Delta x = 0.5$ in Example 10.6 to $\Delta x = 0.43$ in the present case. The eight function values are presented in Table 10.2 for the data points shown in Figure 10.14.

Table 10.2 Eight values of a function for integration using Simpson one-third rule.

n	1	2	3	4	5	6	7	8
x_n	0.5	0.93	1.36	1.79	2.21	2.64	3.07	3.5
$y = y(x_n)$	31.25	54.69	72.3	81.89	81.85	71.54	51.68	25.41

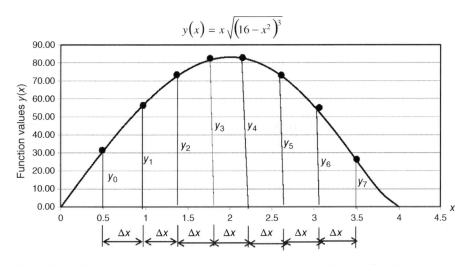

Figure 10.14 Numerical integration using Simpson's one-third rule with eight function values.

We may thus insert the function values presented in Table 10.2 into Equation 10.11 and evaluate the integral of the function:

$$
\begin{aligned}
I &= \int_{0.5}^{3.5} y(x)\,dx = \int_{0.5}^{3.5} x\sqrt{(16 - x^2)^3}\,dx \\
&\approx \frac{\Delta x}{3}(y_0 + 4y_1 + 2y_2 + 4y_3 + 2y_4 + 4y_5 + 2y_6 + y_7) \\
&= \frac{0.43}{3}(31.25 + 4 \times 54.69 + 2 \times 72.30 + 4 \times 81.89 + 2 \times 81.85 + 4 \times 71.54 \\
&\qquad + 2 \times 51.68 + 25.41) \\
&= 186.45
\end{aligned}
$$

This approximate solution of 186.45 of the integral has about 2.6% error from the exact solution of 191.45.

10.4.3 Numerical Integration by Gaussian Quadrature

Numerical integration of functions using the trapezoidal rule (Section 10.4.1) and Simpson's one-third rule (Section 10.4.2) enables us to determine approximate values of integrals of continuous functions $f(x)$ over a range of $(x_b - x_a)$ into equal parts with increment of the variable Δx (or h) as shown in Figures 10.7 and 10.9. This process allows us to select the sampling points and evaluate the integral in terms of the discrete values of the function at these points. These methods usually work well with well-behaved functions in integrals such as the ones used in Examples 10.5 to 10.8. However, neither the trapezoidal rule nor Simpson's one-third rule offers any guidance on the selection of the size of the increment of the variable in these numerical integration methods. There are times when engineers are expected to find numerical values of integrals involving functions that have drastic changes of shape over the range of the required integration. The two methods already discussed do not yield good approximations of the numerical values of these integrals because of improper selection of sampling points. Thus, it is desirable to have a numerical integration method that offers criteria for optimal sampling points in numerical integration.

The Gaussian integration method was established on the basis of strategically selected sampling points. The normal form of a Gaussian integral can be expressed as

$$
I = \int_{-1}^{1} F(\xi)\,d\xi = \sum_{i=1}^{n} H_i F(a_i) \tag{10.12}
$$

in which n is the total number of sampling points, and H_i are the weighting coefficients corresponding to sampling points located at $\xi = \pm a_i$ as given in Table 10.3.

The form of Gaussian integral shown in Equation 10.12 is rarely seen in practice. A transformation of coordinate is required to convert the general form of integration such as shown in Equation 10.6 to the form shown in Equation 10.12, as illustrated in Figure 10.15.

The transformation of coordinates from $y(x)$ in the x-coordinate to the function $F(\xi)$ in the coordinate ξ may be accomplished using the relationship

$$
x = \frac{1}{2}(x_b - x_a)\xi + \frac{1}{2}(x_b + x_a) \tag{10.13}
$$

which leads to the following expression:

$$
\int_{x_a}^{x_b} y(x)\,dx = \frac{x_b - x_a}{2} \int_{-1}^{1} F(\xi)\,d\xi \tag{10.14}
$$

Table 10.3 Weight coefficients of the Gaussian quadrature formula in Equation 10.12 (Kreyszig, 2011; Zwillinger, 2003).

n	$\pm a_i$	H_i
2	$a_1 = 0.577\ 35$	$H_1 = 1.000\ 00$
	$a_2 = -0.577\ 35$	$H_2 = 1.000\ 00$
3	$a_1 = 0.0$	$H_1 = 0.888\ 88$
	$a_2 = 0.774\ 59$	$H_2 = 0.555\ 55$
	$a_3 = -0.774\ 59$	$H_3 = 0.555\ 55$
4	$a_1 = 0.339\ 98$	$H_1 = 0.652\ 14$
	$a_2 = -0.339\ 98$	$H_2 = 0.652\ 14$
	$a_3 = 0.861\ 13$	$H_3 = 0.347\ 85$
	$a_4 = -0.861\ 13$	$H_4 = 0.347\ 85$
5	$a_1 = 0.0$	$H_1 = 0.568\ 88$
	$a_2 = 0.538\ 46$	$H_2 = 0.478\ 62$
	$a_3 = -0.538\ 46$	$H_3 = 0.478\ 62$
	$a_4 = 0.906\ 17$	$H_4 = 0.236\ 92$
	$a_5 = -0.906\ 17$	$H_5 = 0.236\ 92$
6	$a_1 = 0.238\ 61$	$H_1 = 0.467\ 91$
	$a_2 = -0.238\ 61$	$H_2 = 0.467\ 91$
	$a_3 = 0.661\ 20$	$H_3 = 0.360\ 76$
	$a_4 = -0.661\ 20$	$H_4 = 0.360\ 76$
	$a_5 = 0.932\ 46$	$H_5 = 0.171\ 32$
	$a_6 = -0.932\ 46$	$H_6 = 0.171\ 32$
7	$a_1 = 0.0$	$H_1 = 0.417\ 95$
	$a_2 = 0.405\ 84$	$H_2 = 0.381\ 83$
	$a_3 = -0.405\ 84$	$H_3 = 0.381\ 83$
	$a_4 = 0.741\ 53$	$H_4 = 0.279\ 70$
	$a_5 = -0.741\ 53$	$H_5 = 0.279\ 70$
	$a_6 = 0.949\ 10$	$H_6 = 0.129\ 48$
	$a_7 = -0.949\ 10$	$H_7 = 0.129\ 48$
8	$a_1 = 0.183\ 43$	$H_1 = 0.362\ 68$
	$a_2 = -0.183\ 43$	$H_2 = 0.362\ 68$
	$a_3 = 0.525\ 53$	$H_3 = 0.313\ 70$
	$a_4 = -0.525\ 53$	$H_4 = 0.313\ 70$
	$a_5 = 0.796\ 66$	$H_5 = 0.222\ 38$
	$a_6 = -0.796\ 66$	$H_6 = 0.222\ 38$
	$a_7 = 0.960\ 28$	$H_7 = 0.101\ 22$
	$a_8 = -0.960\ 28$	$H_8 = 0.101\ 22$

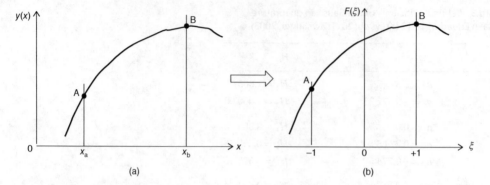

Figure 10.15 Transformation of coordinates for Gaussian integration. (a) With the original coordinates. (b) After transformation of coordinates.

with

$$F(\xi) = y(x)|_{x=\frac{1}{2}(x_b-x_a)\xi+\frac{1}{2}(x_b-x_a)}$$

and

$$d\xi = \left[\frac{1}{2}(x_b - x_a)\right]^{-1} dx$$

We will obtain the expression for the required evaluation of the integral in Equation 10.6 using Gaussian quadrature by substituting the relationship in Equation 10.12 into Equation 10.14.

$$I = \int_{x_a}^{x_b} y(x)\,dx = \frac{x_b - x_a}{2} \sum_{i=1}^{n} H_i F(a_i) \tag{10.15}$$

Numerical values of the weight coefficients H_i with sampling points a_i are given in Table 10.3.

One needs to realize that the term $F(a_i)$ on the right-hand-side of Equation 10.15 denotes the function $y(x)$ evaluated at the sampling points a_i after it has been transformed to the function $F(\xi)$ with integration limits of $(-1$ to $+1)$ as in Figure 10.15b.

Example 10.9

Evaluate the following integral using Gaussian quadrature as shown in Equation 10.14:

$$I = \int_0^{\pi} \cos x\,dx \tag{a}$$

Solution:

We have the function $y(x) = \cos x$ over the integration limits $x_a = 0$ and $x_b = \pi$. The transformation of coordinates makes use of the relationship $x = (\pi/2)\xi + \pi/2$ from Equation 10.13, from which we obtain $dx = (\pi/2)\,d\xi$, and the functions are related as

$$y(x) = \cos x = F(\xi)$$
$$= \cos\left(\frac{\pi}{2}\xi + \frac{\pi}{2}\right)$$
$$= \sin\left(\frac{\pi}{2}\xi\right)$$

from Equation 10.14 with the use of the trigonometric relationships

$$\sin\left(\frac{\pi}{2} + \theta\right) = \cos\theta \quad \text{and} \quad \cos\left(\frac{\pi}{2} + \theta\right) = \sin\theta$$

We thus arrive at the following expression for integral I in Equation (a) using Gaussian quadrature:

$$I = \int_0^\pi \cos x \, dx = \int_{-1}^1 \left[\sin\left(\frac{\pi}{2}\xi\right) \left(\frac{\pi}{2} d\xi\right) \right]$$

$$= \frac{\pi}{2} \int_{-1}^1 \sin \frac{\pi}{2}\xi \, d\xi$$

$$= \frac{\pi}{2} \sum_{i=1}^n H_i \sin\left(\frac{\pi}{2}a_i\right) \tag{b}$$

Let us take, for example, three sampling points (i.e., $n = 3$ from Table 10.3):

$$a_1 = 0 \qquad a_2 = +0.77459 \qquad a_3 = -0.77459$$
$$H_1 = 0.88888 \qquad H_2 = 0.55555 \qquad H_3 = 0.55555$$

Substituting the above values into Equation (b) gives

$$I = \frac{\pi}{2} \left[0.88888 \sin(0) + 0.55555 \sin\left(\frac{\pi}{2} \times 0.77459\right) + 0.55555 \sin\left(-\frac{\pi}{2} \times 0.77459\right) \right]$$

$$= \frac{\pi}{2}[0.55555 \sin(1.2167) - 0.55555 \sin(1.2167)] = 0 \tag{c}$$

Example 10.10

Evaluate the following integral using the Gaussian quadrature as in Equation 10.15:

$$I = 2 \int_0^\pi \sin x \, dx \tag{a}$$

Solution:

The integral in Equation (a) is similar to that in Example 10.9 but with $y(x) = \sin x$. We may thus use a similar derivation to obtain the value of the integral in Equation (a) as

$$I = 2 \int_0^\pi \sin x \, dx$$

$$= 2 \int_{-1}^1 \left[\cos\left(\frac{\pi}{2}\xi\right) \left(\frac{\pi}{2} d\xi\right) \right]$$

$$= 2 \times \frac{\pi}{2} \int_{-1}^1 \cos \frac{\pi}{2}\xi \, d\xi$$

$$= \pi \sum_{i=1}^n H_i \cos\left(\frac{\pi}{2}a_i\right) \tag{b}$$

and the result of using the same three sampling points with $n = 3$ leads to a similar expression to that in Equation (c) in Example 10.9:

$$I = 2 \times \frac{\pi}{2} \left[0.88888 \cos(0) + 0.55555 \cos\left(\frac{\pi}{2} \times 0.77459\right) \right.$$

$$\left. + 0.55555 \cos\left(-\frac{\pi}{2} \times 0.77459\right) \right]$$

$$= \pi[0.88888 + 0.55555 \cos(1.2167) + 0.55555 \cos(-1.2167)] = 4.002497$$

The exact solution is 4.0 for the integral in Equation (a).

Example 10.11

Evaluate the following integral from Example 10.8 using the Gaussian quadrature method:

$$I = \int_{0.5}^{3.5} y(x)\, dx = \int_{0.5}^{3.5} x\sqrt{(16 - x^2)^3}\, dx \qquad (a)$$

Solution:

The function $y(x)$ in the integral in Equation (a) is

$$y(x) = x\sqrt{(16 - x^2)^3} \qquad (b)$$

with integration limits $x_a = 0.5$ and $x_b = 3.5$.

We will first derive the expression of the function $F(\xi)$ for the function $y(x)$ in Equation (a) from the integration limits of $(0.5, 3.5)$ to the limit $(-1, +1)$ as required in Gaussian quadrature.

The transformation of coordinate systems as illustrated in Figure 10.15 begins with the transformation of the variable from x to ξ using the relationship in Equation 10.13, leading to the following relationship between the variables x and ξ:

$$x = \frac{1}{2}(x_b - x_a)\xi + \frac{1}{2}(x_b + x_a) = 1.5\xi + 2$$

from which we obtain

$$dx = 1.5\, d\xi \qquad (c)$$

The integral in Equation (a) in the $y(x)$ vs. x coordinates can thus be transformed to the $F(\xi)$ vs. ξ coordinates by using Equation 10.14, yielding the following expression:

$$I = \int_{0.5}^{3.5} y(x)\, dx = \int_{0.5}^{3.5} x\sqrt{(16 - x^2)^3}\, dx$$

$$= \frac{x_b - x_a}{2} \int_{-1}^{1} F(\xi)\, d\xi$$

$$= \frac{3.5 - 0.5}{2} \int_{-1}^{1} y(x)\big|_{x=1.5\xi+2}\, d\xi \qquad (d)$$

We may thus evaluate the integral from Equation (d) as

$$I = 1.5 \int_{-1}^{1} (1.5\xi + 2)\sqrt{[16 - (1.5\xi + 2)^2]^3}\, d\xi$$

$$= 1.5 \int_{-1}^{1} (1.5\xi + 2)\sqrt{(-2.25\xi^2 - 6\xi + 12)^3}\, d\xi$$

from which we have the function

$$F(\xi) = (1.5\xi + 2)\sqrt{(-2.25\xi^2 - 6\xi + 12)^3} \qquad (e)$$

We may then use Equation 10.12 for the value of the integral I in Equation (a):

$$I = 1.5 \sum_{i=1}^{n} H_i F(a_i) \qquad (f)$$

Let us choose three sampling points for the integral in Equation (f), i.e., $n = 3$ in Equation (e) with a_i and H_i ($i = 1, 2, 3$) from Table 10.3, as tabulated below:

i	a_i	H_i
1	0	0.888 88
2	0.774 59	0.555 55
3	−0.774 59	0.555 55

By substituting the above numbers into Equation (f), we will obtain the value of the integral I as

$$I = 1.5[H_1 F(0) + H_2 F(0.7746) + H_3 F(-0.7746)] \tag{g}$$

We may evaluate the function values at a_i using Equation (e) as

$$F(0) = 83.1384, \quad F(0.774\,59) = 46.4981, \; F(-0.774\,59) = 50.1454 \tag{10.16}$$

Substituting these function values into Equation (f) will give the integral I in Equation (a) as

$$I = \int_{0.5}^{3.5} y(x)\,dx = \int_{0.5}^{3.5} x\sqrt{(16 - x^2)^3}\,dx$$
$$= 1.5(0.8888 \times 83.1384 + 0.5555 \times 46.4981 + 0.5555 \times 50.1454)$$
$$= 191.3684$$

This value of the integral in Equation (a) obtained by Gaussian quadrature with three sampling points is remarkably close to the exact value of 191.45 obtained from a table of integrals (Zwillinger, 2003).

10.5 Numerical Methods for Solving Differential Equations

Differential equations frequently appear in engineering analysis, as described in Section 2.5. Many differential equations that engineers use for their analyses are "linear equations," such as those presented in Chapters 7, 8, and 9, which can be solved by classical solution methods. There are, however, occasions in which engineers need to solve either highly complicated linear differential equations or nonlinear differential equations; in such cases numerical solution methods become viable alternative methods for finding solutions.

Numerical solution methods for differential equations relating to two types of engineering analysis problems: (1) "initial value" problems, and (2) "boundary value" problems. Solution of initial value problems involves a starting point with the variable of the function, say at x_0, that is a specific value of variable x for solution $y(x)$. With the solution given at this starting point, one may find the solutions at $x = x_0+h, x_0+2h, x_0+3h, \ldots, x_0+nh$, where h is the selected "step size" in the numerical computations and n is an integer number of steps used in the analysis. The number of steps n in the computation can be as large as is required to cover the entire range of the variable in the analysis. Numerical solution to boundary value problems is more complicated; function values are often specified at a number of variable values, and the selected steps for solution values may be restricted by the specified values at these variable points.

There are numerous numerical solution methods available for the solutions of differential equations. It is not possible to cover all these solution methods in this section. What we will learn from this section is the principle of converting "differential equations" to "difference equations," followed by numerical computation of solutions at discrete times or locations of the domain in the analysis. We will also confine our coverage to selected numerical solution methods involving only initial value problems. Readers are referred to reference books that have extensive coverage of numerical solutions for differential equations of both ordinary and partial differential equations, such as Ferziger (1998) and Chapra (2012).

10.5.1 The Principle of Finite Difference

We have learned in Chapter 2 that differential equations are equations that involve derivatives. Physically, a derivative represents the rate of change of a physical quantity represented by a function with respect to the change of its variable(s).

Referring to Figure 10.16, we have a continuous function $f(x)$ that has values f_{i-1}, f_i, and f_{i+1} corresponding to the three values of its variable x at x_{i-1}, x_i, and x_{i+1}, respectively. We may also write the three function values at the three x-values as

$$f_i = f(x) \tag{10.16a}$$

$$f_{i+1} = f(x + \Delta x) \tag{10.16b}$$

$$f_{i-1} = f(x - \Delta x) \tag{10.16c}$$

The derivative of the function $f(x)$ at point A with $x = x_i$ in Figure 10.16 is graphically represented by the tangent line A″–A′ to the curve representing function $f(x)$ at point A. Mathematically, we may express the derivative as given in Equation 2.9, or in the form

$$\frac{df(x)}{dx} = \lim_{\Delta x \to 0} \frac{\Delta f}{\Delta x} \approx \frac{\Delta f}{\Delta x} \tag{10.17}$$

where Δx is the increment used to change the values of the variable x, and dx is the increments of variable x with infinitesimally small sizes.

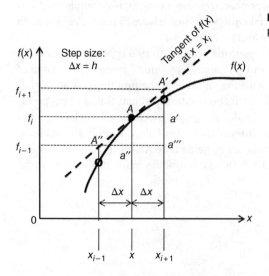

Figure 10.16 Function $f(x)$ evaluated at three positions.

One may observe from Equation 10.17 the important relation that the derivative may be approximated by finite increments of Δf and Δx as indicated in Equation 10.18:

$$\frac{df(x)}{dx} \approx \frac{\Delta f}{\Delta x} \tag{10.18}$$

if the condition on the increment $\Delta x \to 0$ is removed in the evaluation of the rate of change of the function $f(x)$ in Equation 10.17.

Thus from Equation 10.17 we see that derivatives of functions may be approximated by adopting finite—but not infinitesimally small—increments of the variable x. A formulation with such an approximation is called "finite difference".

10.5.2 The Three Basic Finite-difference Schemes

There are three basic schemes that one may use to approximate a derivative: (1) the "forward difference" scheme; (2) the "backward difference" scheme; and (3) the "central difference" scheme. Mathematical expressions of these difference schemes are given below.

The forward difference scheme

In the "forward difference scheme," the rate of change of the function $f(x)$ with respect to the variable x is accounted for between the function value at the current value $x = x_i$ and the value of the same function at the next step, that is, $x_{i+1} = x + \Delta x$ in the triangle $\Delta A'Aa'$ in Figure 10.16. The mathematical expression of this scheme is given in Equation 10.19:

$$\begin{aligned}
\nabla f_i = \frac{df(x)}{dx}\Big|_{x=x_i} &\approx \frac{\Delta f}{\Delta x}\Big|_{x=x_i} \\
&= \frac{f_{i+1} - f_i}{x_{i+1} - x_i} \\
&= \frac{f_{i+1} - f_i}{\Delta x} \\
&= \frac{f_{i+1} - f_i}{h}
\end{aligned} \tag{10.19}$$

in which $h = \Delta x$ is the "step size."

The derivative of the function $f(x)$ at other values of the variable x in the positive direction can be expressed following Equation 10.19 as

$$\nabla f_{i+1} = \frac{f_{i+2} - f_{i+1}}{h}$$
$$\nabla f_{i+2} = \frac{f_{i+3} - f_{i+2}}{h}$$
and so on $\tag{10.20}$

The second-order derivative of the function $f(x)$ at x can be obtained according to the following procedure:

$$\begin{aligned}
\nabla^2 f_i = \frac{d}{dx}\left(\frac{df(x)}{dx}\right)\Big|_{x=x_i} &= \lim_{\Delta x \to 0} \frac{\nabla f_{i+1} - \nabla f_i}{\Delta x} \\
&\approx \frac{\nabla f_{i+1} - \nabla f_i}{\Delta x} = \frac{\nabla f_{i+1} - \nabla f_i}{h} \\
&= \frac{\frac{f_{i+2} - f_{i+1}}{h} - \frac{f_{i+1} - f_i}{h}}{h} \\
&= \frac{f_{i+2} - 2f_{i+1} + f_i}{h^2}
\end{aligned} \tag{10.21}$$

The backward difference scheme

In this difference scheme, the rate of the change of the function with respect to the variable x is accounted for between the current value at $x = x_i$ and the step backward, that is, $x_{i-1} = x - \Delta x$ in the triangle $\triangle AA''a''$ in Figure 10.16. The mathematical expression of this scheme is given in Equation 10.22:

$$\nabla f_i = \lim_{\Delta x \to 0} \frac{f_i - f_{i-1}}{\Delta x}$$

$$= \lim_{\Delta x \to 0} \frac{f(x) - f(x - \Delta x)}{\Delta x}$$

$$\approx \frac{f_i - f_{i-1}}{\Delta x}$$

$$= \frac{f_i - f_{i-1}}{h} \tag{10.22}$$

Following a similar procedure as in the forward difference scheme, we may express the second-order derivative in the following form:

$$\nabla^2 f_i \approx \frac{f_i - 2f_{i-1} + f_{i-2}}{h^2} \tag{10.23}$$

The central difference scheme:

The rate of change of function $f(x)$ in this finite-difference scheme includes the function values between the preceding step at $(x - \Delta x)$ and the step ahead, that is, $(x + \Delta x)$. The triangle involved in this difference scheme is $\triangle A'A''a'''$ in Figure 10.16. We have the first-order derivative as in Equation 10.24:

$$\nabla f \approx \frac{f_{i+1} - f_{i-1}}{x_{i+1} - x_{i-1}} = \frac{f_{i+1} - f_{i-1}}{2h} \tag{10.24}$$

Note that in this finite-difference scheme a much larger step of size $2h$ is used in the first-order derivative as given in Equation 10.24. These "coarse" steps will compromise the accuracy of the values of the derivatives. A better central difference scheme is to employ for "half" steps in both directions. In other words, if we define

$$f_{i+\frac{1}{2}} = f\left(x + \frac{\Delta x}{2}\right) \qquad \text{and} \qquad f_{i-\frac{1}{2}} = f\left(x - \frac{\Delta x}{2}\right) \tag{10.25}$$

we will then have the derivative of the function $f(x)$ using this modified central difference scheme as

$$\nabla f_i \approx \frac{f_{i+\frac{1}{2}} - f_{i-\frac{1}{2}}}{h} \tag{10.26}$$

Example 10.12

Solve the following differential equation using the finite-difference method.

$$\frac{d^2 x(t)}{dt^2} + x(t) = 0 \tag{a}$$

with specified initial conditions

$$x(0) = 1 \tag{b}$$

$$\dot{x}(0) = 0 \tag{c}$$

Solution:

Let us use the *forward difference scheme* according to Equations 10.19 and 10.21 with

$$\frac{dx(t)}{dt} \approx \frac{x(t + \Delta t) - x(t)}{\Delta t} \tag{d}$$

and

$$\frac{d^2x(t)}{dt^2} \approx \frac{x(t + 2\Delta t) - 2x(t + \Delta t) + x(t)}{(\Delta t)^2} \tag{e}$$

Substitution of Equation (e) into Equation (a) results in the following finite-difference form of the differential equation:

$$\frac{x(t + 2\Delta t) - 2x(t + \Delta t) + x(t)}{(\Delta t)^2} + x(t) = 0$$

Upon rearranging the terms in the above equation, we get the following "recurrence relation" for the approximate solution of Equation (a):

$$x(t + 2\Delta t) - 2x(t + \Delta t) + [1 + (\Delta t)^2]x(t) = 0 \tag{f}$$

Applying the finite-difference operator to the initial conditions in Equations (b) and (c) gives

$$x(0) = 1 \tag{g}$$
$$x(\Delta t) = x(0) = 1 \tag{h}$$

We are now ready to solve for $x(t)$ in Equation (a) using the finite-difference operator by repeated use of the recurrence relation in Equation (f). The solution of $x(t)$ will be on the incremental steps of Δt chosen by the user. By referring to the initial conditions in Equations (g) and (h) we may obtain $x(t)$ at all subsequent steps.

Choice of solution steps, Δt:

Let us assume that a step size $\Delta t = 0.05$ is chosen for the solution. The corresponding solution becomes, from Equation (f) with $\Delta t = 0.05$:

$$x(t + 0.1) - 2x(t + 0.05) + 1.0025x(t) = 0 \tag{j}$$

and from Equation (h):

$$x(0.05) = 1$$

Now, at $t = 0$, we get from Equation (j):

$$x(0.1) - 2x(0.05) + 1.0025x(0) = 0$$

Because $x(0) = 1$ is the initial condition in Equation (g), the above relation yields

$$x(0 + 0.1) - 2x(0 + 0.05) = -1.0025$$

But since $x(0.05) = 1$ from Equation (h), we have

$$x(0.1) = -1.0025 + 2x(0.05) = 0.9975 \tag{k}$$

We may now move to the next time point by letting $t = t + \Delta t = 0 + 0.05 = 0.05$ and substituting $t = 0.05$ into Equation (f) to get

$$x(0.05 + 0.1) - 2x(0.05 + 0.05) + 1.0025x(0.05) = 0$$

or

$$x(0.15) - 2x(0.1) + 1.0025x(0.05) = 0 \tag{m}$$

Since we have already obtained $x(0.1) = 0.9975$ from Equation (k) and $x(0.05) = 1$ from Equation (h), we will thus have another solution point from Equation (m):

$$x(0.15) = 2 \times 0.9975 - 1.0025 \times 1 = 0.9925 \tag{p}$$

We move to the next time point, that is, $t = t + \Delta t = 0.05 + 0.05 = 0.1$. Substituting $t = 0.1$ into Equation (f) we get

$$x(0.1 + 0.1) - 2x(0.1 + 0.05) + 1.0025x(0.1) = 0$$

or

$$x(0.2) - 2x(0.15) + 1.0025x(0.1) = 0$$

But with $x(0.15) = 0.9925$ from the last step in Equation (p) and $x(0.1) = 0.9975$ from Equation (k), we have

$$x(0.2) = 2 \times 0.9925 - 1.0025 \times 0.9975 = 0.9850 \tag{q}$$

Thus, following the same procedure as illustrated above, we may obtain solution of Equation (a) at all time points with an increment $\Delta t = 0.05$.

The results obtained from the above exercise are summarized in the tabulation below, with comparison to the exact solution of $x(t) = \cos\ t$.

Variable, t	Solution by the finite-difference method	Exact solution	% Error
0	1	1	0
0.05	1	0.999 996	≈ 0
0.10	0.9975	0.995 0041	0.25
0.15	0.9925	0.988 77	0.38
0.20	0.9850	0.980 066	0.503

One will observe from the tabulated values that the percentage error of the results obtained from the finite-difference method increases with the increase of the variable t. *One may also show that the accuracy of the finite-difference method improves with smaller increments of the variables—the Δt in the present example.*

10.5.3 Finite-difference Formulation for Partial Derivatives

Partial derivatives along a single dimension are computed in the same fashion as for ordinary derivatives (Chapra 2012) illustrated in Section 10.5.2. For example, the central difference scheme for function $f(x,y)$ can be shown to have the following expressions:

$$\frac{\partial f(x,y)}{\partial x} \approx \frac{f(x + \Delta x, y) - f(x - \Delta x, y)}{2\Delta x} \tag{10.27a}$$

$$\frac{\partial f(x,y)}{\partial y} \approx \frac{f(x, y + \Delta y) - f(x, y - \Delta y)}{2\Delta y} \tag{10.27b}$$

The error of the above approximation is of order $\Delta x = h$.

For higher-order partial derivatives, such as $\partial^2 f(x, y)/\partial x \partial y$:

$$\frac{\partial^2 f(x, y)}{\partial x \partial y} = \frac{\partial}{\partial x}\left[\frac{\partial f(x, y)}{\partial y}\right]$$

$$\approx \frac{\dfrac{\partial f(x + \Delta x, y)}{\partial y} - \dfrac{\partial f(x - \Delta x, y)}{\partial y}}{2\Delta x} \tag{10.28}$$

Evaluating each of the partial derivatives in Equation 10.28 will lead to the following final expression:

$$\frac{\partial^2 f(x, y)}{\partial x \partial y} \approx \frac{f(x + \Delta x, y + \Delta y) - f(x + \Delta x, y - \Delta y)}{4\Delta x \Delta y}$$

$$+ \frac{-f(x - \Delta x, y + \Delta y) + f(x - \Delta x, y - \Delta y)}{4\Delta x \Delta y} \tag{10.29}$$

The error of the approximation of second-order differentiation is of order $(\Delta x)^2$.

10.5.4 Numerical Solution of Differential Equations

There are a number of numerical solution methods available for solving differential equations relating to both types of initial value and boundary value problems (Hoffman, 1992; Bronson, 1994; Ferziger, 1998; Chapra, 2012). We will present only the expressions used in the classic fourth-order Runge–Kutta method to illustrate the power of numerical methods for solving differential equations for initial value problems.

The simplest method of solving differential equations is to convert the derivatives in differential equations to the forms of "finite differences" as presented in Section 10.5.2 and Example 10.12. This method is straightforward but usually has significant accumulation of errors in the solution, as indicated in the numerical illustration in Example 10.12. There are several alternative versions available for solutions with better accuracies in the references cited above. We will include the Runge–Kutta method in this section for numerical solution of differential equations using initial value processes.

The Runge–Kutta methods are integrative methods for approximation of solutions of differential equations. These method, with several versions of the technique, were developed around 1900s by German mathematicians C. Runge and M.W. Kutta. The essence of the Runge–Kutta methods involves numerical integration of the function in a differential equation by using a trial step at mid-point of an interval—within a step Δx or h—using numerical integration techniques such as the trapezoidal or Simpson's rules as presented in Section 10.4. The numerical integrations will allow the cancellation of low-order error terms for more accurate solutions. Several versions of Runge–Kutta methods with different orders for solving differential equations have been developed over the years.

10.5.4.1 The Second-order Runge–Kutta Method

This is the simplest form of the Runge–Kutta method, with the formulation for the solution of first-order differential equation in the following form:

$$y'(x) = f(x, y) \tag{10.30}$$

with a specified solution point corresponding to one specific condition for Equation 10.30. The solution points of this differential equation can be expressed as

$$y_{i+1} = y_i + k_2 + O(h^3) \tag{10.31}$$

where $O(h^3)$ is the order of error of the step h^3, and

$$k_2 = h\left(x_i + \frac{1}{2}h, \; y_i + \frac{1}{2}k_1\right) \tag{10.32a}$$

$$k_1 = hf(x_i, y_i) \tag{10.32b}$$

Example 10.13
Use the second-order Runge–Kutta method shown in Equations 10.31 and 10.32a,b to solve the following first-order ordinary differential equation similar to that in Example 7.4:

$$\frac{dy(x)}{dx} + 2y(x) = 2 \tag{a}$$

with a specified condition $y(0) = 2$.

We will solve Equation (a) with condition $y(0) = 2$ in Example 7.4 with an exact solution of $y(x) = 1 + e^{-2x}$.

Solution:
Let us first rearrange Equation (a) in the form

$$\frac{dy(x)}{dx} = y'(x) = 2 - 2y(x) = f(x, y)$$

from which we have

$$f(x, y) = 2 - 2y \tag{b}$$

and the specified solution point $y(0) = y_0 = 2$.

We are now ready to determine the first solution point using Equations 10.30 to 10.32a,b.

Step 1: With $i = 0$ and selected increment $h = 0.1$

$$y_1 \approx y_0 + k_2$$

$$k_1 = 0.1(2 - 2y_0) = 0.1(2 - 2 \times 2) = -0.2$$

$$k_2 = 0.1f\left(x_0 + \frac{h}{2}, \; y_0 + \frac{1}{2}k_1\right)$$

$$= 0.1\left[2 - 2\left(2 + \frac{-0.2}{2}\right)\right] = -0.18$$

We thus have a solution point

$$y_1 = y_0 + k_2 = 2 - 0.18 = 1.82$$

(the exact solution is $y_1 = 1.8187$). We may move the solution point forward to $i = 1$, $h = 0.1$ with $y_1 = 1.82$.

We should have the solution point: $y_2 = y_1 + k_2$ as in Equation 10.31, with

$$k_1 = hf(x_1, y_1) = 0.1(2 - 2y_1)$$

$$= 0.1(2 - 2 \times 1.82) = 0.164$$

$$k_2 = hf\left(x_1 + \frac{1}{2}h, \; y_1 + \frac{1}{2}k_1\right)$$

$$= 0.1\left[2 - 2\left(1.82 + \frac{1}{2}0.164\right)\right] = -0.1804$$

Hence the second solution point y_2 is

$$y_2 = y_1 + k_2 = 1.82 - 0.1804 = 1.6396$$

(the exact solution is $y_2 = 1.67$). We observe that the error of numerical solutions accumulates from 0.07% for y_1 to 1.82% for y_2.

10.5.4.2 The Fourth-order Runge–Kutta Method

This is the most popular version of the Runge–Kutta method for solving differential equations in initial value problems. Formulation of this solution method is similar to that of the second-order method.

The differential equation is similar to that shown in Equation 10.30:

$$y'(x) = f(x, y)$$

with the solution point given by the following formula:

$$y_{i+1} = y_i + \frac{h}{6}(k_1 + 2k_2 + 2k_3 + k_4) \tag{10.33}$$

where

$$k_1 = f(x_i, y_i) \tag{10.34a}$$

$$k_2 = f\left(x_i + \frac{h}{2}, y_i + \frac{k_1 h}{2}\right) \tag{10.34b}$$

$$k_3 = f\left(x_i + \frac{h}{2}, y_i + \frac{k_2 h}{2}\right) \tag{10.34c}$$

$$k_4 = f(x_i + h, y_i + k_3 h) \tag{10.34d}$$

Example 10.14

Use the Runge–Kutta fourth-order method to solve the same differential equation as in Example 10.13 but only for the second solution point y_2.

Solution:

The differential equation we will solve is

$$\frac{dy(x)}{dx} + 2y(x) = 2$$

with the given condition $y(0) = 2$.

Translating this equation into the form of the differential equation using the Runge–Kutta solution method will give $y'(x) = f(x,y)$ in Equation 10.30 with

$$f(x, y) = 2 - 2y \tag{a}$$

and $y_0 = 2$.

Example 10.13 has already solved the first solution point with $y_1 = 1.82$ with a chosen step size of $h = 0.1$. We are required to find the next solution point at y_2 with $i = 1$ and $h = 0.1$.

Using the expression in Equation 10.33 for the fourth-order Runge–Kutta method, we may express the solution at solution point 2 for the differential equation in Example 10.13 as

$$y_2 = y_1 + \frac{0.1}{6}(k_1 + 2k_2 + 2k_3 + k_4) \tag{b}$$

We may obtain the constants k_1, k_2, k_3, and k_4 in Equation (b) by using Equations 10.34a, b, c, and d, respectively:

$$k_1 = (x_1, y_1) = 2 - 2y_1 = 2 - 2 \times 1.82 = -1.64 \tag{c}$$

$$k_2 = f\left(x_1 + \frac{h}{2}, y_1 + \frac{k_1 h}{2}\right) = 2 - 2\left[1.82 + \frac{(-1.64)0.1}{2}\right] = -1.476 \tag{d}$$

$$k_3 = f\left(x_1 + \frac{h}{2}, \ y_1 + \frac{k_2 h}{2}\right) = 2 - 2\left[1.82 + \frac{(-1.476)0.1}{2}\right] = -1.4924 \tag{e}$$

$$k_4 = f\left(x_1 + \frac{h}{2}, y_1 + k_3 h\right) = 2 - 2(1.82 - 1.4924 \times 0.1) = -1.34152 \tag{f}$$

We will evaluate the solution y_2 by substituting the constants k_1, k_2, k_3, and k_4 in Equations (c), (d), (e), and (f) into Equation (b), resulting in

$$y_2 = 1.82 + \frac{0.1}{6}[-1.64 + 2(-1.476) + 2(-1.4924) + (-1.34152)] = 1.6714$$

We notice that the solution y_2 using the fourth-order Runge–Kutta method is remarkably close to the exact solution of $y_2 = 1.67$. This is a much more accurate result than that obtained using the second-order Runge–Kutta method as illustrated in Example 10.13.

10.5.4.3   Runge–Kutta Method for Higher-order Differential Equations

We have seen that Runge–Kutta method can solve differential equations often with remarkable accuracy as demonstrated in Example 10.14. Unfortunately, most textbooks offer the application of this valuable method only for solving first-order differential equations. Its use for solving higher- order differential equations requires the conversion of higher-order differential equations to the first-order-equivalent forms such as that shown in Equation 10.30. The solution of the converted higher-order differential equations can be obtained using expressions such as that given in Equation 10.33 for the fourth-order Runge–Kutta formulation. We will present the following formulation to illustrate how the fourth-order Runge–Kutta method can be used to solve second-order ordinary differential equations (the treatment is derived from an online tutorial http://www.eng.colostate.edu/~thompson/Page/CourseMat/Tutorials/CompMethods/Rungekutta.pdf).

We will use the Runge–Kutta method to solve a second-order ordinary differential equation of the form

$$\frac{d^2 y(x)}{dx^2} = f\left(x, y, \frac{dy(x)}{dx}\right) = f(x, y, y') \tag{10.35}$$

The left-hand side of Equation 10.35 may be expressed as $dy'(x)/dx$, which thus converts the second-order differential equation in (10.35) into a first-order differential equation in the form

$$\frac{dy'(x)}{dx} = f[x, y, y'(x)] \tag{10.36a}$$

with

$$\frac{dy(x)}{dx} = y'(x) = F[x, y, y'(x)] \tag{10.36b}$$

Solution of the second order differential equation in Equation 10.35 may be obtained by the solutions of Equations 10.36a and 10.36b using the fourth-order Runge–Kutta formulations given in Equations 10.37 and Equation 10.38:

$$y_{i+1} = y_i + \frac{1}{6}(F_1 + 2F_2 + 2F_3 + F_4)h \tag{10.37}$$

and

$$y'_{i+1} = y_i + \frac{1}{6}(f_1 + 2f_2 + 2f_3 + f_4)h \tag{10.38}$$

We note that the expression in Equation 10.38 is similar to that in Equation 10.33 for the converted first-order differential equation in Equation 10.36a.

Table 10.4 Coefficients in the fourth-order Runge–Kutta method for solving second-order differential equations.

$f_1 = f(x_i, y_i, y_i')$	$F_1 = y_i'$
$f_2 = f\left(x_i + \frac{1}{2}h, y_i + \frac{1}{2}F_1h, y_i' + \frac{1}{2}f_1h\right)$	$F_2 = y_i' + \frac{1}{2}f_1h$
$f_3 = f\left(x_i + \frac{1}{2}h, y_i + \frac{1}{2}F_2h + \frac{1}{4}f_1h^2, y_i' + \frac{1}{2}f_2h\right)$	$F_3 = y_i' + \frac{1}{2}f_2h$
$f_4 = f\left(x_i + h, y_i + F_3h + \frac{1}{2}f_2h, y_i' + f_3h\right)$	$F_4 = y_i' + f_3h$

The coefficients f_1, f_2, f_3, and f_4 in Equations 10.37 and 10.38 can be obtained from the expressions given in Table 10.4

Example 10.15
Use the fourth-order Runge–Kutta method to solve the following second-order differential equation with specified conditions:

$$\frac{d^2y(x)}{dx^2} - 2\frac{dy(x)}{dx} + y(x) = x^2 - 4x + 2 \tag{a}$$

with

$$y(x)|_{x=0} = y(0) = y_0 = 0 \tag{b1}$$

$$\frac{dy(x)}{dx}\bigg|_{x=0} = y'(0) = y_0' = 0 \tag{b2}$$

Solution:
By comparing Equation (a) with Equation 10.35, we obtain the following expression for the function $f(x,y,y')$ in Equation 10.35:

$$f(x, y, y') = (x^2 - 4x + 2) - y + 2y' \tag{c}$$

with the specified conditions $y_0 = 0$ and $y'_0 = 0$ for our subsequent numerical solution of the differential equation.

We will select step sizes $h = 0.2, 0.1,$ and 0.4 for the three case illustrations using a procedure starting with the variable at $x = 0$. We begin the numerical solution for Equation (a) by letting $i = 0$ in Equation 10.37 with $x_0 = 0$ for the first solution point at $x = h = 0.2$:

We obtain the following coefficients by using the expressions for the coefficients given in Table 10.4:

$$F_1 = y_0' = 0 \quad \text{(a given condition)}$$

$$f_1 = f(x_0, y_0, y_0') = x_0^2 - 4x_0 + 2 - y_0 + 2y_0' = 0 - 0 + 2 - 0 + 2 \times 0 = 2$$

$$F_2 = y_0' + \frac{1}{2}f_1h = 0 + \frac{1}{2}2 \times 0.2 = 0.2$$

$$f_2 = f\left(x_0 + \frac{h}{2}, y_0 + \frac{1}{2}F_1h, y_0' + \frac{1}{2}f_1h\right)$$

$$= \left(x_0 + \frac{0.2}{2}\right)^2 - 4\left(x_0 + \frac{0.2}{2}\right) + 2 - \left(y_0 + \frac{1}{2} \times 0 \times 0.2\right) + 2\left(y_0' + \frac{1}{2} \times 2 \times 0.2\right)$$

$$= 2.01$$

$$F_3 = y_0' + \frac{1}{2}f_2 h = 0 + \frac{1}{2}(2.01) \times 0.2 = 0.201$$

$$f_3 = f\left(x_0 + \frac{h}{2}, y_0 + \frac{1}{2}F_2 h, y_0' + \frac{1}{2}f_2 h\right)$$

$$= \left(x_0 + \frac{0.2}{2}\right)^2 - 4\left(x_0 + \frac{0.2}{2}\right) + 2$$

$$\quad - \left(y_0 + \frac{1}{2}0.2 \times 0.2\right) + 2\left(y_0' + \frac{1}{2}(2.01) \times 0.2\right)$$

$$= 1.992$$

$$F_4 = y_0' + f_3 h = 0 + 1.992 \times 0.2 = 0.3984$$

$$f_4 = f(x_0 + h, y_0 + F_3 h + \frac{1}{2}f_2 h, y_0' + f_3 h)$$

$$= [(0+0.2)^2 - 4(0+0.2) + 2] - (0 + 0.201 \times 0.2 + \frac{1}{2} \times 2.01 \times 0.2) + 2(0 + 1.992 \times 0.2)$$

$$= 1.8078$$

We are ready to find the numerical solution of the differential equation in Equation (a) by substituting the values of f_1, f_2, and f_3 into Equation 10.37, to obtain a solution point y_1 with $i = 0$ and $h = 0.2$:

$$y_1 = y(0.2) = y_0 + y_0' \times 0.2 + \frac{1}{6}(F_1 + 2F_2 + 2F_3 + F_4)h$$

$$= 0 + 0 + \frac{1}{6}(0 + 2 \times 0.2 + 2 \times 0.201 + 0.3984) \times 0.2$$

$$= 0.4001333 \tag{d}$$

The exact solution of Equation (a) is $y(x) = x^2$, which yields an exact solution of $y(0.2) = 0.04$. The numerical solution in Equation (d) has a 0.033% error from the exact solution.

We will also use Equation 10.38 to approximate the value of the first-order derivation $y'(0.2)$ as

$$y_1' = y'(0.2) = y_0 + \frac{1}{6}(f_1 + 2f_2 + 2f_3 + f_4)h$$

$$= 0 + \frac{1}{6}(2 + 2 \times 2.01 + 2 \times 1.992 + 1.8078)0.2$$

$$= 0.3937 \tag{e}$$

One may use the same procedure to obtain the solution of Equation (a) at point $x = x + h$. For instance, the next solution point is at $x = 0.2 + h = 0.4$, or $y(0.4)$ by letting $i = 1$ with $x_1 = 0.2$, $y_1 = 0.0400133$ and $y_1' = 0.3937$.

Like all other numerical solution methods for solving differential equations, the error of the approximated solutions depends largely on the step size h chosen by the analyst. We will demonstrate the effects of the chosen increment size h for the same differential equation as in Equation (a) in two other cases with h-values of $h = 0.1$ and 0.4. The results of these two cases, together with the case $h = 0.2$, are summarized in Table 10.5.

One may observe from the above tabulation that the step size indeed has a significant effect on the accuracy of the numerical solutions of differential equations. It is no surprise that the larger step size $h = 0.4$ resulted in the largest error in the solution, whereas the smaller step size $h = 0.1$ appeared to give the closest solution to the exact solution. This latter particular step size appears to be the optimal size for the accurate solution for this particular differential equation.

Table 10.5 Solutions of a differential equation by Runge–Kutta methods with three different increment sizes.

	Case 1	Case 2	Case 3
x_0	0	0	0
h	0.2	0.1	0.4
F_1	0	0	0
F_2	0.2	0.1	0.4
F_3	0.201	0.100125	0.408
F_4	0.3984	1.999537	0.7904
f_1	2	2	2
f_2	2.01	2.0025	2.04
f_3	1.992	1.99775	1.976
f_4	1.8078	1.999537	1.9446
$y'(x)$	0.3937	0.2000625	0.79844
Approximated $y(x_0 + h)$	0.0400133	0.0100004	0.160427
Exact $y(x_0 + h)$	0.04	0.01	0.16
% Error on $y(h)$	**0.033**	**0.0042**	**0.2667**
% Error on y'(h)	**0.005**	**0.000313**	**0.195**

The three solution points at $x = 0.2, 0.1$, and 0.4 obtained by using the Runge–Kutta method appeared tedious and time-consuming. The same differential equation with the same specified conditions was solved using the software package MatLAB, with the input/output information included in Case 3 of Appendix 4. The results so obtained were remarkably accurate, with solution at the same three points being the exact values as shown tabulated above. MatLAB also offers graphical output such as that shown in Figure 10.17. An overview description of MatLAB software will be presented in Section 10.6.2.

Example 10.16
Use the fourth-order Runge–Kutta method to obtain the solution of the function $x(t)$ at $t = 0.2$ in the second-order differential equation in Example 10.12.

Solution:
The differential equation in Example 10.12 is

$$\frac{d^2x(t)}{dt^2} + x(t) = 0 \tag{a}$$

with the conditions $x(0) = 1$ and $x'(0) = 0$. Since we do not have the term $x'(t)$ in Equation (a), we will not need to evaluate F_1, F_2, F_3, and F_4 in Table 10.5.

We will use the same step size $h = 0.05$ as was used in Example 10.12. The numerical solution at the previous step at $x(0.15) = 0.9925$ will be computed in this example. Since the exact solution of Equation (a) with the specified conditions is $x(t) = \cos t$, we may obtain the first-order derivative of $x(t)$ at $x = 0.15$ as $x'(0.15) = -\sin(0.15) = -0.149\,44$. We thus have the following converted equation and conditions for the present case:

$$\frac{d^2x(t)}{dt^2} = x(t) = f(t, x, x')$$

$$\frac{dx(t)}{dt} = x'(t)$$

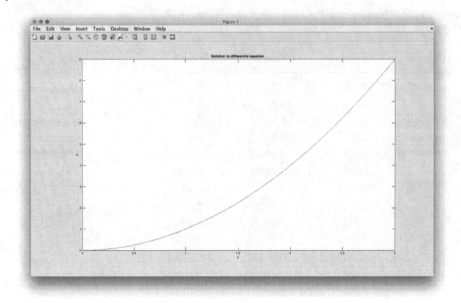

Figure 10.17 Graphical solution of a second-order ordinary differential equation by MatLAB.

and $x_0 = 0.9925$ (the solution obtained in Example 10.12) and $x'_0 = -0.14944$.

We will use the fourth-order Runge–Kutta method as shown in Equations 10.33 and 10.38 and the coefficients given in Table 10.4 to obtain the solution at $x(0.2)$ as follows. Let $i = 0$, $h = 0.05$, $x_0 = 0.9925$, and $x'_0 = -0.14944$ for the solution at $t = 0.2$, or

$$x_1 = x(0.2) = x_0 + x'_0 h + \frac{1}{6}(f_1 + f_2 + f_3)h^2 \tag{b}$$

We evaluate the coefficients f_1, f_2, f_3, and f_4 from Table 10.4 as follows:

$$f_1 = f(t_0, x_0, x'_0) = x_0 = -0.14944$$

$$f_2 = x'_0 + \frac{1}{2}f_1 h = -0.14944 + \frac{(-0.14944) \times 0.05}{2} = -0.153176$$

$$f_3 = x'_0 + \frac{1}{2}f_2 h = -0.14944 + \frac{(-0.153176) \times 0.05}{2} = -0.153694$$

$$f_4 = x'_0 + f_3 h = -0.14944 - 0.153694 \times 0.05 = -0.15710347$$

According to Equation 10.38, we have the solution

$$\begin{aligned}
x_1 = x(0.2) &= x_0 + x'_0 h + \frac{1}{6}(f_1 + f_2 + f_3)h^2 \\
&= 0.9925 + (-0.14944) \times 0.05 + \frac{1}{6}(-0.14944 - 0.153176 - 0.153694)(0.05)^2 \\
&= 0.9848
\end{aligned}$$

This numerical solution has an error of 0.48% from the exact solution, and it is more accurate than that obtained from the simple forward difference scheme in Example 10.12.

10.6 Introduction to Numerical Analysis Software Packages

We have demonstrated the power, and thus the value of numerical methods in solving many problems in engineering analysis involving nonlinear equations, integrations, and differential equations in this chapter. These methods typically require significant time and efforts in arriving at approximate, not exact, solutions of the problem, and the solutions obtained are only available at discrete solution points. More accurate solutions are obtainable with small increment step sizes but with correspondingly more computational effort.

Since almost all the numerical methods involve massive computational effort and the solutions are available only at discrete solution points, sophisticated symbolic manipulation computer packages such as the popular commercially available Mathematica and MatLAB have proven to be valuable tools for engineers in engineering analyses associated with tedious computational processes. In this section, we will briefly survey these two numerical analysis packages, in particular their capabilities in solving engineering problems. Readers are referred to the literature for some excellent references (Malek-Madani, 1998; Chapra, 2012) providing more detailed descriptions and guidance on the effective use of these packages.

10.6.1 Introduction to Mathematica

Mathematica is a computational software program based on symbolic mathematics. It is used in many scientific, engineering, mathematical, and computing fields. The programming languages used in Mathematica are the Wolfram Language by Stephen Wolfram, together with C, C++, and Java. This software package has been in the marketplace since June 1988. It is capable of handling the following major functions and features in engineering analysis:

1) Determining roots of polynomial equations of cubic or higher orders.
2) Integrating and differentiating complicated expressions. For instance, the integral of the function $\sin(x^3)$, for which $\int_0^1 \sin(x^3)\,dx = 0.2338$; in contrast, similar integrals with the simple function $\sin(x)$ but with higher power of the function in the integral, such as $\int_0^1 \sin^3(x)\,dx = 0.3063$, are available from integration tables in mathematical handbooks (Zwillinger, 2003).
3) Solving linear and nonlinear differential equations.
4) Elementary and special mathematical function libraries.
5) Matrix and data manipulation tools.
6) Numeric and symbolic tools for discrete and continuous calculus.

It can also solve the following common analytical engineering problems:

1) Determining Laplace and Fourier transforms of functions.
2) Generating graphics in two and three dimensions.
3) Simplifying trigonometric and algebraic expressions.

According to the Wolfram Language and Systems Documentation Center, Mathematica has the following features and capabilities that are of great value in advanced engineering analyses:

- Support for complex number, arbitrary precision, interval arithmetic, and symbolic computation.
- Solvers for systems of equations, Diophantine equations, ODEs, PDEs, and so on.
- Multivariate statistics libraries including fitting, hypothesis testing, and probability and expectation calculations on over 140 distributions.

- Calculations and simulations on random processes and queues.
- Computational geometry in 2D, 3D, and higher dimensions.
- Finite-element analysis including 2D and 3D adaptive mesh generation.
- Constrained and unconstrained local and global optimization.
- Toolkit for adding user interfaces to calculations and applications.
- Tools for 2D and 3D image processing and morphological image processing including image recognition.
- Tools for visualizing and analyzing directed and undirected graphs.
- Tools for combinatorics problems.
- Data mining tools such as cluster analysis, sequence alignment, and pattern matching.
- Group theory and symbolic tensor functions.
- Libraries for signal processing including wavelet analysis on sounds, images, and data.
- Linear and nonlinear control systems libraries.
- Continuous and discrete integral transforms.
- Import and export filters for data, images, video, sound, CAD, GIS, documents, and biomedical formats.
- Database collection for mathematical, scientific, and socioeconomic information and access to Wolfram alpha data and computations.
- Technical word processing including formula editing and automated report generation.
- Tools for connecting to DLL, SQL, Java, .NET, C++, Fortran, CUDA, OpenCL, and http based systems.
- Tools for parallel programming.
- Mathematica language in notebook computers when connected to the Internet.

The last of the features listed is of particular value to engineers. For example, in Example 10.3 we were required to find the root of the cubic equation $L^3 + 70.3L^2 + 1647.39L - 18656.72 = 0$. A meaningful root of this equation found by the Newton–Raphson method was $L = 8.15$ as shown in Example 10.3. A similar solution of $L = 8.1566$ was obtained by the solution method offered via the internet at the Wolfram/Alpha Widgets website (www.wolframalpha.com/widgets/) with user input of the coefficients of this equation. It offered an instant solution and with an excellent user interface feature.

10.6.2 Introduction to MATLAB

MATLAB is an acronym of "**mat**rix **lab**oratory." This numerical analysis package was designed by Cleve Moler in the late 1970s with an initial release to the public in 1984. The latest version, Version 8.6 was released in September 2015.

MATLAB provides a multiparadigm numerical computing environment and fourth-generation programming language, a proprietary programming language developed by MathWorks. It allows matrix manipulations, plotting of functions and data, implementation of algorithms, and creation of user interfaces that include interfacing with programs written in other languages, including C, C++, Java, Fortran, and Python. It is a popular numerical analysis package mainly because of it has graphics and graphical user interfacing programming capability.

Like Mathematica, MATLAB is capable of handling the following common problems in engineering analysis (Malek-Madani, 1998):

1) Finding roots of polynomials, summing series, and determining limits of sequences.
2) Symbolically integrating and differentiating complicated expressions.

3) Plotting graphics in two and three dimensions.
4) Simplify trigonometric and algebraic expressions,
5) Solving linear and nonlinear differential equations.
6) Determining the Laplace transforms of functions.

Additionally it can handle a variety of other mathematical operations.

Operation of MATLAB requires the user to input simple programs for the solution of the problems. These programs usually consists of three "windows:" (1) the "command window" for the user to enter commands and data; (2) the "graphics window" to display the results in plots and graphs; and (3) the "edit window" to create and edit the M-files, which provide alternative ways of performing operations that can expand MATLAB's problem-solving capabilities.

Detailed instructions for using MATLAB for solving a variety of mathematical problems are available in *MATLAB Primer* published by MathWorks, Inc.(see www.mathworks.com) and two excellent references (Malek-Madani, 1998 Chapra, 2012). Appendix 4 will present the procedure for the input/output (I/O) of three cases of engineering analysis using the MatLAB package:

Case 1: Graphic solution of the amplitudes with "beats" offered by the solution in Equation 8.40 for the near-resonant vibration of a metal stamping machine.

Case 2: The numerical solution with graphic representations of the amplitudes and mode shapes of a flexible rectangular pad subjected to transverse vibration as presented in Equation 9.76.

Case 3: The solution with graphic representation of a nonhomogeneous second-order differential equation.

These cases will demonstrate the value of the MatLAB software package in solving complicated engineering analysis problems.

10.7 Problems

10.1 Use the Newton–Raphson method to solve the following nonlinear equation in Example 10.1: $x^4 - 2x^3 + x^2 - 3x + 3 = 0$.

10.2 A measuring cup illustrated in Figure 10.18a has the overall dimensions shown in Figure 10.18b. (a) Determine the overall volume of the cup. (b) Derive the equation to locate the mark for the volume of 150 ml at the height L from the bottom of the cup, and use the Newton–Raphson method to solve this equation.

10.3 In Example 8.10, we derived a nonlinear equation for the amplitude of the mass attached to a spring in resonant vibration as Equation (m) in that example of the form

$$y(t) = \frac{11}{20} \sin 10t - \frac{t}{2} \cos 10t$$

where $y(t)$ is the amplitude of the vibrating mass at time t. Use the Newton–Raphson method to find the time t_e, at which the spring reaches the breaking stretching extent of 0.005 m.

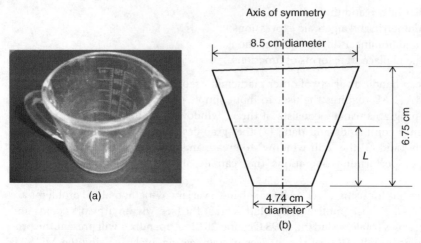

(a)

Axis of symmetry

8.5 cm diameter

4.74 cm diameter

(b)

6.75 cm

L

Figure 10.18 Design of a measuring cup. (a) A measuring cup. (b) The overall dimensions of the cup.

10.4 Use the three numerical integration methods—the trapezoidal rule, Simpson's one-third rule, and Gaussian quadrature with two-sampling points—to determine the values of the following three integrals:

a)
$$I_1 = \int_0^1 \sin x^3 \, dx$$

b)
$$I_2 = \int_0^1 \sin^3 x \, dx$$

c)
$$I_3 = \int_0^1 e^{2x} \sin^3 x \, dx$$

10.5 Use the three numerical integration methods—the trapezoidal rule, Simpson's one-third rule, and Gaussian quadrature with two-sampling points—to determine the plane area of a plate in the form of an ellipse as shown in Figure 10.19.

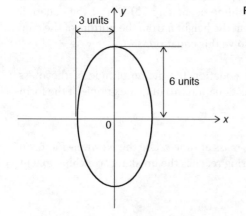

3 units

y

6 units

0

x

Figure 10.19 Plane area of an ellipse.

10.6 Derive the second-order derivative of the function $f(x)$ as shown in Equation 10.23 for the "backward difference" scheme, and also for the "central difference scheme."

10.7 Use the backward and central difference schemes to solve the differential equation in Example 10.12.

10.8 Use the forward difference scheme to solve the differential equation derived for the modeling of the vibration of a metal stamping machine in Example 8.9. Compare your results obtained from this finite-difference approximation and the exact solution given in the example.

10.9 Use both second-order and fourth-order Runge–Kutta methods to solve the first-order differential equation, Equation 7.13, for the instantaneous water level $h(t)$ in a straight-sided cylindrical tank:

$$\frac{dh(t)}{dt} = -\sqrt{2g}\left(\frac{d^2}{D^2}\right)\sqrt{h(t)}$$

with $g = 32.2$ ft/s^2, $D = 12$ inches, $d = 1$ inch, and initial water level $h(0) = h_0 = 12$ inches. Compare your result at two solution points with that from Equation 7.14.

10.10 Use the forward difference scheme to solve the following differential equation with a step size of $\Delta t = 0.1$ and with prescribed conditions $y(0) = 0$ and $y'(0) = 10$.

$$2\frac{d^2y(t)}{dt^2} + 32\frac{dy(t)}{dt} + 100y(t) = 0$$

You need to present the formulations for each solution step for a total of three steps.

10.11 Use the fourth-order Runge–Kutta method to solve the same equation in Problem 10.10 with the same step size. Compare your results obtained from these two methods.

10.12 Use the fourth-order Runge–Kutta method to solve the following differential equation in Equation (a) in Example 8.10:

$$\frac{d^2y(t)}{dt^2} + 16\frac{dy(t)}{dt} + 100y(t) = 10\sin 10t$$

with $y(0) = y_0 = 0$ and

$$\left.\frac{dy(t)}{dt}\right|_{t=0} = y'(0) = 5 \ \text{m/s}$$

Compare your result with the exact solution of $y(2) = 0.25$ cm obtained from the analytical solution of the same equation as indicated in Example 8.10.

11

Introduction to Finite-element Analysis

Chapter Learning Objectives

- Learn the principle of finite-element method for engineering analyses.
- Learn the concept of discretization of continua for approximation solutions.
- Become familiar with the steps in general finite-element analysis.
- Learn to derive interpolation functions for simplex elements.
- Learn the variational principle in deriving element equations.
- Learn to derive element equations using the Rayleigh–Ritz method and the Galerkin method.
- Understand the input/output in general finite-element analysis.
- Learn to assemble element equations to the overall stiffness equations.
- Learn to solve for primary unknown quantities from overall stiffness equations.
- Learn to relate the primary unknown quantities obtained from the finite-element method to other required secondary unknown quantities.
- Learn the use of general-purpose finite-element analysis codes adopted by industry in solving complex real-world problems.

11.1 Introduction

In Chapter 1 we learned the four stages involved in general engineering analyses, with Stage 2 being idealizing actual physical situations for the subsequent mathematical analysis in Stage 3, and Stage 4 being interpretation of analytical results. Idealization of actual physical problems is necessary in most cases of engineering analysis so that analytical tools that engineers have acquired and are available can be used. Idealization of geometry and loading and boundary conditions is frequently carried out in many engineering analyses, and in many of these cases it results in unrealistic solutions from the subsequent mathematical analysis because of possible oversimplification of the real situations in the problems. For instance, in Example 1.2 on the design analysis of a bridge structure some of the assumptions that we made were unrealistic; for example, (1) that the two steel I-beams that support the bridge are "weightless"; (2) that the weight of the bridge surface pavements is also negligible; and (3) that the weight of the passing vehicle is evenly distributed between the front and rear wheels. We made these seemingly unrealistic assumptions in our analysis in order to justify the use of the formula derived from simple beam theory in computing the maximum bending and shearing stresses given in Equations 1.2 and 1.3. This simple analysis with unrealistic idealizations would likely lead to inaccurate analytical results in the maximum bending and sharing stresses that are used to ensure the safety of the bridge structure in meeting the design objectives. Consequently, there was a need to apply

Applied Engineering Analysis, First Edition. Tai-Ran Hsu.
© 2018 John Wiley & Sons Ltd. Published 2018 by John Wiley & Sons Ltd.
Companion Website: www.wiley.com/go/hsu/applied

"safety factors" to those analytical results in Stage 4 in the interpretation of the analytical results to make up for the deficiency in the analyses that involved unrealistic idealizations of some key physical conditions.

Safety factors (SF) in structural analysis, as we learned in Section 1.6, involve "discounting" the material strength in engineering analysis. For instance, a SF of 4 used in the design analysis of unfired pressure vessels that is stipulated in the design code by the American Society of Mechanical Engineers means that the design analysis utilizes only one-quarter, or 25%, of the strength of the material. This underuse of material strength in engineering analysis results in significant increases in materials and production costs, but such a high safety factor is necessary because of the simple formula used in computing the induced stresses in the vessel and the significant simplification of the structure geometry. The time and effort to determine the required thickness of the pressure vessel become trivial in the design of these unfired pressure vessels using the simple formula available in the design code, and the induced errors in the analysis are made up by this high safety factor.

We thus appreciate that engineers need to achieve proper balance between the use of simple methods in design analyses with unrealistic idealizations to reduce the time and effort required in the analyses and sophisticated analyses with little or no idealization of the real physical conditions. The latter approach obviously requires much more time and resources; the benefit, however, is that it allows engineers to use very low safety factors and thus maximize the use of the material strength in the analysis. The finite-element method presented in this chapter offers such benefits because it is a viable analytical tool for sophisticated engineering analysis for engineers and industry.

A number of excellent books on this subject have been published since the 1970s (Zienkiewicz, 1971; Bathe and Wilson, 1976; Segerlind, 1976; Hsu, 1986; Knight, 1993; Logan, 2017). There have also been many valuable technical papers published in archive journals, conference proceedings, and monographs that deal with the theory and application of this technique, including the very first publication on the principles of the finite-element analysis (Turner *et al* 1959).

Generally speaking, there are two approaches for developing finite-element analysis. The first is to place emphasis on the accuracy of the result. This approach requires the development of sophisticated algorithms, such as for special elements with higher-order interpolation functions. Most of such work is tailored to special applications. The second approach is to develop commercially available general-purpose programs such as ANSYS, NASTRAN, ABACUS, etc.. Overview of these commercial codes will be presented in the end of this chapter.

Since the purpose of this chapter is to provide readers with an extended overview of this powerful analytical technique, it is necessary restrict coverage of the detailed derivations of the finite-element analysis. Topics such as element library, in-depth treatment of the variational principle, numerical integration schemes, convergence criteria, and so on will be omitted. These topics and detailed derivations of the relevant expressions and equations are adequately covered in the references cited above, and also in the users' manuals of a number of commercial finite-element codes.

This chapter will therefore present only the fundamental principles of the finite-element method. The essence of the discretization concept of the finite-element analysis will be described in Section 11.3. Steps involved in general finite-element analysis will be presented in such a way as to include the formulation of key equations toward the numerical solutions. Readers are expected to develop a firm grip of this technique and its application in various branches of the engineering discipline with the extended overview provided by this chapter.

Mathematical formulations for physical quantities involving multiple components in this chapter will be expressed in matrix forms with the matrix operations described in Chapter 4.

Matrices with curvy brackets ({}) will designate the column matrices representing such quantities, and those with square brackets ([]) will designate rectangular matrices. Superscript "T" of matrices designates the transpose of the same matrices.

11.2 The Principle of Finite-element Analysis

The essence of the finite-element method can be summarized in the simple phrase "divide and conquer." This phrase is often used to describe a strategy for achieving political or military control, which, of course, is not what the finite-element technique intends to achieve. However, the core strategy of the finite-element method is indeed to "divide" the original continuum of complicated geometry with infinite numbers of degrees-of-freedom (DOF) in the solutions into a number of subdivisions of the continuum with specific simple geometry termed "elements." These elements are interconnected at specific points, either on the sides of the elements and/or at the corners called "nodes" in a discretized model. "Element equations" are derived for each of the elements in the discretized model based on the appropriate physical theories and principles. An "overall structural equation" is then derived by assembling all the element equations in the discretized model, upon which the specified loading and boundary conditions of the original continuum are applied. Desired solutions on the unknown quantities are solved from these "overall structural equations" at every element and nodes using the techniques of solving simultaneous linear equations such as the Gaussian elimination method or its derivatives as presented in Section 4.7.3.

It must be appreciated that the desired solutions are available only at the finite number of elements (and nodes) in the discretized model, not everywhere in the original continuum. In other words, the finite-element method provides solutions only at the elements and nodes of the discretized continuum. It thus reduces the original infinite number of DOF associated with the original continua to a finite number of DOF after the discretization in the finite-element analysis. The concept of "divide and conquer" can thus be viewed as the fundamental principle of this method.

11.3 Steps in Finite-element Analysis

Finite-element analysis (FEA) is now used in many engineering disciplines. It is used to compute stress and temperature distributions in solid structures of complex geometries subject to complicated loading and boundary conditions. It can also be used to predict the pattern of a fluid flowing in channels of complicated geometry. The engineering behavior of solids or fluids with unusual thermophysical and mechanical properties can also be assessed using the FEA. The principles of the application of the FEA in many other engineering disciplines have been illustrated in books such as that of Desai (1979). It is not possible to establish a set of standard procedures for all the computations involved in the problems mentioned. However, as a general guideline, most finite-elements analyses follow eight steps, as will be described below.

Step 1: Discretization of the Real Structures

As already stated, discretization of continua in engineering analyses is the foundation for the formulation of the finite-element analysis. We will present the mathematical expressions that illustrate the principle of the finite-element method by dividing the continuum subjected to the actions of a system of forces, or heat fluxes, or pressures as shown in Figure 11.1a into a

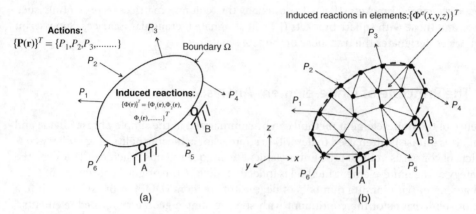

Figure 11.1 Discretization of a continuum subjected to loads. (a) Before discretization. (b) After discretization in a rectangular coordinate system.

number of "elements" of specific geometry interconnected at the nodes (shown as black dots) in Figure 11.1b.

By virtue of the actions representing by: $\{\mathbf{P}\}^{\mathrm{T}} = \{P_1, P_2, P_3, \dots\}$ in Figure 11.1a, there exist the induced reactions $\{\Phi(\mathbf{r})\}^{\mathrm{T}} = \{\Phi_1 \ \Phi_2 \ \Phi_3 \ \dots \dots\}$ (such as deflections, strains, stresses) to the applied forces or pressures $\{P(\mathbf{r})\}^{\mathrm{T}} = \{P_1 \ P_2 \ P_3 \ \dots \ \dots \ \dots\}$ in the case of a stress analysis of solid structures.

One may observe that the discretized model in Figure 11.1b offers an approximation to the geometry of the original continuum in Figure 11.1a. One noticeable difference is that the curved boundary of the original medium is now represented by cords of straight edges for the discretized medium in the finite-element analysis.

Figure 11.2 presents a number of commonly used simplex elements in finite-element analysis. *Simplex elements* are the elements whose physical quantities vary according to linear polynomial functions of the coordinates, and which have the associated nodes at the apexes of the elements.

Of the elements shown in Figure 11.2, the bar element is the simplest form of all elements. It is used in stress analysis of truss structures with long and slender members. Bar elements are also used in beam structures that have similar geometry to the truss members but are able to resist bending loads. Both triangular and quadrilateral elements are used for plate structures with planar geometry. The toroidal elements are used in FE models for axisymmetric solids such as pressure vessels and solids of axisymmetric geometry with cross-sectional areas vary along the axis of symmetry. The elements termed tetrahedral and hexahedral (the former with a geometry similar to a pyramid, but usually skewed in shape, and the latter being six-sided solids with eight apexes) are used to establish FE models for solids of complex three-dimensional geometry such as are illustrated in Figure 11.3.

A major effort in discretizing a continuum for finite-element analysis is involved in registering the discretized FE model to the FE program that has already been installed on the computer. This task requires the user to first set a coordinate system for the discretized model (we will refer to the "FE model" hereafter) and assign numbers following the chronological sequence to all elements and nodes in the model, as in the example of a tapered bar subjected to an in-plane load illustrated in Figure 11.4a. Figure 11.4b is the FE model for the tapered bar with assigned numbers for the 18 hybrid triangular and quadrilateral elements interconnected at 25 nodes.

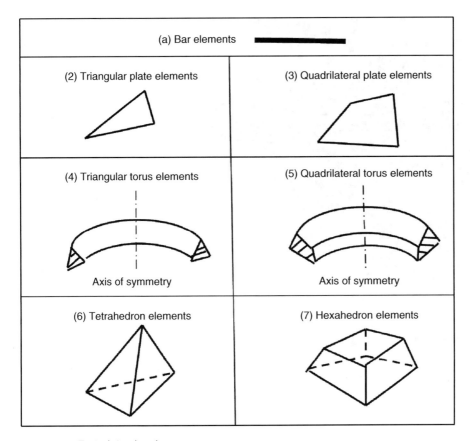

(a) Bar elements

(2) Triangular plate elements

(3) Quadrilateral plate elements

(4) Triangular torus elements

Axis of symmetry

(5) Quadrilateral torus elements

Axis of symmetry

(6) Tetrahedron elements

(7) Hexahedron elements

Figure 11.2 Typical simplex elements.

The FE model is situated in an x–y coordinate system, with node 1 being conveniently located at the origin of the coordinates. For illustrative purposes, we let the tapered bar in Figure 11.4a have dimensions of $H = 8$ cm, $L_1 = 9$ cm, $L = 18$ cm, and $r = 2$ cm.

Upon finalizing the FE model with the set coordinates for all nodes, the user needs to input the description of nodes to the FE program by entering their coordinates and constraints in displacements and the applied forces as indicated in Table 11.1. Additionally, the user needs to input the element description in terms of the associated nodes of each element in the model, as shown in Table 11.2.

One may note from Table 11.1 that not all the nodes in the FE model in Figure 11.4b have been included in the input to the FE program. For example, the coordinates of nodes 2 to 6, and nodes 9 to 13 do not appear in the table. These omissions were possible because these nodes are equally space in the x-coordinates in the FE model, and the FE program is able to interpolate the x-coordinate given the x-coordinates of the two terminal nodes: that is, node 1 and node 7 for all the in-between nodes, and node 8 and node 14 for the x-coordinates of the in-between nodes 9 to node 13. These omissions can significantly lighten the user's effort in inputting the nodal descriptions to the FE program.

The user also needs to input the element descriptions to the FE program as shown in Table 11.2. In this description, the associated nodes of all elements in the FE model need to be specified, as well as the materials of the elements. Since the FE model for the tapered bar in

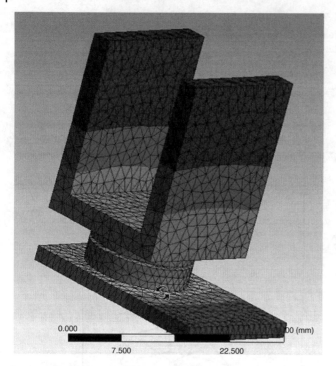

Figure 11.3 FE model for a three-dimensional solid. (Courtesy of ANSYS, Inc., Pittsburg, PA, USA)

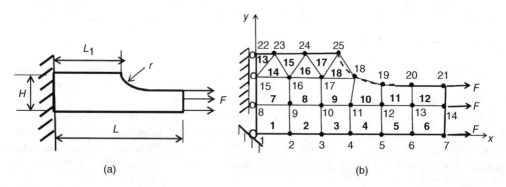

Figure 11.4 FE model for a tapered bar subjected to tensile forces. (a) The tapered bar. (b) An FE model.

Figure 11.4b consists of hybrid triangular and quadrilateral plate elements, we will include the numbers of all four nodes for the quadrilateral elements, and for all the triangular elements will repeat the number of the last node, as shown in Table 11.2. There is a rule that the numbers of the nodes for each element are presented in counterclockwise order. One will also note from Table 11.2 that descriptions of Element 2 to 5, and Element 9 to 11 are missing in the table. These were omitted because the nodes that are related to these elements are automatically provided by the built-in interpolation of the missing node numbers by the FE program, as with the description of nodes in Table 11.1.

The FE program would thus have full information for the nodes and elements in the user's FE model registered and recorded with the inputs as listed in Tables 11.1 and 11.2.

Table 11.1 Nodal description of an FE model of a tapered bar.

Node no.	x-Coordinate (cm)	y-Coordinate (cm)	Nodal constraints in displacements	Applied nodal force
1	0	0	$u_x = u_y = 0$	
7	18	0		F
8	0	3	$u_x = 0$	
14	18	3		F
15	0	6	$u_x = 0$	
17	6	6		
18	10	6		
19	12	4		
21	18	4		F
22	0	8	$u_x = 0$	
23	1.5	8		
25	9	8		

Table 11.2 Element description of an FE model of a tapered bar.

Element no.	Associated node numbers	Material characterized by input material number
1	1, 2, 9, 8	1
6	6, 7, 14, 13	1
7	6, 9, 16, 15	1
8	9, 10, 17, 16	1
12	13, 14, 21, 20	1
13	15, 23, 22, 22	1
14	15, 16, 23, 23	1
15	16, 24, 23, 23	1
16	16, 17, 24, 24	1
17	17, 25, 24, 24	1
18	17, 18, 25, 25	1

Step 2: Identify the Primary Unknown Quantities for the Analysis

The "primary unknown" quantity in finite-element analysis is the first, and usually the principal unknown quantity to be obtained in the analysis. The primary unknown will vary from case to case according to the nature of the problem. Frequently used primary unknowns in FE analysis in mechanical engineering include the following:

- Displacements with $\{U\}^T = \{U_x \quad U_y \quad U_z\}$ in the elements, in which U_x, U_y, and U_z are the displacement components along the x-, y-, and z-coordinates, respectively.
- Temperature T in elements in heat conduction analysis.
- Velocity $\{V\}^T = \{V_x \quad V_y \quad V_z\}$ in the elements in fluid dynamic analysis, in which V_x, V_y, and V_z are the velocity components along the x-, y-, and z-coordinates, respectively.

Other related unknown quantities, known as the "secondary unknowns" may be obtained from the primary unknown quantity. For example in stress analysis, the primary unknowns are displacements in elements, but the secondary unknowns include strains in the elements that can be obtained by the "strain–displacement relations," and other secondary unknown quantities such as stresses in the elements can by computed from the element strains using the stress–strain relations, known as Hooke's law, derived from the theory of elasticity.

Step 3: Derive the Interpolation Functions

Interpolation functions in finite-element analysis relate the primary unknown quantities in the elements and those at the associated nodes of a given element. In general, this function is expressed in mathematical form as in Equation 11.1.

$$\{\Phi(\mathbf{r})\} = [N(\mathbf{r})]\{\varphi\} \tag{11.1}$$

where $\Phi(\mathbf{r})$ is the primary unknown quantities in the element with $\mathbf{r}$ being the position vector representing the corresponding function $\Phi(x, y, z)$ in rectangular coordinate system, or (r,θ,z) in cylindrical polar coordinate system. The quantity $[N(\mathbf{r})]$ is the interpolation function in the form of a rectangular matrix and the column matrix $\{\varphi\}$ is the same primary quantities at the associated nodes of the element.

Interpolation functions such as the ones shown in Equation 11.1 often are referred to as "shape functions" because the form of this function varies with the shape of the elements shown in Figure 11.2. For example, each of the four different shapes of elements in Figure 11.2 has a different number of associate nodes as specified below:

1) Tetrahedral elements for 3D solids with 4 nodes
2) Triangular plate elements for 2D plane solids with 3 nodes
3) Bar elements for 1D solids with 2 nodes
4) Axisymmetric triangular elements for solids with circular geometry with 3 nodes
5) Other element shapes: hexahedral elements with 6 sides and 8 nodes; quadrilateral plate elements with 4 sides and 4 nodes.

As a result, the interpolation function will have the different forms for the different shapes of elements: three-dimensional function $[N(x,y,z)]$ for tetrahedral and hexahedral elements; two-dimensional function $[N(x,y)]$ for triangular and quadrilateral plate elements; and simple one-dimensional function $\{N(x)\}^{\mathrm{T}}$ for bar elements.

Derivation of interpolation functions is an important step in the finite-element analysis because the primary unknown quantities obtained with the formulations at individual nodes of elements in discretized FE models such as shown in Figure 11.4b must be related to the elements using this function. Besides, the nodes are the points at which all elements in the discretized elements are connected to the effect of the approximation between the original continuum and its discretized geometry, as illustrated in Figures 11.1b and 11.4b. Interpolation functions which relate the solution of primary quantities in individual elements to their associate nodes that serve as "linkages" of all elements in the discretized finite-element model thus play a vital role in the finite-element analysis.

11.3.1 Derivation of Interpolation Function for Simplex Elements with Scalar Quantities at Nodes

Derivation of interpolation functions for a particular geometry of the elements begins with assumption of a form of the function that describes the variation of the primary quantity in the elements. The simplest function that one may use for this purpose is a linear polynomial function. We will derive such an interpolation function for a relatively simple triangular plate element with three nodes, as illustrated in Figure 11.5 The interpolation function $[N(x,y)]$ that

Figure 11.5 Relating element values with those at nodes.

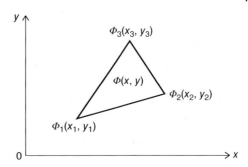

we will derive will relate the primary unknown quantity $\Phi(x,y)$ in the element to the same unknown quantity at the associated nodes with set coordinates φ_1 at (x_1,y_1), φ_2 at (x_2,y_2), and φ_3 at (x_3,y_3).

We assume the function that represents the unknown quantity in the element to be a linear polynomial function that has a mathematical form:

$$\Phi(x, y) = \alpha_1 + \alpha_2 x + \alpha_3 y = \{1 \ \ x \ \ y\} \begin{Bmatrix} \alpha_1 \\ \alpha_2 \\ \alpha_3 \end{Bmatrix} = \{R\}^T\{\alpha\} \tag{11.2}$$

in which the matrix $\{R\}^T = \{1 \ \ x \ \ y\}$, and the coefficients α_1, α_2 and α_3 are constants that can be determined by substituting the specified coordinates of the nodal values φ_1, φ_2, and φ_3, to give

$$\varphi_1 = \alpha_1 + \alpha_2 x_1 + \alpha_3 y_1$$
$$\varphi_2 = \alpha_1 + \alpha_2 x_2 + \alpha_3 y_2$$
$$\varphi_3 = \alpha_1 + \alpha_2 x_3 + \alpha_3 y_3$$

or in a matrix form:

$$\{\varphi\} = [A]\{\alpha\} \ \text{ and } \ \{\alpha\} = [A]^{-1}\{\varphi\} = [h]\{\varphi\} \tag{11.3}$$

where the matrix

$$[A] = \begin{bmatrix} 1 & x_1 & y_1 \\ 1 & x_2 & y_2 \\ 1 & x_3 & y_3 \end{bmatrix}$$

and the inverse of this matrix $[A]^{-1} = [h]$ is obtained with the result

$$[h] = \frac{1}{|A|} \begin{bmatrix} x_2 y_3 - x_3 y_2 & x_3 y_1 - x_1 y_3 & x_1 y_2 - x_2 y_1 \\ y_2 - y_3 & y_3 - y_1 & y_1 - y_2 \\ x_3 - x_2 & x_1 - x_3 & x_2 - x_1 \end{bmatrix} \tag{11.4}$$

in which the determinant $|A|$ can be evaluated to

$$|A| = (x_1 y_2 - x_2 y_1) + (x_2 y_3 - x_3 y_2) + (x_3 y_1 - x_1 y_3) = 2A$$

where $A =$ the plane area of the triangular plate element

Substituting Equations 11.3 and 11.4 into Equation 11.2, the function $\Phi(x,y)$ in the element can be evaluated by the three nodal quantities φ_1, φ_2, φ_3, or $\{\varphi\}$ to be

$$\Phi(x, y) = \{R\}^T[h]\{\varphi\} \tag{11.5}$$

The interpolation function, as defined in Equation 11.1 takes the form

$$[N(x, y)] = \{R\}^{T}[h] \tag{11.6}$$

where the matrix $\{R\}^{T} = \{1 \quad x \quad y\}$ and the matrix $[h]$ is given in Equation 11.4.

11.3.2 Derivation of Interpolation Function for Simplex Elements with Vector Quantities at Nodes

Primary unknown quantities in triangular plate elements with three fixed nodes in a discretized finite-element model may contain only one component as shown in Figure 11.5. However, in general, they contain two components such as $\Phi_x(x,y)$ along the x-coordinate and for the other component as $\Phi_y(x,y)$ along the y-coordinate, as illustrated in Figure 11.6. Consequently, there are two corresponding components of the same primary quantity at each of the three associated nodes with $\varphi_{1x}(x_1,y_1)$ along the x-direction and $\varphi_{1y}(x_1,y_1)$ along the y-coordinate at node 1 situated at (x_1,y_1), and so on.

According to Figure 11.6, the primary unknowns in the element can be represented by two linear polynomial functions as expressed in Equation 11.7:

$$\{\Phi(x, y)\} = \begin{Bmatrix} \Phi_x(x, y) \\ \Phi_y(x, y) \end{Bmatrix} = \begin{Bmatrix} \alpha_1 + \alpha_2 x + \alpha_3 y \\ \alpha_4 + \alpha_5 x + \alpha_6 y \end{Bmatrix} \tag{11.7}$$

where $\alpha_1, \alpha_2, \alpha_3, \ldots, \alpha_6$ are constants to be determined by the nodal values at the specific coordinates of the nodes.

The unknown quantities in the element in Equation 11.7 may be related to those at the three associated nodes as in Equation 11.8:

$$\{\Phi(x, y)\} = \begin{Bmatrix} \Phi_x(x, y) \\ \Phi_y(x, y) \end{Bmatrix} = \begin{Bmatrix} \alpha_1 + \alpha_2 x + \alpha_3 y \\ \alpha_4 + \alpha_5 x + \alpha_6 y \end{Bmatrix}$$

$$= [N(x, y)]\{\varphi\}$$

$$= \{N_1(x, y) \ N_2(x, y) \ N_3(x, y)\} \begin{Bmatrix} \varphi_1(x_1, y_1) \\ \varphi_2(x_2, y_2) \\ \varphi_3(x_3, y_3) \end{Bmatrix} \tag{11.8}$$

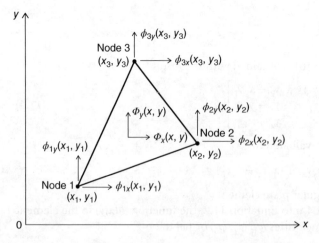

Figure 11.6 Triangular plate element with two unknown components.

in which $N_1(x,y)$, $N_2(x,y)$ and $N_3(x,y)$ are components of the interpolation function $[N(x,y)]$ associated with the unknown nodal quantities φ_1, φ_2, and φ_3 at nodes 1, 2, and 3 respectively.

We realize that each of the three nodes of the element in Figure 11.6 consists of two components:

$$\{\varphi_1\} = \left\{ \begin{array}{c} \varphi_{1x}(x_1, y_1) \\ \varphi_{1y}(x_1, y_1) \end{array} \right\} \qquad \{\varphi_2\} = \left\{ \begin{array}{c} \varphi_{2x}(x_2, y_2) \\ \varphi_{2y}(x_2, y_2) \end{array} \right\} \qquad \{\varphi_3\} = \left\{ \begin{array}{c} \varphi_{3x}(x_3, y_3) \\ \varphi_{3y}(x_3, y_3) \end{array} \right\}$$

By following a similar procedure to that in deriving the interpolation function in the proceeding case, we may derive the three components of the interpolation function, $N_1(x,y)$, $N_2(x,y)$, and $N_3(x,y)$ in Equation 11.8, as in the following expressions (Segerlind, 1976):

$$N_1(x, y) = \frac{1}{A}[(x_2 y_3 - x_3 y_2) + (y_2 - y_3)x + (x_3 - x_2)y] \tag{11.9a}$$

$$N_2(x, y) = \frac{1}{A}[(x_3 y_1 - x_1 y_3) + (y_3 - y_1)x + (x_1 - x_3)y] \tag{11.9b}$$

$$N_3(x, y) = \frac{1}{A}[(x_1 y_2 - x_2 y_1) + (y_1 - y_2)x + (x_2 - x_1)y] \tag{11.9c}$$

where

$$A = \frac{1}{2}[(x_1 y_2 - x_2 y_1) + (x_2 y_3 - x_3 y_2) + (x_3 y_1 - x_1 y_3)]$$

is the area of the triangular element in Figure 11.6.

Example 11.1

Derive the interpolation function for a bar element by following the procedures outlined for simplex elements in Sections 11.3.1 and 11.3.2. The bar element has two nodes at A and B as illustrated in Figure 11.7 with assigned coordinates for node 1 at A as $x = x_1$ and the coordinate of node 2 at B as $x = x_2$. The bar element is subjected to a longitudinal deformation $U(x)$ represented by a linear polynomial function as shown in Figure 11.7.

Solution:

We assume that the primary unknown in this bar element is the displacement along the length of the bar represented by the function $U(x)$, and that this is a linear polynomial function of the form

$$U(x) = \alpha_1 + \alpha_2 x \tag{a}$$

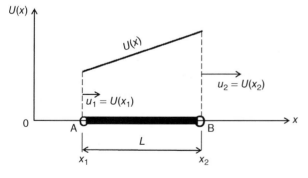

Figure 11.7 Interpolation function of bar elements.

where α_1 and α_2 are two constants that can be evaluated by the nodal displacements u_1 of node 1 and u_2 of node 2 as follows:

$$u_1 = \alpha_1 + \alpha_2 x_1 \tag{b1}$$

$$u_2 = \alpha_1 + \alpha_2 x_2 \tag{b2}$$

Solve for α_1 and α_2 from Equations (b1) and (b2) with:

$$\alpha_1 = -\frac{x_2}{x_1 - x_2} u_1 + \frac{x_1}{x_1 - x_2} u_2 \tag{c1}$$

and

$$\alpha_2 = \frac{1}{x_1 - x_2} u_1 - \frac{1}{x_1 - x_2} u_2 \tag{c2}$$

Hence, we have the element displacement $U(x)$ related to the nodal displacements u_1 and u_2 by the following relationship:

$$U(x) = \left\{ \frac{x - x_2}{x_1 - x_2} \quad -\frac{x - x_1}{x_1 - x_2} \right\} \left\{ \begin{matrix} u_1 \\ u_2 \end{matrix} \right\} \tag{d}$$

The interpolation function $N(x)$ for the bar element is defined in a similar way to that in Equation 11.1, or in this particular case as:

$$U(x) = \{N(x)\} \left\{ \begin{matrix} u_1 \\ u_2 \end{matrix} \right\} \tag{e}$$

with the matrix $\{N(x)\}$ in Equation (e) being the interpolation function for a uniaxially deformed bar element in the form of a row matrix (9.10)

$$N(x) = \left\{ \frac{x - x_2}{x_1 - x_2} \quad -\frac{x - x_1}{x_1 - x_2} \right\} \tag{9.10}$$

Example 11.2

Derive the interpolation function of a simplex triangular planar element involving an unknown prime quantity $T(x,y)$, and with three nodes located in given nodal coordinates as shown in Figure 11.8.

We will use the triangular plate element in Figure 11.5 with three nodes $\{T\}$: T_1, T_2, T_3 located at coordinates (2,1), (6,4), and (5,8), respectively, as shown in Figure 11.8. The interpolation function is derived from a linear polynomial function as shown in Equation 11.2 in the form $T(x, y) = \alpha_1 + \alpha_2 x + \alpha_3 y$ in which α_1, α_2, and α_3 are constants.

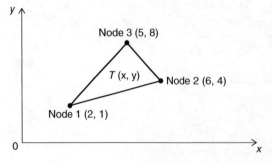

Figure 11.8 Interpolation function for a triangular plate element with specified nodal coordinates.

Solution:

We will use Equation 11.6 to determine the interpolation function $[N(x,y)]$ for the triangular plate element in Figure 11.8, with the given nodal coordinates: $x_1 = 2$, $x_2 = 6$, $x_3 = 5$, and $y_1 = 1$, $y_2 = 4$, and $y_3 = 8$.

The unknown quantity $T(x,y)$ in the element may be related to its nodal values according to Equation 11.1 as follows:

$$T(x,y) = [N(x,y)]\{T\}$$

$$= \begin{bmatrix} N_1(x,y) & N_2(x,y) & N_3(x,y) \end{bmatrix} \begin{Bmatrix} T_1 \\ T_2 \\ T_3 \end{Bmatrix} \tag{a}$$

In which $N_1(x,y)$, $N_2(x,y)$, and $N_3(x,y)$ are the components of the interpolation function $[N(x,y)]$, and T_1, T_2, and T_3 are the components of the quantity $T(x,y)$ in the element at node 1, node 2, and node 3, respectively.

Following the procedure for deriving the interpolation function for scalar quantities in elements with substitution of the three given set of nodal coordinates into Equation 11.4, we compute the determinant $|A| = 19$, which leads to the determination of the matrix $[h]$ in Equation 11.4 as

$$[h] = \frac{1}{19} \begin{bmatrix} 28 & -11 & 2 \\ -2 & 7 & -3 \\ -1 & -3 & 4 \end{bmatrix}$$

We can thus express the interpolation function $[N(x,y)]$ using Equation 11.6, resulting in a row matrix:

$$[N(x,y)] = \{R\}^T[h] = \{1 \ x \ y\} \left(\frac{1}{19}\right) \begin{bmatrix} 28 & -11 & 2 \\ -2 & 7 & -3 \\ -1 & -3 & 4 \end{bmatrix}$$

$$= \begin{Bmatrix} 1.4737 - 0.1053x - 0.0526y & -0.5789 + 0.3684x - 0.1579y \\ 0.1053 - 0.1579x + 0.2105y \end{Bmatrix}$$

We express the components of the interpolation function in Equation (a) with the above expression and obtain

$$N_1(x,y) = 1.4737 - 0.1053x - 0.0526y \tag{b1}$$

$$N_2(x,y) = -0.5789 + 0.3684x - 0.1579y \tag{b2}$$

$$N_3(x,y) = 0.1053 - 0.1579x + 0.2105y \tag{b3}$$

for Equation (a)

Step 4: Define the Relationship between Actions and Induced Reactions

The laws of physics usually provide relationships between the actions on the continuum and the induced reactions from the same substance. Table 11.3 indicates some of these physical quantities frequently involved in engineering analyses.

The laws or equations that relate the actions and induced reactions in Table 11.3 for finite-element analysis include minimization of the potential energy that is produced by the applied forces $\{F\}$ and the induced nodal displacements $\{u\}$ in deformed solid structures; the Fourier law that relates heat supply from heat sources $\{Q\}$ and the induced temperature T in the solids in heat conduction analysis; Bernoulli's law for the relationship between the applied

Table 11.3 Typical actions and induced reactions.

	Actions {P}	Induced reactions {Φ}
Stress analysis	Forces {F}	Displacement {u}
		Strains {ε}
		Stresses {σ}
Heat conduction	Thermal forces {Q}	Temperature {T}
Fluid flow	Pressure or Head {p}	Velocity {V}
Electricity	Voltage {V}	Current flow {i}

pressure {P} and the induced velocity {V} of fluids in motion; and Kirkoff's law that relates the applied voltage {V} and induced current {i} in electric circuitry analysis. These are the laws that are used to develop the finite-element equations for the solutions of the induced reactions from applied actions.

Step 5:Derive Element Equations

Element equations relate the applied actions to the discretized continua and the induced reactions in the elements. There are generally two distinct methods that can be used to derive these equations in finite-element analyses: (1) the Rayleigh–Ritz method, and (2) the weighted residue method, or Galerkin method. The Rayleigh–Ritz method is based on the principle of variational calculus and the Galerkin method is based on the principle of minimizing the difference between the results obtained from the original continua and those from their discretized models. The Galerkin method is used to derive the element equations if a set of differential equations with boundary conditions is available for the solution of the problem. Many heat conduction-convection and fluid mechanics problems can be handled by this method. We will present an overviews of these methods in the following paragraphs.

The Rayleigh–Ritz method

This method is applied to the variation of a suitable functional $\chi(\Phi^e)$ that can characterize the continuum in the analysis, with Φ^e being the unknown primary quantity in the elements of a discretized model as shown in Figure 11.1. (A *functional* is a function of functions). This functional can be expressed in the form of differential quantities such as shown in Equation 11.11.

$$\chi(\Phi^e) = \int_v f\left(\{\Phi^e\}, \frac{\partial\{\Phi^e\}}{\partial r}, \dots\right) dv + \int_s g\left(\{\Phi^e\}, \frac{\partial\{\Phi^e\}}{\partial r}, \dots\right) ds \tag{11.11}$$

where dv and ds are the respective volume and surface of the elements in the discretized continuum in Figure 11.1b.

The variational principle requires that the rate of change of the functional $\chi(\Phi^e)$ in Equation 11.11 with respect to its variable Φ^e be kept in minimum, as shown in Equation 11.12.

$$\frac{\partial \chi(\Phi^e)}{\partial \Phi^e} = \begin{Bmatrix} \dfrac{\partial \chi}{\partial \Phi^e_1} \\ \dfrac{\partial \chi}{\partial \Phi^e_2} \\ \dfrac{\partial \chi}{\partial \Phi^e_3} \\ \vdots \\ \vdots \end{Bmatrix} = \{0\} \tag{11.12}$$

where Φ_1^e, Φ_2^e, Φ_3^e, ... are the primary quantities in the respective elements 1, 2, 3,... in the discretized medium, as shown in Figure 11.1b.

Equation 11.12 will lead to the following equations for each element in the discretized medium:

$$\frac{\partial \chi}{\partial \Phi_1^e} = 0 \quad \text{for element 1} \tag{11.13a}$$

$$\frac{\partial \chi}{\partial \Phi_2^e} = 0 \quad \text{for element 2} \tag{11.13b}$$

$$\frac{\partial \chi}{\partial \Phi_3^e} = 0 \quad \text{for element 3} \tag{11.13c}$$

and so on. Equations 11.13a,b,c,... may lead to the "element equations" for all the elements in the discretized continuum in Figure 11.1b.

The Galerkin method

This method is sometimes referred to as the "weighted residue method," in which the weighting functions are assumed to be identical to the interpolation functions used in the discretized models of the continua.

Consider a physical problem that can be described mathematically in the form of a differential equation:

$$D(\varphi) = 0 \text{ in domain } v, \text{ and } B(\varphi) = 0 \text{ on boundary } s$$

The system can be replaced by an integral equation:

$$\int_v W\, D(\varphi)\, dv + \int_s \overline{W}\, B(\varphi)\, ds = 0 \tag{11.14}$$

where W and $\overline{W}$ are arbitrary weighting functions.

For a discretized systems, the primary unknown quantity in the element is $\Phi \approx \sum N_i \varphi_i$ with N_i and φ_i being the interpolation functions and the primary unknown nodal values, respectively. Equation 11.14 can thus be approximated as

$$\int_v W_j D\left(\sum N_i \varphi_i\right) dv + \int_s \overline{W}_j B\left(\sum N_i \varphi_i\right) ds = \text{Re} \tag{11.15}$$

in which W_j and $\overline{W}_j$ are discretized weighting functions and Re on the right-hand side of Equation 11.15 is the residue of the difference between the original continuum in Figure 11.1a and that of the discretized state in Figure 11.1b. A good discretization system, of course, should make the residue Re a minimum, or Re→0.

The Galerkin method with identical weighting functions (W_j and $\overline{W}_j$) and with both these weighting functions equal to the interpolation function (N_i) is the most popular form used in finite-element analyses, notwithstanding that it has been proposed that other forms may provide even more stable solutions (Heinrich *et al.*, 1977). Thus, replacing the weighting functions with the interpolation function in Equation 11.15, one arrives at the following formulation for the element equations:

$$\int_v N_i D\left(\sum N_i \varphi_i\right) dv + \int_s N_i B\left(\sum N_i \varphi_i\right) ds \to 0 \tag{11.16}$$

The left-hand side of Equation 11.16 can be made to be the functional $\chi(\Phi)$ both for inside the element (v) and for the actions across the boundaries (s). Satisfaction of this equation requires the satisfaction of Equation 11.12, or for discretized finite-element models:

$$\frac{\partial \chi(\Phi^e)}{\partial(\Phi)} = \sum \frac{\partial \chi^e(\Phi^e)}{\partial(\Phi^e)} = 0$$

The right-hand side of the above expression will lead to the typical form of element equations as shown in Equation 11.13.

We will find that the element equations derived according to the above two distinct methods can be made to take the following form:

$$[K_e]\{\varphi\} = \{F\} \tag{11.17}$$

where $[K_e]$ is the element coefficient matrix, $\{\varphi\}$ is the vector of primary unknown quantities normally expressed by the quantities at the nodes that link all elements in the discretized model, and $\{F\}$ on the right-hand side of Equation 11.17 are the applied force vectors at the nodes.

Element equations are considered to be the fundamental fabric of the finite-element analysis; let us take a look at the triangular plate element illustrated in Figure 11.9 in which the primary unknown quantity in the element is $\Phi(x,y)$—usually a displacement in a deformed solid, consisting of two components: $\Phi_x(x,y)$ and $\Phi_y(x,y)$ corresponding to the two components along the x- and y-coordinate, respectively. The same primary unknown quantities at the three associated nodes are φ_{1x} and φ_{1y} at node 1 located at (x_1,y_1); φ_{2x} and φ_{2y} at node 2 located at (x_2,y_2), and φ_{3x} and φ_{3y} at node 3 located at (x_3,y_3). We note that node 1 is completely fixed at (x_1,y_1), whereas node 2 is constrained from movement in the y-direction. A force F is applied at node 3 in the direction shown in Figure 11.9. We thus realize that there are a total of six unknown displacements associated with the three nodes, which accounts for six DOF of the triangular plate element in Figure 11.9.

Let us assume that the elements of the coefficient matrix $[K_e]$ of the element in Figure 11.9 are obtained using the Rayleigh–Ritz method as K_{ij}, with $i, j = 1, 2, \ldots, 6$. The element equation in Equation 11.17 may be shown to take the following form, including the specified boundary

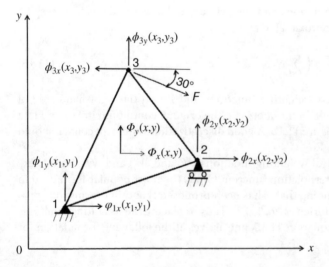

Figure 11.9 Triangular plate element with nodal constraints and force.

conditions of $\varphi_{1x} = \varphi_{1y} = 0$ and $\varphi_{2y} = 0$ and the specified loading of $F_{3x} = F\cos 30°$ and $F_{3y} = -F\sin 30°$:

$$
\begin{bmatrix} k_{11} & k_{12} & k_{13} & k_{14} & k_{15} & k_{16} \\ k_{21} & k_{22} & k_{23} & k_{24} & k_{25} & k_{26} \\ k_{31} & k_{32} & k_{33} & k_{34} & k_{35} & k_{36} \\ k_{41} & k_{42} & k_{43} & k_{44} & k_{45} & k_{46} \\ k_{51} & k_{52} & k_{53} & k_{54} & k_{55} & k_{56} \\ k_{61} & k_{62} & k_{63} & k_{64} & k_{65} & k_{66} \end{bmatrix} \begin{Bmatrix} \varphi_{1x} = 0 \\ \varphi_{1y} = 0 \\ \varphi_{2x} \\ \varphi_{2y} = 0 \\ \varphi_{3x} \\ \varphi_{3y} \end{Bmatrix} = \begin{Bmatrix} F_{1x} = 0 \\ F_{1y} = 0 \\ F_{2x} = 0 \\ F_{2y} = 0 \\ F_{3x} = F\cos 30° \\ F_{3y} = -F\sin 30° \end{Bmatrix}
$$

Step 6: Derive Overall Stiffness Equations

The overall stiffness equations for the entire discretized continua in finite-element analyses are derived by assemblies of all individual element equations in the discretized model. These equations allow engineers to determine the unknown quantiles at all the nodes in the discretized model. The overall stiffness equations for all unknown nodal quantities $\{\varphi\}$ have the form

$$[K]\{\varphi\} = \{R\} \tag{11.18}$$

where $[K]$, the overall coefficient matrix, equals $\sum\limits^{M}[K_e]$, with M the total number of elements in the discretized model, and $\{R\}$ is the matrix with assembly of the resultant applied actions at nodes, such as $\{F\}$in Equation 11.17 for individual elements.

It is common in finite-element analyses that the element coefficient matrix $[K_e]$ in Equation 11.17 is shared by more than one element in the discretized model. For instance, several elements, such as Elements 1, 2, 7, 8, share common node 9 in the finite-element model in Figure 11.4b. The same node 16 is share by Elements 7, 8, 14, 15, and 16. The values such nodes as nodes 9 and 16, involve contributions from all the elements that share these nodes. This fact must be considered in summing the related element stiffness matrices $[K_e]$ and the force matrix $\{F\}$ in Equation 11.17 in the proper places in the overall coefficient matrix $[K]$ and the resultant action matrix $\{R\}$in Equation 11.18. We will illustrate this process by Example 11.3.

Example 11.3

A thin plate of quadrilateral shape shown in Figure 11.10 is subjected to a set of in-plane forces. The plate is subdivided into two triangular elements, with Element 1 consisting of nodes 1, 3, and 4, and Element 2 having nodes 2, 4, and 3. Nodes 3 and 4 are shared by both Elements 1 and 2. Draw a map for the assembly of these two element equations to form the overall stiffness equation for the plate structure if the element equations for both Elements 1 and 2 are expressed in Equations (a) and (b) respectively:

$$
\begin{bmatrix} \Delta & \Delta & \Delta & \Delta & \Delta & \Delta \\ \Delta & \Delta & \Delta & \Delta & \Delta & \Delta \\ \Delta & \Delta & \Delta & \Delta & \Delta & \Delta \\ \Delta & \Delta & \Delta & \Delta & \Delta & \Delta \\ \Delta & \Delta & \Delta & \Delta & \Delta & \Delta \\ \Delta & \Delta & \Delta & \Delta & \Delta & \Delta \end{bmatrix} \begin{Bmatrix} u_{1x} \\ v_{1y} \\ u_{3x} \\ v_{3y} \\ u_{4x} \\ v_{4y} \end{Bmatrix} = \begin{Bmatrix} F_{1x} \\ F_{1y} \\ F_{3x} \\ F_{3y} \\ F_{4x} \\ F_{4y} \end{Bmatrix} \tag{a}
$$

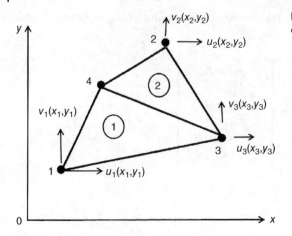

Figure 11.10 Finite-element model of a quadrilateral plate structure.

for Element 1, and

$$
\begin{bmatrix} \bullet & \bullet & \bullet & \bullet & \bullet & \bullet \\ \bullet & \bullet & \bullet & \bullet & \bullet & \bullet \\ \bullet & \bullet & \bullet & \bullet & \bullet & \bullet \\ \bullet & \bullet & \bullet & \bullet & \bullet & \bullet \\ \bullet & \bullet & \bullet & \bullet & \bullet & \bullet \\ \bullet & \bullet & \bullet & \bullet & \bullet & \bullet \end{bmatrix} \begin{Bmatrix} u_{2x} \\ v_{2y} \\ u_{3x} \\ v_{3y} \\ u_{4x} \\ v_{4y} \end{Bmatrix} = \begin{Bmatrix} F_{2x} \\ F_{2y} \\ F_{3x} \\ F_{3y} \\ F_{4x} \\ F_{4y} \end{Bmatrix}
$$

(b)

for Element 2, in which the entries of the element matrix in Equation (a) are denoted by open triangles Δ of real numbers for Element 1 and the entries of the element matrix in Equation (b) are denoted by closed circles of real numbers for Element 2.

Solution:
We notice that the thin plate structure in Figure 11.10 is discretized into two triangular plate elements interconnected at nodes 3 and 4. We further realize that each of the four nodes in the discretized plate structure has two unknown components (e.g., the displacements) along the respective x- and y-coordinates, which accounts for two degrees-of-freedom (DOF). The structure in the present example thus has a total of $2 \times 4 = 8$ DOFs which leads to the size of the assembled overall coefficient matrix [K] in Equation 11.18 to be 8×8.

We may logically assign the entries to this 8×8 overall coefficient matrix [K] in the following way: Row and Column numbers 1 and 2 for the two unknown quantities of node 1, Row and Column numbers 3 and 4 for the two unknown quantities of node 2, Row and Column numbers 5 and 6 for the unknown quantities of node 3, and the Row and Column numbers 7 and 8 for the unknown quantities of node 4. However, because both nodes 3 and 4 are shared by Elements 1 and 2 as shown in Figure 11.10, entries to Row and Column numbers 5 to 8 need to be the summations of the unknown quantities of nodes 3 and 4 from both Elements 1 and 2 as illustrated in Figure 11.11.

One will note that the entries in Row and Column numbers 5 to 8, which relate to nodes 3 and 4 of the discretized structure, have the sum of the corresponding values from both Element 1 and Element 2 in the above assembled overall stiffness matrix.

Step 7: Solve for Primary Unknown Quantities

It is apparent that Equation 11.18 represents a set of simultaneous linear equations. The total number of equations is the same as the total number of DOFs in the analysis.

Column number

Row number	1	2	3	4	5	6	7	8
1	Δ	Δ			Δ	Δ	Δ	Δ
2	Δ	Δ			Δ	Δ	Δ	Δ
3			●	●	●	●	●	●
4			●	●	●	●	●	●
5	Δ	Δ	●	●	Δ+●	Δ+●	Δ+●	Δ+●
6	Δ	Δ	●	●	Δ+●	Δ+●	Δ+●	Δ+●
7	Δ	Δ	●	●	Δ+●	Δ+●	Δ+●	Δ+●
8	Δ	Δ	●	●	Δ+●	Δ+●	Δ+●	Δ+●

Figure 11.11 Map for assembling element coefficient matrices.

Depending on the size of the overall stiffness matrix [K], and thereby the number of simultaneous equations to be solved, there are generally two methods that can be used to solve for the nodal unknown quantities $\{\varphi\}$ in Equation 11.18. These are (1) the Gaussian elimination method or its derivatives, and (2) the matrix inversion method. Both these methods are described in Chapter 4.

In practice, however, it is desirable to partition the overall stiffness matrix [K] in Equation 11.18 by re-arranging the terms in this equation into the following partitioned form:

$$\begin{bmatrix} K_{aa} & K_{ab} \\ K_{ba} & K_{bb} \end{bmatrix} \begin{Bmatrix} \phi_a \\ \phi_b \end{Bmatrix} = \begin{Bmatrix} R_a \\ R_b \end{Bmatrix} \tag{11.19}$$

where $\{\varphi_a\}$ are the specified primary unknown quantities at the nodes with specified boundary conditions. For instance, both displacement components at node 1 in the finite-element model in Figure 11.4b are zero, and the displacement components along the x-direction are zero for nodes 8, 15, and 22. In other words, $\{\varphi_a\}$ in Equation 11.19 are the unknown nodal quantities that need not to be determined in the analysis. The values of $\{R_b\}$ in Equation 11.19 are the specified (known) applied nodal resultant forces, such as the forces $\{F\}$ applied at nodes 7, 14, and 21 along coordinate x in the case illustration in Figure 11.4b.

The unknown primary quantities at the nodes, i.e., $\{\varphi_b\}$ in Equation 11.19 can be computed from this partitioned overall stiffness equation by the following expression:

$$\{\varphi_b\} = [K_{bb}]^{-1}(\{R_b\} - [K_{ba}]\{\varphi_a\}) \tag{11.20}$$

The reader is reminded that the partitioning of the finite-element formulation for the overall discretized substances requires interchanging rows in the [K], $\{\varphi\}$, and {R} matrices. Any interchange of rows of the aforementioned matrices must be followed by the interchange of the corresponding columns in the same matrices.

Example 11.4

Use the matrix partitions in Equation 11.19 and the solution in Equation 11.20 to solve the unknown value of φ_3 from the following overall stiffness matrix equation involving four nodes at φ_1, φ_2, φ_3, and φ_4 with specified values of: $\varphi_1 = 2$, $\varphi_2 = 3$, and $\varphi_4 = 4$.

$$\begin{bmatrix} 1 & 2 & -3 & 4 \\ 2 & -3 & 1 & 1 \\ 3 & 2 & 1 & -2 \\ -4 & 1 & -2 & 3 \end{bmatrix} \begin{Bmatrix} \varphi_1 \\ \varphi_2 \\ \varphi_3 \\ \varphi_4 \end{Bmatrix} = \begin{Bmatrix} 21 \\ 0 \\ 5 \\ 5 \end{Bmatrix} \tag{a}$$

where

$$[K] = \begin{bmatrix} 1 & 2 & -3 & 4 \\ 2 & -3 & 1 & 1 \\ 3 & 2 & 1 & -2 \\ -4 & 1 & -2 & 3 \end{bmatrix}$$

is the overall coefficient matrix;

$$\{\varphi\} = \begin{Bmatrix} \varphi_1 \\ \varphi_2 \\ \varphi_3 \\ \varphi_4 \end{Bmatrix}$$

is the unknown matrix; and

$$\{R\} = \begin{Bmatrix} 21 \\ 0 \\ 5 \\ 5 \end{Bmatrix}$$

is the resultant force matrix.

Solution:

Since the unknown quantity that we need to solve is φ_3 in Row number 3 in the overall coefficient matrix, we need to interchange Rows 3 and 4 in the [K] matrix as follows:

After interchanging Row 3 and Row 4:

$$[K] = \begin{bmatrix} 1 & 2 & -3 & 4 \\ 2 & -3 & 1 & 1 \\ -4 & 1 & -2 & 3 \\ 3 & 2 & 1 & -2 \end{bmatrix}$$

followed by interchanging Column 3 and Column 4:

$$[K] = \begin{bmatrix} 1 & 2 & 4 & -3 \\ 2 & -3 & 1 & 1 \\ -4 & 1 & 3 & -2 \\ 3 & 2 & -2 & 1 \end{bmatrix}$$

We will thus have Equation (a) being modified to the form that is compatible to Equation 11.19:

$$\left[\begin{array}{ccc:c} 1 & 2 & 4 & -3 \\ 2 & -3 & 1 & 1 \\ -4 & 1 & 3 & -2 \\ \hdashline 3 & 2 & -2 & 1 \end{array}\right] \begin{Bmatrix} \phi_1 \\ \phi_2 \\ \phi_4 \\ \hdashline \phi_3 \end{Bmatrix} = \begin{Bmatrix} 21 \\ 0 \\ 5 \\ \hdashline 5 \end{Bmatrix} \tag{b}$$

and the following submatrices derived in the form of Equation 11.20 from Equation (b):

$$
[K_{aa}] = \begin{bmatrix} 1 & 2 & 4 \\ 2 & -3 & 1 \\ -4 & 1 & 3 \end{bmatrix} \quad [K_{ab}] = \begin{Bmatrix} -3 \\ 1 \\ -2 \end{Bmatrix} \quad \{\varphi_a\} = \begin{Bmatrix} \varphi_1 = 2 \\ \varphi_2 = 3 \\ \varphi_4 = 4 \end{Bmatrix} \quad \{R_a\} = \begin{Bmatrix} 21 \\ 0 \\ 5 \end{Bmatrix}
$$

$[K_{ba}] = \{3 \quad 2 \quad -2\}$, $\{K_{bb}\} = 1$, and $\{R_b\} = 5$.

The unknown value of φ_3 can thus be computed using Equation 11.20 as

$$
\{\varphi_b\} = \varphi_3 = [K_{bb}]^{-1}(\{R_b\} - [K_{ba}]\{\varphi_a\})
$$

$$
= (1)^{-1} \left(5 - \{3 \ 2 \ -2\} \begin{Bmatrix} 2 \\ 3 \\ 4 \end{Bmatrix} \right)
$$

$$
= 5 - (3 \times 2 + 2 \times 3 - 2 \times 4) = 1
$$

Step 8: Solve for Secondary Unknown Quantities

As was mentioned early in Step 2 of finite-element analysis, the selection of primary quantities for the analysis may vary from case to case. For instance, element displacements are denoted as $\Phi(x, y) = \{U\}^T = \{U_x \quad U_y\}$ where U_x and U_y are the two components of the unknown displacement U in the element along the x- and y-coordinate, respectively. These unknown primary quantities in the elements of the discretized model may be related to the quantities at the associated nodes, designated by $\{\varphi\}$ using the adopted interpolation functions. These unknown nodal quantities are later computed by the overall stiffness equations of the discretized model in Equation 11.18. In reality, however, there is the need to determine other quantities required in the analysis. For instance, in stress analysis of machine structures such as the case illustrated in Figure 11.4, the "stress" and the "maximum stresses" in the structures were required to determine whether the structures would be strong enough to withstand the applied forces F at nodes 7, 14, and 21. We thus need to determine the stress distributions and the magnitude and locations of maximum stress in the structure from the finite-element analysis. The "stresses" and the related "strains" are referred to as the required "secondary unknown quantities" in finite-element analysis.

The secondary unknown quantities can be determined from the primary quantities computed in finite-element analysis by using appropriate laws of physics. For instance, the strain components in the elements may be determined by the element displacement using the "displacement–strain relations" derived in the theory of linear elasticity, and the "element strains" may be related to the element "stress components" by use of Hooke's law. Similar formulations are available for heat transfer analysis, in which Fourier's law of heat conduction and Bernoulli's law for fluid dynamics analysis are frequently used to determine the secondary quantities from the primary unknown quantities obtained from finite-element analyses.

11.4 Output of Finite-element Analysis

Results obtained from finite-element analyses may take different forms, which in general may include (1) tabulated data; (2) contour maps of the computed unknowns in the elements or at nodes; (3) different zones of both the primary and secondary unknown quantities, with different color designations over the input discretized model; and (4) visual animation of selected

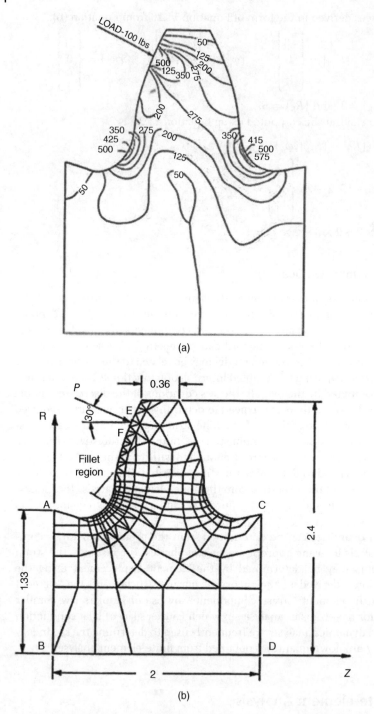

Figure 11.12 Contour output of a finite-element analysis. (a) Isoclinic stress contour. (b) Finite-element model.

unknown quantities in the discretized model. Many commercially available finite-element analysis codes such as the ANSYS code have a "post processor" to produce graphic outputs.

Tabulation of output usually includes the user's input of material properties of input elements, descriptions of elements and nodes in the discretized model with specified boundary conditions and loadings, as illustrated in Step 1 of the finite-element analysis. There are, of course, the computed nodal displacements and stresses in all elements. The element stresses in the output are expressed as von Mises stress for multiaxially loaded structures. von Mises stresses are expressed as in Equations 11.21a and 11.21b:

$$\overline{\sigma} = \frac{1}{\sqrt{2}} \sqrt{(\sigma_1 - \sigma_2)^2 + (\sigma_2 - \sigma_3)^2 + (\sigma_1 - \sigma_3)^2} \qquad (11.21a)$$

where σ_1, σ_2, and σ_3 are three principal stresses in elements, or the von Mises stress in a solid defined by a rectangular coordinate system of (x,y,z) takes the following form:

$$\overline{\sigma} = \frac{1}{\sqrt{2}} \sqrt{(\sigma_{xx} - \sigma_{yy})^2 + (\sigma_{yy} - \sigma_{zz})^2 + (\sigma_{xx} - \sigma_{zz})^2 + 6(\sigma_{xy}^2 + \sigma_{yz}^2 + \sigma_{xz}^2)} \qquad (11.21b)$$

in which σ_{xx}, σ_{yy}, ..., σ_{xy}, σ_{xz}, etc. are the stress components as will be defined subsequently in Section 11.5.1.

von Mises stress is used to show the output stress of elements in finite-element stress analysis because it is regarded as the "effective" stress in multiaxial stress situations, which is frequent in the case of finite-element analyses. Readers will find that the von Mises stress as expressed in Equation 11.21b can be deduced to a single stress component for a uniaxial stress situation with $\overline{\sigma} = \sigma_{xx}$ or $\overline{\sigma} = \sigma_{yy}$, or $\overline{\sigma} = \sigma_{zz}$ in uniaxial stress situations with $\sigma_{xy} = \sigma_{yz} = \sigma_{xz} = 0$.

Output in the form of contour maps of the computed unknowns in the elements or at nodes is useful to the users because it offers them the results of the finite-element analysis at a glance. Figure 11.12a shows the isoclinic (principal) stress contours of a gear tooth obtained by a finite-element analysis, with the its discretized model illustrated in Figure 11.12b (Hsu, 1986). The same contours were observed in model testing using a photoelasticity technique. Another useful form of graphic output in of ranges of the unknown quantities in different zones is shown in Figure 11.13.

Many commercial finite-element analysis codes offer animation of analytical results, such as continuous deformation of a solid structure subject to continuously varying loads. Video clips that record such animation provide great value to users in visualizing the effects on structures in motion, or specific processes in manufacturing or production processes.

11.5 Elastic Stress Analysis of Solid Structures by the Finite-element Method

In this section we will present key formulations for computing induced deformations and stresses in solid structures resulting from externally applied forces. Structural integrity is the number one concern in the design analysis of any machine/device. Finite-element analysis is a valuable analytical tool for stress analysis of machines or devices of complex geometry subject to complex loading and boundary conditions.

The formulations for finite-element analysis presented in this section are excerpts from the author's earlier books (Hsu, 1986; Hsu and Sinha, 1992) Readers are reminded that the physical quantities of multiple components in elastic stress analysis of structures such as displacements, strains, and stresses in deformed solids will be expressed in matrices, and the mathematical operations of the matrices presented in Chapter 4 will be extensively used in deriving the element equations and other formulae in the finite-element analysis.

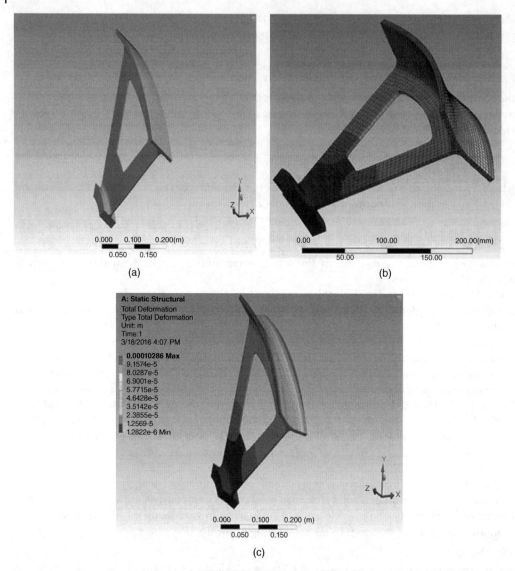

Figure 11.13 Contours of deformation of a wheel section by finite-element analysis. (a) The solid for FE modeling. (b) Deformation with the FE model. (c) Contours of deformation. (Courtesy of ANSYS Inc., Pittsburg, PA, USA).

11.5.1 Stresses

Two physical responses occur in a homogeneous solid with isotropic material properties when it is subjected to a set of applied forces {P} as illustrated in Figure 11.14a. In these responses (1) the solid deforms into a new shape, and (2) the solid develops internal resistance to the applied forces while it deforms into new shapes. The induced internal resistance by the solid eventually reaches a state of equilibrium with the applied forces, and the solid ceases further deformation under this new equilibrium condition.

The intensity of the induced resistance to the applied force in the solid is termed the "stress" at any given location. These stresses may be oriented in different directions in the space defined

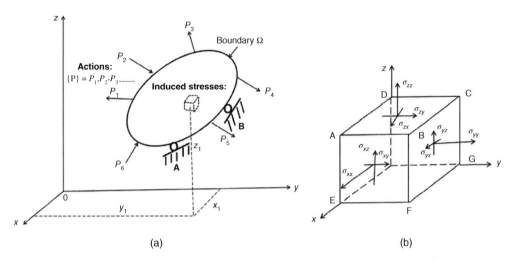

Figure 11.14 Induced stress components in a solid due to external loads. (a) The solid subjected to external loads. (b) Induced stress in the solid.

by the x-, y-, and z-coordinates in Figure 11.14a. The magnitude, location, and orientation of the induced stresses can be denoted by the special notation system described next.

Let us consider an infinitesimally small cubic element located at (x_1, y_1, z_1) with its edges parallel to the three coordinate axes as shown in Figure 11.14b. We envisage that there are six faces of this infinitesimally small cubic element, and there exist stress components with some normal to the faces of the cube, and some others on the faces. A commonly accepted way to denote these stress components is to attach two subscripts to the symbol σ, which denotes the stress magnitude, with the first subscript indicating the coordinate normal to the specific face of the cube, and the second subscript indicating the direction of the stress component. We can thus account for the nine stress components on each of the six faces of the small cubic element inside the deformed solid with the following matrix expression:

$$[\sigma] = \begin{bmatrix} \sigma_{xx} & \sigma_{xy} & \sigma_{xz} \\ \sigma_{yx} & \sigma_{yy} & \sigma_{yz} \\ \sigma_{zx} & \sigma_{zy} & \sigma_{zz} \end{bmatrix} \tag{11.22a}$$

where σ_{xx}, σ_{yy}, and σ_{zz} are the stress components on the faces perpendicular to the respective x-, y-, and z-coordinates, and oriented along the same respective coordinates. We will call these stress components the *normal stresses* in the deformed solid. The other six stress components with different subscripts—σ_{xy}, σ_{xz}, and so on—are called *shearing stress* components; for example, σ_{xy} acts on the cube face that is normal to the x-coordinate and points in the y-direction, and σ_{xz} is another shearing stress on the same cube face but pointing in the z-direction.

We may account a total of nine stress components on the small cubic element as included in Equation 11.22a. This number is reduced to six in Equation 11.22b via the relationships $\sigma_{xy} = \sigma_{yx}$, $\sigma_{yz} = \sigma_{zy}$, and $\sigma_{xz} = \sigma_{zx}$ because of the new equilibrium condition of the solid in the deformed condition.

We thus have the six independent stress components in deformed solid in three-dimensional stress analyses:

$$[\sigma] = \begin{bmatrix} \sigma_{xx} & \sigma_{xy} & \sigma_{xz} \\ & \sigma_{yy} & \sigma_{yz} \\ \text{SYM} & & \sigma_{zz} \end{bmatrix} \tag{11.22b}$$

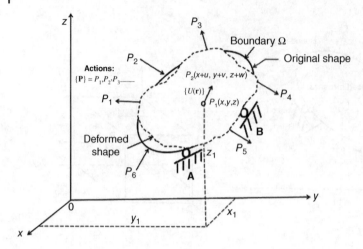

Figure 11.15 Displacements in a deformed solid.

We may also express these stress components in the form of a column matrix in our subsequent derivations:

$$\{\sigma\}^{\mathrm{T}} = \begin{Bmatrix} \sigma_1 & \sigma_2 & \sigma_3 & \sigma_4 & \sigma_5 & \sigma_6 \end{Bmatrix} \tag{11.23}$$

in which $\sigma_1 = \sigma_{xx}$, $\sigma_2 = \sigma_{yy}$, $\sigma_3 = \sigma_{zz}$, $\sigma_4 = \sigma_{xy}$, $\sigma_5 = \sigma_{yz}$, $\sigma_6 = \sigma_{xz}$

One will note that the stress components with identical subscripts are the normal stresses, whereas those with different subscripts are the shearing stresses.

11.5.2 Displacements

The change of shape of the solid due to externally applied forces is illustrated in Figure 11.15, in which the change of the positions of the points inside and on the boundary of the solid is represented by the movement of point P_1 at position (x,y,z) before the loading to point P_2 at a new location $(x + u, y + v, z + w)$ after the deformation. The net movement of point P_1 to P_2 is represented by a vector **U** with components u, v, and w along the coordinates x, y, and z, respectively. The displacement vector of a point located at $\mathbf{r}(x,y,z)$ can thus be expressed as

$$\{U(\mathbf{r})\}^{\mathrm{T}} = \begin{Bmatrix} u(\mathbf{r}) & v(\mathbf{r}) & w(\mathbf{r}) \end{Bmatrix} \tag{11.24}$$

where the bold-face **r** is the position vector representing the coordinates as defined in Chapter 3.

11.5.3 Strains

As we have defined stress in a deformed solid to be the intensity of internal resistance by the solid induced by applied forces, it is appropriate to define the induced strain to be the rate of change of displacement of the solid. Like stresses, strain components can be classified as *normal strain* components and *shearing strain* components. However, the reader should bear in mind that normal strain components result in changes in dimensions (and thus the *size*) of the solid, whereas the shearing strain components cause change only of the *shape* of the solid, which change is given by the angles of the sides of the cubic faces of the cubic element, which is 90° in the undeformed cubic element before the deformation and the angles between the same faces after the deformation, with a unit of "radians." Strain components are usually expressed as column matrices corresponding to the six independent stress components, as show in Equation 11.25:

$$\{\varepsilon\}^T = \{\varepsilon_1 \ \varepsilon_2 \ \varepsilon_3 \ \varepsilon_4 \ \varepsilon_5 \ \varepsilon_6\} \tag{11.25}$$

with $\varepsilon_1 = \varepsilon_{xx}$, $\varepsilon_2 = \varepsilon_{yy}$, $\varepsilon_3 = \varepsilon_{zz}$, $\varepsilon_4 = \varepsilon_{xy}$, $\varepsilon_5 = \varepsilon_{yz}$, $\varepsilon_6 = \varepsilon_{xz}$.

11.5.4 Fundamental Relationships

The three essential physical quantities of stress, strain, and displacement in stress analysis of solid structures are mutually interrelated by mathematical expressions derived from the theory of elasticity. We will present only those formulations that are relevant to the subsequent derivations of equations and formulae required for finite-element analysis.

11.5.4.1 Strain–Displacement Relations

As illustrated in Figure 11.15, the displacements $\{U(\mathbf{r})\}$ from point P_1 at (x,y,z) to point P_2 at $(x + u, y + v, z + w)$ with $u = u(x, y, z)$ for the net movement of the point along the x-coordinate, $v = v(x, y, z)$ for the net movement of the same point along the y-coordinate, and $w = w(x, y, z)$ for the net movement of the point along the z-coordinate. As we have defined the strain components that account for the rate of change of the displacements along the coordinates that define the deformed solid, and the strain components that exist in the same cubic element as stresses in Figure 11.14b with $\varepsilon_{xx} = \varepsilon_{xx}(x, y, z)$, $\varepsilon_{yy} = \varepsilon_{yy}(x, y, z)$, $\varepsilon_{zz} = \varepsilon_{zz}(x, y, z)$, $\varepsilon_{xy} = \varepsilon_{xy}(x, y, z)$, $\varepsilon_{yz} = \varepsilon_{yz}(x, y, z)$, and $\varepsilon_{xz} = \varepsilon_{xz}(x, y, z)$, we may relate the various strain components to the related displacement components as shown in the following equations:

$$\varepsilon_{xx}(x, y, z) = \frac{\partial u(x, y, z)}{\partial x} \tag{11.26a}$$

$$\varepsilon_{yy}(x, y, z) = \frac{\partial v(x, y, z)}{\partial y} \tag{11.26b}$$

$$\varepsilon_{zz}(x, y, z) = \frac{\partial w(x, y, z)}{\partial z} \tag{11.26c}$$

$$\varepsilon_{xy}(x, y, z) = \frac{\partial v(x, y, z)}{\partial x} + \frac{\partial u(x, y, z)}{\partial y} \tag{11.26d}$$

$$\varepsilon_{yz}(x, y, z) = \frac{\partial w(x, y, z)}{\partial y} + \frac{\partial v(x, y, z)}{\partial z} \tag{11.26e}$$

$$\varepsilon_{xz}(x, y, z) = \frac{\partial w(x, y, z)}{\partial x} + \frac{\partial u(x, y, z)}{\partial z} \tag{11.26f}$$

The relationships in Equations 11.26a to 11.26f may be expressed in the following matrix form:

$$\begin{Bmatrix} \varepsilon_{xx} \\ \varepsilon_{yy} \\ \varepsilon_{zz} \\ \varepsilon_{xy} \\ \varepsilon_{yz} \\ \varepsilon_{xz} \end{Bmatrix} = \begin{bmatrix} \frac{\partial}{\partial x} & 0 & 0 \\ 0 & \frac{\partial}{\partial y} & 0 \\ 0 & 0 & \frac{\partial}{\partial z} \\ \frac{\partial}{\partial y} & \frac{\partial}{\partial x} & 0 \\ 0 & \frac{\partial}{\partial z} & \frac{\partial}{\partial y} \\ \frac{\partial}{\partial z} & 0 & \frac{\partial}{\partial x} \end{bmatrix} \begin{Bmatrix} u(x, y, z) \\ v(x, y, z) \\ w(x, y, z) \end{Bmatrix} \tag{11.27}$$

We may further condense the relationship in Equation 11.27 into the following form for the subsequent derivation of element equations:

$$\{\varepsilon\} = [D]\{U\} \tag{11.28}$$

where matrix [D] takes the form shown in Equation 11.29 and the element displacement matrix {U} is given in Equation 11.24.

$$[D] = \begin{bmatrix} \dfrac{\partial}{\partial x} & 0 & 0 \\ 0 & \dfrac{\partial}{\partial y} & 0 \\ 0 & 0 & \dfrac{\partial}{\partial z} \\ \dfrac{\partial}{\partial y} & \dfrac{\partial}{\partial x} & 0 \\ 0 & \dfrac{\partial}{\partial z} & \dfrac{\partial}{\partial y} \\ \dfrac{\partial}{\partial z} & 0 & \dfrac{\partial}{\partial x} \end{bmatrix} \tag{11.29}$$

11.5.4.2 Stress–Strain Relations

In Figure 11.14 we represented the six independent stress components existing at any point (represented by infinitesimally small cubic elements in the figure) located at (x,y,z). We realize that there are equal numbers of corresponding strain components associate with these stress components. The relationship between these stress and strain components can be expressed by the generalized Hooke's law, as given, fore example, in Hsu (1986). Equation 11.30 shows the matrix form of such a generalized Hooke's law:

$$\begin{Bmatrix} \sigma_{xx} \\ \sigma_{yy} \\ \sigma_{zz} \\ \sigma_{xy} \\ \sigma_{yz} \\ \sigma_{xz} \end{Bmatrix} = \dfrac{E}{(1+v)(1-2v)} \begin{bmatrix} 1-v & v & v & 0 & 0 & 0 \\ & 1-v & v & 0 & 0 & 0 \\ & & 1-v & 0 & 0 & 0 \\ & & & \dfrac{1-2v}{2} & 0 & 0 \\ & & & & \dfrac{1-2v}{2} & 0 \\ & \text{SYM} & & & & \dfrac{1-2v}{2} \end{bmatrix} \begin{Bmatrix} \varepsilon_{xx} \\ \varepsilon_{yy} \\ \varepsilon_{zz} \\ \varepsilon_{xy} \\ \varepsilon_{yz} \\ \varepsilon_{xz} \end{Bmatrix} \tag{11.30}$$

or in a compact form:

$$\{\sigma\} = [C]\{\varepsilon\} \tag{11.31}$$

in which E is the Young's modulus and v is the Poisson's ratio of the material. The matrix [C], called the elasticity matrix, has the form:

$$[C] = \dfrac{E}{(1+v)(1-2v)} \begin{bmatrix} 1-v & v & v & 0 & 0 & 0 \\ & 1-v & v & 0 & 0 & 0 \\ & & 1-v & 0 & 0 & 0 \\ & & & \dfrac{1-2v}{2} & 0 & 0 \\ & & & & \dfrac{1-2v}{2} & 0 \\ & \text{SYM} & & & & \dfrac{1-2v}{2} \end{bmatrix} \tag{11.32}$$

11.5.4.3 Strain Energy in Deformed Elastic Solids

The above formulations are derived from a postulate that in elastic solids, the deformations, stresses, and strains induced by external forces applied to the solid in the equilibrium state will disappear altogether after removal of the applied forces. This postulate implies that there must be a mechanism that restores the solid to its original state after removal of the applied loads. This mechanism is referred to as the "strain energy." This energy is induced in the solid during the deformation process, and it is "stored" in the deformed solid at the equilibrium state of the solid under external loading. It will be released to restore the deformed solid to its original state upon removal of the applied actions. Since this energy is created by the deformation of the solid expressed by the strains induced by the externally applied forces in the form of the induced stresses, we may mathematically express this energy in the following form:

$$U(\{\varepsilon\}, \{\sigma\}) = \frac{1}{2} \int_v (\sigma_{xx}\varepsilon_{xx} + \sigma_{yy}\varepsilon_{yy} + \sigma_{zz}\varepsilon_{zz} + \sigma_{xy}\varepsilon_{xy} + \sigma_{xz}\varepsilon_{xz} + \sigma_{yz}\varepsilon_{yz})\,dv \tag{11.33}$$

or in a compact version:

$$U(\{\varepsilon\}, \{\sigma\}) = \frac{1}{2} \int_v \{\varepsilon\}^T \{\sigma\}\,dv \tag{11.34}$$

11.5.5 Finite-element Formulation

We will follow the formulations described in Step 5 of Section 11.3 in deriving the element and overall stiffness equations for stress analysis of elastic solid continua. Figure 11.16 illustrates how a cam-shaft assembly may be discretized by a finite number of tetrahedral elements interconnected at nodes. These elements are typically used for discretization of solids with complicated geometry in three-dimensional finite-element analyses. Shapes of these elements are illustrated in Figure 11.2.

It was mentioned earlier that hexahedral elements are commonly used in the finite-element modeling of solids of complicated geometry in three dimensions, but since each hexahedral element can be subdivided into four or five tetrahedral elements in actual computations, we will present the subsequent finite-element formulations in terms of typical tetrahedral elements. As illustrated in Figure 11.17, tetrahedral elements have shapes that are similar to pyramids with

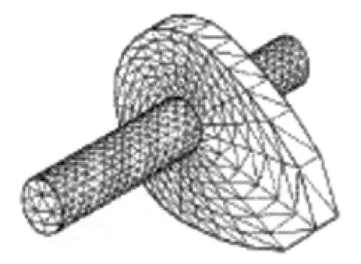

Figure 11.16 Finite-element model of a cam-shaft assembly. (*Source*: Hsu and Sinha, 1992.)

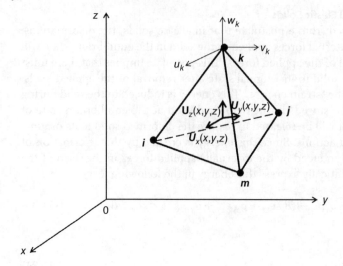

Figure 11.17 Typical tetrahedral elements.

four apexes (or nodes) for each element. These nodes may be designated by nodes i, j, k, and m as indicated in Figure 11.17.

The primary unknown quantity in typical finite-element analysis in stress analysis is the displacements of the element $U(\mathbf{r})$ with $\mathbf{r}$ being the position vector in a discretized model representing the coordinates of the solution point (usually located at the center of gravity) in the element. The element displacement vector is expressed as $U(x,y,z)$ in the finite-element formulation.

The displacement in a deformed solid is a vector quantity with components along the three coordinates that define the space in which the element is situated. Mathematically, we may represent the element displacement as

$$\{U(x,y,z)\} = \begin{Bmatrix} U_x(x,y,z) \\ U_y(x,y,z) \\ U_z(x,y,z) \end{Bmatrix} = \begin{Bmatrix} U(x,y,z) \\ V(x,y,z) \\ W(x,y,z) \end{Bmatrix}$$

for the element, in which the components U_x, U_y, and U_z along the coordinates x, y, and z are often denoted by $U(x,y,z)$, $V(x,y,z)$, and $W(x,y,z)$, respectively.

The discretization process in the finite-element analysis requires a relationship between the displacements in the element and those at the associated nodes according to Equation 11.1 or as expressed in Equation 11.35 for the current situation:

$$\{U(x,y,z)\} = \begin{Bmatrix} U_x(x,y,z) \\ U_y(x,y,z) \\ U_z(x,y,z) \end{Bmatrix} = [N(x,y,z)]\{u\} \tag{11.35}$$

for the element displacement in a discretized finite-element model with $\{u\}$ being the displacement components of the associated nodes at fixed coordinates in the discretized model. $N(x,y,z)$ is the interpolation function for the tetrahedral element in Figure 11.17. The displacement components at each of the four nodes of the element in Figure 11.17 consist of three components as in the following:

$$\{u\}|_{node\,i}^{T} = \{u_i \ v_i \ w_i\},$$

$$\{u\}|_{\text{node}\,j}^{\text{T}} = \{u_j \; v_j \; w_j\},$$

$$\{u\}|_{\text{node}\,k}^{\text{T}} = \{u_k \; v_k \; w_k\}$$

$$\{u\}|_{\text{node}\,m}^{\text{T}} = \{u_m \; v_m \; w_m\}$$

These nodal displacements are point values in the form of real numbers. The total number of displacement components of the four nodes of a tetrahedral element in Figure 11.17 is thus equal to 12, as expressed in the following matrix form:

$$\{u\}^{\text{T}} = \{u_i \; v_i \; w_i \; u_j \; v_j \; w_j \; u_k \; v_k \; w_k \; u_m \; v_m \; w_m\} \tag{11.36}$$

where the subscripts identify the nodes: for example, the subscripts i, j, k, and m denote nodes i, j, k, and m, respectively. The notations u, v, and w in Equation 11.36 represent the displacement components along the x-, y-, and z-directions, respectively.

Referring to the tetrahedral element in Figure 11.17, Equation 11.35 can be expanded to give

$$\{U(x,y,z)\} = \begin{Bmatrix} U_x(x,y,z) \\ U_y(x,y,z) \\ U_z(x,y,z) \end{Bmatrix}$$

$$= [N(x,y,z)]\{u\}$$

$$= \{N_i(x,y,z) \; N_j(x,y,z) \; N_k(x,y,z) \; N_m(x,y,z)\}\{u\}$$

or, in an compact form:

$$\{U(x,y,z)\} = \{N_i u_i \; N_j u_j \; N_k u_k \; N_m u_m\} \tag{11.37}$$

where N_i, N_j, N_k, and N_m are the components of the interpolation function associated with node i, j, k, and m, respectively, in Figure 11.17.

Substituting Equation 11.35 into Equation 11.28 results in the relationship between the element strains and nodal displacements:

$$\{\varepsilon(x,y,z)\} = [B(x,y,z)]\{u\} \tag{11.38}$$

where the matrix [B] has the form:

$$[B(x,y,z)] = [D][N(x,y,z)] \tag{11.39}$$

The above expression indicates that the matrix $[B(x,y,z)]$ may be obtained from the derivatives of $[N(x,y,z)]$ with respect to the coordinates x-, y-, and z-, respectively. The matrix [D] is given in Equation 11.29.

The element stresses and the nodal displacement components are related by substituting Equation 11.38 into Equation 11.31 to give Equation 11.40:

$$\{\sigma(x,y,z)\} = [C][B(x,y,z)]\{u\} \tag{11.40}$$

And finally, the strain energy in the element is obtained by substituting Equations 11.38 and 11.40 into Equation 11.34, resulting in the following equation for the strain energy of the deformed element:

$$U(\{u\}) = \frac{1}{2}\int_v \{u\}^{\text{T}}[B(x,y,z)]^{\text{T}}[C][B(x,y,z)]\{u\}\,dv \tag{11.41}$$

The strain energy $U(\{u\})$ in Equation 11.41 is the energy stored in the deformed solid (in the elements in this case). It is induced by the deformation of the elements in a discretized finite-element models by the applied forces at the nodes of the elements.

Strain energy in the deformed elements constitutes an important part of the potential energy of discretized solid continua such as shown in Figure 11.1b. The potential energy in a deformed solid is the difference between the induced strain energy in the deformed solid and the work done on the solid by applied external forces. For a continuum such as shown in Figure 11.1a, this difference should be zero; or conversely, the induced strain energy in the deformed solid should be equal to the work done on the solid by the applied forces. In the discretized solid model consisting of individual elements such as that illustrated in Figure 11.1b, however, the equality of the induced strain energy in elements $U\{u\}$ in Equation 11.41 and the work performed on the same element by applied forces on the surface in the form of surface tractions $\{t\}$ and body forces $\{f\}$ may likely not hold. To ensure that the discretized solid is stable and in mechanical equilibrium, such a difference must be kept to a minimum.

The formulation of the potential energy (Π) in a deformed discretized model such as shown in Figure 11.1b can thus be expressed mathematically in the form

$$\Pi(\{u\}) = U(\{u\}) - W_p$$

where the work produced by applied forces in the deformed element by the body forces $\{f\}$ and surface tractions $\{t\}$ can be expressed as

$$W_p = \int_v \{U(x,y,z)\}^T \{f\}\, dv + \int_s \{U(x,y,z)\}^T \{t\}\, ds$$

in which $\{U(x,y,z)\}$ are the displacement components of the element, v and s are the respective volume and surface of the element, and $\{f\}$ and $\{t\}$ are the respective body force components and surface tractions applied to the element.

The potential energy in the element can thus be expressed in terms of nodal displacement with the interpolation function $[N(x,y,z)]$ in the following expression:

$$\Pi(\{u\}) = \frac{1}{2}\int_v \{u\}^T [B(x,y,z)]^T [C][B(x,y,z)]\{u\}\, dv$$

$$- \int_v \{u\}^T [N(x,y,z)]^T \{f\}\, dv - \int_s \{u\}^T [N(x,y,z)]^T \{t\}\, ds \qquad (11.42)$$

By applying the principle of variational calculus, the potential energy function in Equation 11.42 has either a maximum or minimum value by solving the following equation:

$$\frac{\partial \Pi(\{u\})}{\partial \{u\}} = 0$$

with nodal displacement components $\{u\}$ to be "point values" which are independent to the coordinates in differentiation. This variation will thus lead to the following equality:

$$\left(\int_v [B(x,y,z)]^T [C][B(x,y,z)]\, dv\right)\{u\}$$

$$- \int_v [N(x,y,z)]^T \{f\}\, dv - \int_s [N(x,y,z)]^T \{t\}\, ds = 0 \qquad (11.43)$$

One may also demonstrate the second-order derivative of the potential energy function $\Pi(\{u\})$ to be positive:

$$\frac{\partial \Pi^2(\{u\})}{\partial \{u\}^2} = \int_v [B(x,y,z)]^T [C][B(x,y,z)]\, dv > 0$$

We may thus ensure that the solution of the induced nodal displacements $\{u\}$ by applied loads as obtained from Equation 11.43 will result a minimum value of the potential function in Equation 11.42.

Equation 11.43 may be expressed in the following form, called the "element equations":

$$[K_e]\{u\} = \{p\} \tag{11.44}$$

in which

$[K_e]$ = Element stiffness matrix

$$= \int_v [B(x, y, z)]^{\mathrm{T}}[C][B(x, y, z)]\, dv \tag{11.45}$$

with the matrix $[B(x,y,z)]$ as in Equation 11.39 and the matrix $[C]$ available from Equation 11.32. The matrix $\{p\}$ has the form:

$\{p\}$ = Nodal force matrix

$$= \int_v [N(x, y, z)]^{\mathrm{T}}\{f\}\, dv + \int_s [N(x, y, z)]^{\mathrm{T}}\{t\}\, ds \tag{11.46}$$

and the matrix $[N(x,y,z)]$ is the interpolation function defined in Equation 11.1

In a strict sense, a deformed solid is in equilibrium only when the potential energy Π in Equation 11.42 in all elements in the discretized model is minimum. Consequently, we postulate that the potential energy of the entire discretized solid can be obtained by the summation of potential energies in all individual elements in the discretized solid. This postulate legitimizes our using the following equation for the entire discretized structure:

$$[K]\{u\} = \{P\} \tag{11.47}$$

where

$$[K] = \sum_{m=1}^{M} [K_e^m] = \text{overall coefficient matrix} \tag{11.48}$$

with

$$[K_e] = \int_v [B(x, y, z)]^{\mathrm{T}}[C][B(x, y, z)]\, dv$$

as in Equation 11.45; M is total number of elements in the discretized model, and $\{p\}$ represents the total nodal forces as expressed in Equation 11.46.

The assembly of the element stiffness matrices for the overall stiffness matrix in Equation 11.47 entails following the rule stipulated in Step 6 in Section 11.3 and illustrated in Example 11.3.

11.5.6 Finite-element Formulation for One-dimensional Solid Structures

We will derive the element equation of a bar element such as shown in Figure 11.7 in Example 11.1 to illustrate the general procedure for developing finite-element formulations.

We will begin by referring to the bar element illustrated in Figure 11.7, made of a material with Young's modulus E. We locate the bar element in a plane defined by the x–y coordinate system as shown in Figure 11.18.

The interpolation function $[N(x)]$ that relates the displacement in the bar element to its two nodes may be derived from a simple polynomial function as we did in Example 11.1. This function has the form shown in Equation 11.10:

$$N(x) = \left\{ \frac{x - x_2}{x_1 - x_2} \quad -\frac{x - x_1}{x_1 - x_2} \right\} \tag{11.10}$$

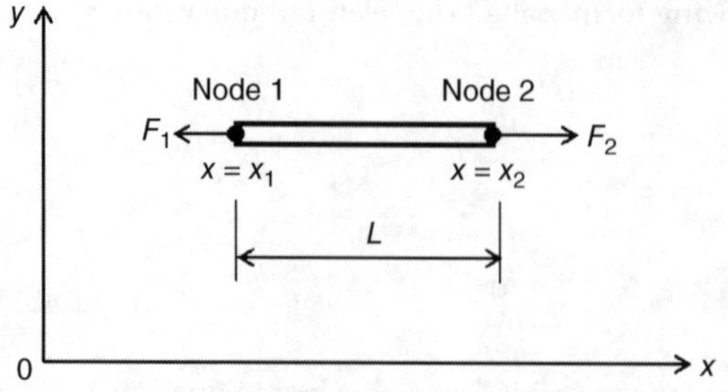

Figure 11.18 An axially deformed bar.

The matrix $[B(x)]$ that relates the element strain and the nodal displacement may be obtained using Equation 11.39:

$$[B(x)] = \frac{d}{dx}[N(x)]$$

$$= \frac{d}{dx}\left\{\frac{x-x_2}{x_1-x_2} \quad -\frac{x-x_1}{x_1-x_2}\right\}$$

$$= \left\{\frac{1}{x_1-x_2} \quad -\frac{1}{x_1-x_2}\right\} \tag{11.49}$$

Since the bar element is subjected to two applied forces with $-F_1$ at node 1 and $+F_2$ at node 2 and both these forces act along the longitudinal direction, that is, along the x-coordinate of the bar, the only induced stress component in the bar is σ_{xx} along the x-coordinate. Likewise, the only induced strain component is ε_{xx} along the same direction. The relationship between the induced stress and strain obeys Hooke's law with $\sigma_{xx} = E\varepsilon_{xx}$ where E is the Young's modulus of the bar material. We thus obtain the elasticity matrix $[C] = E$.

By substituting the above expression for $[N(x)]$ from Equation 9.10, $[B(x)]$ in Equation 11.49, and with $[C] = E$ and $dv = A\,dx$, where A is the cross-sectional area of the bar element and $x_1 - x_2 = -L$ into Equations 11.42 and (11.43), we obtain the element stiffness matrix from Equation 11.45:

$$[K_e] = \int_0^L \left\{-\frac{1}{L} \quad \frac{1}{L}\right\}^T E \left\{-\frac{1}{L} \quad \frac{1}{L}\right\} A\,dx$$

$$= \frac{EA}{L}\begin{bmatrix} 1 & -1 \\ -1 & 1 \end{bmatrix} \tag{11.50}$$

The nodal forces matrix in this case is

$$\{p\}^T = \{-F_1 \quad F_2\}$$

From Equation 11.44 the element equation of the bar element thus has the form

$$\frac{EA}{L}\begin{bmatrix} 1 & -1 \\ -1 & 1 \end{bmatrix}\begin{Bmatrix} u_1 \\ u_2 \end{Bmatrix} = \begin{Bmatrix} -F_1 \\ F_2 \end{Bmatrix} \tag{11.51}$$

Example 11.5
Use the finite-element formulation in Section 11.5 to determine the displacements at the joint of a compound bar made of two rods subjected to a uniaxial force P. The dimensions of the rods are given in Figure 11.19. The compound bar has a cross-sectional area of $A = 650$ mm^2 and the Young's moduli of copper and aluminum are $E_{cu} = 10\,300$ MPa and $E_{al} = 69\,000$ MPa, respectively.

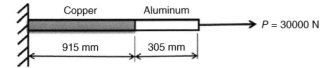

Figure 11.19 A uniaxially loaded compound bar made of copper and aluminum.

Figure 11.20 Finite-element model for a uniaxially loaded compound bar.

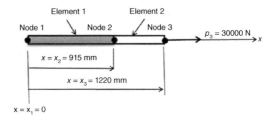

Solution:

Since the compound rod is uniaxially loaded by a force P as shown in Figure 11.19, it is expected to elongate along the same direction along the x-coordinate. The finite-element model for this problem is illustrated in Figure 11.20, with two elements and three nodes, with node 1 located at $x = 0$, node 2 at $x = 915$ mm, and node 3 at $x = 1220$ mm. The displacements at these nodes are expressed as: $\{u\}^T = \{u_1 \quad u_2 \quad u_3\}$, with a fixed boundary condition at node 1 of $u_1 = 0$.

Our solution begins with developing the element equations as shown in Equation 11.44 using the element equation that we derived in Example 11.1 and the element stiffness matrices for a bar element in Equation 11.50 as follows.

For Element 1 made of copper:

$$
\begin{aligned}
\left[K_e^1\right] &= \frac{E_1 A_1}{L_1} \begin{bmatrix} 1 & -1 \\ -1 & 1 \end{bmatrix} \\
&= \frac{10\,300 \times 650}{915 \times 10^{-3}} \begin{bmatrix} 1 & -1 \\ -1 & 1 \end{bmatrix} \\
&= \begin{bmatrix} 7.317 & -7.317 \\ -7.317 & 7.317 \end{bmatrix} \times 10^6 \quad \text{N/m}
\end{aligned}
\tag{a}
$$

in which we have used the given values of Young's modulus, $E_1 = 10\,300$ MPa for copper, the cross-sectional area of the rod, $A_1 = 650$ mm^2, and the length of the element, $L_1 = 915$ mm.

Likewise, we may compute the stiffness matrix for Element 2 made of aluminum:

$$
\begin{aligned}
\left[K_e^2\right] &= \frac{E_2 A_2}{L_2} \begin{bmatrix} 1 & -1 \\ -1 & 1 \end{bmatrix} \\
&= \frac{69\,000 \times 650}{305 \times 10^{-3}} \begin{bmatrix} 1 & -1 \\ -1 & 1 \end{bmatrix} \\
&= \begin{bmatrix} 147.05 & -147.05 \\ -147.05 & 147.05 \end{bmatrix} \times 10^6 \quad \text{N/m}
\end{aligned}
\tag{b}
$$

The element equations for both Elements 1 and 2 thus take the following forms in Equation 11.44, expanded to give

$$
\begin{bmatrix} 7.317 \times 10^6 & -7.317 \times 10^6 \\ -7.317 \times 10^6 & 7.317 \times 10^6 \end{bmatrix} \begin{Bmatrix} u_1 \\ u_2 \end{Bmatrix} = \begin{Bmatrix} p_1 \\ p_2 \end{Bmatrix}
\tag{c}
$$

for Element 1, and

$$\begin{bmatrix} 147.05 \times 10^6 & -147.05 \times 10^6 \\ -147.05 \times 10^6 & 147.05 \times 10^6 \end{bmatrix} \begin{Bmatrix} u_2 \\ u_3 \end{Bmatrix} = \begin{Bmatrix} p_2 \\ p_3 \end{Bmatrix} \tag{d}$$

for Element 2.

We are now ready to assemble the element stiffness matrices in Equations (c) and (d) for the overall stiffness matrix of the structure following the rule of assembly as described in Example 11.3 and Figure 11.11:

$$[K] = 10^6 \begin{bmatrix} 7.317 & -7.317 & 0 \\ 0 & (7.317 + 147.05) & -147.05 \\ 0 & 0 & 147.05 \end{bmatrix}$$

$$= 10^6 \begin{bmatrix} 7.317 & -7.317 & 0 \\ 0 & 154.367 & -147.05 \\ 0 & 0 & 147.05 \end{bmatrix} \text{N/m} \tag{e}$$

We may thus use the above finite-element equation for the overall stiffness equation in Equation 11.47 as

$$10^6 \begin{bmatrix} 7.317 & -7.317 & 0 \\ 0 & 154.367 & -147.05 \\ 0 & 0 & 147.05 \end{bmatrix} \begin{Bmatrix} u_1 \\ u_2 \\ u_3 \end{Bmatrix} = \begin{Bmatrix} p_1 \\ p_2 \\ p_3 \end{Bmatrix}$$

The above equation may be expressed in Equation (f) with specified nodal constraint and applied nodal forces as

$$10^6 \begin{bmatrix} 7.317 & -7.317 & 0 \\ 0 & 154.367 & -147.05 \\ 0 & 0 & 147.05 \end{bmatrix} \begin{Bmatrix} u_1 = 0 \\ u_2 \\ u_3 \end{Bmatrix} = \begin{Bmatrix} p_1 = 0 \\ p_2 = 0 \\ p_3 = 30\,000 \end{Bmatrix} \tag{f}$$

after inserting the applicable boundary condition $u_1 = 0$ at node 1, and the loading conditions $P_1 = 0$ at node 1, $P_2 = 0$ at node 2, and $P_3 = 30\,000$ N at node 3. Equation (f) is partitioned following Equation 11.19 in the form

$$10^6 \begin{bmatrix} 7.317 & -7.317 & 0 \\ 0 & 154.367 & -147.05 \\ 0 & 0 & 147.05 \end{bmatrix} \begin{Bmatrix} u_1 = 0 \\ u_2 \\ u \end{Bmatrix} = \begin{Bmatrix} P_1 = 0 \\ P_2 = 0 \\ P_3 = 30000 \end{Bmatrix}$$

We will thus require to solve the following equations:

$$10^6 \begin{bmatrix} 154.367 & -147.05 \\ 0 & 147.05 \end{bmatrix} \begin{Bmatrix} u_2 \\ u_3 \end{Bmatrix} = \begin{Bmatrix} 0 \\ 30\,000 \end{Bmatrix} \tag{g}$$

Equation (g) offers the solution of displacement of node 3 with $u_3 = 30\,000/(147.05 \times 10^6) = 204 \times 10^{-6}$ m or 2.04 mm. With $u_3 = 2.04$ mm, we have the following relationship from the same equation:

$$10^6(154.367u_2 - 147.05u_3) = 0$$

from which we may solve for $u_2 = 1.94$ mm.

The total elongation of the compound rod is thus $u_{total} = u_2 + u_3 = 1.94 + 2.04 = 3.98$ mm.

The induced strains in the two elements are obtained by using the relationship $\{\varepsilon\} = [B]\{u\}$ in Equation 11.38 with

$$[B(x)] = \frac{d}{dx}[N(x)] = \left\{ \frac{1}{x_1 - x_2} \quad -\frac{1}{x_1 - x_2} \right\}$$

Thus, for Element no. 1 with $x_1 - x_2 = -L_1 = -915$ mm,

$$[B_1] = \left\{ -\frac{1}{915} \quad \frac{1}{915} \right\}$$

resulting to the strain in Element 1 as

$$\varepsilon_{xx}^1 = \left\{ -\frac{1}{915} \quad \frac{1}{915} \right\} \left\{ \begin{matrix} u_1 = 0 \\ u_2 \end{matrix} \right\} = \frac{1}{915}u$$

$$= \frac{1.94}{915} = 0.21\%$$

The [B] matrix for Element 2 with $x_1 - x_2 = 915 - 1220 = L_2 = -305$ mm results in

$$[B_2] = \left\{ -\frac{1}{305} \quad \frac{1}{305} \right\}$$

leading to the induced strain in Element 2 as

$$\varepsilon_{xx}^2 = \left\{ -\frac{1}{305} \quad \frac{1}{305} \right\} \left\{ \begin{matrix} u_2 \\ u_3 \end{matrix} \right\} = -\frac{u_2}{305} + \frac{u_3}{305}$$

$$= -\frac{1.94}{305} + \frac{2.04}{305}$$

$$= \frac{0.1}{305} = 0.033\%$$

The corresponding stress in both elements can be computed using Hooke's law in Equation 11.31 with $[C_1] = E_{cu} = 10\,300$ MPa, and $[C_2] = E_{al} = 69\,000$ MPa.

We thus have the induced stresses in both elements:

$$\sigma_{xx}^1 = [C_1]\varepsilon_{xx}^1 = E_{cu}\varepsilon_{xx}^1 = 10\,300 \times 0.0021 = 1.03 \ \text{MPa}$$

in Element 1, and

$$\sigma_{xx}^2 = [C_2]\varepsilon_{xx}^2 = E_{al}\varepsilon_{xx}^2 = 69\,000 \times 0.000\,33 = 22.77 \ \text{MPa}$$

in Element 2.

11.6 General-purpose Finite-element Analysis Codes

The incredible versatility of the finite-element method in solving complicated but closer to real engineering problems has resulted in wide acceptance of this method by members of the scientific and engineering communities for handling a great variety of problems with great accuracy and reliability well beyond the capabilities of traditional analytical tools. The broad acceptance of this method has also prompted the commercialization of finite-element analysis software package, or "codes" by a number of reputable companies. Finite-element analysis codes such as ANSYS, ABACUS, COSMOS, and others have served the industry well in the past half a century. Updated versions of these codes with improved accuracy and expanded capabilities were made available to users continually. It is not possible to describe here all of these commercially available codes: what will be presented in this section is merely an overview.

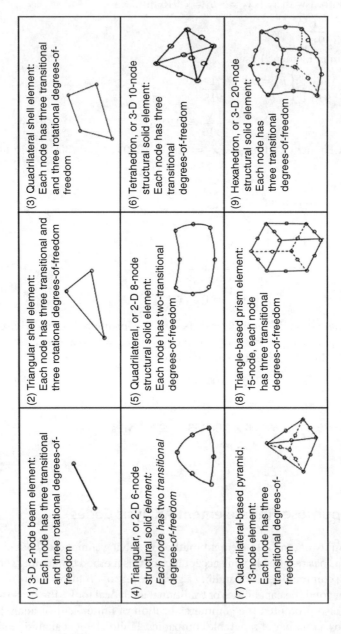

(1) 3-D 2-node beam element:
Each node has three transitional
and three rotational degrees-of-
freedom

(2) Triangular shell element:
Each node has three transitional and
three rotational degrees-of-freedom

(3) Quadrilateral shell element:
Each node has three transitional
and three rotational degrees-of-
freedom

(4) Triangular, or 2-D 6-node
structural solid element:
*Each node has two transitional
degrees-of-freedom*

(5) Quadrilateral, or 2-D 8-node
structural solid element:
Each node has two-transitional
degrees-of-freedom

(6) Tetrahedron, or 3-D 10-node
structural solid element:
Each node has three
transitional
degrees-of-freedom

(7) Quadrilateral-based pyramid,
13-node element:
Each node has three
transitional degrees-of-
freedom

(8) Triangle-based prism element:
15-node, each node
has three transitional
degrees-of-freedom

(9) Hexahedron, or 3-D 20-node
structural solid element:
Each node has
three transitional
degrees-of-freedom

Figure 11.21 Advanced elements for finite-element models.

11.6.1 Common Features in General-purpose Finite-element Codes

Finite-element programs that can be used to solve a large class of engineering problems go under the generic name of "general-purpose programs." Common desirable features of such a program are described below.

1) The program should offer extensive capabilities of solving engineering problems ranging from static and dynamic elastic to elastoplastic stress analysis of solid structures made either of traditional materials or of unusual materials such as composites, biomedical materials, and so on, with applied loads from nontraditional sources such as piezoelectric and electro-static forces as well as molecular forces in the stress analysis of structures at micrometer and nanometer scales.

2) It must have a large element library with many more element types to choose from than the simplex elements offered in Figure 11.2. There are other types of elements that can make the finite-element analysis more efficient in terms of setting discretization models and that pro-duce more accurate analytical results. Figure 11.21 shows some of these elements available in the element library of the ANSYS code (Lee, 2015). One will observe from Figure 11.21 that the first of the three listed types of elements is for beam structures using bar elements and the second and third types are for modeling plates of planar geometry. Thin-shell ele-ments are available for structures such as pipes and pressure vessels. One may also note that the last six types of elements have additional nodes at the middle of the edges of the elements. These are the elements normally called the "serendipity" elements, with many such elements having additional nodes on selected locations along the edges of the ele-ments (Bathe and Wilson, 1976; Hsu, 1986). The functions that represent the variation of the primary unknowns in these elements are higher-order polynomial functions, such as the Lagrange polynomials. There are many advantages of using these elements and high-order polynomial interpolation functions, with one obvious advantage of better fitting the element shapes to solids of curved geometry. The higher order of the polynomial functions used in the computations can yield more accurate results from the analyses. The additional nodes in the elements may result in the use of fewer elements in the discretized model for the finite-element analysis. There are many other types of elements that are often offered by general-purpose finite-element codes, such as "spring elements," "contact elements," and "interface elements" with unusually large aspect ratio (defined as the ratio of the longest to shortest edges of the element). Other special elements may include the "birth and death ele-ments," such as are found in the element library of the ANSYS code. These elements are used for cases simulating the "growth" of solid substances involving welding, or the "disappear-ance" of solid substances such as in etching of structures in microfabrication process, or the growth and propagation of cracks in solid structures in fracture mechanics analyses.

3) The package must have user-friendly pre-processor with sufficient versatility. A good pre-processor should include, in addition to extensive material database, easy-to-use automatic mesh generation for users to establish the required discretized model with user-selected density and configurations of finite-element meshes, such as illustrated in Figures 11.3 and 11.13b.

4) It must have a comprehensive and easy-to-use post-processor that can display the results of the analysis with visual outputs, either in static graphical form or in animated graphical output for continuous action if so desired. Users should be offered ample options in choosing the desired form for outputs of their analyses.

5) It is desirable to offer cost-effective ways to use the code in terms of minimal effort in learning the use of the code, minimal human input, and high computational efficiency. Most com-mercial finite-element codes adopt highly efficient solution methods.

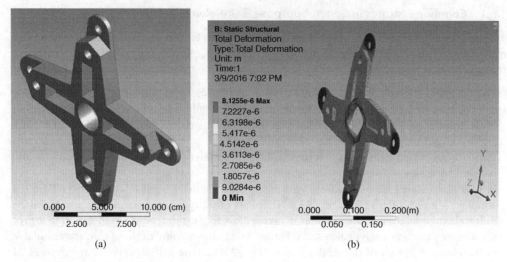

Figure 11.22 Finite-element analysis on imported solid geometry. (a) Imported solid geometry. (b) Finite-element analytical results. (Courtesy of ANSYS Inc. , Pittsburg, PA, USA).

6) Effective interfacing of computer-aided design (CAD) with the finite-element analysis (FEA) is also a desirable feature of general-purpose FE code. It allows the user to perform finite-element analysis using the imported solid model geometry from a CAD package. Figure 11.22a shows a solid model of a coupler from a computer-aided design package to be automatically translated to the ANSYS code for finite-element stress analysis; and Figure 11.22b shows the variation of deformation of the structure under applied load. There are also codes that offer interfacing of CAD/FE with computer-aided-manufacturing (CAM), with the solid models of structures being downloaded to a CAM package ready for mechanical analysis as well as coding for computer-numerically controlled (CNC) machine tools after satisfactory finite-element analysis.

11.6.2 Simulation using general-purpose finite-element codes

A relatively new application of general-purpose finite-element code is to simulate critical engineering processes, in particular those involved in new product development. Such simulations may result in significant cost saving for the development process, and also reduce the time to market for these products. While the process of product development may vary from case to case, Figure 11.23 illustrates the general procedure with various stages in such a development.

A recent article on "What does product development really cost" by K. Douglass [http://www .pivotint.com] reports some resounding outcomes, such as that a 12-month in reduction in time-to-market (TTM) would result in 92% increase in "internal rate of return" (IRR) on the product development investment. A 63% increase in IRR was achieved with a 9-month reduction in TTM. Further observation indicates that other than the costs of personnel and management, and the market research activities, the most costly activities in new product development are the design, production, and testing of the two prototypes, as indicated in Figure 11.23. Not only are activities relating to the prototyping costly, they also prolong time-to-market.

General-purpose finite-element code such as ANSYS has developed such capability that it can simulate many of the activities involved in product development that are shown in Figure 11.23. It can produce high-quality simulation based on expected functions of products of complex geometry subjected to all conceivable loading conditions. It also provides graphic simulations

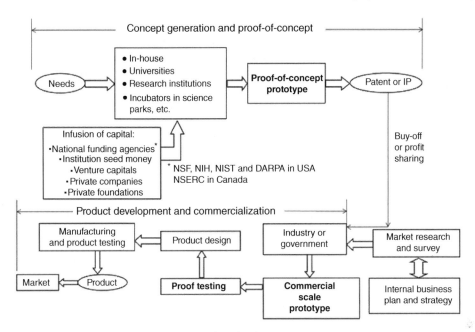

Figure 11.23 General processes in new product development.

of what "real" physical prototypes can achieve under all conceivable conditions and beyond, which results not only in substantial savings in costs of building the real prototypes but also in substantial time saving in production and testing of the product using the virtual prototypes.

Figure 11.24 illustrates a case of simulation by the ANSYS code of various vibrational modes of the disk-and-pad assembly of an automobile braking system. The design objective of this

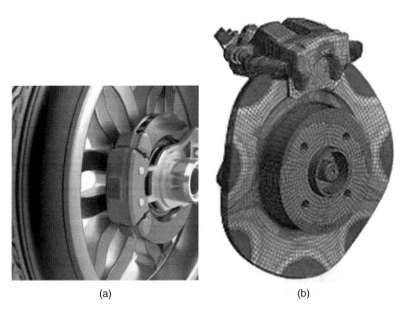

(a)   (b)

Figure 11.24 Simulation of the vibration of a disk-coupler during braking of a vehicle. (a) New design of braking pad. (b) Mode shape of disk. (Courtesy of ANSYS Inc. , Pittsburg, PA, USA).

system was to identify the onset of instability in braking using a modal analysis. The ANSYS solution from simulating the performance of a new design of a braking system such as shown in Figure 11.24 enabled engineers to achieve a first-time virtual design of a cost-effective, quiet braking system for multiple platforms in far less time than would have been possible with conventional methods using physical prototyping and field testing.

11.7 Problems

11.1 Use no more than 25 words in the answers to each of the following questions: (1) Why does the discretized FE model used in FEA lead to approximate but not the exact solutions to the real problems? (2) What is the major error induced in FEA with simplex elements using linear polynomial functions for the interpolation functions in the FEA? (3) How do you get more accurate results from the FEA involving simplex elements in question (2)? (4) What is the principal reason for using general-purpose FE codes to simulate new product development?

11.2 Show that the nodal force vector in stress analysis of an element may be expressed as

$$\{p\} = \int_v [B(\mathbf{r})]^T \{\sigma\} \, dv$$

in which matrix $[B(\mathbf{r})]$ is given in Equation 11.39 with $\mathbf{r}$ being the coordinates, $\{\sigma\}$ is the stresses in the element, and v is the volume of the element.

11.3 Derive the interpolation function for a triangular toroidal element as illustrated in Figure 11.2. The cross-section of the element is made of nodes i, j, and k located at (r_i, z_i), (r_j, z_j), and (r_k, z_k), respectively, in the r–z coordinate system, with the z-coordinate being the axis of symmetry. Assume that the element displacement components in the radial and axial directions obey linear polynomial functions $U_r(r, z) = b_1 + b_2 r + b_3 z$, and $U_z(r, z) = b_4 + b_5 r + b_6 z$ in which $b_1, b_2, b_3, \ldots, b_6$ are arbitrary constants.

11.4 Find the total elongation and the displacement at the nodes that join the three different rods in a compound bar as illustrated in Figure 11.25. The structure is similar to that in Example 11.5 but now with a third rod made of steel with Young's modulus $E_{st} = 200$ GPa.

11.5 Solve Problem 11.4 with the rods of copper and steel swapped in their respective positions in the compound bar in Figure 11.25. Compare the results in elongations of the three rods with those computed by formulae given in books of classical mechanics of materials.

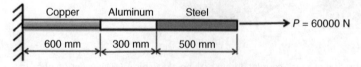

Figure 11.25 A compound bar made of three different materials.

11.6 Use the finite-element technique with a linear polynomial function for the interpolation function to compute the stresses and strains in the triangular plate element in Figure 11.26. Also find the displacement at the three nodes. The plate element is subjected to a force acting at node B. Use the material properties $E = 70$ GPa and $v = 0.3$ in your computations. The thickness of the plate element is 20 mm.

11.7 Solve the displacements, strains, and stresses in a triangular plate similar to that in Problem 11.6 but with the applied force at node B with an angle of 30° downward from the x-axis.

11.8 Use the finite-element method to compute the following quantities for the triangular plate in Figure 11.27:
a) Displacement components at the corners in the x- and y-directions.
b) The stresses and strains in the plate.
c) Reactions at the supports at nodes A and C.
The plate is made of aluminum alloy with Young's modulus $E = 70$ GPa and Poisson's ratio $v = 0.3$.
Hint: The distributed pressure loading on the edge of an element of the finite-element model may be treated as concentrated forces equally acting on the two adjacent nodes of the edge over the plane area of the edge. The plate is 20 mm thick.

Figure 11.26 A triangular plate subjected to a nodal force.

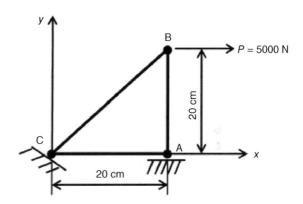

Figure 11.27 A triangular plate subjected to uniform pressure loading.

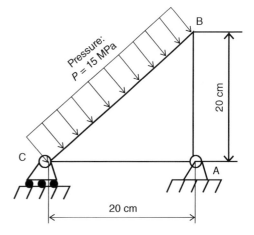

11.9 Show the overall finite-element equation similar to Equation 11.45 for a plane structure defined by an x–y coordinate system as an assembly of two elements with the individual element equations being as follows:

For Element no. 1:

$$
\begin{bmatrix}
1 & 2 & 3 & 4 & 5 & 6 \\
 & 1 & 2 & 3 & 4 & 5 \\
 & & 1 & 2 & 3 & 4 \\
 & \text{SYM} & & 1 & 2 & 3 \\
 & & & & 1 & 2 \\
 & & & & & 1
\end{bmatrix}
\begin{Bmatrix}
u_1^1 \\ v_1^1 \\ u_2^1 \\ v_2^1 \\ u_4^1 \\ v_4^1
\end{Bmatrix}
=
\begin{Bmatrix}
p_{1x}^1 \\ p_{1y}^1 \\ p_{2x}^1 \\ p_{2y}^1 \\ p_{4x}^1 \\ p_{4y}^1
\end{Bmatrix}
$$

and for Element no. 2:

$$
\begin{bmatrix}
2 & 3 & 4 & 5 & 6 & 7 \\
 & 2 & 3 & 4 & 5 & 6 \\
 & & 2 & 3 & 4 & 5 \\
 & & & 2 & 3 & 4 \\
 & \text{SYM} & & & 2 & 3 \\
 & & & & & 2
\end{bmatrix}
\begin{Bmatrix}
u_2^2 \\ v_2^2 \\ u_3^2 \\ v_3^2 \\ u_4^2 \\ v_4^2
\end{Bmatrix}
=
\begin{Bmatrix}
p_{2x}^2 \\ p_{2y}^2 \\ p_{3x}^2 \\ p_{3y}^2 \\ p_{4x}^2 \\ p_{4y}^2
\end{Bmatrix}
$$

11.10 Use the finite-element method to solve the same problem as in Problem 11.6 with two elements in the discretized model as illustrated in Figure 11.28. The finite-element model in Figure 11.28 involves an additional node (node D) situated half-way between nodes B and C, and the numbers in circles are the element numbers in the model. Show what lesson you have learned from solving Problems 11.6 and 11.10.

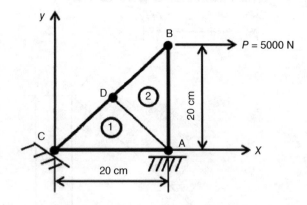

Figure 11.28 Finite-element model of a triangular plate with two elements.

12

Statistics for Engineering Analysis

Chapter Learning Objectives

On completing this chapter readers will have learned about the following topics in the application of statistics in engineering analysis and the introduction of statistical process control in mass production of products:

- The use of statistics in engineering practice.
- Common terminology in statistical analysis.
- Standard deviation and its physical significance.
- The normal distribution of statistical data and its physical significance.
- The normal distribution function, a mathematical model for statistical analysis.
- The Weibull distribution function for probabilistic engineering design.
- The concept of statistical process control.
- The application of statistical process control using control charts for quality assurance in mass production environments.

12.1   Introduction

Statistics is the science of decision making in a world full of uncertainties.

The world in which we live is indeed full of uncertainties. The following are just a few examples of sources of such uncertainty:

- A person's daily routine, career, and health conditions.
- The life expectancies of citizens of different countries of the world.
- Fluctuations of stock markets.
- Weather forecast for regions within countries and the whole world.
- National and global politics.
- Epidemics and natural disasters around the world.

The broad application of statistical methods has made this branch of science a stand-alone major discipline of research and academic programs. A great many archival papers and books have been published in this specialty. It is not possible for this chapter to cover all of the important topics in statistics. The effort will be focused on a small subset, including the most commonly used terminology in statistical analysis and introducing topics that are relevant to engineering analysis with specific applications in probabilistic design analysis and in process control for quality control and assurance of industrial products in a mass production environment. For example, probabilistic design of structures made of brittle materials such as ceramics and cast iron is frequently employed for machines and structures

Applied Engineering Analysis, First Edition. Tai-Ran Hsu.
© 2018 John Wiley & Sons Ltd. Published 2018 by John Wiley & Sons Ltd.
Companion Website: www.wiley.com/go/hsu/applied

operating in high-temperature environments. Extensive treatments of the use of statistics for solving engineering problems are available in numerous books (see, for example: Stout, 1985; Vardeman, 1994; Leemis, 1995; Rosenkrantz, 1997; Morrison, 2009; Navidi, 2013).

Statistical methods are playing increasingly important roles in nearly all phases of human endeavor (Spiegel, 1961). The reason why statistical methods are becoming more popular is that the world in which we live has become ever more complicated in many aspects that involve increasing numbers of uncertainties, and the traditional methods based on deterministic concepts can no longer deal with these complex issues and problems. The use of statistics has accordingly expanded to encompass nearly all aspects of human activity. The following are some of these areas of application.

- *Agriculture:* According to the U.S. Department of Agriculture, statistical analyses offer key indicators, outlook analysis, and a wealth of data on the food and agricultural system, along with information on farming practices, structure, and performance. They also produce data on such diverse topics as farm and rural households, commodity markets, food marketing, agricultural trade, diet and health, food safety, food and nutrition assistance programs, natural resources and the environment, and the rural economy.
- *Biology:* There are many applications of statistics in studying the cell theory of structural growth, evolution, and taxonomy of living species and organs. Statistics is also a powerful tool in the study of the physiology of tissues and organs, as well as in genetics, involving such critical areas as DNA/RNA conversion. Bioinformatics is a relatively new interdisciplinary field of science that combines computer science, statistics, mathematics, and engineering to study and process biological data.
- *Business:* Statistics is a powerful tool for developing inferences about certain characteristics of a population in the business domain, which may include people, objects, or collections of information. Such inferences may lead to critical actions in financial analysis, econometrics, auditing, production and operations including service improvements, and marketing research. Statistics is also widely used by business communities in performance management and strategic planning that includes critical components such as alternative scenarios.
- *Economics:* The data of concern to economic statisticians may include those of an economy of region, country, or group of countries. Analyses within economic statistics both make use of and provide the empirical data needed in economic research, whether descriptive or econometric. They are a key input for decision making as to economic policy. The subject includes statistical analysis of topics and problems in microeconomics, macroeconomics, business, finance, forecasting, data quality, and policy evaluation. It also includes such considerations as what data to collect in order to quantify some particular aspect of an economy and how best to collect it in any given instance.
- *Electronics and communication:* A major application of statistical methods in electronics involves noise in electronics circuits and systems. Noise and frequency modulations are major problems in the design of sound systems for communication. Another major area of application of statistical methods is in the market evaluation of consumer electronic products, which is essential to developing strategies for new products.
- *Healthcare:* Statistical methods are used to provide information for understanding, monitoring, improving, and planning the use of resources for the benefit of people's lives, thereby facilitating the provision of services and promoting their wellbeing. Health statistical data help us to understand the impacts of health on people and to work for their betterment. As we monitor the health of a population, we enhance our understanding of strategies to promote its health.
- *Medicine:* Statistical methods are used in clinical trials, epidemiological studies, and new drug discovery and delivery.

- *Physics:* Probability theory, a major branch of statistics, is used to deal with large populations and approximations in many physical problems. It can describe a wide variety of fields with an inherently stochastic nature. In particular, statistical mechanics develops the phenomenological results of thermodynamics from a probabilistic examination of the underlying microscopic systems. Historically, one of the first topics in physics where statistical methods were applied was the field of mechanics, which is concerned with the motion of particles or objects when subjected to forces.

- *Politics and social sciences:* Major application areas involve policy development in circumstances of uncertain geopolitics and public opinion, sociology in population growth, actuarial studies, censuses, crime statistics, demography for social welfare planning, and numerous other fields such as climate change and environmental studies.

These broad applications of statistics have established it as a specialized academic discipline in many higher educational institutions. It is not possible to treat even a small fraction of the territory of statistical analysis in this chapter. We will cover only a narrow span of this vast field of statistics with the introduction of commonly used terminology in statistical analyses followed by applications in statistical process control, with a couple of control charts for quality control of mass-produced products. An introductory section will be included on reliability design concepts using the Weibull distribution for engineering systems involving materials with random strength data.

12.2   Statistics in Engineering Practice

In Chapter 1 we learned that one of the three principal functions of engineers is "decision making." Proper decisions were indicated that were made on sound engineering principles through credible analyses. Such an approach may have worked well in the past. However, as technologies have evolved at an unprecedented pace in recent times, the demand for accurate results in all aspects of engineering practices has become ever stronger—yet practical engineering activities are full of uncertainties as will be demonstrated in the following description. The statistical approach, in sharp contrast to the traditional deterministic approach, has offered engineers a viable alternative in handling problems involving many uncertainties.

There are indeed many uncertainties in almost all engineering activities. The following are just a few examples; there are many uncertainties in systems design and analysis. Section 1.4 outlined four distinct stages involved in many engineering analyses. Of these four stages in the analysis, Stage 2 relates to idealization of real physical situations that engineers have to deal with. The areas that need idealizing include the geometry and loading and support conditions that engineers would not have the available analytical tools to deal with. There are other areas in which the engineers again may not be certain in their analyses, including the following.

- *Uncertainties in material properties.* Among the most common inputs to any engineering design analysis are the thermophysical properties of the materials to be used in the intended engineering system. A simple stress analysis of a uniaxially loaded bar requires the Young's modulus (E) of the material. An additional property called the Poisson's ratio (v) needs to be included in the generalized Hooke's law for multiaxially loaded solids, as illustrated in Chapter 11. Thermal conductivity (k) is a required property in heat conduction analysis (Equation 7.24), and other properties such as permittivity (ε) are required in electromagnetic analysis in Equation 3.33a. Most of these properties are available in professional handbooks (Avallone *et al.*, 2006).

However, one must realize that all the material properties that are available in textbooks and handbooks are the "average" values of the measured data from testing of many samples, with the assumption that the sample materials are homogeneous (and isotropic, as in many cases). The reality, however, is that no material can be as homogeneous as one would assume it to be. One would find that the sample materials for property measurements are full of minute voids and cavities, and with grains oriented in random directions, as can be observed under electron microscopes. These randomly occurring micro voids with random orientations can make the measured material properties uncertain in many ways, and this uncertainty will be amplified many-fold in the design analysis of engineering systems at micro- and nanoscales (Hsu, 2008), in which device structures are of molecular or atomic scale. Any uncertainty in material homogeneity and purity will produce significant error in engineering analyses.

- *Methodology of engineering analysis.* Many engineering analyses involve mathematical modeling as described in Chapter 2. Derivation of the mathematical model adopted in an engineering analysis would involve a number of assumptions and hypotheses such as were illustrated in Chapters 7, 8, and 9. Additionally, there are times at which engineers need to use numerical techniques such as the finite-difference or finite-element methods as described in Chapters 10 and 11; one will readily appreciate that these numerical methods, though popular and powerful, are used to obtain only approximate solutions due to the many assumptions required in their formulation and that they also include accumulative rounding-off and truncations errors in computation. Thus, particular analytical methods also introduces uncertainties into the analyses.

- *Fabrication methods.* Once a system is beyond the design stage, it will be subject to another major source of uncertainty in the fabrication of the designed systems into products. Huge uncertainty may arise in fabrication or process methods, in selection of machine tools or process procedures, and in quality assurance techniques applied to the finished products. Human factors will also come into play because operators involved in the fabrication process will have varying levels of experience and skill.

These uncertainties that can arise in design methodologies, material properties, and fabrication techniques constitute a lively example of the well-known "Murphy's Law," which stipulates that "*Anything that can go wrong will go wrong.*" Many of the factors in engineering systems can indeed "go wrong."

12.3   The Scope of Statistics

Statistics is concerned with developing and applying scientific methodology in the following major activities.

Collecting relevant information

Methods are required for collecting relevant information for statistical analysis according to the nature and the extent of the analysis. It is a costly activity but it is a critical part of the process because the data collected for analysis can affect the accuracy and credibility of the analysis. The sizes of the datasets and the sources from which the data are collected may significantly affect the outcome of the analysis.

There are generally four sources from which the analysts may collect relevant data for their analyses: (a) from the entire population; (b) from a subset of the population; (c) from controlled studies undertaken to understand cause-and-effect relationships; and (d) from whatever existing studies attempting to understand the cause-and-effect relationships on the subject of interest. The latter source offers the least costly option in collecting data, but the analyst does not have any control of the populations that generated the data or the conditions that dictated the generation of the collected data.

Organizing the collected information

Once relevant information for statistical analysis has been collected, the analyst must determine the best strategy for organizing and analyzing the collected information by developing databases.

There are two types of databases commonly used in statistical analysis: (a) quantitative databases for the information collected in numerical form, such as that on the frequency of specific behaviors; and (b) qualitative databases for nonnumerical information, such as responses gathered through interviews, observations, focus groups, or survey questionnaires. Surveys on market demands for certain products by business and industry are in this category.

Datasets in quantitative databases usually involve a set of numbers in either ascending or descending order of magnitude: for example, the measured temperature in an experiment at specific instants, or a particular dimension of a machine component manufactured by mass production.

Three elements are required for well-organized databases: (a) they should carry a unique identifier; (b) there should be prescreening and exclusion of incorrect information according to the analyst's best judgment; (c) the collected information should be entered into the database in a consistent format.

Summarizing and presenting the information to concerned parties

Databases constructed according to the above criteria are the basis for the subsequent analyses that will allow conclusions or proper decision making by senior managers of business and industry. Often the analyst may be required to present the essential information that the database itself expresses without a detailed analysis using the collected data. It is thus important that the analyst be able to *summarize* the collected information and present this information to the concerned parties. Key items the analyst may present to characterize a quantitative database include the "frequency distributions," the "central tendency," and "variability" of the dataset. Presentation of the findings from information collected on qualitative data is more of a challenge to the analyst. The analyst would present the findings that appear to correspond to the choice of which information should be emphasized, minimized, or omitted in the analysis.

Analyzing data to generate valid conclusions and allow reasoned decision making on the basis of such analysis

A major task of the analyst at this stage is performing an inferential analysis of the database that he or she has established. The objective of this effort is to establish the "significance" in the relevant aspect offered by the dataset. "Statistical significance" derived from probability theory is often used to assess the validity of the dataset. This indicates whether a result is more probable than what would have been expected due to random error. For the dataset to be significant, it must have high probability that the results were not produced by chance. Common statistical tests such as "chi-square" are often used to assess the "goodness of fit" of the collected data with what is expected, such as the normal distribution, as will be elaborated later in the chapter.

Example 12.1

A local firm in Silicon Valley fabricates a batch of microchips. The quality assurance engineer took 5 samples from a bin containing mass-produced chips, and measurements of the dimensions of these chips were made at three predetermined locations on each sample. The measured data that the engineer collected and recorded are tabulated below:

Sample 1:	2.15	2.35	1.95
Sample 2:	2.70	1.83	2.25
Sample 3:	1.97	2.03	2.13
Sample 4:	2.06	2.70	2.15
Sample 5:	2.03	1.75	1.85

All measured data have the same unit of millimeters. Organize the collected data in ascending order for further analysis.

Solution:
The set of data is organized in an ascending order as

1.75, 1.83, 1.85, 1.95, 1.97, 2.03, 2.03, 2.06, 2.13, 2.15, 2.15, 2.25, 2.35, 2.70, 2.70

12.4   Common Concepts and Terminology in Statistical Analysis

The following terminology is frequently used in statistical analysis using databases created by quantitative data collection. For illustration purpose, we present only cases with small sample sizes with small numbers of data, which does not normally represent in reality.

12.4.1   The Mode of a Dataset

The *mode* of a dataset is that value (or more than one value) that occurs with the greatest frequency: that is, it is the most common value or values in the dataset. There are cases such that the mode may not exist in a dataset. In other cases there may be more than one mode for the dataset, as will be demonstrated in the following examples.

Example 12.2
Find the mode in each of the following datasets:

a) 2, 2, 5, 7, 9, 9, 9, 10, 10 11, 12, 18.
b) 3, 5, 8, 10, 12, 15.
c) The dataset that appears in Example 12.1.

Solution:

a) The mode of this dataset is 9 because this number appears three times in the dataset, more frequently than any other datum in the set.
b) This dataset has no mode because there is no number in the set that appears more than once.
c) The dataset appears in Example 12.1 has triple modes of 2.03, 2.15, and 2.70 since each of these numbers appears twice in the dataset.

12.4.2   The Histogram of a Statistical Dataset

A *histogram*—often called a "frequency distribution diagram"—is most commonly used by statisticians and engineers for expressing the physical sense of a set of data. General rules for establishing this type of diagram are outlined as follows:

1) Identify the largest and smallest numbers in the entire dataset, and thus the overall range of the collected data.

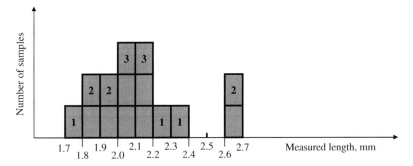

Figure 12.1 Histogram of measured lengths of chips.

2) Establish a convenient number of intervals of the same size (value) within the overall range of the collected data.
3) Determine the number of the data falling into each of the set intervals. These numbers will be the "frequency" of the data in each of the set range in the histogram.

Example 12.3
Establish a histogram of the measured dimensions of the microchips presented in Example 12.1.

Solution:
We have conveniently divided the range of the measured length between 1.75 mm and 2.7 mm into 10 intervals as tabulated below:

Intervals (mm)	1.7–1.8	1.8–1.9	1.9–2.0	2.0–2.1	2.1–2.2	2.2–2.3	2.3–2.4	2.4–2.5	2.5–2.6	2.6–2.7
Sample number in each interval	1	2	2	3	3	1	1	0	0	2

The histogram corresponding to the tabulated numbers is plotted in Figure 12.1, in which the vertical axis represents the frequency of the data in the dataset.

Example 12.4
Establish a histogram of the following tabulated marks that the students in a class earned in their final examination:

Test scores	45–49	50–54	55–59	60–64	65–69	70–74	75–79	80–84	85–89	90–94
Frequency	1	2	2	5	11	9	10	5	4	2

Solution:
The corresponding histogram of the above mark distribution is illustrated in Figure 12.2.

The data presented in Example 12.4 resulted in an approximately "symmetrical bell-shaped" histogram as shown in Figure 12.2. The corresponding data that form the histogram are referred to as a "normally distributed dataset." So-called *normal distribution* of datasets is common for many physical phenomena, and the resulting bell-shaped histogram produced by these datasets is referred to as the *normal distribution* by statisticians. *Most statistical data from large sample sizes fit well with bell-shaped distribution curves.*

12.4.3 The Mean

The *mean* of a dataset is the arithmetic average of all data in the set. It is a popular measure of "central tendency" of the dataset.

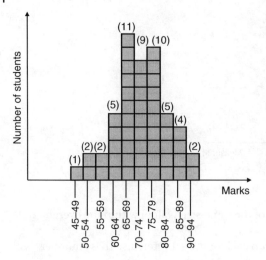

Given a data set of n numbers represented by $x_1, x_2, x_3, \ldots, x_n$, we may express the summation of all numbers in the dataset as $\sum_{i=1}^{n} x_i$. The arithmetic mean, $\bar{x}$ of the dataset may be obtained by the following expression:

$$\bar{x} = \frac{\sum_{i=1}^{n} x_i}{n} \tag{12.1}$$

Example 12.5
Determine the mean of the dataset that was given in Example 12.1:

1.75, 1.83, 1.85, 1.95, 1.97, 2.03, 2.03, 2.06, 2.13, 2.15, 2.15, 2.25, 2.35, 2.70, 2.70

Solution:
We find the sum of the 15 numbers ($n = 15$) in the dataset to be 31.9. Thus, for the mean of this set data Equation 12.1 gives

$$\bar{x} = \frac{31.9}{15} = 2.127$$

Use of the mean in statistical analysis has many advantages:

1) It includes *all* the data in the set.
2) It always exists.
3) It is usually reliable in representing the "central tendency" of the dataset.

A disadvantage of using the mean in statistical analysis is that "significant error" of use of the mean to represent the central tendency of the dataset can occur with some extreme outlier numbers appearing in any part of the dataset, as will be demonstrated in the following example.

We may readily compute the mean of the dataset: 2, 3, 5, 7, 9, 11, 13 to be 7.14 which is close to the "central value" of the dataset. However, the same dataset but with the value of the last datum in the dataset replaced by a large number, say 73, will have a mean of 15.71 instead. This value is far from being the "central value" of the dataset with the last datum of 73. We thus observe that the central tendency of a dataset cannot be represented by the mean of a dataset of the latter type. We must recognize that although the mean may represent the central tendency of a dataset in most cases, there are instances when other methods need to be used to establish the "central tendency" of the dataset.

12.4.4 The Median

Like the mean, the *median* of a dataset is also used to represent the central tendency of a dataset. The "central" datum is readily identified in datasets with an odd number of data. For datasets with an even number of data, the median is the average of the two central data. For example, the median of the dataset 5, 9, *11, 14*, 16, 19 is $(11+14)/2 = 12.5$.

The difference between the mean and median is that the latter represents the data at the "center" when the dataset is arranged in either ascending or descending order. For example, the median value $(7+9)/2 = 8$ of the ambiguous dataset 2, 3, 5, 7, 9, 11, 13, 73 that we used earlier may represent the central tendency of the dataset much better than the mean does.

The median is frequently used to represent the central tendency of large datasets involving a few extreme data values, or of datasets with large fluctuations. Examples are such things as the median income of a population or the median price of houses in places like Santa Clara Valley in California, where the median but not the mean house price is more representative in a market with prices ranging from close to \$1 000 000 for the majority houses to a smaller number of houses valued at many millions of dollars.

12.4.5 Variation and Deviation

The degree to which numerical data deviate from the "average" value (the mean or median of a dataset) is called the "variation" or "dispersion" of the data in the dataset. Deviation is one of the measures of variation in a statistical analysis. It is a measure of how the data in a set deviates from its mean value of $\bar{x}$. In the world of statistics, small deviation represents a more uniform dataset with less "scatter" from the mean. Thus, for a dataset of n data with $x_1, x_2, x_3, \ldots, x_n$, the variation of individual data can be determined by simply computing the individual variations: $(x_1 - \bar{x})$, $(x_2 - \bar{x})$, $\ldots$, $(x_n - \bar{x})$. The total deviation of the entire dataset will be the sum of all the individual deviations. Unfortunately, one will find that the deviations sum to zero:

$$\sum [(x_1 - \bar{x}) + (x_2 - \bar{x}) + (x_3 - \bar{x}) + \cdots + (x_n - \bar{x})] = 0$$

which leads to an unrealistic implication that there is *no* overall variation of the data in the dataset. A close inspection of the components in the above summation will indicate that half of the individual components in the summation carry negative sign whereas the other half of the terms carry positive sign. The net value of the summed components becomes zero, as expected.

To avoid this situation of zero overall deviation as demonstrated in the above summation, one may use an alternative expression using the squares of the deviations to represent the overall deviation of the dataset:

$$\sqrt{\sum_{i=1}^{n} (x_i - \bar{x})^2} \tag{12.2}$$

One will observe in Equation 12.2 that all individual (*squared*) terms of variation in Equation 12.2 will carry positive sign, and the result of the square root of the summed variations will be a real nonzero number.

12.5 Standard Deviation (σ) and Variance (σ^2)

12.5.1 The Standard Deviation

The standard deviation (σ) for a dataset is a measure that is used to quantify the amount of variation or *dispersion* of the data values in the dataset. A standard deviation close to 0 indicates that the data points tend to be very close to the mean (the central tendency, also called the *expected value*) of the set, while a high value of standard deviation indicates that the data points are spread out over a wider range of values. Mathematically, the standard deviation of a dataset may be computed from Equation 12.3:

$$\sigma = \sqrt{\frac{\sum_{i=1}^{n}(x_i - \bar{x})^2}{n-1}} \tag{12.3}$$

where n is the number of data in the dataset.

12.5.2 The Variance

Like standard deviation, variance is also used as a measure of scatter of data in a datasets; it has the following properties:

1) It is proportional to the scatter of the data (small variance means the data are clustered together, and a large variance results from a widely scattered dataset).
2) It is independent of the number of values in the dataset.
3) It is independent of the mean (since now we are only interested in the spread of the data, not its central tendency).

The mathematical relation for the variance is derived from Equation 12.2 and is given by the following expression:

$$\sigma^2 = \frac{\sum_{i=1}^{n}(x_i - \bar{x})^2}{n-1} \tag{12.4}$$

If a very large number of data are involved in the dataset and only a partial dataset of n values is used to determine the standard deviation, we use the following modified expression for standard deviation:

$$\sigma = \sqrt{\frac{n\left(\sum_{i=1}^{n}x_i^2\right) - \left(\sum_{i=1}^{n}x_i\right)^2}{n(n-1)}} \tag{12.5}$$

where n is the selected number of data used in the computation.

Example 12.6

Determine the standard deviation and the sample variance of the dataset:

5, 9, 11, 14, 19

Solution:

We see that there are five numbers in the dataset, so $n=5$. The mean value is calculated to be $\bar{x}=11.6$ using Equation 12.1. The standard deviation of the dataset can be obtained using

Equation 12.3 as

$$\sigma = \sqrt{\frac{(5-11.6)^2 + (9-11.6)^2 + (11-11.6)^2 + (14-11.6)^2 + (19-11.6)^2}{5-1}}$$

$$= 5.27$$

The sample variance is $\sigma^2 = (5.27)^2 = 27.8$.

12.6 The Normal Distribution Curve and Normal Distribution Function

The "normal distribution" is the most frequently encountered statistical distribution pattern, appearing in many natural circumstances, such as the age distribution of the citizens of a country, the annual temperature variations in a region, the performance of certain machines and devices, and so on. The data in overwhelming numbers of datasets, when plotted as histograms such as shown in Figure 12.2, will have envelopes that are the bell-shaped curves known as "normal curves," or "normal distribution curves." These distribution curves are usually presented with the mean of the dataset, such as that shown in Figure 12.2, relocated to the center of the distribution of the data.

The common occurrence of data in bell-shaped distributions motivated mathematicians and statisticians to develop mathematical models that can be used to represent this common statistical data distribution. This development evolved from the Gaussian distribution function that often appears in probability theory. Probability theory is a branch of mathematics concerned with the analysis of random phenomena. The outcome of a random event (such as the academic performance of the class in Example 12.2) cannot be determined before it occurs, but it may be any one of several possible outcomes. The actual outcome is considered to be determined by chance. In this theory, the normal (or bell-shaped) distribution is a very common *continuous* probability distribution of random variables, and the Gaussian function that is symmetrical about the central tendency of the dataset is widely used to assess the likelihood of observing the expected outcomes.

The *probability density* of a normal distribution in probability theory is given by:

$$f(x|\mu, \sigma) = \frac{1}{\sigma\sqrt{2\pi}} e^{-(x-\mu)^2/2\sigma^2} \tag{12.6}$$

(see https://en.wikipedia.org/wiki/Probability_density_function). Here, μ is the *mean* or *expectation* of the distribution (and it can also equal the median and mode of a statistical dataset). The parameter σ is its standard deviation, with its variance being σ^2. A random variable with a Gaussian distribution is said to be *normally distributed* and is called a *normal deviate*.

Thus, if we plot the histogram in Figure 12.2 for the academic performance of students in a class by moving the vertical axis for the population density to coincide with the central value of the dataset—the mean—and connect the peaks of each individual attribute of mark range, we will have the modified distribution in a continuous curve expressed as solid line that is a close approximation of the bell-shape, as illustrated in Figure 12.3.

Because the bell-shaped normal distribution represents the statistical variation of the overwhelming number of natural phenomena, it can be graphically represented by what is termed the "normal curve" as shown in Figure 12.4, with a corresponding mathematical function that is called the "normal distribution function" as follows.

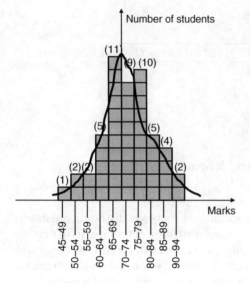

Figure 12.3 Approximated normal distribution of mark distribution of a class.

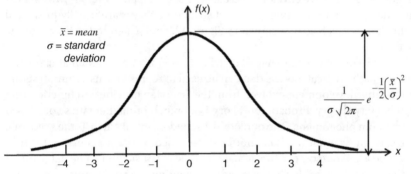

Figure 12.4 Normal distribution curve.

The mathematical expression that represents the normal curve or the normal distribution function takes the form

$$f(x) = \frac{1}{\sigma\sqrt{2\pi}}e^{-(1/2)[(x-\bar{x})/\sigma]^2} \tag{12.7}$$

in which σ is the standard deviation of the dataset as given in Equation 12.3, and $\bar{x}$ is the mean of the dataset as in Equation 12.1.

We note that the normal distribution function in Equation 12.7 is effectively identical to the Gaussian distribution function in Equation 12.6.

There are many advantages of having the normal distribution curve represented by the normal distribution function of Equation 12.7, as many useful mathematical models for scientific and engineering observations can be developed using this function. For example, the following important properties of observed phenomena have been derived using the normal distribution function of Equation 12.7:

1) The normal curve (or normal distribution) is symmetrical about the mean $\bar{x}$.
2) Being a credible mathematical model with a continuous function that is valid for a large number of natural phenomena the normal distribution function serves as a basis for mathematical modeling of other statistical analyses.

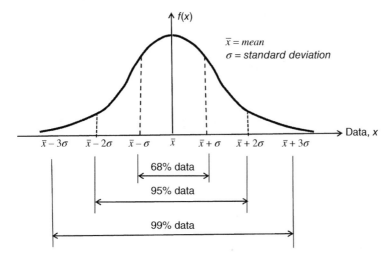

Figure 12.5 Properties of the normal distribution.

3) With the function in Equation 12.7, the standard deviation, σ, of a dataset that fits a normal distribution can be interpreted in the following interesting and extremely valuable way: 68.26% data of the set are within the value of the mean $\bar{x} \pm 1$ standard deviation (σ). 94.40% data of the set are within the value of the mean $\bar{x} \pm 2\sigma$. 99.73% data of the set are within the value of the mean $\bar{x} \pm 3\sigma$.

These properties are illustrated in Figure 12.5.

Example 12.7
A tire manufacturing company supplies tires for 200 000 cars each year. Of these cars, 15%, or 30 000 cars, were used to evaluate the life of its tires. Test results indicated that the average life of the tires involved in these tests was 45 000 miles with a standard deviation of 4000 miles. Determine the lives of the tires produced by the company.

Solution:
The mean value of the lives of the tires in the tests was $\bar{x}=$ 45 000 miles with a standard deviation $\sigma = 4000$ miles. We assume that all measured lives of the tires fit a normal distribution (or normal distribution curve) as shown in Figure 12.4. We may then interpret the test results for the tire lives in the following way:

68.26% cars had tire life of $\bar{x} \pm \sigma = 45\ 000 \pm 4000$ miles,
94.40% cars had tire life of $\bar{x} \pm 2\sigma = 45\ 000 \pm 8000$ miles
99.73% cars had tire life of: $\bar{x} \pm 3\sigma = 45\ 000 \pm 12\ 000$ miles.

12.7 Weibull Distribution Function for Probabilistic Engineering Design

Like the normal distribution function (or Gaussian distribution function in probabilistic analysis), the Weibull distribution function is another continuous function that can be used for physical (or engineering) phenomena that are significantly skewed from the normal distribution illustrated in Figure 12.4. It is particularly useful for dealing with engineering analyses with significant scatter of analytical parameters such as material property inputs to a design analysis. We will focus on its application in this section.

12.7.1 Statistical Approach to the Design of Structures Made of Ceramic and Brittle Materials

There has been increasing use of ceramic or cermet (composites of metal and ceramics) by the aerospace and nuclear industries for components operating at very high temperature but requiring light weight. The ceramic SiC has a mass density of 3200 kg/m^3, which is only 41% of that of structural steel, but has a Young's modulus of 375 000 MPa, that is almost twice that of steel. SiC also has a melting point of 4263 K vs. 1703 K for structural steel. Not only does the much higher melting point of SiC provide much higher strength at high temperature, but SiC is also much less vulnerable to creep deformation, which normally occurs at an operating temperature above half of a material's homologous melting point (that is, half of the melting point on the Kelvin scale).

Despite the attractive strength of SiC at high temperature, it is brittle, and with randomly varying material properties—in particular, its fracture strength—presents major challenges to engineers using well-established deterministic mathematical models in their design analysis, such as was illustrated in Figure 1.4. Table 12.1 shows the fracture strength of SiC produced by 4-point bending tests at the author's laboratory at the Whiteshell Nuclear Research Establishment of Atomic Energy of Canada Ltd. in 1970. It shows that the fracture strength of the Carborundum KT grade SiC specimens had wide scatter in the measured fracture strength, ranging from 15 ksi (103.5 MPa) to 21 ksi (147 MPa).

Table 12.1 Fracture strength of silicon carbide in 4-pont bending tests.

Order	Ordered fracture strength (ksi)	Survival probability, R	Order	Ordered fracture strength (ksi)	Survival probability, R
1	15.28	0.9678	16	18.67	0.4838
2	15.35	0.9355	17	18.86	0.4516
3	16.96	0.9032	18	19.04	0.4194
4	17.15	0.8710	19	19.30	0.3871
5	17.65	0.8387	20	19.57	0.3548
6	17.85	0.8065	21	19.59	0.3226
7	18.08	0.7742	22	19.72	0.2903
8	18.17	0.7419	23	19.96	0.2581
9	18.36	0.7097	24	20.26	0.2258
10	18.39	0.6774	25	20.68	0.1935
11	18.43	0.6451	26	21.11	0.1613
12	18.45	0.6129	27	21.18	0.1290
13	18.50	0.5806	28	21.31	0.0968
14	18.63	0.5484	29	21.55	0.0645
15	18.65	0.5161	30	21.57	0.0323

Survival probability $R = 1 - i/(n+1)$ where i is the order number and $n =$ total number of specimens $= 30$.
(Hsu and Gillespie, 1971.)

Like many other ceramic materials, the fracture strength of KT-SiC also shows strong variation with the regions that are subjected to tensile stresses; above all, the properties vary significantly with the volume of the structure, which leads to strong size effects in design analysis. In sharp contrast, metallic materials such as structural steel exhibits much less scatter in the strength values. A report (Barnett and McGuire, 1966) indicated that an investigation involving 30 000 tons of structural steel showed only two out of 3124 specimens to yield below the minimum specified value of 33 ksi (227.7 MPa), and they fell no lower than 31 ksi (21.4 MPa). A definitive allowable strength for structural steel is therefore possible.

Similar inconsistent fracture strength also occurs in conventional metallic materials that are embrittled at low temperature. The highly inconsistent mechanical strength of ceramics, cermet and brittle materials alike, with the fracture strength of KT-SiC as in Table 12.1, is mainly attributed to the lack of plastic deformation in these materials. Consequently, sharp stress concentration near the tip of inherent cracks inside the material prompts these cracks to propagate through the specimens or the structure because of lack of plastic yielding of the materials. This type of structural failure fits the statistic model referred to as "failure by the weakest link," meaning that the failure of the weakest of the many links that make up the structure is what causes the overall structure to fail. The random nature of inherent cracks in brittle materials creates weakest links inside materials of this type, which leads to the failure in random fashion of specimens or structures made of brittle materials.

The significant inconsistency of material strength of ceramic and brittle materials, like that of KT-SiC as shown in Table 12.1, precludes the use of the conventional deterministic theory in engineering design analysis such as illustrated in Examples 1.1 and 1.2 in Section 1.5 of Chapter 1. A radically different approach to design analysis of structures made of these materials is thus required.

The statistical approach for the design of structures made of ceramic and brittle materials is derived using the Weibull distribution function, with the strength of the materials measured by special techniques, and with the specimen geometry being such as described by Barnett and McGuire (1966). The "interpretation of results" for the design analysis described in the Stage 4 of engineering analysis in Chapter 1 is no longer to keep the maximum stress in a structure below the maximum allowable stress, because no definitive allowable stresses can be established with this type of material. Rather, one will use the term "reliability" of the structure as the design criterion instead. This approach of using statistical methods is referred to as *probabilistic analysis.*

12.7.2 The Weibull Distribution Function

Statistician Waloddi Weibull proposed the well-known eponymous distribution function for structures subjected to applied load with induced stress σ (Weibull, 1939). The mathematical expression of this distribution function is given in Equation 12.8:

$$P = 1 - e^{(-r)} \tag{12.8}$$

where P is the probability of fracture of the material, and r is the risk of rupture.

The risk of rupture r of a given material in Equation 12.8 is obtained by an integration as in Equation 12.9:

$$r = \int_v f(\sigma) \, dv \tag{12.9}$$

in which v is the volume of the material under tensile load, and $f(\sigma)$ is a function of stress distribution (or variations) in the material of volume v. We note that the Weibull distribution function is constructed on the hypothesis that compressive stresses in material will not result in the failure of the structure.

The original form of $f(\sigma)$ in the proposed Weibull distribution function has two parameters, as in the Equation 12.10:

$$f(\sigma) = \left(\frac{\sigma}{\sigma_0}\right)^m \tag{12.10}$$

in which m is the "Weibull modulus," which is a measure of the scatter of the collected data, and σ_0 is a "Weibull parameter." In general, this parameter is related to the gap between the lower bound and the mode of the dataset.

The Weibull distribution function for the probability of rupture of a material with the "risk of rupture r" is illustrated in Figure 12.6. In the figure, a probability of fracture $P = 1$ indicates a 100% likelihood of fracture of the sample specimen or structure.

One should be aware that Weibull did not attempt to justify his distribution function theoretically. However, the distribution function of Equation 12.8 did show good fit to experimental strength data of many materials ranging from cotton fibers to steel.

In 1951, Weibull proposed to add a lower bound σ_u to his distribution function in Equation 12.8 giving the following modified distribution function (Weibull, 1951):

$$P = 1 - \exp\left[-KV\left(\frac{\sigma - \sigma_u}{\sigma_0}\right)^m\right] \tag{12.11}$$

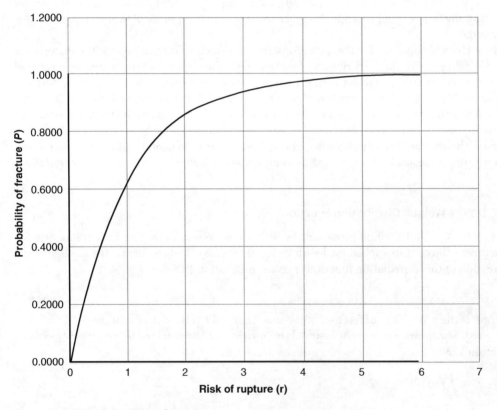

Figure 12.6 Weibull distribution function.

Figure 12.7 Four-point bend test specimens for brittle materials.

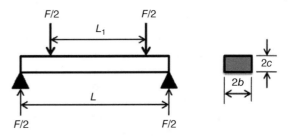

where K is the load factor and V is the volume of the specimen or the structure with induced tensile stresses due to externally applied loads. The stress σ_u is referred to as the "zero probability of rupture strength," or the "threshold stress," below which no fracture of specimens (or structure) occurs.

The value of the load factor K in Equation 12.11 is determined by how the specimens used to determine the material fracture strength are loaded (Gregory and Sprill, 1962). For instance, $K = 1$ is for a specimen subjected to uniform tensile load, and

$$K = \frac{1}{6(m+1)} \left(\frac{2}{m+1} + 1 \right)$$

for 4-point third span (the center span) bending tests as illustrated in Figure 12.7.

Fracture strength testing for ceramic or brittle materials such as KT-SiC is often performed in 4-point bending tests with a specimen configuration shown in Figure 12.7. The loading arrangement shown in this figure results in "pure bending" of the central portion (the third span) of the beam. This mode of beam bending makes it easy to identify the region that is subject to "pure" tensile load. And more importantly, this arrangement can mitigate the parasitic stresses on the specimen induced by gripping the specimen both ends.

The following measurements were included in producing the fracture strength data for KT-SiC in Table 12.1:

Length of the specimen $(L) = 1.75$ inches
Length of the third span $(L_1) = 1.0$ inch
Volume of the specimen $(V) = 2c \times 2b \times L = (2 \times 0.1259)(2 \times 0.01383) \times 1.75 = 0.0541$ in^3
Loading factor (K):

$$K = \frac{1}{2(m+1)} \left(\frac{0.4286}{m+1} + 0.5714 \right)$$

12.7.3 Estimation of Weibull Parameters

Three methods are available for engineers to determine the Weibull parameters appearing in Equation 12.11. These are (1) the log-log plot method; (2) the method of statistical moments (Gregory and Sprinill, 1962); and (3) the least mean squares approximation (Gregory and Sprinill, 1962).

We will present the log-log plot method here for its simplicity. This process begins with re-writing Equation 12.11 in the following form as a linear relationship:

$$\ln \ln \left(\frac{1}{1-P} \right) = m \ln (\sigma - \sigma_u) - m \ln (\sigma_0) + \ln (KV) \qquad (12.12)$$

One may compute the probability of fracture of each specimen from the testing data listed in Table 12.1.

The following are the major steps in determining the Weibull parameters in Equation 12.11:

Step 1: Assuming an arbitrary value of the threshold stress σ_u.

Step 2: Compute $\sigma_i - \sigma_u$; that is, $\sigma_1 - \sigma_u, \sigma_2 - \sigma_u, \sigma_3 - \sigma_u, \sigma_4 - \sigma_u, \ldots$.

Step 3: Compute $\ln(\sigma_i - \sigma_u)$ obtained in Step 2.

Step 4: Plot $\ln(\sigma_i - \sigma_u)$ against $\ln \ln[1/(1 - P)]$ and obtain one curve plot (name this as "plot A")

Step 5: If the curve plot is not a straight line as expected, begin the iteration with another assumed value of σ_u and repeat Steps 1 to 4 until a reasonably straight line plot is obtained (call this curve plot "plot B"). The last assumed value of σ_u is the desired value.

Step 6: The Weibull modulus m is then obtained as the slope of the latest straight line in plot B with the last chosen value of σ_u.

Step 7: The Weibull parameter σ_0 may be obtained by referring to the sketch in Figure 12.8, in which we pick up the stress σ_s corresponding to $\ln(KV)$ on the vertical axis, representing $\ln \ln[1/(1 - P)]$ with the known values of load factor K and the volume of the specimen V. By virtue of Equation 12.12, we may reach the relationship $m \ln(\sigma_s - \sigma_u) = m \ln(\sigma_0)$, which leads to $(\sigma_s - \sigma_u) = (\sigma_0)$. The value of σ_0 can thus be determined as $\sigma_0 = \sigma_s - \sigma_u$ with the newly determined value of σ_s.

Example 12.8

Use the log-log plot method to determine the Weibull parameters in Equation 12.12 for the KT-S$_i$C specimens whose fracture strength under 4-point bending is shown in Table 12.1. The 30 specimens have an average volume of 0.0541 in^3.

Solution:

The parameters in Equation 12.11 were determined by following the seven steps outlined in Section 12.7.3 with four arbitrarily assumed values of the threshold stresses of $\sigma_u = 11$ ksi in case (1), 7.0 ksi in case (2), 5.0 ksi in case (3), and 0 ksi in case (4). The results for all four cases of the assumed value σ_u are plotted in Figure 12.9. We observe that case (4) with the assumed $\sigma_u = 0$ results in a straight line in Figure 12.9. The Weibull modulus m was obtained by the computing the slope of this line to be $m = 13.33$. The parameter σ_0 was computed to be 11.82 ksi following the procedure stipulated in Step 7. The Weibull distribution function for the fracture strength of the 30 KT-SiC specimens listed in Table 12.1 thus has the following expression:

$$P = 1 - \exp\left[-0.0013\left(\frac{\sigma}{11.82}\right)^{13.3}\right] \qquad (12.13)$$

The Weibull distribution function in Equation 12.13 obtained by the log-log plot method is shown in Figure 12.10. Also shown in the figure are author's attempts at evaluating the Weibull parameters by other two methods: the 3-moment method with an assumed threshold fracture strength $\sigma_u = 11.86$ ksi, resulted in $m = 5.2$ and $\sigma_0 = 2.54$ ksi; and the least mean square

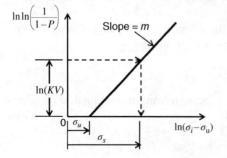

Figure 12.8 Determination of Weibull parameters by the log-log plot method.

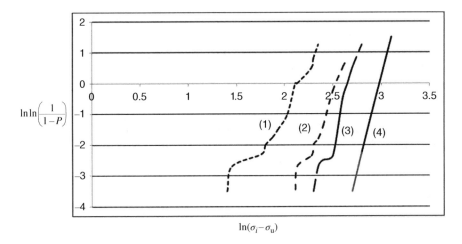

Figure 12.9 Estimation of Weibull parameters for KT-SiC by the log-log method with chosen values of σ_u.

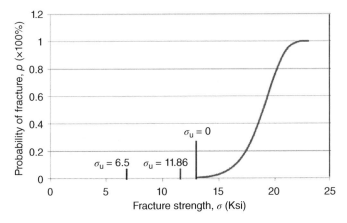

Figure 12.10 Weibull distribution function for the fracture strength of KT-SiC.

approximation method with $\sigma_u = 6.5$ ksi resulted in $m = 8.15$ and $\sigma_0 = 6.08$ ksi. The procedures of determining the Weibull parameters by the latter two methods are available in Gregory and Sprinill (1962). It is remarkable that all three methods used in determining Weibull parameters fit well to the P–σ curve plot as shown in Figure 12.10.

12.7.4 Probabilistic Design of Structures with Random Fracture Strength of Materials

In view of the randomness of the fracture strength of many ceramic and brittle materials and the significance of the size of the structure, conventional deterministic approaches involving maximum allowable stress with "safety factors," as described in Section 1.6 in Chapter 1, have little meaning in the design analyses. A radically different concept and corresponding theory need to be adopted for the analyses of structures made of these materials. Probabilistic design analysis using Weibull distribution functions is considered a viable alternative.

This radically different concept is to design the structure with random strength based on the "probability P of fracture of the structure" or "reliability R of the structure," with $R = 1 - P$. The design criterion P or R depends on the nature and purpose of the structure; for instance a value of $P = 10^{-6}$, meaning a one-millionth chance of structural failure, is considered to be desirable for entities that require extremely high reliability and safety such as nuclear reactor cores or crucial components of aerospace equipment. The low probability of fracture P is used because the consequences of failure of these structures under anticipated loads in both normal operation and accident conditions are infinitely serious.

The volume effect is another major factor to be included in this type of analysis. Conventional wisdom indicates that engineering materials have inherent voids and minute flaws in reality. It is logical to envisage that the larger the volume of the material the more voids and flaws will be present inside the volume. This scenario leads to anticipation that larger structures are vulnerable to higher chances of structural failure. This "size effect" is often ignored in traditional design analysis, but it is an important parameter in the Weibull distribution function as expressed in Equation 12.11. The Weibull distribution function with volume V in Equation 12.11 leads to the following relationship for this purpose:

$$\frac{\sigma_1}{\sigma_2} = \left(\frac{V_2}{V_1}\right)^{1/m} \tag{12.14}$$

where the subscripts on σ denote the tensile stress in materials with respective given volumes V, and m is the Weibull modulus

The following steps are to be followed in the design analysis for reliability of structures made of materials with random and widely scattered fracture strength distributions:

Step 1: Establish the Weibull distribution of $P–\sigma$ from the measured fracture strength data with Weibull parameters such as illustrated in Figure 12.10.

Step 2: Perform a detailed stress analysis on the structure either using classic theories or by numerical methods such as the finite-element method presented in Chapter 11.

Step 3: Divide the structure into n volumes; V_i with $i = 1, 2, 3, \ldots$, with no volume of the structure smaller than the gage volumes of the test specimens used to generate the Weibull parameters for the Weibull distribution function in Step 1. Also, try to choose the individual volumes in the structure with approximately uniform stress distributions within the volume.

Step 4: Determine the "worst" or the highest stress condition in each volume V_i if the stress distribution in the unit volume is not uniformly. Also assume that the corresponding principal stresses S_1, S_2, and S_3 act uniformly throughout the volume. Principal stresses in a tube or pipe are the stress in the radial, tangential, and longitudinal directions.

Step 5: Determine the reliability of each volume element by first finding the reliability of the gage volume of the specimens under each of the principal stresses S_1, S_2, and S_3 using the Weibull distribution function in Equation 12.13 for KT-SiC with $1 - P_{gv}(S_1)$, $1 - P_{gv}(S_2)$, and $1 - P_{gv}(S_3)$, in which $P_{gv}(S_i)$, with $i = 1, 2, 3$, denotes the probability of failure of "gage volume" of the test specimens under principal stress S_i. From these $P_{gv}(S_i)$ values, we may determine the reliability of the gage element subject to simultaneous application of these principal stresses as

$$P_{gv}(S_i) = [1 - P_{gv}(S_1, S_2, S_3)]$$
$$= [1 - P_{gv}(S_1)][1 - P_{gv}(S_2)][1 - P_{gv}(S_3)]$$

where $P_{gv}(S_1, S_2, S_3) = $ probability of failure of the volume element subject to simultaneous principal stress S_1, S_2, and S_3.

The reliability of the volume element of the structure with different volume than the gage volume of the specimens can thus be determined by the following expression:

$$(1 - P_{iv}) = (1 - P_{gv})\frac{V_i}{V_g} \tag{12.15}$$

where P_{iv} = probability of failure of the volume element of the structure V_i and V_g is the volume of the test specimens.

We have the reliability of the gage volume of the specimens from the testing for the material's fracture strength, represented by the Weibull distribution function

$$R_{gv} = 1 - P_{gv} = \exp\left[-V_g\left(\frac{\sigma - \sigma_u}{\sigma_0}\right)^m\right]$$

We may substitute the relationship in Equation 12.15 into the above expression to obtain the reliability of each volume element of the structure:

$$R_{iv} = 1 - P_{iv} = \exp\left[-V_i\left(\frac{\sigma - \sigma_u}{\sigma_0}\right)^m\right] \tag{12.16}$$

Step 6: The reliability of the overall structure using the weakest link theory can thus be obtained from the following expression:

$$R = \prod_{i=1}^{n} R_{iv} \tag{12.17}$$

where n is the total number of volume elements in the structure and the R_{iv} are the reliability of each volume element in the structure computed from Equation 12.16.

Example 12.9

Determine the reliability of a ceramic tube made of KT-SiC with fracture strength measured from 30 specimens as listed in Table 12.1. The dimension of the tube is 6 inches OD × 4 inches ID × 4 inches long. The tube is subjected to a uniform internal pressure $P_i = 4000$ psi. The Weibull parameters for the measured fracture strength are determined by a least squares approximation method, and they fit the Weibull distribution function

$$P_{gv} = 1 - \exp\left[-0.3378\left(\frac{\sigma - 6.5}{6.083}\right)^{8.15}\right] \tag{a}$$

Assume that the tube is at room temperature and the end effects of the pressurized tube are neglected.

Solution:

The physical situation of the pressurized tube is illustrated in Figure 12.11. The tube is free to deform in the radial direction because the end constraints are assumed to be negligible. Consequently, only two stress components need to be included in the analysis. These are the hoop (or tangential) stress designated by $\sigma_{\theta\theta}$ and the radial stress σ_{rr} in the tube wall. From advanced strength of materials textbooks (Volterra and Gaines, 1971) or mechanical engineering handbooks (Avallone, 2006), one may compute these stresses with the applied internal pressure P_i for thick-walled tubes and pipes using the following equations:

$$\sigma_{rr}(r) = \frac{P_i r_i^2}{r_0^2 - r_i^2}\left(1 - \frac{r_0^2}{r^2}\right) \tag{b}$$

$$\sigma_{\theta\theta}(r) = \frac{P_i r_i^2}{r_0^2 - r_i^2}\left(1 + \frac{r_0^2}{r^2}\right) \tag{c}$$

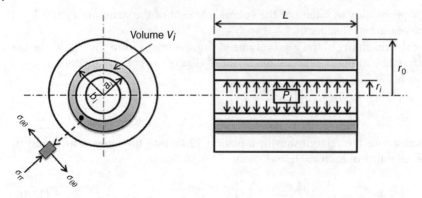

Figure 12.11 A ceramic tube subject to internal pressure loading.

Table 12.2 Reliability of a pressurized tube made of KT-SiC ceramic.

At $P_i = 4000$ psi

i	a_i (in)	b_i (in)	V_i (in^3)	$\sigma_{\theta\theta}$ (ksi)	$-V_i\left(\dfrac{\sigma_{\theta\theta}-6.5}{6.083}\right)^{8.15}$	$1-P_{iv}=\exp\left[-V_i\left(\dfrac{\sigma_{\theta\theta}-6.5}{6.083}\right)^{8.15}\right]$
1	2	2.25	13.3518	10.4	-0.35649	0.700 13
2	2.25	2.5	14.9226	8.889	-0.00734	0.992 68
3	2.5	2.75	16.4934	7.808	-5.99×10^{-5}	0.999 94
4	2.75	3	18.0642	7.008	-2.89×10^{-8}	1

At $P_i = 3500$ psi

				9.1	-0.013085	0.987
				7.7778	-0.000045	0.999 95
				6.832	0	1
				6.132	$\sigma_{\theta\theta} < \sigma_u$	1

From Equations (a) and (b) one may conclude that the radial stress in Equation (b) is compressive, which is not a concern in the analysis. We will thus only be concerned with the tangential stress $\sigma_{\theta\theta}$, which is tensile at all times. This stress component may be computed using Equation (c).

Let us arbitrarily divide the tube wall into four concentric elements with dimensions shown in Table 12.2, with a_i and b_i being the inside and outside diameters of element i with volume V_i as shown in Figure 12.11.

The reliability of the four volume elements under $P_i = 4000$ psi are computed, with the results tabulated in the right-hand column of Table 12.2. The reliability of the overall structure, that is, the KT-SiC tube, can be computed using Equation 12.17 to give

$$R = \prod_{i=1}^{n} R_{iv} = 0.700\,13 \times 0.992\,68 \times 0.999\,94 \times 1.0 = 0.694\,96$$

or $R = 69.5\%$ for the KT-SiC tube subjected to an internal pressure $P_i = 4000$ psi.

The computation has also been performed for the same tube subjected to a reduced internal pressure loading at 3500 psi, at which the reliability of the tube rises to 98.7%, as indicated in

Table 12.2. It makes sense to expect the tubular structure to have higher reliability (thus lower probability of fracture) at a lower applied pressure load, as computed in this analysis.

12.8 Statistical Quality Control

Conventional wisdom indicates that a product cannot succeed in the marketplace unless it offers consumers competitive price and good quality. The quality of a product commonly relates to its finished appearance and its reliability in performance. It also relates to a number of other factors that are related to good quality in industrial products. These factors may include the fitting of the components in an engineering system; sustainability of the product in the expected environmental and/or operating conditions; delivery of the intended performance; and persistence to the life expectancy of the designed objectives. These quality requirements are critical concerns by both the producers and consumers. Poor or unreliable quality of a product not only jeopardizes its success in the marketplace but, more seriously, it also causes liability that may cost the producer millions of dollars in recall of the products, or lawsuits and other severe penalties resulting from litigation. Recalls of massive numbers of products, such as have happened in automobile industry, are mainly due to the poor quality of the products.

Poor quality of a product in relation to engineering practice may be attributed to many causes as outlined below:

1) Poor design in setting the dimensions and tolerances, in surface finishing, in improper selection of materials, and so on.
2) Manufacturing and fabrication processes relating to improper machining, assembly, testing, and inspection.
3) Improper condition of machine tools and fabrication process control.
4) Poor workmanship in all of the above production processes.

The cost of poor product quality can be significant to producers, as can be seen from the following typical example. In addition to incurring substantial costs in recall of defective products already in the marketplace, firms may expend a significant portion of their earned revenues in the costs related to such quality-related areas as inspection of the product at various stages in production, scrappage and rework, prevention, and warranty.

Quality of industrial products is closely related to the rigor and quality of engineering design, to the fabrication equipment, and to the knowledge, experience, and work ethic of operators and inspectors. The exorbitant costs associated with quality of products could be mitigated by better design, better manufacturing processes, better workmanship, and so on. All improvements in quality of industrial products through more sophisticated design process—e.g., setting proper tolerances for assembled engineering systems—better fabrication equipment, and employment of more experienced and skillful operators, as well as implementing more thorough inspections on finished products will incur costs for the improvement. Naturally one will expect the value of the product to increase with the improvement in quality, with associated better customer receptivity for and satisfaction with the products. It will also result in significant reduction in the costs related to product warranty, scrappage, and rework. It thus comes to an issue of reaching an optimal "balance" between the *improved quality* from and the *required cost* for such improvements. A qualitative relationship between these two factors is illustrated in Figure 12.12 (Stout, 1985).

In Figure 12.12, the solid line curve represents the cost associated with improvement of quality of a product and the broken line curve represents the increase of product value. For a specific improvement of product quality, the associated cost and product value can be identified as

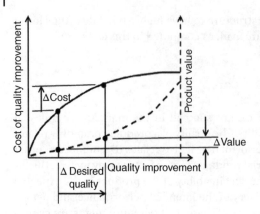

Figure 12.12 Cost and product value associated with quality improvement.

shown in the figure. It is apparent that the planned quality improvement of the product is worthwhile only if the resulting increase of product value (ΔValue) is significantly larger than the associated increase of cost (ΔCost) in Figure 12.12. The particular situation illustrated in Figure 12.12 is not a desirable one according to the analysis of cost vs. benefits outlined above.

12.9 Statistical Process Control

12.9.1 Quality Issues in Industrial Automation and Mass Production

Many would consider that industrial automation began with the introduction of assembly lines in the Ford Motor Company in early 1900. Massive industrial automation did not materialize, however, until shortly after World War II in response to the strong demand for the production of life-sustaining products such as foods and construction materials in large quantities after the war. Industrial automation became a way of producing consumer electronics in the 1980s, followed by information technology equipment a decade later. This mode of production has become commonplace in many other industries, for example for production of electronics, automobiles, and information technology and communication equipment. Industrial automation has now been extended to the healthcare industry, in particular in drug manufacturing and production. A major issue arising from mass production of products by industrial automation is that of quality assurance and quality control during production.

The need for control of quality of products produced by mass production embraces many fields that involve cultural aspects and technical management; these include design, manufacture, functional testing, and inspection through an understanding of sampling procedures and sampling theory by both employees and the management that has overall responsibility for the maintenance of quality. These aspects of quality control and assurance are major challenges to engineers and managers involved in such production.

Most would agree that frequent and thorough inspection of finished products is the key factor contributing to good quality assurance of items or components produced in a mass production environment. Vigorous inspections of the quality of these items or components during the production stage will certainly result in better quality of the finished products. In theory, the ideal way to control the quality of finished products is to thoroughly inspect *every* piece that is produced by a machine or by an automated production process in order to ensure not only that the dimensions and tolerances set for all pieces are consistent with the design specifications but also that all pieces will perform the function(s) that are designed for. This idea is obviously not practicable in mass production situations: the amount of time and the costs associated with thorough inspection of every piece of finished product would be prohibitive to any producer. Nor is it practicable to inspect the quality of every finished or quasi-finished item in batches of millions of items that have been produced. A more realistic scheme is to perform inspections on a limited number of randomly selected samples and employ statistical techniques to be able to ensure the quality of all products produced by the same manufacturing process. Key questions involved in such practice are "How many samples" from a production batch does one needs to inspect, and "How many measurements" does one needs to perform on each of the selected sample items in order to be sure that "enough is enough?"

The following illustration will present the procedure that a quality engineer might follow in the quality assurance and quality control of a mass-produced shaft used as the guide rail for printer head cartridges. A machine or a process is used to produce 10 000 circular shafts each day. These shafts in the form of long circular rods must have uniform and precise diameter along the length to allow smooth sliding of the cartridge head under the guidance of high-precision motion controls. The quality assurance engineer who is responsible for such production must determine the number of sample shafts that he or she will randomly pick from the batches of the finished products to assure that the finished sample rods comply with the set standards. Additionally, the engineer also needs to determine the number of measurements of the rod diameters on each of the selected sample shaft that are required in order to be sure that the diameters measured will have the least discrepancies in value at the measurement locations.

For the purpose of illustration, let us assume that the engineer has randomly picked 100 sample rods from a batch of finished rods and chosen six diameter measurement stations along the length of the selected shafts as indicated in Figure 12.13.

The readers is now faced with the critical question of "How confident is the quality assurance engineer that selecting only six measurement stations on each of the 100 samples taken would lead to a 'credible and robust' assurance of the quality for 'all' of the 10 000 rods produced by the same machine or process?" Conversely, the question may be put "How valid are the results of the inspections on a *limited number of sample products* from the batch with *limited number of inspections on each of the limited number of sample products* for all the products produced by the same machine or process?"

The *statistical process control* (SPC) technique may provide answers to this critical question, as will be presented in the following section.

12.9.2 The Statistical Process Control Method

The statistical process control (SPC) method widely used by industry in current times was pioneered by Walter A. Shewhart of Bell Laboratories in the early 1920s. He developed the control charts for statistical quality control in 1924. This concept was further modified and improved for practical applications in modern-day statistical quality control of industrial products by W. Edwards Deming (1900–1993), an American statistician, who is regarded as the pioneer of this powerful tool for quality control of mass-produced products. The wide adoption of SPC by

Figure 12.13 A rod sample with six measurement stations.

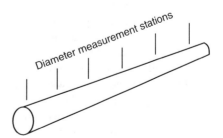

Japanese industry after WWII is viewed as the principal reason for Japan's enormous success in consumer electronics and automobiles in the global marketplace. One of Deming's followers, Dr. Genichi Taguchi, is considered the architect of applied SPC for industrial production. SPC is widely used in the semiconductor industry in this country.

The major advantages of using statistical process control (SPC) are outlined below:

1) It offers assurance that those items or components remaining in production after the application of SPC will have satisfactory quality.
2) If the procedures are applied correctly, the defect rate attributed to manufacturing will rarely exceed 1%. As a result, more parts with consistent quality will be produced, and scrappage, reworking, and repair will be reduced to a minimum.
3) When effective machines are used, manufacture will be trouble-free. This control method will also identify ineffective machines being used in production.
4) When machines of marginal capability are used in production, then the number of defective components produced by these machines will be kept to the unavoidable minimum.
5) The procedures will improve shopfloor personnel participation in quality control.
6) The procedures will have the effect of reducing company scrap rates during manufacture and assembly, which will reflect advantageously in total quality control cost.

In short, SPC emphasizes early detection of defective products and the fabrication processes that produce defective products. It also increases the rate of production of quality-assured finished products.

One will appreciate that, it being such an essential tool in modern industrial production, full coverage of the subject of SPC would require considerably more space and time than is available in this one chapter. What we will do here is introduce one commonly used technique known as "control charts." These charts allow the quality assurance engineer to control the functions and capabilities of the machines or processes that are used in mass production of parts and goods.

12.10 The "Control Charts"

Control charts are established by quality assurance engineers using statistical methods to provide the *upper* and *lower bounds* within which the measured parameters from randomly selected samples are deemed to represent satisfactory quality. These control charts are derived on the assumption that all the measured parameters from the sampled products fit the bell-shaped distribution that was illustrated in Figure 12.4, so that the mathematical model for such a distribution—Equation 12.7 for the normal distribution function—can be used as a theoretical basis for such derivation.

Figure 12.14 shows the form of a typical control chart consisting of a lower control limit (LCL) and an upper control limit (UCL). These limits serve as the lower and upper bounds that determine acceptance of the parts subjected to the same manufacturing and inspection processes after the establishment of the control chart. Once the chart is established, quality assurance engineers will randomly pick samples from batches of parts manufactured by the same process for further inspection by measurement of the same parameters. The manufacturing process is considered sound and healthy if the values of the measured parameters in these additional samples are within the upper and lower control limits of the control chart. On the other hand, the quality assurance engineer will order an immediate halt of production should he or she finds the value of the measured parameters on any of these additional samples to be outside these control limits. The engineer will immediately launch an investigation to identify the sources of the problem, whether they arise from malfunction of the manufacturing process or whether

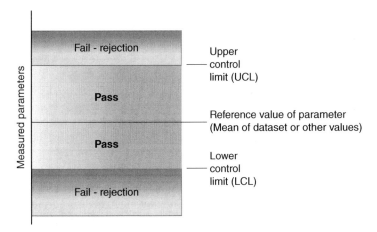

Figure 12.14 Typical control chart for quality control in mass production.

there is human error on the part of the operators. Production resumes after these problems are resolved.

We will include in later subsections two particular types of control charts that are popular for SPC in mass production environment. We will require collection of the following data for establishing the control charts of both types:

1) The sample size, k (the number of randomly selected samples from a specific manufacturing process)
2) The number of measured parameters of each sample, n.

Thus the total number of measurements for all randomly selected samples in the SPC analysis will be $k \times n$.

12.10.1   Three-Sigma Control Charts

The term "three-sigma" derives from the description of the physical meaning of standard deviation that was given in Section 12.6 and implies that 99.73% of the collected data (the measurements of the specific parameter) will be included in the analysis. For k samples and n measurements on each sample taken from a production batch, the mean of the measured value from the total $k \times n$ measured parameters is denoted by $\bar{x}$.

If the mean of all $k \times n$ measurements is $\bar{x}$, with standard deviation σ, obtained from Equation 12.3, we will have the bounds of the "upper control limit" and "lower control limit" from the following expressions:

Lower control limit:

$$\mathrm{LCL}_{\bar{x}} = \bar{x} - \frac{3\sigma}{\sqrt{n}} \tag{12.18a}$$

Upper control limit:

$$\mathrm{UCL}_{\bar{x}} = \bar{x} + \frac{3\sigma}{\sqrt{n}} \tag{12.18b}$$

According to the definition of standard deviation, σ, as a measure of variation of data from the mean value, "3σ" that appears in Equations 12.18a and 12.18b implies that there is a 99.73%

Figure 12.15 Three-sigma control chart.

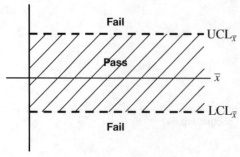

probability of the measured data fitting within the value of $\bar{x} \pm 3\sigma$. Consequently, we may say that there is only a $1 - 0.9973 = 0.0027$, or 0.27%, probability that further mean measured values of the parameter would fall outside the two control limits.

The above situation may be graphically illustrated in Figure 12.15. We may readily see that any average measured control parameter on *each* additional sample ($\bar{x}$) that falls in the shaded zone would mean acceptance of the product as being of good quality, whereas any further sample whose value of $\bar{x}$ falls outside the shaded zone will be rejected. In such case, the quality assurance engineer needs to investigate the causes for the defective product and impose whatever remedial action is necessary.

A necessary word of caution is that the above SPC control chart is built on a fundamental assumption that all the measured data fit the "normal distribution." Engineers thus need to be sure that large datasets with large numbers of samples are included in the establishment of the control charts. The limited numbers of data used in the subsequent examples may not offer correct answers for this reason. These limited numbers of data are used in these examples solely for illustrative and didactic convenience.

Example 12.10

A firm in Silicon Valley produces integrated circuit (IC) chips for a customer. The quality assurance engineer measured the output voltage from three outlets of each of the five sample chips that he randomly selected from a production station. The measured voltages are tabulated below:

Sample 1:	2.25	3.16	1.80
Sample 2:	2.60	1.95	3.22
Sample 3:	1.75	3.06	2.45
Sample 4:	2.15	2.80	1.85
Sample 5:	3.15	2.76	2.20

All measured data have the unit of millivolts.

Establish the upper and lower control limits using the "three-sigma" control chart method and the tabulated data.

Solution:

We have $k = 5$ (five samples) and $n = 3$ (three measurements on each sample). The total number of measurements is $k \times n = 5 \times 3 = 15$. One can determine the mean of all 15 measurements to be $\bar{x} = 2.477$ millivolts, and from Equation 12.3 using the total number of data of $n = 15$, the standard deviation to be $\sigma = 0.5281$ millivolt.

Figure 12.16 A three-sigma control chart for IC chip output.

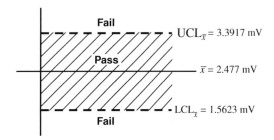

Thus, using Equations 12.18a and 12.18b, we have the lower and upper control limits:

$$\text{LCL}_{\bar{x}} = \bar{x} - \frac{3\sigma}{\sqrt{n}} = 2.477 - \frac{3 \times 0.5281}{\sqrt{3}} = 1.5623 \quad \text{millivolts}$$

$$\text{UCL}_{\bar{x}} = \bar{x} + \frac{3\sigma}{\sqrt{n}} = 2.477 + \frac{3 \times 0.5281}{\sqrt{3}} = 3.3917 \quad \text{millivolts}$$

Graphically, the above results provide the quality assurance engineer with the control chart illustrated in Figure 12.16.

The quality control engineer will pick further samples during the continuing production run and perform similar measurements of output voltages on the same leads as he did with the sample chips. Should the average the measured output (i.e., $\bar{x}$ = average of the three measurements) from any additional sample chip be beyond these bounds in Figure 12.16, he will not only scrap this particular chip but will also order an immediate halt of the production and investigate the causes for the defective chip. Production will be resumed after all the problems are resolved.

12.10.2 Control Charts for Sample Ranges (the R-Chart)

We learned in Section 12.10.1 about establishing three-sigma control charts with the mean values ($\bar{x}$) of the parameter from each of the selected measurement stations on the selected samples. There is another type of control chart, called the R-chart, that is established for the *range of measured parameter* on each of the selected samples.

Again, if we let k = the number of selected samples and n = the number of measurements stations on each sample, the recorded measurements may be expressed in the format presented in Table 12.3.

Table 12.3 Working sheet for establishing the R-chart

Sample	Measured parameters							Sample mean, $\bar{x}$	Sample range, R
$k=1$	x_1	x_2	x_3	x_4	x_5	...	x_n	$\bar{x}_{k=1}$	$x_{max} - x_{min}$
$k=2$	•	•	•	•	•	...	•	$\bar{x}_{k=2}$	•
•	•	•	•	•	•	...	•	•	•
•	•	•	•	•	•	...	•	•	•
•	•	•	•	•	•	...	•	•	•
$k=k$	•	•	•	•	•	...	•	•	•
						MEAN:		$\bar{x}$	$\bar{R}$

One sees from the bottom line of Table 12.3 that the "mean" of the Range, $\overline{R}$ may be obtained by the arithmetic average of all the sample ranges in the last column of the table. However, the same mean value of the sample range $\overline{R}$ can be obtained theoretically from the expression

$$\overline{R} = d_2\sigma \tag{12.19}$$

in which σ is the standard deviation of *all measurements*—that is, from $(k \times n)$ measurements. The factor, d_2 may be determined on the basis of the value of $n =$ the number of measurements of the parameter on each sample from Table 12.4 (Rosenkrantz, 1997).

This value of mean sample range $\overline{R}$ obtained by the measured average of all samples should approach the $\overline{R}$ value computed from Equation 12.19 if the sample size is large enough for the collected data to fit well with the normal distribution as illustrated in Figure 12.4 and represented by Equation 12.7, with which the $\overline{R}$-value can be calculated using Equation 12.19.

The discrepancy in the $\overline{R}$ values determined by these two methods is not significant because the number of measured data in a real-life quality assurance process would be large enough to have all the measured data closely fitting the normal distribution curve illustrated in Figure 12.4 and justify the use of Equation 12.19 for determining the $\overline{R}$ values in establishing the R-control charts.

The lower and upper control limits based on the range of the measured data may be computed from the following equations:

$$\textit{Lower control limit:} \quad \text{LCL}_R = D_1\sigma \tag{12.20a}$$

$$\textit{Upper control limit:} \quad \text{UCL}_R = D_2\sigma \tag{12.20b}$$

where the coefficients, D_1 and D_2 can be found from Table 12.4.

Table 12.4 Factors for estimating. $\overline{R}$ and lower and upper control limits (Rosenkrantz, 1997)

No. of measurements on each sample, n	Factor d_2	Coefficient D_1	Coefficient D_2
2	1.128	0	3.69
3	1.693	0	4.36
4	2.059	0	4.70
5	2.326	0	4.92
6	2.534	0	5.08
7	2.704	0.20	5.20
8	2.847	0.39	5.31
9	2.970	0.55	5.39
10	3.075	0.69	5.47
11	3.173	0.81	5.53
12	3.258	0.92	5.59
13	3.336	1.03	5.65
14	3.407	1.12	5.69
15	3.472	1.21	5.74

Figure 12.17 Control chart using sample ranges (the R-chart).

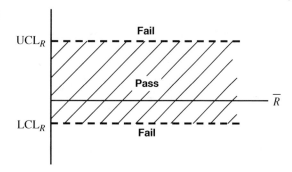

A similar control chart to that of 3σ-method in Figure 12.16 may be established for the control chart based on the range of the measured data as in Figure 12.17.

Example 12.11

Establish an R-chart using the data presented in Example 12.10.

Solution:

We will first establish the working chart similar to that shown in Table 12.3 for the present case as shown below:

Sample	Measured voltage (mV)			Mean value, $\bar{x}$	Sample Range
$k=1$	2.25	3.16	1.80	2.4033	1.36
2	2.60	1.95	3.22	2.5900	1.27
3	1.75	3.06	2.45	2.4200	1.31
4	2.15	2.80	1.85	2.2667	0.95
5	3.15	2.76	2.20	2.7033	0.95
Total $k=5$		Total $n=3$		Mean, $\mu=2.477$	$\bar{R}=1.168$
					(from dataset)

We may use Equation 12.3 to calculate the standard deviation from the dataset with the $k \times n = 15$ measured data and calculate $\sigma = 0.5281$.

We may use Equation 12.19 to compute the value of $\bar{R} = d_2\sigma = 1.693 \times 0.5281 = 0.8941$, using $d_2 = 1.693$ obtained from Table 12.4 with $n = 3$. This value obviously is very different from what we calculated from the average sample range of $\bar{R} = 1.168$ as indicated in the above table. This discrepancy is to be expected because the total number of 15 measurements in this example is hardly enough for these measured data to fit a normal distribution.

However, given that the control limits given in Equations 12.20a and 12.20b are derived from the normal distribution, we will adopt the theoretical value $\bar{R} = 0.8941$ in determining these two control limits for the present case, notwithstanding the unrealistic number of sample data.

Thus, using Equations 12.20a and 12.20b, we may compute the control limits for the R-chart to be

$$\text{LCL}_R = D_1\sigma = 0 \times 0.5281 = 0$$
$$\text{UCL}_R = D_2\sigma = 4.36 \times 0.5281 = 2.3025$$

with $D_1 = 0$ and $D_2 = 4.36$ obtained from Table 12.4 with $n = 3$.

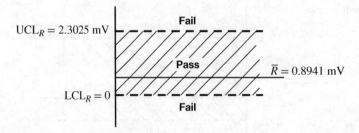

Figure 12.18 The R-control chart for quality control of chips.

Graphically the R-chart in this case may be shown as in Figure 12.18.

The control chart shown in Figure 12.18 will be used for quality control for inspecting chips in further production. Any range of the measured output voltages on any chip that is within the bounds of the control limits will be deemed to be of acceptable quality. Those chips that fail to satisfy this condition will be rejected, and the quality assurance engineer should investigate the causes for the unacceptable quality of the product and seek remedial actions.

12.11   Problems

12.1   A farm-implement manufacturing company in Midwest U.S.A. purchases steel castings from a local foundry. Thirty castings were selected at random and weighed, and their masses were recorded to the nearest kilogram, as shown below:

235	232	228	228	240	231
225	220	218	230	222	229
217	233	222	221	228	228
238	232	230	226	236	226
227	227	229	229	224	227

Do the following:
a) Group the measurements into a frequency distribution table having six equal classes from 215 to 244.
b) Construct a histogram of the distribution.
c) Determine the median, mode, and mean of the dataset

12.2   Approximations of missile velocities were recorded over a predetermined fixed distance and are presented in the following dataset. Each value in the dataset is rounded to the nearest 10 m/s:

980	960	950	1010
930	880	870	960
850	1020	970	940
970	900	1030	950
1000	940	970	600

Do the following:
a) Group these measurements into a frequency distribution table having six equal classes that range from 500 to 1100.
b) Construct a histogram of the distribution.
c) Determine the median, mode, and mean of the data

12.3 Using the data given in Problem 12.2, calculate the following:
a) The standard deviation of all data.
b) The variance of the dataset.

12.4 Find the mean $\bar{x}$, the median x_d and the standard deviation σ from the following set of measurements on the diameter of a shaft in centimeters:
2.5, 2.45, 2.48, 2.75, 2.32, 2.55, 2.47, 2.25

How do you interpret the standard deviation (σ) in a physical sense? If this dataset fits a normal distribution, what implications are for the 1σ, 2σ and 3σ that you obtained in the computation?

12.5 Find the mean, $\bar{x}$, and the standard deviation, σ, of the fracture strength of KT-SiC tabulated in Table 12.1. Interpret the physical meaning of the standard deviation that you have computed.

12.6 Find the reliability of a tube made of KT-SiC with the wall thickness indicated in Example 12.9. The tube is subjected to internal pressures of 3000 and 2500 psi.

12.7 The thickness of an epitaxial layer on a silicon wafer is specified to be 14.5 ± 0.5 μm. Assume that the manufacturing process is in statistical control, where the distribution of the thickness is normal with the mean $\bar{x} = 14.5$ μm and standard deviation $\sigma = 0.4$ μm. To monitor the production process, the mean thickness of four wafers is determined at regular time intervals. Compute the lower and upper control limits for a three-sigma control chart for the process mean.

12.8 A Silicon Valley company produces computer chips in large quantities. A particular microfabrication process for the production involves thin dioxide films on silicon sub-strates. The measured mean thickness of deposited thin film from 50 samples with 8 measurements on each sample was 15.4 micrometers (μm). The total sum of the range of measured thickness on each sample is 1.25 μm. If one assumes the measured data fit normal distribution, do the following:
a) Determine the standard deviation of the measured data, σ, and its physical meaning in relation to the measured data.
b) Determine the variance of the measured thickness.
c) Construct the 3σ control chart for the measured thickness.
d) Construct the control chart for the sample range of the measured thickness.
e) Indicate the use of these control charts for quality control in the production.

12.9 The diameters (measured in mm) of ball bearings are monitored by $\bar{x}$ and R-charts. After 30 samples of size $n = 6$ were taken, the following values were recorded: the sum of the mean of all samples $\sum_i \bar{x} = 150$ mm and the sum of the range of the measured values of all samples $\sum_i \bar{R}_i = 12$ mm. Assuming the process is in statistical control and the measured values fit the normal distribution, calculate the $\bar{x}$ and R control limits.

12.10 A precision machine shop is involved in mass production of shafts for a special electromechanical system. A quality assurance engineer is assigned the responsibility for establishing quality control charts. She took five samples from the batch of the shafts produced by one lathe. The diameters that she measured from three selected locations along the length of each of these samples are listed below:

Sample no.	Diameter at position 1	Diameter at position 2	Diameter at position 3
1	2.25	3.16	1.85
2	2.60	2.28	3.10
3	1.85	2.35	2.20
4	1.95	2.15	2.75
5	2.15	2.85	3.10

a) Determine the mode of all the measured data.
b) Determine the median and mean of all measured data.
c) Determine the standard deviation and indicate the physical meaning of this standard deviation.
d) Construct the control chart based on three-sigma method.
e) Construct the control chart based on sample ranges.
f) How would the quality assurance engineer use these charts in controlling the quality of shafts produced by that particular lathe?

12.11 A batch of microchips is fabricated by a local firm in Silicon Valley. The quality assurance engineer took five samples from a bin and measured the length of these chips at three fixed locations on each sample. The measured data that the engineer recorded are:

Sample 1:	2.15	2.35	1.95
Sample 2:	2.70	1.83	2.25
Sample 3:	1.97	2.03	2.13
Sample 4:	2.06	2.70	2.15
Sample 5:	2.03	1.75	1.85

All measured data have the unit of mm.
Determine the upper and lower control limits using:
a) The three-sigma method
b) The control chart for the ranges.

12.12 A quality assurance engineer randomly took 20 samples out of a batch of 2000 high-precision shafts manufactured by a special process. She measured the diameter of the shaft at six different locations on each sample shaft. She recorded the following: $\sum_i \bar{x}_i = 52$ mm and $\sum_i \bar{R}_i = 5$ mm, in which $\bar{x}_i$ and $\bar{R}_i$ are the sample mean and sample range of the measured values from each sample, respectively. On the assumption that all measured data fit the normal distribution, establish the following control charts for quality assurance purposes: (a) the 3σ control chart, and (b) the control chart for sample range (the R-chart) for the measured data

Bibliography

Avallone, E.A., Baumeister III, T., and Sandegh, A., "Marks' Standard Handbook for Mechanical Engineers," 11th edition, McGraw-Hill, New York, 2006, ISBN 0071428674.

Ayres, F., Jr., "Differential and Integral Calculus," Schaum's Outline Series in Mathematics, 2nd edition, McGraw-Hill Book Company, New York, 1964.

Barnett, R.L. and McGuire, R.L., "Statistical Approach to Analysis and Design of Ceramic Structures," Ceramic Bulletin, Vol. 45, No. 6, 1966, pp. 595–602.

Bathe, K-J and Wilson, E.L., "Numerical Methods in Finite Element Analysis," Prentice Hall, Inc., Englewood Cliffs, NJ, 1976, ISBN 0-13-627190-1.

Beer, F.P., Johnston Jr., E.R., and Clausen, W.E., "Vector Mechanics for Engineers – Dynamics," 7th edition, McGraw-Hill Higher Education, Boston, 2004, ISBN 0-07-230492-8.

Bishop, R.H., editor, "The Mechatronics Handbook," CRC Press, Inc., Boca Raton, 2002, ISBN 0-8493-0066-5.

Bronson, R., "Differential Equations," Schaum's Outlines, 2nd edition, McGraw-Hill Companies, New York, 1994, ISBN 0-07-008019-4.

Burden, R.L. and Faires, J.D., "Numerical Analysis," 9th edition, Brooks/Cole, Australia, 2011, ISBN-13: 978-0-538-73351-9.

Chapra, S.C., "Applied Numerical Methods with MATLAB," 3rd edition, International edition, McGraw-Hill Companies, New York, 2012, ISBN 978-007-108618-8.

Desai, C.S., "Elementary Finite Element Method," Prentice Hall, Englewood, NJ, 1979.

Ferziger, J.H., "Numerical Methods for Engineering Application," 2nd edition, John Wiley & Sons, Inc., New York, 1998, ISBN 0-471-11621-1.

Friend, E. and Idelchik, I.E., "Flow Resistance: A Design Guide for Engineers," Taylor & Francis Ltd., Philadelphia, PA, 1989. ISBN 1-56032-487-2 (paperback).

Gad-el-Hak, M., editor, "The MEMS Handbook," The CRC, Inc., Boca Raton, 2002, ISBN 0-8493-0077-0.

Gibilisco, S., editor, "The Illustrated Dictionary of Electronics," McGraw-Hill, New York, 1997, ISBN 0-07-024186-4.

Gray, A. and Mathews, G.B. "A Treatise on Bessel Functions and Their Applications to Physics," 2nd edition, Dover Publications, Inc., NY, 1966.

Gregory, L.D. and Sprinill, C.E. "Structural Reliability of Re-entry Vehicles Using Brittle Materials in Primary Structure," IAS Aerospace Systems Reliability Symposium, 1962.

Hagen, K.D., "Introduction to Engineering Analysis," 4th edition, Pearson Education, Inc., London, 2014.

Heinrich, J.C., Huyakorn, P.S., Zienkiewicz, O.C., and Mitchell, A.R. "An 'upwind' finite element scheme for two-dimensional convective transport equation," International Journal for Numerical Method for Engineering, Vol. 11, 1977, pp. 131–143.

Applied Engineering Analysis, First Edition. Tai-Ran Hsu.
© 2018 John Wiley & Sons Ltd. Published 2018 by John Wiley & Sons Ltd.
Companion Website: www.wiley.com/go/hsu/applied

Hibbler, R.C., "Dynamics," 11th edition, Pearson Prentice Hall, Inc., Upper Saddle River, NJ, 2007, ISBN 0-13-221504-7.

Hoffman, J.D., "Numerical Methods for Engineers and Scientists," McGraw-Hill, Inc., New York, 1992, ISBN 0-07-029213-2.

Hogben, L., "Elementary Linear Algebra," West Publishing Company, St. Paul, Minnesota, 1987.

Hsu, T.R. and Gillespie, G.E., "Thermal Shock on Ceramic Tubes," Nuclear Engineering and Design, Vol. 16, 1971, pp. 45–58.

Hsu, T.R. and Bertel, A.W.M., "Propagation and Opening of a Through Crack in a Pipe subject to Combined Cyclic Thermo-Mechanical Loadings," Journal of Pressure Vessel Technology, ASME, Transactions, Vol. 98, No. 1. February, 1976, pp. 17–25.

Hsu, T.R., "The Finite Element Method in Thermomechanics," 'Chapter 7 on "Thermofracture Mechanics," Allen & Unwin, Boston, 1986. ISBN 0-04-620013-4.

Hsu, T.R., Chen, G.G., Gong, Z.L., and Sun, N.S., "A Hybrid Experimental-Numerical Approach to the Determination of J-integral in a Leaking Pipe," An Invited Paper in Proceedings of International Conference on Computational Engineering Science, Springer-Verlag, 1988.

Hsu, T.R. and Sinha, D.K. "Computer-Aided Design- an integrated approach," West Publishing Company, St. Paul, MN, 1992, ISBN 0-314-80781-0.

Hsu, T.R., "MEMS & Microsystems Design and Manufacture," McGraw-Hill Higher Education, Boston, 2002, ISBN 0-07-239391-2.

Hsu, T.R., "MEMS and Microsystems Design, Manufacture, and Nanoscale Engineering," 2nd edition, John Wiley & Sons, Inc., Hoboken, NJ, 2008, ISBN 978-0-470-08301-7.

Janna, W.S., "Introduction to Fluid Mechanics," 3rd edition, PWS Publication Company, Boston, 1993, ISBN 0-534-93334-3.

Jeffrey, A., "Advanced Engineering Mathematics," Harcourt Academic Press, San Diego, 2002.

Kim, Y.J. and Hsu, T.R., "A Numerical Analysis on Stable Crack Growth Under Increasing Load," International Journal of Fracture, Vol. 20, 1982, pp. 17–32.

Knight, C.E., "The Finite Element Method in Mechanical Design," PWS-Kent Publishing Company, Boston, 1993, ISBN 0-534-93187-1.

Kreith, F., editor, "The CRC Handbook of Mechanical Engineers," CRC Press, Inc., ISBN 0-8493-9418-X, 1998.

Kreith, F. and Bohn, M.S.,"Principles of Heat Transfer," 5th edition, PWS Publishing Company, Boston, 1997, ISBN 0-534-95420-0.

Kreyszig, E., "Advanced Engineering Mathematics," 10th edition, John Wiley & Sons, Inc., Hoboken, NJ, 2011, ISBN 978-0-470-45836-5.

Landes, J.D., Begley, J.A. and Clarke, G.A., editors, "Elastic-plastic Fracture," STP 668, American Society for Testing and Materials, 1979.

Leemis, L.M., "Reliability-Probabilistic Models and Statistical Methods," Prentice-Hall, Edgewood Cliffs, New Jersey, 1995, ISBN 0-13-720517-1.

LeVeque, R., "Finite Difference Method for Ordinary and Partial Differential Equations," Society for Industrial & Applied Mathematics, Philadelphia, PA, 2007, ISBN: 978-0-898716-290.

Lee, H-H., "Finite Element Simulations, with ANSYS Workbench 16," SDC Publications, 2015, ISBN 13: 978-1-58503-983-8.

Logan, D.L., "A First Course in the Finite Element Method," 6th edition, Cengage Learning, Australia, 2017.

Luxmoore, A.R., Owen, D.R.J., Rajapakse, Y.P.S., and Kanninen, M.F., editors, "Numerical Methods in Fracture Mechanics," Proceedings of the 4th International Conference, San Antonio, TX, USA, 1987.

Malek-Madani, R., "Advanced Engineering Mathematics-with Mathematica and MATLAB," Addison Wesley Longman, Inc., Reding, MA, Vol. 1 and 2, 1998, ISBN 0-201-59881-7 (Vol. 1), ISBN 0-201-32549-7 (Vol. 2).

Meriam, J.L. and Kraige, L.G., "Dynamics," 6th edition, John Wiley & Sons, Inc., Hoboken, NJ, 2007, ISBN 13 978-0-471-73931-9.

Moody, L.F., "Friction Factors for Pipe Flows," Transactions of the American Society of Mechanical Engineers, Vol. 66, No. 8, 1944, pp. 671–684.

Morrison, J., "Statistics for Engineers," John Wiley & Sons, 2009, ISBN 978-0-470-74556-4.

Navidi, W., "Statistics for Engineers and Scientists,"4th edition, McGraw-Hill, Columbus, OH 2013, ISBN 978-0-07340-1331.

Norton, R.L., "Machine Design: An Integral Approach," 5th edition, Prentice Hall, 2013, ISBN 978-0133-56717.

Ozisik, M.N., "Boundary Value Problems of Heat Conduction," International Textbook Company, Cranton, PA, 1968. ISBN 978-0-486-78286-7.

Rosenkrantz, W.A., "Introduction to Probability and Statistics for Scientists and Engineers," McGraw-Hill, NY, 1997, ISBN 978–0070539884.

Sauer, T., "Numerical Analysis," 2nd edition, Pearson Education, Ltd., London, 2011, ISBN 978-1-292-02358-8.

Segerlind, L.J., "Applied Finite Element Analysis," John-Wiley & Sons, Inc., NY, 1976, ISBN 0-471-77440-5.

Shigley, J.Ed. and Mischke, C.R., "Mechanical Engineering Design," McGraw-Hill Book Company, New York, 1989, ISBN 0-07-056899-5.

Spiegel, M.R., "Statistics," Schaum Publishing Company, NY, 1961.

Spiegel, M.R., "Theory and Problems of Advanced Calculus," Schaum Publishing Company, NY, 1963.

Stout, K., "Quality Control in Automation," Prentice-Hall, Inc., Englewood Cliffs, NJ, 1985, ISBN 0-13-745159-8.

Turner, M.J., Clough, R.W., Martin, H.C., and Topp, L.J., "Stiffness and Deflection Analysis of Complex Structure," Journal of Aeronautical Science, Vol. 23, 1959, pp. 805–823.

Vardeman, S.B., "Statistics for Engineering Problem Solving," PWS Publishing Company, Boston, MA, 1994, ISBN 0-534-92871-4.

Volterra, E. and Gaines, J.H., "Advanced Strength of Materials," Prentice-Hall, Inc., Englewood, NJ, 1971, ISBN 13-013854-1.

Weibull, W., "A Statistical Theory of Strength of Materials," Ing. Vetenskaps Akad. Hnadl., No. 151, 1939.

Weibull, W., "A Statistical Distribution Function of Wide Applicability," Journal of Applied Mechanics, American Society of Mechanical Engineers, Vol. 18, No. 293, 1951.

Whitaker, J.C., editor, "The Electronics Handbook," The CRC Press, Boca Raton, 1996, ISBN 0-8493-8345-5.

White, F.M., "Fluid Mechanics," 3rd edition, McGraw-Hill, Inc., New York, 1994, ISBN 0-07-911695-7.

Wylie, C.R. and Barrett, L.C., "Advanced Engineering Mathematics," 6th edition, McGraw-Hill, NY, 1995.

Young, W.C., Budynes, R.G., and Sadogh, A.M., "Roark's Formulas for Stress and Strain," 8th edition, McGraw-Hill, New York, 2012, ISBN 978-0-07-174248-1.

Zienkiewicz, O.C., "The Finite Element Method in Engineering Science," McGraw-Hill, NY, 1971, ISBN 07-094138-6.

Zill, D.G. and Cullen, M.R., "Advanced Engineering Mathematics," PWS Publishing Company, Independence, Ky, U.S.A., 1992, ISBN 0-534-92800-5.

Zwillinger, D., "CRC Standard Mathematical Tables and Formulae," 31st edition, Chapman & Hall/CRC, Boca Raton, FL, 2003, ISBN 1-58488-291-3.

Appendix 1

Table for the Laplace Transform

Case no.	$f(t)$	$F(s) = L[f(t)] = \int_0^\infty e^{-st}f(t)dt$
1	1	$1/s$
2	t	$1/s^2$
3	t^{n-1}	$(n-1)!/s^n \ (n = 1,2,3,\ldots.)$
4	$t^{-1/2}$	$\sqrt{\pi}/s$
5	$t^{1/2}$	$\sqrt{\pi}/(2s^{3/2})$
6	t^{k-1}	$\Gamma(k)/s^k \ (k > 0), \ \Gamma(k) = $ Gamma function
7	e^{at}	$1/(s-a)$
8	te^{at}	$1/(s-a)^2$
9	$t^{n-1}e^{at}$	$(n-1)!/(s-a)^n \ (n = 1,2,3,\ldots.)$
10	$t^{k-1}e^{at}$	$\Gamma(k)/(s-a)^k \ (k > 0), \ \Gamma(k) = $ Gamma function
11	$e^{at}-e^{bt}$	$(a-b)/[(s-a)(s-b)] \ (a \neq b)$
12	$ae^{at}-be^{bt}$	$(a-b)s/[(s-a)(s-b)] \ (a \neq b)$
13	Dirac function: $\delta_0(t)$	1
14	$\delta_a(t)$	e^{-as}
15	Unit step function: $u_a(t)$	e^{-as}/s
16	$\ln t$	$\dfrac{1}{s}\left(\ln\dfrac{1}{s} - 0.5772156\ldots\right)$
17	$\sin(\omega t)$	$\omega/(s^2 + \omega^2)$
18	$\cos(\omega t)$	$s/(s^2 + \omega^2)$
19	$\sinh(at)$	$a/(s^2-a^2)$
20	$\cosh(at)$	$s/(s^2-a^2)$
21	$e^{at}\sin(\omega t)$	$\omega/[(s-a)^2 + \omega^2]$
22	$e^{at}\cos(\omega t)$	$(s-a)/[(s-a)^2 + \omega^2]$
23	$t\sin(\omega t)$	$2\omega s/(s^2 + \omega^2)^2$
24	$\omega t - \sin(\omega t)$	$\dfrac{\omega^3}{s^2(s^2 + \omega^2)}$
25	$J_0(at)$	$\dfrac{1}{\sqrt{s^2 + a^2}}$
26	$J_0(2\sqrt{kt})$	$e^{-k/s}/s$

Applied Engineering Analysis, First Edition. Tai-Ran Hsu.
© 2018 John Wiley & Sons Ltd. Published 2018 by John Wiley & Sons Ltd.
Companion Website: www.wiley.com/go/hsu/applied

Appendix 2

Recommended Units for Engineering Analysis

Length	meter (m)
Area (A)	square meter (m^2)
Volume (v)	cubic meter (m^3)
Time (t)	second (s)
Temperature (T)	degree Celsius (°C), or in the scale of Kelvin (K)
Force (F)	Newton (N)
Weight (W)	kilogram (kg$_f$) = 9.81 N
Pressure (P)	Pascal (P) = N/m^2
Mass (m)	gram (g)
Mass density (ρ)	g/cm^3
Work (W_k)	N-m
Energy	Joule (J) = 1 N-m
Power	Watt (W) = 1 J/s = 1 N-m/s
Stress (σ)	Mega Pascal (MPa) = 10^6 Pa = 10^6 N/m^2
Velocity (V)	m/s
Acceleration (a)	m/s^2
Specific heat (c)	J/g-°C
Thermal conductivity (k)	W/m-°C
Heat transfer coefficient (h)	W/m^2-°C
Thermal diffusivity (α)	m^2/s
Coefficient of linear thermal expansion (α)	/°C
Dynamic viscosity (μ)	N-s/m^2

Applied Engineering Analysis, First Edition. Tai-Ran Hsu.
© 2018 John Wiley & Sons Ltd. Published 2018 by John Wiley & Sons Ltd.
Companion Website: www.wiley.com/go/hsu/applied

Appendix 3

Conversion of Units

Length	1 meter (m) = 39.37 inches = 3.28 feet (ft)
	1 centimeter (cm) = 10^{-2} m
	1 millimeter (mm) = 10^{-3} m
	1 micrometer (μm) = 10^{-6} m
	1 kilometer (km) = 0.6212 mile (mi) = 3280 feet (ft)
Area	1 square meter (m^2) = 10.76 ft^2
Volume	1 cubic meter (m^3) = 35.29 ft^3
	1 liter (l) = 1000 cm^3
Temperature	degree Celsius, °C = (5/9) × [degree Fahrenheit (°F) – 32]
	K = °C + 273
Force	1 N = 0.2252 pound force (lb_f) = 1 kg-m/s^2
Weight	1 kilogram force (kg_f) = 2.2 lb_f
Pressure and stress	1 MPa = 145.05 pound per square inch (psi)
	1 Pa = 1 N/m^2 = 10 dynes/cm^2 = 2.089 × 10^{-2} lb_f/ft^2
Mass	1 g = 68.5 × 10^{-6} slug
	1 kg = 1000 g = 2.2 pounds mass (lb_m)
Energy	1 J = 0.2389 calories (cal) = 10^7 ergs =
	0.7376 ft-lb_f = 9.481 × 10^{-4} Btu
	1 kilowatt-hour (kW-h) = 3.6 × 10^6 J = 3413 Btu
Power	1000 watts (W) = 1.341 horsepower (hp) =
	737.6 ft-lb_f/s = 0.9483 Btu/s
	1 W = 1 J/s
Velocity	1 m/s = 3.28 ft/s
	1 km/h = 0.6212 miles/h = 0.9111 ft/s
Specific heat	1 J/g-°C = 0.2394 Btu/lb_m-°F
Thermal conductivity	1 W/m-°C = 13.3816 × 10^{-6} Btu/in-s-°F
Heat transfer coefficient	1 W/m^2-°C = 1.1151 × 10^{-6} Btu/in^2-s-°F
Thermal diffusivity	1 m^2/s = 1545 in^2/s
Coefficient of linear thermal expansion	/°C = 0.5555/°F
Dynamic viscosity	N-s/m^2 = 10 poise = 0.02089 slug/ft-s

Applied Engineering Analysis, First Edition. Tai-Ran Hsu.
© 2018 John Wiley & Sons Ltd. Published 2018 by John Wiley & Sons Ltd.
Companion Website: www.wiley.com/go/hsu/applied

Appendix 4

Application of MATLAB Software for Numerical Solutions in Engineering Analysis

Contributed by *Vaibhav Tank*

As described in Section 10.6.2, the MATLAB software package offered by The MathWorks, Inc. provides a powerful environment for numerical computations, visualization, and programming for engineers in their engineering analyses. Engineers may use this software package to quickly solve complex mathematical problems, analyze data, and develop algorithms, which would otherwise be extremely tedious to solve using traditional methods. This appendix will present three case examples to illustrate how MATLAB can be used to solve different types of problems in engineering analysis with a detailed description of the input/output procedures for using this software package. These selected cases are related to the examples in various chapters of the book.

The author of this book and the contributor of this appendix would like to express their appreciation to The San Jose State University for providing them the access to a late version of MATLAB R2015a used to solve the following three cases:

Case 1: Plot the instantaneous amplitudes of a vibrating mass $y(t)$ in Equation 8.40 in Chapter 8, where t is the time after the inception of the vibration in a near-resonant vibration situation illustrated in Figure 8.24.

The solution of the instantaneous amplitudes of the vibrating mass M subjected to a cyclic force with circular excitation frequency ω is expressed as:

$$y(t) = \left[\frac{2F_0}{M(\omega_0{}^2 - \omega^2)} \right] \sin \varepsilon t \sin \omega t \tag{8.40}$$

where $F_0 = 1000$ N is the maximum value of a cyclic force $F_0\cos(\omega t)$ applied to the elastic bench support of a stamping machine in Figure 8.19 with the mass $M = 1000$ kg and the natural frequency of the vibrating bench $\omega_0 = 5$ rad/s. The frequency ω of the excitation force is 4.95 rad/s.

The process of plotting the required graphic solution in Equation 8.40 with assigned numerical data using the MATLAB software package begins with the following steps:

1) Open the version of MATLAB installed in the computer.
2) Go to File – New – Script (Ctrl + N or ⌘ + N) to create a new script file that MATLAB will run.
3) By looking at the problem statement, we can see that we have all of the variables needed to obtain the graphic solution of Equation 8.40. The only variable in this equation is t (time), so we will list it as such in the beginning. As with any mathematics or physics problem, we will start the solution procedure by listing all available information as:

Applied Engineering Analysis, First Edition. Tai-Ran Hsu.
© 2018 John Wiley & Sons Ltd. Published 2018 by John Wiley & Sons Ltd.
Companion Website: www.wiley.com/go/hsu/applied

```
syms t;           %Define system variables

Fo=1000;          %Excitation force in N
M=1000;           %Mass in kg
omega_o=5;        %Natural frequency of system in rad/s
omega=4.95;       %Excitation frequency in rad/s
```

The statements followed by the percentage symbol (%) are in-line comments. The compiler ignores everything following the symbol on that line. Comments are a great way to keep the user's thoughts organized, so that anyone who might look at the code later can understand what the user was thinking. MATLAB does not require the user to end each statement with a semicolon. In MATLAB, semicolons are used to construct arrays and separate commands on the same line, or to suppress the outputs. In this case, we used the semicolon at the end of each line for latter function, or else the result of that command would return to us through the Command Window. We will observe this later on in the code when we will purposely leave out the semicolon, so that we can actually obtain the result of that line of code.

We note that the value of epsilon (ε) is not explicitly expressed in Equation 8.40. This quantity is given by the following expression in this book:

$$\varepsilon = \frac{\omega_o - \omega}{2} \tag{8.39b}$$

Converting the above relation into the MATLAB code, we get the following:

```
epsilon=(omega_o-omega)/2;    %Circular frequency (epsilon)
                              %calculated from natural and
                              %excitation frequency
```

We also need to define the "vector" for the variable t with a specified step size.

```
t=[0:.01:10];    % array for time values
```

Here, the vector t starts at 0 and continues up to 10 seconds with a selected step size of 0.01 seconds. There is a reason for choosing such a small step size, as will be discussed later. Now that we have all of the required variables defined, we can go ahead and specify the solution of the equation that needs to be plotted graphically. One important thing to keep in mind is that most of the variables we defined so far are scalars, but the array t is treated as a vector. In order to multiply scalars and vectors together, MATLAB uses special notations to signify element-wise multiplication, where one may use a period (.) in front of the operator as shown below:

```
y=(2.*Fo)./(M.*(omega_o^2-omega^2)) .* sin(epsilon.*t)
.*sin(omega.*t); %Nonlinear equation to be graphed
```

This will ensure that MATLAB does not interpret why we are trying to multiply two vectors together. Trying to evaluate the same line above without the periods will display an error message that reads the "Inner Matrix dimensions must agree," since matrices require the inner dimensions to be equal in order to multiply them as required in Chapter 4.

Now that we have defined all the variables as well as the function that needs to be shown in the graph, we can proceed using the built-in *plot()* function to create a 2-D plot and give a title as well as axis labels to it as follows:

```
plot(t,y)          %Graph y(t)
title('Near-resonant vibration situation(Amplitude vs. Time)'),
xlabel('time (seconds)'), ylabel('y(t)')
```

The final input to the code is shown as follows:

```
case1.m  ×  +
1 -   syms t;              %Define system variables
2
3 -   Fo=1000;             %Excitation force in N
4 -   M=1000;              %Mass in kg
5 -   omega_o=5;           %Natural frequency of system in rad/s
6 -   omega=4.95;          %Excitation frequency in rad/s
7 -   epsilon=(omega_o-omega)/2;    %Circular frequency (epsilon) calculated from natural and excitation frequency
8
9 -   t=[0:.01:20];        %array for time values
10
11 -  y=(2.*Fo)./(M.*(omega_o^2-omega^2)).*sin(epsilon.*t).*sin(omega.*t);    %Nonlinear equation to be graphed
12
13 -  plot(t,y)            %Graph y(t)
14 -  title('Near-resonant vibration situation (Amplitude vs. Time)'), xlabel('time (seconds)'), ylabel('y(t)')
```

4) Click on the green "play" icon in the ribbon on the computer screen to run the script.

<u>Output:</u>

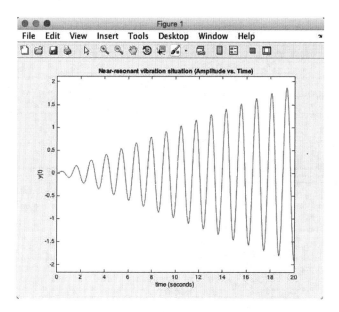

Figure A4.1 Amplitude of the vibrating mass in the first 20 seconds.

We note that two different trigonometric functions are present in Equation 8.40 along with the coefficient in the square brackets. Both these trigonometric functions and the associated coefficient play important roles in the output results. The quantity in the square brackets represents the maximum oscillating amplitude of the vibrating mass. The function $\sin(\varepsilon t)$ *regulates the variation of the amplitudes in* shorter periods as seen in Figure A4.1. However, in order to see how $\sin(\omega t)$ plays a role in this near-resonant vibration situation, we need to zoom out the graphing window and edit line 9 in the script:

```
t=[0:.01:125]; % array for time values
```

We would then run the script again resulting in the following graph:

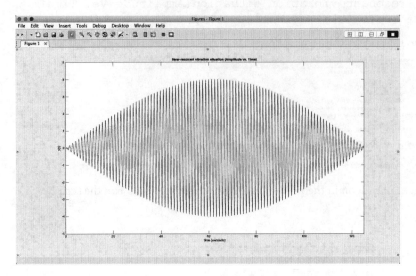

Figure A4.2 Amplitude of the vibrating mass for half cycle.

Increasing the value of t to 250 seconds by editing the same line, one can see the behavior of the entire function.

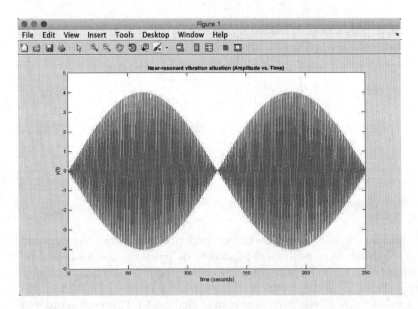

Figure A4.3 Amplitude of the vibrating mass after one complete cycle.

From this graph, we can see that the effect of the second sinusoidal function $\sin(\omega t)$ *produces* fluctuation of the variations of the amplitudes produced by the function $\sin(\varepsilon t)$ *in* a wave-like behavior as shown in Figure A4.3.

While preparing the input to the MATLAB code in this particular case, we made a decision on purposely using a small step size in order to capture every "small cycle" of the vibrating mass (call "beats") in the graphic display. Not having the appropriate step size, also called undersampling, can result in aliasing, which creates a distorted view of the actual results. If we had chosen a step size of $t = 5$ s instead of 0.01, the results would have been as shown in Figure A4.4.

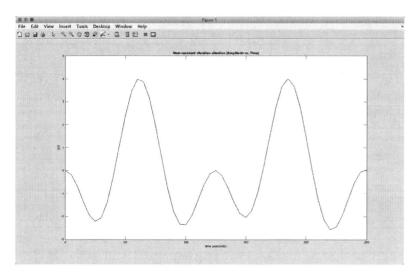

Figure A4.4 Improper graphical results due to aliasing.

The user is thus reminded that undersampling, in this particular case illustration, could lead to improper results that do not resemble those shown in Figures A4.1–A4.3, because we are capturing data points that are too large of an interval than the true behavior of the solution of Equation 8.40. Thus, it is important to choose the correct solution step sizes to ensure that we get the correct representation of the function $y(t)$.

Case 2: On the mode shapes of a vibrating flexible pad

This case deals with the determination of the shapes of Mode 1 and of a vibrating flexible pad from a free vibration analysis. Flexible pads are reasonable approximation of thin plate structures.

We will use the MATLAB software package to generate graphical solution to the free vibration analysis of a flexible pad described in Section 9.8 with the following physical conditions:

The pad has a rectangular shape as illustrated in Figure 9.25 with length $b = 10$ inches and width $c = 5$ inches. The pad has a thickness of 0.185 inch and a mass density of $1.55 \, \mathrm{lb_m/in^2}$. The pad has an initial sag due to its own weight, which can be described by a function $f(x, y) = (10 - x)(5 - y)$, and this shape is sustained by a tensile force $P = 0.5 \, \mathrm{lb_f/in}$.

We performed the following procedures for deriving the required graphical solutions by using the MATLAB software package:

1) Open the version of MATLAB in the computer.

2) Go to File – New – Script (Ctrl + N or ⌘ + N) to create a new script file that MATLAB will run.

3) Since the solution of free vibration of flexible pads is already available as Equation 9.76 in the book, with the induced amplitudes of the vibrating pad expressed in Equation 9.81, we will focus on creating surface plots using the latter equation.

4) We will put an effort to list all the given variables as shown below:

```
syms x y                    % Define variables to be graphed

b = 10;              % Length of pad in inches
c = 5;               % Width of pad in inches
P = 0.5;             % Tension applied to pad in lbf/in
g = 32.2;            % Gravitational constant in ft/s²
ρ = 0.00155;         % Mass density of pad in lbm/in²
m = 1;            % Mode variable; will be changed to 3 for different mode
n = 1;            % Mode variable; will be changed to 3 for different mode
```

From the above given information, we can calculate the frequency ω_{mn} of the system given in Equation (p) in Section 9.8.2 to be:

$$\omega_{mn} = \sqrt{\left(\frac{m\pi}{b}\right)^2 + \left(\frac{n\pi}{c}\right)^2}$$

and the constant coefficient a in Section 9.8.3 to be:

$$a = \sqrt{\frac{Pg}{\rho}}$$

Using the above equations, we can let the MATLAB to perform the calculations with the following lines of code:

```
w_mn = sqrt((m*pi/b)^2+(n*pi/c)^2); % Frequency in rad/s

a = sqrt(P*g*12/ρ);                 % Constant coefficient in
in/s; formula multiplied by 12 to convert from feet to inches
```

In order to calculate the required mode shapes of the flexible plate, we also need to determine the coefficient A_{mn} using the following equation:

$$A_{mn} = \frac{16(bc)^2[1 + (-1)^{n+1}][1 + (-1)^{m+1}]}{m^3 n^3 \pi^6} \tag{9.82}$$

where b = 10 inches and c = 5 inches. The input for evaluating the coefficients A_{mn} to the MATLAB code is:

```
A_mn = (16*(b*c)^2*(1+(-1)^(n+1))*(1+(-1)^(m+1)))/(m^3*n^3*pi^6);
% Calculate coefficient A_mn
```

One reason to use MATLAB scripts is that they make computing tedious and repetitive calculations easy and quick with just a few lines of code. We will create a *while* loop in MATLAB to iterate through different time intervals at which we would like to compute the equation. To begin the loop, we have to initialize the iterating variable T as follows:

```
T = 0;       % Initialize time in seconds
f = 1;       % Initialize figure label counter
```

After initializing the variables, we have to define the boundary conditions at which we would like to end the loop. In this case, we want to run the loop up to time $t = 0.25$ seconds, which should give us a good view on the shapes of the pad immediately after it is subjected to an instantaneous disturbance. The corresponding input coding is:

```
while T <= 1/4    % While loop to create graphs at different time
intervals
```

The goal of this "*while* loop" is to create multiple surface plots at different time intervals. Thus, it is important to create figure labels with the *figure( )* command, as shown below:

```
figure(f)      % Figure label
```

Although this may not seem like a crucial command, not including this one line in the script will overwrite the previous plot as soon as each loop has been executed. Thus, leaving out the *figure( )* command will produce only one graph instead of three, as we intended.

Now that all the variables are accounted for in the input file, we need to define the equation whose results need to be displayed graphically:

$$z(x, y, t) = A_{mn}\left(\sin \frac{m\pi x}{10}\right)\left(\sin \frac{n\pi y}{5}\right)(A_{mn} \cos a\omega_{mn}t) \tag{9.81}$$

where x and y are the coordinates for the length and width of the pad, respectively.

```
    z = A_mn * sin(m*pi*x/10) * sin(n*pi*y/5) * A_mn *
(cos(a*w_mn*(T)));      % Function to be graphed
```

In order to create the contoured surface plots, we will use the built-in *ezsurfc( )* function to perform this task. The input parameters include the function that needs to be displayed graphically, followed by the domain in vector format. The first two parameters in the array correspond to the x domain and the last two parameters correspond to the y domain.

```
 ezsurfc(z,[0, 10, 0, 5]) % Function to create a contoured surface
graph
```

To assign the appropriate plot titles, we will use the built-in string concatenation function *strcat()* to create the correct string *str* that will be used in the titles for each plot. This function concatenates strings horizontally separated by commas as shown below:

```
 str = strcat({'Shape of the pad at '}, {'T = '}, num2str(T), '
seconds', {' in Mode '}, num2str(m)); % Concatenate strings for figure
title
      title(str)
```

An important thing to keep in mind is that the function takes only the arguments of type *string*, which is why we had to convert the variables T and m to strings using *num2str()*. Otherwise, MATLAB will throw an error that reads "Inputs must be cell arrays or strings."

The main part of the loop is now over. However, in this "*while* loop," we need to specify how to increment the iterating variable T as well as f, which is used in the figure labels. Without specifying the increments of T, MATLAB will not know how to proceed, which will create an infinite loop and eventually crash the application. At the end of the "*while* loop," we need to let MATLAB know that we have reached its terminating point by using the *end* command. Thus, the following lines of code are very important:

```
    T = T + 1/8; % Increment time by 1/8 seconds.
    f = f + 1;   % Increment figure number
end
```

The input file to MATLAB in this case is now completed, and we are ready to run the case by the MATLAB code. The following is the input data file for the user to make a final check for its appropriateness.

```
Case3.m  ×  +
1 -    syms x y          % Define variables to be graphed
2
3 -    b = 10;           % Length of pad in inches
4 -    c = 5;            % Width of pad in inches
5 -    P = 0.5;          % Tension applied to pad in lbf/in
6 -    g = 32.2;         % Gravitational constant in ft/s2
7 -    p = 0.00155;      % Mass density of pad in lbm/in2
8 -    m = 1;            % Mode variable
9 -    n = 1;            % Mode variable
10
11 -   w_mn = sqrt((m*pi/b)^2+(n*pi/c)^2);    % Frequency in rad/s
12 -   a = sqrt(P*g*12/p);     % Constant coefficient in in/s
13
14 -   A_mn = (16*(b*c)^2*(1+(-1)^(n+1))*(1+(-1)^(m+1)))/(m^3*n^3*pi^6); % Calculate coefficient A_mn
15
16 -   T = 0;            % Initialize time in seconds
17 -   f = 1;            % Initialize figure label counter
18
19 -   ⊟while T <= 1/4   % While loop to create graphs at different time intervals
20 -   |
21 -         figure(f)
22 -         z = A_mn * sin(m*pi*x/10) * sin(n*pi*y/5) * A_mn * (cos(a*w_mn*(T)));   % Function to be graphed
23
24 -         ezsurfc(z,[0, 10, 0, 5]) % Function to graph
25 -         str = strcat({'Shape of the pad at '}, {'T = '}, num2str(T), ' seconds', {' in Mode '}, num2str(m));
26 -                      % Concatenate strings for figure title
27 -         title(str)
28
29 -         T = T + 1/8;    % Increment time by 1/8 seconds.
30 -         f = f + 1;      % Increment figure number
31 -   └ end
```

4.5 Click on the green "play" icon on the computer screen to run the case after having been ensured with the correct inputs to the problem.

Output:

Mode 1:

Figures A4.5a,b,c show the shape of the pad in Mode 1 at different times. Although the shapes look similar to each other, one can observe the reduction in the maximum amplitude by looking

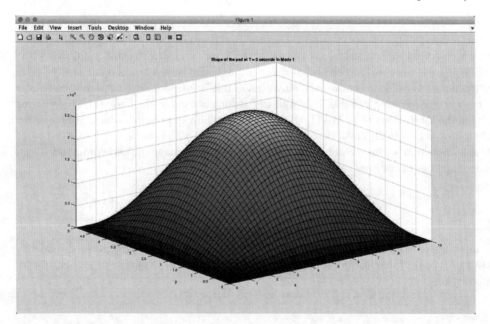

Figure A4.5a Initial shape at $t = 0$ s of the flexible pad.

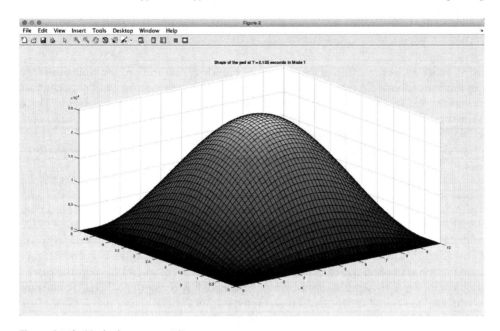

Figure A4.5b Mode shape at $t = 1/8$ s.

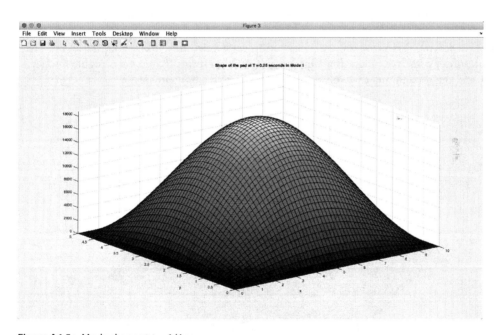

Figure A4.5c Mode shape at $t = 1/4$ s.

at the changes in the scale of the amplitude on the left. There is a drastic drop in the amplitude between $t = 0$ s and $t = 1/4$ s, as shown in Figures 4.5a and 4.5c. In order to see what happens to the pad under Mode 3 vibration, we need to go back to the input file again and change m and n to 3 in lines 8 and 9, respectively.

```
m = 3;      % Mode variable
n = 3;      % Mode variable
```

After pressing the "play" button again, we get the following results:

Mode 3:

Figures A4.6a,b,c show the shape of the flexible pad in Mode 3 vibration. Multiple peaks are present in this mode, both along the x- and y-directions. The figure above overlays a contour plot on a surface plot in order to show the locations where the peak amplitudes occur. These locations are of particular interest to engineers during vibrational analysis and the results were achieved by using the *ezsurfc( )* function of MATLAB.

Case 3: Solve the following ordinary differential equation with graphic output (Example 10.15 in the book):

$$\frac{d^2y(x)}{dx^2} - 2\frac{dy(x)}{dx} + y(x) = x^2 - 4x + 2 \tag{a}$$

with $y(0) = 0$ and $y'(0) = 0$

The reader will find from Example 10.15 in Chapter 10 that a fourth-order Runge–Kutta method was used to solve the ordinary differential equation, Equation (a), but not without lengthy computations of several recurrence relations provided by this method. Here, the reader will find that solution with significant easiness is obtainable by using the MATLAB software package, with the additional capability of the graphic output for the solution. The procedure for obtaining the MATLAB solution of this differential equation is as follows:

1) Open the version of MATLAB in the computer.

2) Go to File – New – Script (Ctrl + N or ⌘ + N) to create a new input (script) file that MAT-LAB will run.

3) To begin solving this problem, we will need to create two functions: one that defines the second-order ordinary differential equations (ODEs) in terms that are acceptable by MAT-LAB, and the second one that solves the equation using a built-in solver.

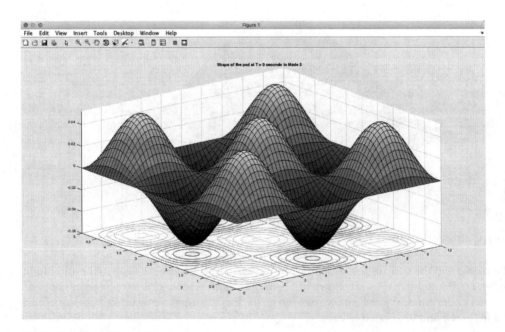

Figure A4.6a Initial state of the pad ($t = 0$ s).

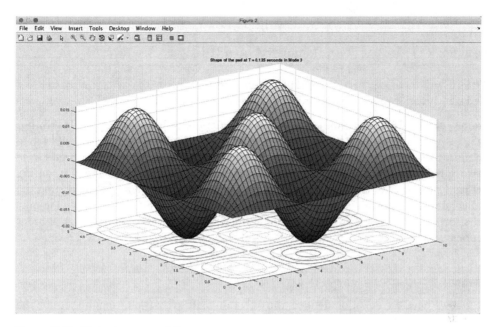

Figure A4.6b Mode shape at $t = 1/8$ s.

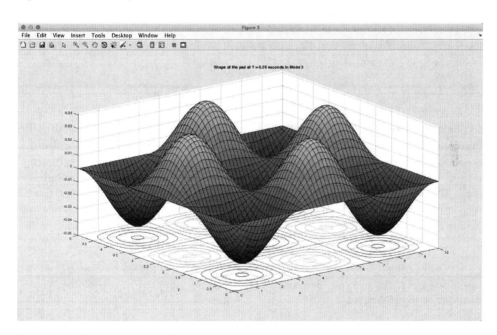

Figure A4.6c Mode shape at $t = 1/4$ s.

We can name the function the way we like; the only restriction is that the filename should match the exact name of the function. This script uses a *main* function that calls the definition of the differential equation. However, the user is free to create separate scripts for each of the functions. We will define an arbitrary range of x-values and list the given initial conditions for this equation as:

```
function case2_main

x=0:0.001:3;              % Range of x-values

initial_y    = 0;         % Initial value of y
initial_dydx = 0;         % Initial value of dy/dx
```

There are several built-in "ordinary differential equation" solvers within MATLAB, but ode45( ) is one of the most versatile and should be the first solver that one should try to solve most problems. For instance, on the left-hand side of Equation (a), one should state the two variables involved in the equation using a vector. The ode45( ) function on the right-hand side takes in the parameters such as the function, the range of values, and the initial conditions. One thing to keep in mind is that this solver is capable of solving only the first-order ordinary differential equations (ODEs), so derivatives in the higher-order equations need to use a special command, as will be seen a little later.

```
[x,y] = ode45( @rhs, x, [initial_y initial_dydx] ); % Differential
equation setup using ode45()
```

Now that we have the equation defined, the only thing left to do is plotting the results; the solver takes care of the rest. Because of the initial conditions defined in the ode45( ) solver, the y-vector returns two columns: $y(t)$ and dy/dt. When creating the plot, we only want to plot the first column of the y vector against the x-axis since that is the solution to the problem. When referencing a matrix in MATLAB, one needs to reference it using the following format: *matrix_name(row, column)*. Because we want to plot graphs for all of the rows of the first column of the y matrix, we use *y(:,1)* as shown below.

```
plot(x,y(:,1));
xlabel('t'); ylabel('x'); title('Solution to differential equation');
```

At this point, the main function is complete. The lines of code below are used to confirm the solutions with that of Example 10.15 in the book which was solved using the fourth-order Runge–Kutta method. We would like to confirm the y-values at specific x-values. In order to do this, we need to find the index of a specific x value first by searching through the x matrix. The next few lines of code search through the x matrix with $x = 0.2$, 0.6, and 1.2, respectively. These index values are stored in their corresponding variable names and are then used to search through the y matrix to find the exact value of the solution at the index as:

```
index_x1 = x==0.2;  % Find index of array element where x = 0.2
y1 = y(index_x1);   % Find corresponding value of y at x = 0.2 by
looking at the index x1

index_x2 = x==0.6;  % Find index of array element where x = 0.6
y2 = y(index_x2);   % Find corresponding value of y at x = 0.6 by
looking at the index x2
```

```
index_x3 = x==1.2;  % Find index of array element where x = 1.2
y3 = y(index_x3);  % Find corresponding value of y at x = 1.2 by
looking at the index x3
```

After storing the exact values found by searching through the arrays, we can print the results on the Command Window by using the sprintf( ) commands. Strings used within this command are enclosed within quotation marks. The user may also use formatting operators as placeholders for variables within a print statement. In this case, we use the *%f* formatting operator because we know that the results from the ode45( ) solver will contain float values. The decimal value (.3) in front of the format specifier indicates the number of decimal places the user wants to print. We chose to print three decimal places to avoid too much clutter. As shown below, the format specifiers are replaced with the variables that follow the commas.

```
sprintf('Solutions: ')
sprintf('y(0.2) = %.3f', y1)
sprintf('y(0.6) = %.3f', y2)
sprintf('y(1.2) = %.3f', y3)

end
```

So far, we have set up the parameters in the ode45( ) solver and plotted them, but the function needs to be defined yet. We start off by naming and creating a function called *rhs* containing the equation that needs to be solved; this will then be used in the ode45( ) solver. Earlier, we mentioned that the solvers can only solve first-order ODEs, so we need to be creative in converting the second-order ordinary differential equations (ODEs) that we want to solve into two first-order ODEs as follows:

We will express Equation (a) in the form of the first-order differential equation by redefining $\frac{dy(x)}{dx}$ as a function $y(2)$ and move all other terms in Equation (a) to the right-hand side of the equation as follows:

$$\frac{d^2y(x)}{dx^2} = 2y(2) - y(1) + x^2 - 4x + 2$$

where the functions $y(1) = y(x)$ and $y(2) = \frac{dy(x)}{dx}$.

The solution to this problem will be represented as a vector $dydx = [y(1); y(2)]$. Keep in mind that only $y(1)$ corresponds to the solution of y in the original equation and the function $y(2)$ corresponds to the derivative of $y(x)$, which can just be ignored as a byproduct of this method. This is the reason why we seek only the graphical solution of the first column of the y-vector in the *plot( )* function. Once this is done in defining the differential equation to the first-order ODE, we need to terminate it with the *end* command.

```
function dydx = rhs(x,y)
    dydx_1 = y(2);
    dydx_2 = 2*y(2) - y(1) + x^2 - 4*x + 2;

    dydx = [dydx_1; dydx_2];
end
```

The above action marks the completion of the entire script. Double-check the cursor to make sure that the displayed script matches the one shown below:

```
1       % The function below solves the ODE function in question.
2     ⊟ function case2_main
3
4  -    x=0:0.001:3;          % Range of x-values
5
6  -    initial_y    = 0;  % Initial value of y
7  -    initial_dydx = 0;  % Initial value of dy/dx
8
9  -    [x,y] = ode45( @rhs, x, [initial_y initial_dydx] ); % Differential equation setup using ode45()
10
11 -    plot(x,y(:,1))
12 -    xlabel('x'); ylabel('y'); title('Solution to differential equation');
13      |
14 -    index_x1 = x==0.2;      % Find index of array element where x = 0.2
15 -    y1 = y(index_x1);       % Find corresponding value of y at x = 0.2 by looking at the index x1
16
17 -    index_x2 = x==0.6;      % Find index of array element where x = 0.6
18 -    y2 = y(index_x2);       % Find corresponding value of y at x = 0.6 by looking at the index x2
19
20 -    index_x3 = x==1.2;      % Find index of array element where x = 1.2
21 -    y3 = y(index_x3);       % Find corresponding value of y at x = 1.2 by looking at the index x3
22
23 -    sprintf('Solutions: ')
24 -    sprintf('y(0.2) = %.3f', y1)
25 -    sprintf('y(0.6) = %.3f', y2)
26 -    sprintf('y(1.2) = %.3f', y3)
27
28 -    └ end
29
30      % The function below defines the ODE function to be
31      % solved. To solve higher-order ODEs, we need to
32      % convert them to systems of first-order ODEs and then
33      % solve those systems. Thus, in this function, we re-write the second-order ODE as a
34      % system of first-order ODEs.
35
36 ⊟ function dydx = rhs(x,y)
37 -          dydx_1 = y(2);
38 -          dydx_2 = 2*y(2) - y(1) + x^2 - 4*x + 2;
39
40 -          dydx = [dydx_1; dydx_2];
41 -    └ end
```

4) Click on the green "play" icon appearing on the computer screen to run the case computation.

Output:

```
>> case2_main
ans =
Solutions:
ans =
y(0.2) = 0.040
ans =
y(0.6) = 0.360
ans =
y(1.2) = 1.440
>>
```

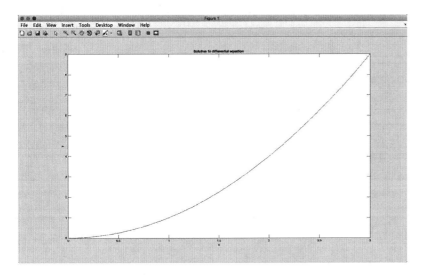

Figure A4.7 Solution to the second-order differential equation.

Index

Applied Engineering Analysis, First Edition. Tai-Ran Hsu.
© 2018 John Wiley & Sons Ltd. Published 2018 by John Wiley & Sons Ltd.
Companion Website: www.wiley.com/go/hsu/applied

Printed and bound by CPI Group (UK) Ltd, Croydon, CR0 4YY